CW01067047

HANDBOOK ON WEALTH AND THE SUPER-RICH

Handbook on Wealth and the Super-Rich

Edited by

Iain Hay

Flinders University, Australia

Jonathan V. Beaverstock

University of Bristol, UK

Cheltenham, UK • Northampton, MA, USA

© Iain Hay and Jonathan V. Beaverstock 2016

All rights reserved. No part of this publication may be reproduced, stored in a retrieval system or transmitted in any form or by any means, electronic, mechanical or photocopying, recording, or otherwise without the prior permission of the publisher.

Published by
Edward Elgar Publishing Limited
The Lypiatts
15 Lansdown Road
Cheltenham
Glos GL50 2JA
UK

Edward Elgar Publishing, Inc.
William Pratt House
9 Dewey Court
Northampton
Massachusetts 01060
USA

A catalogue record for this book
is available from the British Library

Library of Congress Control Number: 2015950293

This book is available electronically in the Elgaronline
Social and Political Science subject collection
DOI 10.4337/9781783474042

ISBN 978 1 78347 403 5 (cased)
ISBN 978 1 78347 404 2 (eBook)

Typeset by Servis Filmsetting Ltd, Stockport, Cheshire
Printed and bound by CPI Group (UK) Ltd, Croydon, CR0 4YY

Contents

Figures

Tables

Contributors

Rowland Atkinson holds the Research Chair in Inclusive Society within the Department of Town and Regional Planning at the University of Sheffield, UK.

Jonathan V. Beaverstock is Professor of International Management at the University of Bristol, UK.

Lucy Budd is Senior Lecturer in Air Transport at Loughborough University, UK.

Roger Burrows is Professor of Cities based in the School of Architecture, Landscape and Planning at Newcastle University, UK.

Louise Crewe is Professor of Human Geography at the University of Nottingham, UK.

Aidan Davison is Senior Lecturer in the Discipline of Geography and Spatial Science at the University of Tasmania, Australia.

Adam D. Dixon is Reader in Economic Geography at the University of Bristol, UK.

Ray Forrest is Professor of Housing and Urban Studies in the Department of Public Policy at City University of Hong Kong.

David R. Green is Professor of Historical Geography at King's College London, UK.

Sarah Hall is Professor of Economic Geography at the University of Nottingham, UK.

Tim Hall is Professor of Interdisciplinary Social Studies at the University of Winchester, UK.

Iain Hay is Matthew Flinders Distinguished Professor of Human Geography at Flinders University, South Australia.

Ilan Kapoor is Professor of Critical Development Studies in the Faculty of Environmental Studies, York University, Toronto, Canada.

Sin Yee Koh is Assistant Professor in Geography in the Institute of Asian Studies at Universiti Brunei Darussalam.

Giovanni Mangraviti is a doctoral student in the Departments of International Politics and Management at City University London, UK.

Amber Martin is a Research Executive at 2CV, a consumer research agency, in London, UK.

Isaac 'Asume' Osuoka is a postdoctoral researcher at the Faculty of Environmental Studies, York University, Toronto, Canada.

Alastair Owens is Professor of Historical Geography at Queen Mary University of London, UK.

Ronen Palan is Professor of International Political Economy at the City University London, UK.

Chris Paris is Emeritus Professor of Housing at the University of Ulster, Northern Ireland.

David Rhodes is Research Fellow in the Centre for Housing Policy at the University of York, UK.

Andrew Sayer is Professor of Social Theory and Political Economy at Lancaster University, UK.

Paul G. Schervish is Professor of Sociology and Director of the Center on Wealth and Philanthropy at Boston College, USA.

Sam Schulz is Lecturer in the Sociology of Education at Flinders University, South Australia.

John Rennie Short is Professor of Public Policy at the University of Maryland, Baltimore County, USA.

Emma Spence is a doctoral student in the School of Planning and Geography at the University of Cardiff, Wales.

Allan Watson is Senior Lecturer in Human Geography at Staffordshire University, UK.

Bart Wissink is an Associate Professor in the Department of Public Policy at City University of Hong Kong.

Michael Woods is Professor in the Department of Geography and Earth Sciences at Aberystwyth University, Wales.

Anna Zalik is an Associate Professor in the Faculty of Environmental Studies, York University, Toronto, Canada.

1. They've 'never had it so good':[1] the rise and rise of the super-rich and wealth inequality

Jonathan V. Beaverstock and Iain Hay

INTRODUCTION

Perhaps one silver lining on the storm clouds of the recent global financial crisis, particularly in Europe and North America, has been the growing attention given to 'super-rich capitalism', igniting debates in both the public sphere and academy about inequitable levels of prosperity and wealth in global society. Public commentators and journalists like Chrystia Freeland (2012), Robert Frank (2007) and John Kampfner (2014) have presented informative and popular writings on the rise of the global super-rich, and how they occupy the most exclusive places and networks on the planet, distanced far from the rest of us. Even public and private television broadcasters have got in on the act, illuminating the prosperous lifestyles and conspicuous consumption of the super-rich, from multimillionaire celebrities to the more outgoing and publicly available of the world's billionaires (e.g., see the British Broadcasting Corporation's *The Super-Rich and Us*, broadcast in 2015, and CNBC's 2014 series *Secret Lives of the Super Rich*).

Thankfully, beyond the perspective on glitzy lifestyles and super-rich consumption offered through the lens of popular culture, members of the social sciences community, from economists to urban studies specialists, have begun to look critically at the excessive growth and consumption, political economy and inequitable wealth creation of the super-rich in global society. To date, the pinnacle of the academy's intervention on wealth and inequality has come from Thomas Piketty's (2014) widely acclaimed *Capital in the Twenty-First Century*, a volume showcased in many of the chapters in this *Handbook*. Despite its considerable merits, Piketty's book cannot of course tell the full story. And since the 1980s there has been a bloom of informative empirical and theoretical analyses, which this *Handbook* joins, on the growth of wealth and accompanying unequal income distributions (e.g., Atkinson et al., 2011; Saez, 2013; Piketty and Saez, 2014); the histories of the rich and affluent in society (Thorndike, 1980; North, 2005); the rise of the super-rich (Lundberg,

1988; Haseler, 1999; Irvin, 2008; Beaverstock et al., 2013; Hay, 2013; Hay and Muller, 2014); and, finally, on the so-called 1 per cent – that small, wealthiest fraction of the global population (Dorling, 2014; Di Muzio, 2015; Sayer, 2015). Moreover, the economically superior position of the super-rich in society has also been the focus of great attention in the 'Occupy' movement, which not only highlighted the outrageous 'bonus' culture of Wall Street and City of London bankers in a context of bank bailouts during the financial crisis, but also rekindled the wider spotlight on income inequality across global society. During 2014, Oxfam's paper 'Working for the few' highlighted very starkly the wealth inequality of the global super-rich in society. As Oxfam (2014, p. 1) noted, 'Economic inequality is rapidly increasing in the majority of countries. The wealth of the world is divided in two: almost half going to the richest one per cent; the other half to the remaining 99 per cent'. It is the richest 1 per cent, or indeed the more rarefied space occupied by the 0.1–0.01 per cent (see Sayer, 2015), that interests us in this *Handbook*, the structure of which we shall turn to shortly. But first we briefly outline our subjects of study: wealth and the super-rich.

WEALTH AND THE SUPER-RICH

While the conspicuous behaviours of the rich were highlighted by Thorstein Veblen ([1899] 1985) in the late nineteenth century in his book *A Theory of the Leisure Class*, perhaps the most widely known and systematic analysis of the rich and wealthy emerged almost a century later in 1982 with the first annual publication of *Forbes* magazine's list of the wealthiest 400 US citizens (see Koh et al., Chapter 2 in this volume for a more detailed discussion of the history of literature on wealth and the super-rich). The first *Forbes* list was populated by those who had inherited wealth from the famous industrialist families, oil barons, bankers and financiers, and tycoons of the first Gilded Age (e.g., Gordon Getty, David Rockefeller Sr), as well as by the new band of entrepreneurs like Sam Walton and Steve Jobs who set the pace for the so-called second Gilded Age (see Smith, 2001; Short, 2013). For the next seven years, the *Forbes* list of the wealthy became the dominating and most widely accepted for identifying and examining – in wonderment, rather than in any critical fashion – the super-rich in American society. This kind of informative and informed, yet uncritical analysis of the super-rich was supplemented in 1989 with the publication of the inaugural *Sunday Times Rich List*. This rich list of the UK's 1000 wealthiest people followed the 'Big Bang' in the City of London in 1986 and Margaret Thatcher's reduction in the top rate of personal income tax from 60 per cent to 40 per cent in the 'giveaway' Budget of 1988. Together these events sowed the seeds for much new wealth, growing numbers of millionaires in the UK's banking and financial services industries, as well as asset booms in London and the south-east of England (see Leyshon and Thrift, 1997). On the heels of these two major rich lists, and their accompanying and often valuable journalistic analyses, has come a wave of rich lists in different jurisdictions. *Forbes* magazine produces lists for many nations (e.g., Brazil, Indonesia, South Africa, Singapore), but there is also a growing number of indigenous lists including the *BRW* (Australia's *Business Review Weekly*) Rich 200; the *Challenges* magazine list of the richest people in France; the *Hurun Report* China Rich List; the *Canadian Business* magazine's list of the 100 richest Canadians; and *National Business Review*'s rich list for New Zealand. Individually and collectively these lists – often compiled by teams of experienced and determined journalists dedicated full-time to their production – provide useful descriptive insights into the remarkable recent growth and extent of super-wealth in specific jurisdictions as well as more broadly.

In the late 1990s and 2000s two authors offered some analytical perspectives on the composition of the super-rich: Haseler (1999) in *The Super-Rich*; and Frank (2007) in *Richi$tan: A Journey Through the 21st Century Wealth Boom and the Lives of the New Rich*. Haseler (1999) looked closely at the wealth make-up of the super-rich and identified a super-rich pyramid. At the pinnacle were the mega-rich billionaires, each with net worth in excess of US$1 billion. In the middle were the multi-millionaires, who according to Haseler (1999, p. 2) in the USA averaged over US$10 million in net worth (and numbered about 500 000 people) in 1995. At the base of the pyramid were the US$ millionaires who numbered about 6 million worldwide in 1996. Based on his professional experience and observations as a *Wall Street Journal* reporter Robert Frank (2007, pp. 6–12) constructed a similar pyramid that provided the basis for his book's 'Richi$tan'. At the top of this pyramid was so-called 'Billionairesville' and those who inhabited this space had a household net worth of more than US$1 billion. Below were, respectively, the residences of 'Upper Richi$tan' (US$100 million to US$1 billion), 'Middle Richi$tan' (US$10 million to US$100 million) and 'Lower Richi$tan' (US$1 million to US$10 million). Those at the lower end of 'Lower Richi$tan' were deemed to be the merely affluent rather than the super-rich.

The banking and professional services sectors of the economy have been publishing definitions and segmented classifications of an individual's personal wealth since the mid-1990s. The *World Wealth Reports* first published from 1996 by the French multinational management consulting corporation Capgemini and Merrill Lynch, the wealth management division of the Bank of America, classify the wealthy and super-rich as a high net worth individual (HNWI) and segment the HNWI market into three distinctive bands: HNWIs with investible assets of US$1 million or more; mid-tier millionaires with investible assets of US$5 million to US$30 million; and ultra-HNWIs with investible assets in excess of US$30 million (CML, 2007). Most banking and professional services firms have similar definitions to define and segment the super-rich (see Beaverstock et al., 2013). For example, International Financial Services London[2] (2009) retained the structure of the Capgemini Merrill Lynch classification, but inserted a 'Mass Affluent' segment below HNWI (with net worth over US$100000 and up to US$1 million). In a similar vein, the Boston Consulting Group (2011), who publish the influential annual *Global Wealth Report*, define the wealth of households as: ultra high net worth (more than US$100 million); established wealthy (more than US$5 million); emerging wealthy (US$1 million–US$5 million); and affluent (US$100000–US$1 million). The key factor in most segmented definitions of the super-rich employed by the wealth management sector is that they take account of liquid, investible assets only (i.e., they exclude primary household residence, collectables and consumer durables; CML, 2007).

Most of the authors contributing to this *Handbook* take some space at the outset of their chapter to comment on growing wealth inequality in global society and some of them draw from commonly accepted definitions of the super-rich, such as those outlined above. We recognize and indeed embrace the variability surrounding understandings of super-wealth, for as Hay and Muller (2012) have observed there is no definitive threshold to identify the super-rich: wealth takes on different meanings depending on the context within which one is located. Nevertheless, for practical purposes – and notwithstanding Koh et al.'s reservations (see Chapter 2 in this volume) – we acknowledge that international definitions of what it takes to be super-rich now typically begin for an *individual* (not a household) with liquid wealth and investible assets in excess of US$1 million – an amount that to some in the wealthy North may seem to constitute a fairly low bar but for billions of other people in the rest of the world represents a vast sum of money. And for most purposes, it is this indicator of wealth that we adopt in this volume to distinguish the super-rich from the rest.

In 2013, the world population of HNWIs (individuals with US$1 million or more) reached 13.7 million according to Capgemini and Royal Bank of Canada Wealth Management (CRBCWM, 2014), which was +14.2 per cent more than the previous year, and a staggering +59.3 per cent (+5.1 million) more than in 2008, during the depths of the global financial crisis (Table 1.1). Similarly, there has been a rebound in the

Table 1.1 *The global population of HNWIs and value of private wealth, 1996–2013*

	Number (Millions)	Change (%)	Wealth (US$ trillions)	Change (%)
1996	4.5	–	16.6	–
1997	5.2	+15.6	19.1	+15.1
1998	5.9	+13.5	21.6	+13.1
1999	7.0	+18.6	25.5	+18.1
2000	7.2	+2.9	27.0	+5.9
2001	7.1	−1.4	26.2	−3.7
2002	7.3	+2.8	26.7	+2.7
2003	7.7	+5.5	28.5	+6.7
2004	8.2	+6.5	30.7	+7.7
2005	8.8	+7.3	33.4	+8.8
2006	9.5	+8.0	37.2	+11.4
2007	10.1	+6.3	40.7	+9.4
2008	8.6	−14.9	32.8	−19.4
2009	10.0	+17.1	39.0	+18.9
2010	10.9	+8.3	42.7	+9.7
2011	11.0	+0.8	42.0	−1.7
2012	12.0	+9.2	46.2	+10
2013	13.7	+14.2	52.6	+13.9

Source: Adapted from CMLGWM (2008, 2009, 2010, 2011); CML (2002, 2007); CRBCWM (2012, 2013, 2014).

value of wealth held by this HNWI population worldwide, increasing from US$32.8 trillion in 2008 to US$52.6 trillion in 2013 (+US$19.8 trillion or +60.4 per cent). This is quite simply a stupendously large figure.[3] CRBCWM data (2014) alone show very dramatically the reappearance of the rich in global society. Since 2008, much of this new wealth and growth in the HNWI population has been created in the Asia-Pacific region (see Beaverstock and Hall, Chapter 20 in this volume), increasing from 2.4 million HNWIs in 2008 to 4.3 million in 2013, +79.2 per cent (+1.9 million). In 2013 the Asia-Pacific had the same stock of HNWIs as North America, 4.3 million, accounting for 31.4 per cent of the total share (Europe had by contrast 3.8 million HNWIs or a 27.7 per cent total share). In terms of wealth distribution, the Asia-Pacific stood at US$14.2 trillion in 2013, just behind North America at US$14.9 trillion, but ahead of Europe at US$12.4 trillion (see CRBCWM, 2014).

Beyond the publishing and banking and professional services sectors of

Table 1.2 The growth rate of personal wealth for the top richest population, 1987–2013

Top Wealth Holders in the Global Population	Number[a]		Wealth (US$ billion)		Average Year Growth Rate (%)[b]
	1987	2013	1987	2013	
The top 1/100 millionth	30	45	3	35	+6.8
The top 1/20 millionth	150	225	1.5	15.0	+6.4
Average world wealth per adult					+2.1
Average world income per adult					+1.4
World adult population					+1.9
World GDP					+3.3

Notes:
a. Out of 3 billion people in the 1980s and 4.5 billion in 2010.
b. Adjusted after inflation rate.

Source: Adapted from Piketty (2014, p.435).

the economy, data on the super-rich and wealth creation and distribution has been generated by other sources, most notably in the form of national governments' data on income (as analysed by Piketty, 2014 and colleagues – Atkinson et al., 2011; Saez, 2013; Piketty and Saez, 2014), which are picked up by various chapters in this volume (e.g., those by Hay and Sayer – Chapters 4 and 5 respectively). One notable table from Piketty's (2014) treatise shows how the top wealth holders in society have benefited disproportionately in the average growth rate of wealth in relation to the average world wealth and income per adult (Table 1.2).

Others generate similar findings. For example, Piketty and Saez (2006) and Reich (2013) have identified and publicized growth in the top incomes and the development of historical U-shaped income distributions in the USA and Europe, partly accounted for by the phenomenon of the 'working rich' – as well as by significant changes in taxation and welfare policies favouring the wealthy (also see Folkman et al., 2007; Philippon and Reshef, 2012; Sayer, 2015). Indeed, Saez (2013) has concluded that between 2009 and 2012 the wealthiest 1 per cent of the USA's population received 95 per cent of that country's financial growth in the wake of the recent crises.

As Saez's observation suggests, increases in wealth have not been shared evenly. Nor – as Sayer (2015) reminds us – have they been distributed fairly amongst those responsible for that wealth's very production. Indeed the distribution of wealth globally is now so disproportionate as to be as outrageous and alarming as it is absurd. For instance in its recent report *Wealth: Having it All and Wanting More*, Oxfam (2015) points out that where it took 388 billionaires to have accumulated the same amount of wealth as the bottom 50 per cent of the population in 2010, by 2014 it took just 80 to match the wealth of 3.5 billion people (including children). As Kasia Moreno (2014), Editorial Director at *Forbes Insights*, remarks: 'Both groups have $US1.7 trillion. That's $20 billion on average if you are in the first group, and $486 if you are in the second group'.

The figures in specific jurisdictions are almost as disheartening, as Hay details in this volume in Chapter 4. For instance, the richest 86 Canadians are reported to have the same wealth as that country's poorest 11.4 million. Expressed in other terms, 0.002 per cent of Canadians have wealth equivalent to that held by 34 per cent of the population (Macdonald, 2014, p. 6). In the past 15 years China has gone from having the world's second most equal distribution of wealth to now having inequality among the world's highest (2012 Gini coefficient of 0.55 compared with 0.45 for the United States) (Dianqing and Xin, 2014). And in Australia, at A$56.2 billion, the 2014 wealth of the country's seven richest people, including Gina Rinehart, Frank Lowy and James Packer, exceeded the approximate wealth (A$54 billion) of the poorest 1.73 million households (Hutchens, 2014).

It is in the context of such astonishing data on the vastly uneven distribution of income and wealth in global society and the rise and rise of the super-rich that we position this timely *Handbook*. We have been purposely eclectic in considering contributors' disciplinary backgrounds for this book's interrogation of the concentration of wealth in the hands of individuals (rather than corporations or the nation-state) and its careful scrutiny of the lifestyles, performances and agency of the super-rich in global society. We believe the diversity of contributions here gives some indication of the different disciplinary powers that can usefully be brought to bear to examine the interwoven subjects of individual wealth and the super-rich. And we firmly believe that it is past time for broader parts of the social sciences and humanities communities to look as critically at the rich and wealth as we have at the poor and poverty in global society. This book is one small part of that uprising.

STRUCTURE

The selection of authors in this volume, each of whom we admire greatly, spans aviation transportation, development studies, education, housing, human geography, management, political economy, public policy, sociology, urban planning and urban studies, and provides a unique multi- and interdisciplinary appraisal of wealth and the super-rich. The authors explore and analyse a broad array of intriguing and significant issues that neither editor has been able to reconnoitre since their own initial forays into 'exposing' the geographies of the super-rich (Beaverstock et al., 2004; Hay and Muller, 2012). We believe the *Handbook's* contributions combine synergistically to yield a multifaceted, yet coherent, analysis of individual wealth and the super-rich, where most certainly, drawing on Aristotle's wisdom, 'the whole is greater than the sum of its parts'. The authors, all experts from North America, Europe and Asia-Pacific, have not only delivered empirical projects with different regional foci, but have also critically evaluated the (re)production and concentration of wealth in the hands of the global few, often berated in the public discourse of the post-global financial crisis as members of the '1 per cent' (Dorling, 2014; Oxfam, 2015; Sayer, 2015).

It is of course impossible to cover every conceivable angle in the study of wealth and the super-rich in contemporary society. Our motives to edit this *Handbook* emanated from our shared desire to encourage and support critical work on the distribution of wealth over time and space; how the wealthy live and express their wealth; how wealth has transformed places; and finally, how the wealthy experience (and enjoy) socioeconomic power relations and lifestyles that are almost completely separated from the remaining 99 per cent of society. We are also concerned that as much attention be given to wealth and the processes that underpin its generation and its maldistribution as is afforded the individuals who are super-rich. Just as good studies of poverty and the poor examine both the ways in which poverty is produced and the experiences of the poor (e.g., Deepa et al., 2000; Banerjee and Duflo, 2012), so we seek to emphasize wealth generation, concentration, maintenance and distribution as well as the experiences of the super-rich. To these ends we have organized the *Handbook* into three distinctive sections entitled: 'Wealth, Self and Society'; 'Living Wealthy'; and 'Wealth and Power'. Each is summarized in turn below, drawing on some of the chapters' distinctive contributions.

However, before those sections there is one additional prefatory chapter by Sin Yee Koh, Bart Wissink and Ray Forrest that we believe usefully reviews the current status of super-rich and wealth studies in the social sciences. While they limit discussion to urban studies to make their case, Koh et al.'s chapter offers a helpful exemplar upon which to consider work in many of the other social sciences. They observe that we need to move beyond questions such as 'Who [and we might add, where] are the super-rich?' and 'What do they do?' to questions about 'What made the super-rich and why?' – questions that are taken up by several of the contributors to this volume. While we do not regard the future of studies of the super-rich and wealth to be quite so clearly the 'either/or' proposition implied by Koh et al., their case does emphasize the point we highlighted earlier in this chapter, and that is that the future of studies in this field demands that we consider simultaneously and from a broad set of standpoints those who are super-rich, and their wealth – its meaning, its production and its (mal)distribution.

Part I, 'Wealth, Self and Society', comprises eight chapters that together interrogate the morality and legitimacy of extreme wealth and the advent of the 'new' super-rich in global society. An important thread that runs through these chapters is an exploration of the (construction of the) character, nature and identities of the super-rich. In the first chapter of this section, Chapter 3, Alastair Owens and David Green provide a historical account of the changing composition and distribution of personal wealth from the 1800s up until the 1930s. The chapter's central focus is on the ways in which institutional developments during industrial capitalism and the emergence of the core-periphery world capitalist system (Wallerstein, 2011), like the development of the banking and financial system, created new opportunities for the accumulation of wealth and assets, and, significantly, unprecedented personal fortunes from both domestic and overseas commercial interests.

The following chapter by Iain Hay, 'On plutonomy: economy, power and the wealthy few in the second Gilded Age' brings us forward to the contemporary period and critically evaluates the economic and power relations of the extreme wealthy in the so-called 'second Gilded Age' (Short, 2013). Central to Chapter 4 is Hay's appraisal of Citigroup reports authored by Ajay Kapur and colleagues (Kapur et al., 2005, 2006) that separated the world into two key blocs: the very wealthy 'plutonomy' and the rest. Hay not only exposes the major arguments fostered by Kapur and his colleagues on the key characteristics and socioeconomic-political power of the super-rich, but also, and dare we say skilfully, links Citigroup's approach to Piketty's (2014) game-changing analysis of the accumulation and uneven reproduction of wealth (rate of return on private capital exceeds the rate of economic growth $-r > g$) over the last hundred years or so.

In a similar vein to Hay, Andrew Sayer's chapter (Chapter 5) critically interrogates the legitimacy and return of the rich in contemporary society, linking that return inextricably to the rollout of neoliberalism and the rise of financialization. Sayer adopts a moral economy approach to evaluate how the rich accumulate their wealth through the growth of unearned income, aided and abetted by the ownership and control of assets, the financialization of everything it seems, and the backdrop of highly favourable personal and corporate taxation regimes. Sayer's analysis of the emergence of rentiership is at the heart of this very fine chapter's moral economic critique of the global super-rich.

In Chapter 6, Ilan Kapoor presents a critical and illuminating account of the celebrity philanthropy and charity work of billionaires Bill Gates and George Soros. Kapoor adopts the noteworthy term, 'decaf capitalism' to explain how such corporate giving is an explicit and hardnosed business practice that helps to rebut the murkier side of capitalism and maintain the status quo of the neoliberal economic order in global society. The case studies of Gates and Soros are used as exemplars to elucidate the wider, deeply problematic and often overlooked issues of 'tycoon philanthropy' (Phillips, 2008) or super-philanthropy (Hay and Muller, 2014).

In Chapter 7, Paul Schervish takes us into the very personal processes by which super-wealth is constructed, thoughtfully unpacking the entrepreneurship and wealth creation of the high net worth individual. Drawn from an intensive interview survey of 49 super-rich entrepreneurs (with a median net worth greater than US$16.5 million), Schervish outlines the economic rules and procedures and personal practices that established these individuals as successful and wealthy entrepreneurs. Significantly, the chapter explains the nature and morality of prosperous entrepreneurship and evaluates their role as 'hyperagents' in society.

Sam Schulz and Iain Hay's chapter (Chapter 8) picks up a gauntlet thrown down by Beaverstock et al. (2004) and later Caletrío (2012), setting out a theoretical case as to why social sciences researchers should study the prosperity and privilege associated with the super-wealthy. The chapter goes on to draw from a careful analysis of the documentary film *Born Rich* made by Johnson & Johnson heir, Jamie Johnson, to show how discourses and practices of silence and unproblematic notions of meritocracy together form vital parts of super-rich subjectivities, fortifying elite power while concentrating the disadvantage of excluded others.

The final chapters in this section on 'Wealth, Self and Society' are written by Allan Watson and Tim Hall and explore two high-profile, licit and illicit economic activities of some of the super-rich. Respectively, these are the multimillionaire dollar business of rap music and the rise of the 'hip-hop mogul', and the ill-gotten super-gains of the criminal and corrupt. In Chapter 9, Watson examines the entrepreneurial success of the self-made hip-hop mogul, drawing on the case studies of the two most successful male moguls, Sean 'Diddy' Combs and Shawn 'Jay-Z' Carter. He also comments on the importance of material excesses of wealth in the hip-hop music culture and tensions that exist for such moguls in their quest to be of the 'street', but as multimillionaires who may have 'sold out' their connections with the grassroots in their quest to become rich.

Turning to the illicit aspects of wealth generation, Tim Hall gives attention to the criminal super-wealthy, whose economic success has been generated through narcotics or financial fraud. Until now, and perhaps not surprisingly, scholarly discourse about the super-rich has included very little on the illegal accumulation of wealth. Chapter 10 takes a big step to correct this lacuna. His fascinating chapter explores the global illicit economy through a biographical approach, discussing the unlawful activities and wealth accumulation of three high-profile super-rich individuals: the gangsters Pablo Escobar and Dawood Ibrahim Kaskar, and the financial fraudster, Bernard Madoff.

Part II of the *Handbook*, 'Living Wealthy', draws on seven chapters to unpack the sociocultural characteristics and lifestyles of the super-rich, especially those associated with their luxurious living spaces, both urban and rural. Rowland Atkinson, Roger Burrows and David Rhodes provide a compelling case study of London's housing market for the super-rich in Chapter 11, investigating closely the prime markets in Kensington and Chelsea, Westminster and Camden. These authors map the wealthiest neighbourhoods across London using the extraordinarily rich geodemographic database MOSAIC (owned by Experian) and provide a critical evaluation of the politics and scale of housing change in London. An interesting highlight is the internationalization of London's housing stock ownership in the super-prime central areas like Westminster, an observation that raises much bigger questions about who London is actually for in British society.

Staying with the subject of housing for the super-rich, and taking up some of its international dimensions, Chris Paris's chapter (Chapter 12) charts the residential spaces of the wealthy across the globe. He draws primarily on super-rich residential housing data compiled by the commercial real estate company Knight Frank, looking in detail at prime housing markets in the world's global cities. He shows that the super-rich have disproportional impacts on national and international housing systems that are far greater than their mere numbers suggest. The types of residential spaces owned and consumed by the super-rich have remade places through the decoupling of prime residential real estate from national housing markets and in zones of exclusive hyperconsumption.

In Chapter 13, and maintaining a focus on real estate, Michael Woods presents a careful historical analysis of rural land ownership and wealth, illustrating the historically embedded relationship between wealth and landownership in rural societies. His chapter focuses on the old wealth of the landed gentry in Europe; the new wealth of gentrifying middle classes; and the global wealth of the super-rich – three distinct forms of wealth that have had an impact on rural places – and reminds us of the uneven power relations of the aristocracy and agricultural land owners, including holders of the vast estates of the UK. Not surprisingly, in the UK context, the invasion by the new wealthy middle classes of the rural idyll was stimulated by wealth creation in new money and favourable fiscal regimes stimulated by successive Conservative governments from 1979. As such gentrification was not only limited to the UK, Woods also includes examples from the USA and mainland Europe.

The next three chapters in this section of the *Handbook* by Emma Spence, Lucy Budd, and Louise Crewe and Amber Martin, examine closely the lifestyles and consumption of the super-rich. First, Emma Spence takes us to Monaco, a place that is almost synonymous with wealth and the super-rich. Her fascinating chapter draws from extensive personal experience on super-yachts and in a yacht brokerage in Monaco to make three main observations. She points first to the prospective value of working with and through intermediaries – such as yacht brokers, luxury goods representatives and air crew – to gain access to the hard-to-reach super-rich. Second, she draws from experience as and with an intermediary to offer simply fascinating insights into the ways in which being super-rich is performed in the Principality. And third, she alerts us to Monaco's dependence on those performances for its continuing success as a destination attractive to the super-rich.

Chapter 15 takes us skyward to examine the elite aeromobilities of the global super-rich. Budd refers to these subjects as the 'aerial elites' and takes us through the luxuries and practices of their flight, embracing scheduled First Class travel with carriers like Etihad as well as private business aviation. The substantive content of the chapter is focused on private business aviation in small, 12+ seater corporate jets built by the likes of Airbus, Boeing and Learjet (Bombardier), and assesses the benefits of such elite mobilities for the super-rich. Budd points out that despite their relative invisibility from scholarly and public debate, compared with arrangements surrounding mass aeromobility, it is important that forms of transport enjoyed by the world's richest people are not omitted from critical assessment of their implications for inequality and sustainability.

In the following chapter, Chapter 16, Crewe and Martin keep us focused on high-end luxury. Their study examines in detail the remarkable resilience of luxury retailers in the face of global recession and faltering consumer spending. They show how global luxury brands such as Prada and Gucci build and fix their value spatially, and they make clear the role that factors such as location, labels, architecture and design play in the making of luxury markets. Crewe and Martin pay particular attention to the largest luxury conglomerates like Burberry, Hermès, Kering, LVMH, charting their growth especially in China, and examining in detail the complexity of luxury retail, like Gucci, in 'uber' exclusive flagship stores located in prestigious sites within selected global cities like Paris, London, New York and Milan.

In the final chapter of this section (Chapter 17), and in an appropriate successor to the preceding chapter's discussion, Aidan Davison presents a critical appraisal of the conspicuous consumption of the super-rich and their resource use. Drawing primarily on the United States given its position as a high consumption society home to many of the world's wealthiest people, Davison considers the position of the super-rich in the (mis)alignment of environmental concerns with neoliberal political economy. His analysis of the resurgence of conspicuous wealth and luxury fever for the super-rich opens wider debates about such consumption, its sustainability and impacts on the environment. Davison quite rightly concludes that the super-rich and their environmental implications require further scrutiny and critical debate.

Part III, 'Wealth and Power', comprises five chapters that discuss the economic supremacy of the super-rich in contemporary capitalism and globalization. A vital narrative that runs through these chapters is the super-rich's ability to exert power and influence over the rest of society, whether economic, political or cultural. John Rennie Short's chapter (Chapter 18) illustrates clearly how the super-rich find it very easy, unlike the rest of us, to adhere to different nation-state immigration policies because of their ease in meeting different country entry requirements based on wealth, income and assets. Drawing on Ley's (2010) ideas about 'millionaire migrants', Short provides a highly informative analysis of national immigration programmes that seek to attract the super-rich, with examples from Canada, the UK, USA, Australia, Singapore and Malaysia. Short quite rightly reminds us that as the super-rich are *the* most highly mobile people in global society, nation-states are in a race to attract them, and immigration programmes have looked favourably on the mobility and multi-residency (and perhaps, citizenship) needs of the wealthy.

The next three chapters by Adam Dixon, Jonathan Beaverstock and Sarah Hall, and Ronen Palan and Giovanni Mangraviti, are drawn from financial geography and international political economy approaches to critically evaluate wealth production and management, and the protective and secretive nature of offshore financial centres, also known as 'tax havens'. Chapter 19 focuses on the production of sovereign wealth funds and the nation-state. Dixon is the only author in the book who focuses overtly on the state, where often the sovereign wealth is under the ownership and control of a 'ruling dynasty'. The chapter considers the rise and geography of sovereign wealth funds and their links to the commodity economy, and in particular, oil, and export-led growth in Asia.

In contrast, Beaverstock and Hall's Chapter 20 explores the establishment of the private wealth management industry, particularly 'offshore', to preserve and accumulate the wealth of the global super-rich. Since the late 1980s the private wealth management industry, comprising private banks, the wealth management division of investment banks, and many professional services like accounting, insurance and law, has grown significantly in the world economy, and has contributed to the re-emergence of the offshore financial centre. Beaverstock and Hall look closely at the development of the private wealth sector in offshore financial centres, and present a case study of Singapore.

Continuing with the tax haven theme taken up in Chapter 20, in Chapter 21 Palan and Mangraviti examine the corporate world's multi-jurisdictional arbitrage and mechanisms for tax efficiency and reduction. After a discussion of troublesome wealth and tax havens, their chapter goes on to discuss wealth protection strategies and the multi-jurisdictional networks and tax footprint reduction of both the private and corporate worlds.

Part III concludes with a chapter on 'black gold', or oil. Isaac 'Asume' Osuoka and Anna Zalik take us in Chapter 22 to the primary commodity that has created unprecedented wealth concentrated in a 'global oil elite' comprising familial and inherited wealth, multimillion US dollar (or equivalent) remuneration levels for the executive and scientific workforce, and the beneficiaries of the financialization of the oil industry. Of significance, these authors draw on examples from North America and West Africa to illustrate the uneven production of wealth gained from oil for a selected global elite. The case study on West Africa shows clearly the cosy relationships that have been established with those multinational energy companies that seek to extract and refine oil and natural gas, and the local, often ruling elites who own land and lay claim over the natural resources.

To conclude, if nothing else, the global financial crisis of the late 2000s intensified what had been a flickering light of attention on the lives of the super-rich and vast wealth they had been accumulating since the 1970s. In the few years that have followed the crisis we have witnessed new levels of public and scholarly scrutiny of super-wealth and the lives of the super-rich, even to the extent of excoriating accounts of their place in contemporary society from the least likely of sources (see, e.g., the recent book by Steve Hilton et al., 2015, recalling that Hilton was a recent former senior adviser to UK Prime Minister David Cameron). To date, however, the level of popular media attention dedicated to wealth and the wealthy has probably exceeded the academic attention given to those subjects. This book is part of a corrective to that imbalance. We hope that by way of its thoughtful scholarly insights to a diverse array of topics such as the moral economy of wealth, environmental implications of super-rich lifestyles, and wealth preservation through means such as professional wealth management and tax havens, this volume's examination of the concentration of extraordinary wealth in the hands of a small number of individuals and its analysis of super-rich lifestyles and super-rich 'selves' makes a useful and timely contribution to the emerging literature on the super-rich and their wealth in contemporary society.

We hope too that this book signals our concern that scholarly attention in this field be given to the processes that underpin super-wealth generation and distribution, and indeed its very meaning, as well as to the lives and lifestyles of those sometimes high-profile individuals who are super-rich and to whom much popular interest gravitates. Where effective studies of poverty and the poor, for example, examine both poverty's production and the wretched ways it is played out in the lives of the impoverished, so future studies of wealth and the super-rich will do well to emphasize wealth generation, concentration, maintenance and distribution as well as the profound significance of lifestyles and experiences of the super-rich.

NOTES

1. Attributed to Harold Macmillan, the British Prime Minister who used this phrase in a speech to the British public in 1957 during the 1950s boom (see Evans, 2010): 'Most of our people have never had it so good'.
2. In 2010 IFSL was merged into an organization known as TheCityUK.
3. To illustrate how large this figure is: if you earned the generous sum of US$1 million per year, you would need to work for 5.26 million years to earn this amount.

REFERENCES

Atkinson, A.B., T. Piketty and E. Saez (2011), 'Top incomes in the long run of history', *Journal of Economic Literature*, **49** (1), 3–71.

Banerjee, A.V. and E. Duflo (2012), *Poor Economics: A Radical Rethinking of the Way to Fight Global Poverty*, New York: Public Affairs.

Beaverstock, J.V., S. Hall and T. Wainwright (2013), 'Servicing the super-rich: new financial elites and the rise of the private wealth management retail ecology', *Regional Studies*, **47** (6), 834–49.

Beaverstock, J.V., P.J. Hubbard and J.R. Short (2004), 'Getting away with it? Exposing the geographies of the super-rich', *Geoforum*, **35** (4), 401–7.

Boston Consulting Group (2011), *Global Wealth 2011*, accessed 2 September 2013 at http://www.rb.ru/skip/upload/admins/files/BCG_Shaping_a_New_Tomorrow_May_2011.pdf.

British Broadcasting Corporation (BBC) (2015), *The Super-Rich and Us* [TV series], accessed 15 May 2015 at http://www.bbc.co.uk/programmes/b04xw2x8.

Caletrío, J. (2012), 'Global elites, privilege and mobilities in post-organized capitalism', *Theory Culture Society*, **29** (2), 135–49.
Capgemini Merrill Lynch (CML) (2002), *World Wealth Report 2001*, accessed 6 January 2010 at www.ml.com.
Capgemini Merrill Lynch (CML) (2007), *World Wealth Report 10th Anniversary 1997–2006*, accessed 6 January 2010 at www.ml.com.
Capgemini Merrill Lynch Global Wealth Management (CMLGWM) (2008), *World Wealth Report 2008*, accessed 6 January 2010 at www.ml.com.
Capgemini Merrill Lynch Global Wealth Management (CMLGWM) (2009), *World Wealth Report 2009*, accessed 6 January 2010 at www.ml.com.
Capgemini Merrill Lynch Global Wealth Management (CMLGWM) (2010), *World Wealth Report 2010*, accessed 6 January 2010 at www.ml.com.
Capgemini Merrill Lynch Global Wealth Management (CMLGWM) (2011), *World Wealth Report 2011*, accessed 13 March 2012 at www.ml.com.
Capgemini and Royal Bank of Canada Wealth Management (CRBCWM) (2012), *World Wealth Report 2012*, accessed 21 June 2012 at http://www.capgemini.com/insights-and-resources/by-publication/world-wealth-report-2012--spotlight/.
Capgemini and Royal Bank of Canada Wealth Management (CRBCWM) (2013), *World Wealth Report 2013*, accessed 2 September 2013 at http://www.capgemini.com/thought-leadership/world-wealth-report-2013-from-capgemini-and-rbc-wealth-management.
Capgemini and Royal Bank of Canada Wealth Management (CRBCWM) (2014), *World Wealth Report 2014*, accessed 25 June 2014 at https://www.worldwealthreport.com/.
CNBC (2014), *Secret Lives of the Super Rich* [TV series], accessed 12 August 2015 at http://www.cnbc.com/secret-lives-of-the-super-rich/.
Deepa, N. with R. Patel and K. Schafft et al. (2000), *Voices of the Poor: Can Anyone Hear Us?* New York: Oxford University Press for the World Bank.
Dianqing, X. and L. Xin (2014), *Income Disparity in China*, Singapore: World Scientific Publishing.
Di Muzio, T. (2015), *The 1% and the Rest of Us. A Political Economy of Dominant Ownership*, London: Zed Books.
Dorling, D. (2014), *Inequality and the 1%*, London: Verso Books.
Evans, M. (2010), 'Harold Macmillan's "never had it so good" speech followed the 1950s boom', *The Telegraph*, 19 November, accessed 11 May 2015 at http://www.telegraph.co.uk/news/politics/8145390/Harold-Macmillans-never-had-it-so-good-speech-followed-the-1950s-boom.html.
Folkman, P., J. Froud and S. Johal et al. (2007), 'Working for themselves? Capital market intermediaries and present day capitalism', *Business History*, **49** (4), 552–72.
Frank, R. (2007), *Richi$tan. A Journey through the 21st Century Wealth Boom and the Lives of the New Rich*, New York: Piatkus.
Freeland, C. (2012), *Plutocrats. The Rise of the New Global Super-Rich*, London: Penguin.
Haseler, S. (1999), *The Super Rich: The Unjust New World of Global Capitalism*, London: St. Martin's Press.
Hay, I. (ed.) (2013), *Geographies of the Super-Rich*, Cheltenham, UK and Northampton, MA, USA: Edward Elgar Publishing.
Hay, I. and S. Muller (2012), '"That tiny, stratospheric apex that owns most of the world" – exploring geographies of the super-rich', *Geographical Research*, **50** (1), 75–88.
Hay, I. and S. Muller (2014), 'Questioning generosity in the golden age of philanthropy. Towards critical geographies of super-philanthropy', *Progress in Human Geography*, **38** (5), 635–53.
Hilton, S., S. Bade and J. Bade (2015), *More Human: Designing a World Where People Come First*, London: W.H. Allen.
Hutchens, G. (2014), 'Wealth of the seven richest Australians exceeds that of 1.73 million households', *Sydney Morning Herald*, 8 July, accessed 14 May 2015 at http://www.smh.com.au/federal-politics/political-news/wealth-of-seven-richest-australians-exceeds-that-of-173-million-households-20140707-3bj0q.html.

Irvin, G. (2008), *Super Rich. The Rise of Inequality in Britain and the United States*, Cambridge, UK: Polity Press.

Kampfner, J. (2014), *The Rich. From Slaves to Super-Yachts. A 2000 Year History*, London: Little Brown Group.

Kapur, A., N. MacLeod and T.M. Levkovich et al. (2005), 'Plutonomy: buying luxury, explaining global imbalances', *Citigroup Global Markets*, 16 October.

Kapur, A., N. MacLeod and T.M. Levkovich et al. (2006), 'The plutonomy symposium – rising tides lifting yachts', *Citigroup Global Markets*, 29 September.

Ley, D. (2010), *Millionaire Migrants: Trans-Pacific Lifelines*, Chichester, UK: Wiley.

Leyshon, A. and N. Thrift (1997), *Money/Space*, London: Routledge.

Lundberg, F. (1988), *The Rich and the Super-Rich*, New York: Citadel Press.

Macdonald, D. (2014), *Outrageous Fortune. Documenting Canada's Wealth Gap*, Canadian Centre for Policy Alternatives, accessed 14 May 2015 at https://www.policyalternatives.ca/sites/default/files/uploads/publications/National%20Office/2014/04/Outrageous_Fortune.pdf.

Moreno, K. (2014), 'The 67 people as wealthy as the world's poorest 3.5 billion', *Forbes Insights*, 25 March, accessed 14 May 2015 at http://www.forbes.com/sites/forbesinsights/2014/03/25/the-67-people-as-wealthy-as-the-worlds-poorest-3-5-billion/.

North, R.D. (2005), *Rich is Beautiful. A Very Personal Defence of Mass Affluence*, London: The Social Affairs Unit.

Oxfam (2014), 'Working for the few', *Oxfam Briefing Paper No. 178*, accessed 25 June 2014 at http://www.oxfam.org/en/policy/working-for-the-few-economic-inequality.

Oxfam (2015), 'Wealth: having it all and wanting more', *Oxfam Issue Briefing*, January, accessed 15 May 2015 at https://www.oxfam.org/sites/www.oxfam.org/files/file_attachments/ib-wealth-having-all-wanting-more-190115-en.pdf.

Philippon, T. and A. Reshef (2012), 'Wages and human capital in the US finance industry: 1909–2006', *The Quarterly Journal of Economics*, **127** (4), 1551–609.

Phillips, M. (2008), 'Tycoon philanthropy: power and the annihilation of excess', in D. Crowther and N. Capaldi (eds), *Research Companion to Corporate Social Responsibility*, Farnham, UK: Ashgate, pp. 249–66.

Piketty, T. (2014), *Capital in the Twenty-First Century* [French edition published 2013 as *Le capital au XXI siècle*, Editions du Seuil], Cambridge, MA: Belknap Press of Harvard University Press.

Piketty, T. and E. Saez (2006), 'The evolution of top incomes: a historical and international perspective', *AEA Papers and Proceedings*, **96** (2), 200–205.

Piketty, T. and E. Saez (2014), 'Inequality in the long-run', *Science*, **344** (6186), 838–42.

Reich, R. (2013), *Inequality for All* [documentary film, directed by J. Kornbuth], distributed by RADiUS-TWC.

Saez, E. (2013), 'Striking it richer: the evolution of top incomes in the United States (updated with 2012 preliminary updates)', accessed 23 June 2014 at http://eml.berkeley.edu/~saez/saez-UStopincomes-2012.pdf.

Sayer, A. (2015), *Why We Can't Afford the Rich*, Bristol, UK: Policy Press.

Short, J.R. (2013), 'Economic wealth and political power in the second Gilded Age', in I. Hay (ed.), *Geographies of the Super-Rich*, Cheltenham, UK and Northampton, MA, USA: Edward Elgar Publishing, pp. 25–43.

Smith, R.C. (2001), *The Wealth Creators: The Rise of Today's Rich and Super-Rich*, New York: Truman Books.

Thorndike, J. (1980), *The Very Rich: A History of Wealth*, New York: Crown.

Veblen, T. ([1899] 1985), *A Theory of the Leisure Class*, London: Allen and Unwin.

Wallerstein, I. (2011), *The Modern World-System, Vol. IV: Centrist Liberalism Triumphant, 1789–1914*, Berkeley, CA: University of California Press.

2. Reconsidering the super-rich: variations, structural conditions and urban consequences

*Sin Yee Koh, Bart Wissink and Ray Forrest**

INTRODUCTION

For a long time, social science research paid little attention to the super-rich. The super-rich were not visible as a social problem and were also hard to locate and study. Instead, research tended to focus on the lower strata of society rather than on its upper echelons, while the middle classes only started to receive serious attention in the last few decades. However, evident processes of extreme and persistent income segregation and the emergence of elite enclaves in major cities have shifted the research agenda. The local effects of transnational real estate investments by the super-rich have now started to receive critical and, increasingly, politically charged attention (see Chapters 12 and 13 by Paris and Woods in this volume). There is, for example, increasing debate on the unbalanced attention to high-end real estate development in cities (Cook, 2010); on impacts on local real estate markets and prices that put properties and neighbourhoods out of reach of local residents (Ley and Tutchener, 2001; Ley et al., 2002; Ley, 2010); and on neighbourhoods where (second-home) houses remain unoccupied most of the year (Paris, 2011). As Lees (2012) highlights, gentrification of world cities and the 'gentrifying global elites' have now become a global problem.

Geographers have been key in setting the agenda for studies of the super-rich (Beaverstock et al., 2004; Hay, 2013b), particularly in urban studies (Pow, 2011) and the wealth management industry (Beaverstock et al., 2011, 2013). Other contributions to this literature come from mobilities studies (Elliott and Urry, 2010; Birtchnell and Caletrío, 2014), macroeconomics (Alvaredo et al., 2013; Piketty, 2014) and sociology (Volscho and Kelly, 2012; Keister, 2014; Keister and Lee, 2014). More recently, Hay and Muller (2014) call for a critical examination of super-rich philanthropy.

While academic research into the super-rich is thus rapidly evolving, we do observe some problems. First and foremost, the literature – academic, commercial and popular – *collectively* contributes towards a normative and actor-centred take on 'the super-rich'. This has been shaped in part by the growing voice of the commercial and popular literature, and in part by the large proportion of existing empirical works describing the social and material practices of the super-rich compared to those adopting structural analyses.[1] Regrettably, such writings can be easily conflated with portrayals of the super-rich in the commercial and popular literature, thus contributing towards the implicit assumption that the super-rich are a pre-existing social group, with limited attention given to the broader forces that structure their production. Fortunately, a rapidly emerging literature is now starting to address the structural basis of the super-rich and their agency (Hay and Muller, 2012; Volscho and Kelly, 2012; Piketty, 2014; Keister, 2014; Sassen, 2014; Sayer, 2014). But clearly more needs to be done to move the debate beyond the questions of 'Who are the super-rich?' and 'What do they do?' to questions of 'What made the super-rich and why?'

Second, and consequently, the academic literature inadvertently contributes towards a dichotomous (i.e., rich versus poor) reading of the super-rich that is already prevalent in the commercial and popular literature. For example, the *Vancouver Sun* reports that 'millionaire migrants. . .are raising housing costs for everyone' (Todd, 2014), making an explicit and literal reference to Ley's (2010) *Millionaire Migrants*. While Ley's book emphasizes the role of the neoliberal state and investor immigration policies that have enabled the transnational real estate purchases of Asian immigrants in Canada in the first place, as well as the challenges experienced by the millionaire migrants themselves, these points were not touched upon in the news article. The titles of academic writings are also not helping the case. For example, while Keister (2014) adopts a structural analysis to understand the super-rich in the USA, the article title 'The one per cent' can be easily taken out of context and conflated with media portrayals of a dichotomous divide between 'the rich and the rest' (Beddoes, 2012, p. 8).

Building on the emerging attention to structural conditions, in this chapter we aim to advance research on the super-rich by highlighting the need to tease out the structural explanations for the expansion of super-rich wealth and the implications of their activities for local, urban contexts. Following this introduction, the next section first reviews how the existing literature (academic, commercial and popular) collectively leads to normative and actor-centred readings of the super-rich. The subsequent three sections suggest ways to counteract that tendency, paying attention to the super-rich as both discursive and material products, and by teasing out the implications for urban studies research. We conclude by calling for a reconsideration of the super-rich in the social sciences generally, and in urban studies specifically.

DEFINING AND DESCRIBING THE SUPER-RICH

Despite increasing attention on the super-rich, there is 'no definitive threshold to meaningfully identify' this group (Hay, 2013b, p. 2). While some texts refer to 'the super-rich' without further explanation, other texts define this group more explicitly. In doing so, authors generally use one or more of four characteristics.[2] The first and most usual characteristic refers to material wealth, typically in the form of net worth (Davis, 1985; Haseler, 2000; Beaverstock and Faulconbridge, 2014). This builds in part on categorizations of the super-rich in the commercial wealth management literature (e.g., Capgemini and RBC Wealth Management, 2013b; Credit Suisse, 2013; Wealth-X Institute and UBS, 2013) and in part on popular literature fuelling the general public's fascination with the super-rich and their wealth (e.g., Simmons, 2011). However, more recent work by Piketty and others (Kopczuk and Saez, 2004; Atkinson and Piketty, 2007; Piketty, 2014) provides a different approach. On the basis of datasets constructed from tax records, they replace definitions of the super-rich in terms of predefined wealth thresholds by an analysis by centiles (and even the top thousandth) regarding income and wealth. These recent works advance an exploration of the top 1 per cent of the population in relation to their key characteristics without the need for any predefined level of wealth or income.

A second closely related characteristic to which various authors refer, is the capacity to generate and accumulate wealth. While earlier studies note the prevalence of intergenerational wealth through family inheritance (Pinçon and Pinçon-Charlot, 1998; Gilding, 2005), it appears that there is a shift towards self-made wealth in recent years. Indeed, an amazing two-thirds of the 1645 super-rich individuals included in the 2014 *Forbes* billionaires list 'built' their own fortunes in various ways, compared to 13 per cent who inherited their wealth (Dolan and Kroll, 2014). This has to do with the expansion of the global economy and technological developments that have enabled some to take advantage of entrepreneurial opportunities in the global arena. Short (2004, p. 112) writes of the 'global super-rich' who are 'key movers and shakers in the global economy. . .whose reach and influence encompass the entire capitalist system'. Whether through family inheritance or entrepreneurship of various kinds, what is crucial here is that the super-rich, more than others, have the capacity to generate and accumulate wealth.

This capacity is closely related to the third characteristic of the super-rich to which the literature regularly refers: their elite positions as 'people of the higher circles' (Mills, 1956, pp. 3–4). Carroll (2010) uncovers connected networks between members of super-rich families and their directorial positions in the world's top 500 (G500) companies, as well as the important roles of 'key non-billionaire networkers [who] serve as *brokers*, connecting billionaire social circles that would otherwise remain disjointed' (p. 150; original emphasis).[3] Urry (2014a) suggests that these networks operate in a more systematic and conscious manner, manipulating tax and regulatory regimes well beyond the evident and disproportionate influence of wealthy elites in political lobbying. More recently, *The Economist* (2014) found that involvement in 'crony' sectors[4] with heavy rent-seeking practices contributed significantly to billionaire wealth. Clearly more needs to be done to examine how various super-rich networks interconnect and operate in relation to the mobilization, deployment and protection of extreme wealth holdings.

Finally, the super-rich are sometimes characterized by their transnational mobility, owning multiple homes (Paris, 2011, 2013) and a variety of investments in world cities and offshore tax havens (Shaxson, 2011; Urry, 2014b). Recently, the Economic Intelligence Unit (EIU) reports on 'internationally-mobile wealthy individuals (IMWIs)', defined as those who 'live or work outside their country of birth or spend more than half of their time outside their home country' (EIU and RBC Wealth Management, 2012, p. 3). In parallel to this, the mobilities literature (Birtchnell and Caletrío, 2014) suggests that super-rich mobility is highly exclusive: the modes of mobility (e.g., private jets and yachts) are only affordable to the super-affluent, while privacy and security are highly sought after by the super-rich due to a desire for comfort and safety (see Budd, Chapter 15 this volume). However, it is easy to overstate the 'residential promiscuity' of the super-rich, as social factors like gender and life course, as well as the strategic locations of core business interests may be important structuring factors. Furthermore, their 'hypermobility' (Paris, 2013) also needs to be contextualized to the increasing ease and norm of transnational travel and migration that have similarly enabled the mobility of other segments of society.

In sum, the existing literature draws on a number of characteristics to describe and therefore define the super-rich. First, the super-rich possess extreme wealth (or super-affluence) relative to the rest of society. Second, and in relation to this, they have the capacity to generate and accumulate wealth. Third, this capacity is closely related to their positions in the socio-economic hierarchy and their ([il]legal) networked connections (see Hall, Chapter 10 this volume). Finally, some super-rich individuals enjoy hyper-mobility, living transnational lives across multiple homes and travelling in highly exclusive styles.

These four characteristics, implicitly and explicitly invoked by much of the existing literature, have contributed towards the normative and actor-centred image of 'the super-rich' in current debates. We argued earlier that, notwithstanding the growing attention to the structural conditions of the super-rich and their agency, parts of the academic literature have been complicit with the commercial and popular literature in perpetuating a certain image of the super-rich – associated with financial wealth, luxurious lifestyles, elite status, power and transnational mobility. Furthermore, as their 'conspicuous' spending and luxurious lifestyles are also made visible by the super-rich literature (Paris, 2011; Featherstone, 2014a, 2014b), it then becomes easy to point fingers at the super-rich as the prime targets in debates around the social problems of wealth inequality.[5]

However, we argue that this is problematic as it shifts attention away from two critical tasks. First, we must build on recent work to interrogate the structural forces that have contributed to the rise of the super-rich. As Nathwani (2014; original emphasis) observes, attacks on the 1 per cent have been on 'bourgeois *personhood* rather than on economic, cultural, and social structures that permit the person her place of privilege at the expense of the 99 per cent'. Second, it is important to go beyond the uniformly enviable facade of the super-rich that has been discursively sustained by parts of the super-rich literature, to examine the stratification and differentiations within, and between, 'the super-rich'. This goes beyond taking for granted the super-rich as actors with agency – or an uncritically assumed *super*-agency due to their wealth, mobility and elite status – in order to position the super-rich with regard to broader structural and historical developments, as well as to understand how and why segmentations of the super-rich relate to those forces.

THE SUPER-RICH AS A DISCURSIVE PRODUCT

An analysis of the structural conditions behind the rise of the super-rich should have a dual focus, embracing both the discursive and material production of the super-rich. The first of these tasks includes analysing how and under which conditions the super-rich came to be perceived as a distinct social group. After all, as research in cultural studies shows, social groups are 'fluid and often shifting, but nevertheless real' (Young, 1990, p. 9). Ahistorical definitions of the super-rich are problematic as they contribute to the illusion that the super-rich are a pre-existing social group, obscuring the socio-political work and related (political) aims and assumptions underlying their constitution. After all, '[c]ommunication is. . .always ideological in that it not only (1) constitutes identities and relationships of power, and (2) reproduces dominant systems of belief (or

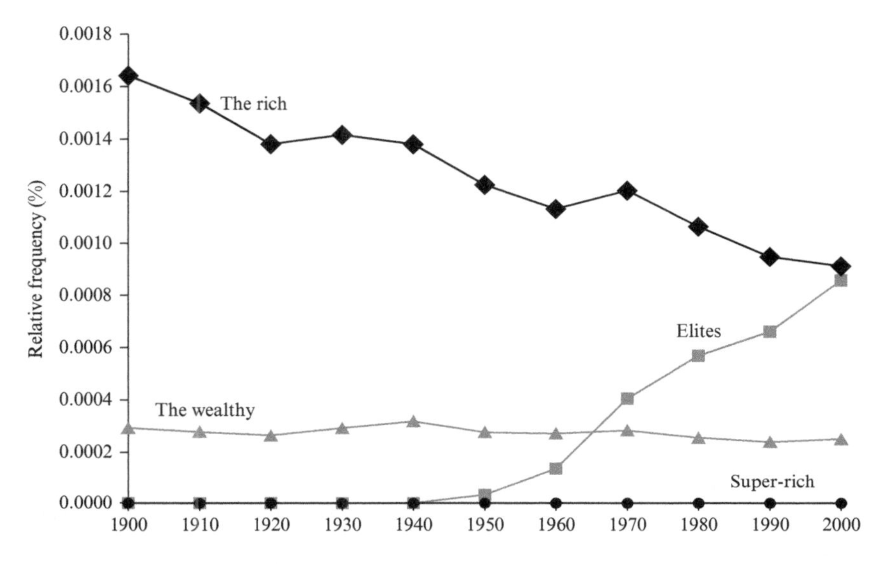

Source: Michel et al. (2011).

Figure 2.1 Trends of the usage of selected wealth-related terms in English-language books, 1900–2000

"good taste"), but also because it (3) maintains structures of inequality and privilege' (Thurlow and Jaworski, 2012, pp. 490–91).

The term 'super-rich' was not a popular term used to describe the rich and wealthy before the 1970s (Figures 2.1 and 2.2).[6] Instead, the (English) literature used terms such as 'the rich', 'the wealthy', and 'elites'. Writings on the 'rich' and 'wealthy' started to emerge in the early nineteenth century. During the 1940s, the 'curious literary genre' (Pessen, 1990, p. 310) of the 'rich lists' such as *Wealth and Biography of the Wealthy Citizens of New York City* (Beach, 1845) started to flourish. These were not so much concerned with the amount of wealth possessed by the wealthy, but instead emphasized individual biography and genealogy. However, these early 'rich lists' were factually flawed, containing highly subjective comments. Bailey (1954, p. 25), for example, notes that the biographical comments 'are not reliable and are sometimes derogatory'. Nonetheless, they capture the materialistic and capitalistic spirit of that era.

Attention to the rich and wealthy continued to grow in relation to the second Industrial Revolution in America and Britain (late nineteenth to early twentieth century). It was during this first Gilded Age that some individuals and families were able to amass financial assets and other

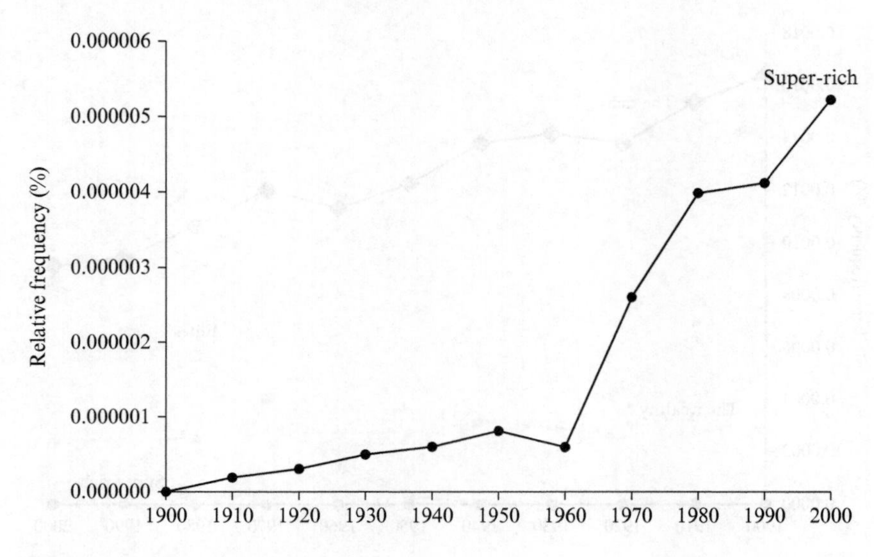

Source: Michel et al. (2011).

Figure 2.2 Trends of the usage of 'super-rich' in English-language books, 1900–2000

capital (cf. Short, 2013). Rapid technological advances, industrialization and urbanization led to the emergence of an urban middle class. Pioneers in industries such as manufacturing, oil, rail transport and newspaper publishing saw a rapid expansion in their financial wealth. This increasing wealth and disposable income related to an emerging culture of 'conspicuous consumption' – a term Veblen ([1899] 1992) coined to describe public displays of affluence by the emerging rich, in his mind primarily motivated by greed and desire.

While the fascination with the rich continued well into the twentieth and twenty-first centuries, there has been a notable shift. Before the 1980s, rich people were called 'the wealthy' (e.g., Katona and Lansing, 1964; Wooster, 1967), or 'wealth holders' (Rubinstein, 1981). Arguably, it was only from the late 1970s and early 1980s that term 'the rich' gained popularity in conjunction with the emergence of what some describe as the second Gilded Age (Short, 2013). This suggests a shift from attention on 'the wealthy' – more specifically, wealth that is tied to genealogy and family inheritance – to 'the rich' – which alludes to wealth that is tied to new money as a result of the expansion of the global economy. This shift from 'the wealthy' to 'the rich' can be traced through the development of two parallel literatures: rich lists and research monographs on the wealthy and elites.

Forbes published rich lists as early as 1918, but it was only from 1982 that it started to publish the annual *Forbes 400*, ranking the 400 richest individuals in America by their net worth (Hesseldahl, 2002). In Australia, the *Business Review Weekly* introduced *The BRW Rich 200*, a list of the top 200 wealthiest individuals and families in 1984. In the UK, *The Sunday Times* followed with the yearly *The Sunday Times Rich List* as a magazine supplement in 1989. Listing the 1000 wealthiest individuals or families who reside predominantly in Britain, this list was even more ambitious than its predecessors. Reflecting China's emergence as an economic world power, the *Hurun Report* started its China Rich List in 1999. While the early rich lists focused predominantly on individual countries, there soon followed a shift to the global stage. *Forbes*, for example, started publishing the *World's Billionaires* from 1987, and has also listed the rich in specific countries including Hong Kong, India, Japan, Africa and China. Bloomberg introduced the *Bloomberg Billionaires Index* in 2012, and it is updated on a daily basis. In a different type of rich list, Beresford and Rubinstein (2007) present an historical overview of the 250 wealthiest Britons since the year 1066. It might be true, as *The Huffington Post* reports, that 'we live in a society that has long had a fascination with the ultra rich' (El-Erian, 2012), but this emergence of a daily updated rich list suggests a new 'rich list obsession' (Day, 2011), constituting an increasing visibility of the rich and their wealth in the popular media.

In addition to these rich lists, there is also a specialized literature on the rich and their wealth. Ferdinand Lundberg's *America's 60 Families* (1937) and *The Rich and the Super-Rich* (1968) for instance, discuss the concentration of wealth in the hands of selected individuals and families. Edward Pessen (1990) presents 'the rich and elite' in major American cities in *Riches, Class, and Power: America Before the Civil War*. William Rubinstein (1981) examines the accumulation of Victorian British wealth in *Men of Property*, and he continues to write books on the history of wealthy individuals in Britain (2009). In *Grand Fortunes*, Pinçon and Pinçon-Charlot (1998) discuss wealthy families in France, while Michael Gilding (1999) describes 'Superwealth in Australia: entrepreneurs, accumulation and the capitalist class', and more recently the *Secrets of the Super Rich* (2002). These works build upon earlier writings by economists, sociologists and historians on wealth and income distribution since the late nineteenth century (see Rubinstein, 1981, pp. 19–22). Additionally, there are also studies on powerful corporate individuals such as Michael Useem's (1984) *The Inner Circle*, and Charles Wright Mills's (1956) *The Power Elite* on individuals who hold prominent military, economic and political decision-making positions in America. Recently, Robert Frank ([1999] 2010) published *Luxury Fever*, arguing that there is a culture of wasteful consumption as the general public blindly follows the lifestyles and consumption practices of the super-rich. These works suggest that the emergence of the wealthy and powerful in American and British society have been of academic, socio-political and public interest since the mid-twentieth century.

The term 'the super-rich' can be attributed to Lundberg's (1968) *The Rich and the Super-Rich*. While his first book (1937) examined how political and media power was held by a small group of wealthy families, this second book argued that the majority of the 'new rich' were inheritors of the big fortunes concentrated amongst the 60 families he had examined earlier. Indeed, the rise of family fortunes in mid-twentieth-century America saw a corresponding emergence of family offices beginning in the late 1980s. Companies such as Family Office Exchange (FOX) (since 1989), the Institute for Private Wealth Investors (since 1991) and the CCC Alliance (since 1994) were established and offered investment and private wealth management education, as well as discreet peer-to-peer networking opportunities for ultra high net worth (UHNW) clients. Services offered by these family offices are predominantly focused on the clients' financial capital, and range from centralized management of investments, tax planning, estate planning, succession and inheritance planning, philanthropic planning, to lifestyle and concierge services.

The wealth management industry is an important complicit contributor to the discursive production of the super-rich as a distinct social group. Through the publication of annual wealth reports, press releases, infographics and news bites, the research arms of this industry help to discursively produce the clientele that it serves. For example, the Wealth-X Institute (2013) published a press release outlining the 'statistical profile of the Hong Kong UHNWI'. According to this document, on average, the Hong Kong super-rich individual is 57 years old, male, married, and has an average net worth of US$150 million. Such typical profiles contribute towards the construction of the super-rich individual in concrete ways – we know his or her age, gender, marital status and asset class. Indeed, reports on wealthy lifestyles, luxury spending and investment preferences (e.g., Wealth-X Institute, 2012; Capgemini and RBC Wealth Management, 2013a; Wilkinson, 2013; Knight Frank, 2014; Wealth-X Institute and Arton Capital, 2014) have us believe that the super-rich *do* exist.

In addition to the wealth management industry literature, news and popular media also work to produce the super-rich. In general, recent media and popular writings on the super-rich can be categorized into three groups. The first focuses on depicting the enviable lifestyles of the super-rich. For example, reporting on a study by the Boston Consulting Group, *The Guardian* notes that the super-rich 'shift their thrills from luxury goods to costly experiences' (Wood, 2014). A second group describes the rise of the super-rich, particularly in the American context. Crucially, this literature contributes towards dichotomizing the super-rich and the non-super-rich. For example, Freeland's (2012) popular book *Plutocrats* is subtitled *The Rise of the New Global Super-Rich and the Fall of Everyone Else*. A third group takes a normative stand in blaming the super-rich. For example, *The Guardian* reports that the Equality Trust think-tank found that inequality 'costs Britain £39 [billion] a year' (McVeigh, 2014). And another *Guardian* article, drawing on an Oxfam report, highlights that 'Britain's five richest families now own more wealth than the poorest 20 per cent of the population' (Elliott, 2014). In doing so, this literature also produces a moral segregation between the 1 per cent and the 99 per cent. Consequently, social problems such as wealth inequality, gentrification and urban segregation are easily blamed upon the super-rich, their wealth and their 'conspicuous' lifestyles.

There are important socio-political implications if we proceed to analyse the super-rich without paying attention to their discursive production. Attention is shifted away from the way in which the super-rich are constituted as a social group in relation to specific historical and material settings. This runs the risk of producing social science research that privileges the present without necessarily accounting for the historical hinterlands that have shaped the discourse of the subject matter in question, and related power inequalities. Furthermore, research along these lines also fails to acknowledge possible variations amongst the super-rich and their practices, as well as variations between the super-rich in different places in the world.

The discursive construction of the super-rich as a distinct group also structures normative attitudes towards this group that are easily reproduced in research and popular understanding. Dichotomous views of 'the super-rich' versus 'everyone else', for example, can influence the entry points through which researchers tackle issues related to the super-rich; and this in turn can easily result in a negative reading of 'the super-rich'. Thus, there are numerous studies of elite enclaves, elite mobilities and super-rich lifestyles (including in this volume), but a relative lack of empirical studies on their social integration and relations to others (e.g., productive comparative research into the super-rich versus the non-super-rich). While there could certainly be a case for giving special attention to this group (e.g., by virtue of their characteristics, status and activities), there is the more important task of getting to the 'deeper systemic dynamics that articulate much of what now appears as unconnected' (Sassen, 2014, p. 5).

THE SUPER-RICH AS A MATERIAL PRODUCT

Research into the constitution of the super-rich should also focus on the mechanisms of their material production (and perhaps even their moral constitution as Schervish discusses in this volume in Chapter 7). After all, the growth and increasing visibility of individuals and households with extraordinary levels of wealth is inextricably linked to the restructuring and reshaping of contemporary capitalism over the last three to four decades. Early conceptions of these structural transformations were typically described as 'polarization' – a tendency for an increasing concentration of households towards either ends of the income/wealth spectrum or a growing rift between the work rich and work poor. Needless to say, this view of the changing social structure, which originated in the early 1980s, generated considerable debate and disagreement (Hamnett, 1994, 1996; Van Kempen, 1994). Moreover, there was a lack of empirical precision and persistent vagueness about whether the polarization referred to income or wealth, or both. Also, unsurprisingly, initially there was a strong focus on the impoverished end of the spectrum, on notions of an underclass or disconnected marginal poor (e.g., Auletta, 1982; Murray, 1990). Pahl's (1988) thoughtful piece on this subject did conceive of an onion-shaped structure in which both the top end (the very wealthy) and the bottom end (the very poor) were becoming increasingly divorced from the middle mass. As intimated earlier, academic and popular concern focused on the problems of extreme poverty rather than extreme wealth.

This neglect of the changes occurring in the upper echelons is easy to understand because of the relative lack of reliable empirical data on the extremely wealthy but also, and more importantly, because of the pervasive belief in continued social mobility in the early phases of the neoliberal era. The middle class escalator was still believed to be recruiting increasing numbers to the ranks of what Piketty (2014) refers to as the patrimonial middle class – essentially a middle mass of residential property owners. Who cares if the very rich are getting even richer, if most of us are also doing better, or have realistic prospects of doing so?

Economic shocks and crises have, however, created a very different set of expectations and assumptions bolstered by the evidence of increasing income inequality. The middle class has taken a growing share of the tax burden, of so-called austerity measures – the poor are too poor and the rich are too elusive, too offshored and well served by tax advisors (Urry, 2014b).

More regressive tax regimes across a wide range of societies (most notably, generally lower marginal rates), extreme salary inflation among the senior managerial class, policies of privatization and outsourcing are all interconnected features of the reshaping of contemporary capitalist societies. In combination, and in different ways in different societies, these processes have produced different factions of the super-rich. For example, the 'shock doctrine' (Klein, 2007) restructuring of the Russian economy produced the oligarchs who have amassed prodigious wealth from the disposal of state assets at knockdown prices. And sharply, lower top tax rates have both encouraged the payment of super-high corporate salaries in the USA and also enabled those who receive them to retain a substantial proportion. A sophisticated and complex structure of financial intermediation plays the crucial role of lubricating the machinery of financialized, neoliberal capitalism. In this context Davies (2014) refers to 'the tyranny of intermediaries', the power and influence of those 'who sit in the crevices between businesses, enabling the market system to work at all' (p. 1). Davies's broader point is that a fixation with the super-rich or 1 per cent risks a neglect of the nuts and bolts of the system, the panoply of actors and institutions that enable and construct the super-rich. The system seems to be on remote control, lacking critical targets. As he further observes, 'shaming and taking down such individuals leaves a zombie system still unharmed. Targeting the "1 per cent" only confirms the frustration of our current predicament: the culprits are so few as to be virtually invisible' (p.1).

Building on extensive research with several others, Piketty's (2014) more recent and substantial intervention shows that in absolute terms the 1 per cent are more visible than has been assumed and can represent a significant number of people in a particular society as well as globally. His extensive historical analysis[7] provides a systematic and sober framing of the current fascination with the super-rich. Whilst emphasizing throughout his analysis that debates around the distribution of income and wealth in societies are, and should be, matters of political and democratic discussion, his empirical work aims to provide unassailable evidence of structurally driven growing income and wealth inequalities. Essentially, his argument is that the erosion of income from capital, the decline in power and influence of the capitalist class, roughly between 1910 and 1970, was a product of two World Wars and the associated economic and social turmoil they generated. This period saw a general trend towards more progressive income structures and wealth distributions. But historically, the rate of return on capital exceeds the rate of growth of output and income and that historical dynamic is now reasserting itself. The consequence is growing inequality with the evident increase of the super-rich. Put simply, from a longer historical perspective, the current rapid increase in inequalities appears to mark a return to business as usual, albeit with some clear differences in relation to the shaping and composition of the rich elite of the twenty-first century compared to their counterparts of the early twentieth century. For example, the highly paid supermanagers are a distinctive contemporary phenomenon. And those who derive their income primarily from capital are a very select group within the top 1 per cent (see detailed discussion in Piketty, 2014, pp. 276–7).

FROM SPACES OF THE SUPER-RICH TO SUPER-RICH SPATIALITY

What are the implications of these arguments for the urban literature on the super-rich? For a long time urban studies paid little attention to the rich although there are notable exceptions such as Zorbaugh's (1929) *The Gold Coast and the Slum*, which provided a detailed sociological analysis of the 'social game' through which social climbers tried to become part of Chicago's elite (dubbed the 'Four Hundred'). A decade later, Lynd and Lynd (1937) argued that the so-called X family was central to the control of the business class of local politics in *Middletown*, a pseudonym for Muncie, Indiana. Their analysis laid the groundwork for a tradition of community power structure studies, discussing the influence – or lack thereof – of local elites on decision-making (cf. Hunter, 1953; Dahl, 1961; Domhoff, 1967, 2014; Stone, 1989).

Notable as these texts are, they were an exception in an urban studies literature that, like most other social sciences, generally paid little attention to elites.[8] Much urban research was understandably more committed to addressing social problems of the poor and deprived. This changed radically in the early 1970s with much greater attention given to the formative influence of capitalist elites in the development of cities (Castells, 1972; Harvey, 1973). Five years after the publication of *Whose City?* in 1970, Ray Pahl (1975, p. 1), argued that '[o]ne does not have to be very astute now in order to answer the question "Whose City?": quite evidently the capitalists own British cities and up to 1973 they grew fat on their rents and the revaluations of their portfolios'. This was recently echoed and radicalized by Andy Merrifield's (2013) bold assertion that 'parasites' own our cities:

> In parasitic cities, social wealth is consumed through conspicuously wasteful enterprises, administered by parasitic elites, our very own aristocracy (the 1%) who squander generative capacity by thriving exclusively from unproductive activities: they roll dice on the stock market, profit from unequal exchanges, guzzle at the public trough; they filch rents and treat land as a pure financial and speculative asset, as a form of fictitious capital.

Capitalist elites were never popular with social science commentators and this did not get any better after the 2007–08 global financial crisis.

The financial crisis directly relates to changes in capitalism that started to take shape in the 1970s (Harvey, 2010). The resulting new period of 'late' capitalism saw the advent of transnational corporations that were increasingly strategic about the location of their activities (Dicken, 2011). National governments shifted away from investments in domestic industrial production towards the higher returns of service industries. And neoliberal city governments replaced a managerial approach focused on redistribution and social investment with an entrepreneurial style of governance aiming to 'win' intercity competition by appeasing business interests in the 1980s (Harvey, 1989; Jessop and Sum, 2000).

Ushering in new 'spaces of neoliberalism' (Brenner and Theodore, 2002), these transformations proved to have radical consequences for cities, theorized and critiqued in various related strands of urban research. Global city research showed how some cities were transformed into key nodes of the global economy, thus producing new urban hierarchies and uneven geographies – both between cities and inside cities (Smith, 1984; Sassen, 1991; Castells, 1996). Theories of neoliberalism revealed how elites controlled wealth and power through new strategies of accumulation by dispossession (Harvey, 2004). Gentrification research focused on how elites took over and transformed residential neighbourhoods (Smith, 1996) and research into the privatization of public space questioned the transformation of generally accessible and mixed public spaces into exclusionary collective spaces aimed at consumption (Sorkin, 1992). Research into gated communities spoke of affluent residents who increasingly separated themselves spatially from the rest of society (Blakely and Snyder, 1997; Atkinson and Blandy, 2005). And Graham and Marvin (2001) suggested that the privatization of infrastructures was one of the driving forces behind this 'splintering urbanism'.

Encompassing a plethora of topics, these theories are inevitably quite diverse. But against the background of the discussion about the super-rich, they nonetheless have some striking characteristics in common. First, while they present economic globalization and neoliberalism as the institutional setting for urban transformations, some theories remain rather unclear about the driving actors within that setting. In Sorkin's (1992) analysis, for instance, the acting agents never really become clear, while Davis (1992) speaks cryptically about 'inexorable forces'. But implicitly or explicitly, many authors see elites as the main driving actors. For Harvey (2010), it is the 'capitalist class' that is driving spatial transformation, while Smith (1996) points at 'yuppies and real estate magnates'. In gated communities research, the consumption practices and production interests of elites are presented as implicit factors in the rapid emergence of urban enclaves.

At the other end of the spectrum – and this is a second characteristic – these theories discern the underprivileged as the losers in urban transformations. Again, this group is mostly treated as a coherent body. Living within structural constraints, the underprivileged are pushed away into residual places.[9] Third, urban transformations are presented as generic developments, taking place in cities around the world. Thus, Sorkin (1992, p. xv) argues that the 'sites discussed are representative; they do not simply typify the course of American urbanism but are likely to be models for urban development throughout the world'. There is little attention to differences between urban enclaves in different places, or for the specific ways in which these relate to and are influenced by local contexts (Douglass et al., 2012). And fourth and directly related, these transformations to the city are interpreted negatively (Merrifield, 1996; Crawford, 1999; Amin and Thrift, 2002), as they threaten community, solidarity and democracy.

Our previous discussion points to a number of unanswered empirical questions that do not automatically come into view within the existing literature: Who are these super-rich, and how do they perceive themselves? How do they behave, both producing and consuming cities? Which variations can we observe amongst the super-rich? How are their behaviours structured through networks of intermediaries? How is this different in various cities with their own unique institutional networks and spatial settings? And how does the rest of society perceive their agency? Answering these questions on the one hand acknowledges that the super-rich are not one coherent group, and they are not the same everywhere. For instance, they might have diverse interests, related to the source of their wealth. Or they might participate in different social practices, related to the location of their residence. On the other hand, these questions acknowledge that the super-rich operate in a structured setting, which generates and forms their agency. Both adjustments replace mere descriptions of lifestyle and practices of the super-rich, and of specific spaces of the super-rich, by a detailed analysis of the localized practices of the super-rich and of the structural settings within which they operate. Informed by the arguments in this chapter, we suggest four particular lines of research inquiry.

First, there is an obvious need for research into the influence of the super-rich in the production of space. On the one hand, this research could focus on the agency of specific individuals in development processes: see, for instance, the role of Cédric Naudon in *La Jeune Rue* development in Paris (Rodriguez, 2014). How does the personal involvement of super-rich entrepreneurs influence the outcomes of these projects? On the other hand, this research should also highlight the structural conditions under which these entrepreneurs operate, with special attention to the intermediaries that structure practices of production (Davies, 2014). Through comparative research, attention should also be placed on the mediation of global power in local urban settings, and to the possible local variations in space that result from this.

Second, there is a need for research to extend into the situated connections of the super-rich within cities and outside of cities, both with the super-rich and with others. This should include a comparative analysis of the residential segregation of the super-rich in different cities, as well as an inquiry into their *everyday* mobilities including both visits to exclusionary and mundane spaces. Groups of super-rich may be very different; and their spatial practices may also differ between cities. It is important, for example, not to overstate the residential 'promiscuity' of the super-rich.

Third and related to this, building on existing research (e.g., Paris, 2013) we advocate additional research into the various investment strategies regarding residential properties by the super-rich. Residential properties might play a crucial role in the life worlds of the super-rich, anchoring their mobilities, or they might have a specific function in relation to securing citizenship and residential statuses, facilitating wealth generation and capital accumulation, or accessing desired schools and colleges for children. These rationales may not be apparent if residential properties of the super-rich are solely seen as investment vehicles.

Fourth, we suggest undertaking research into the discursive construction of the super-rich and their agency by diverse groups in the city. Which controversies invoke a discussion of super-rich agency? How are the super-rich and the effects of their actions perceived? And which (re)actions are suggested? As there could be considerable differences in super-rich narratives in different cities, this research should again have a clear comparative component.

CONCLUSIONS – RECONSIDERING THE SUPER-RICH

As Neumayer (2004, p. 795) notes, '[t]he accumulation of great fortunes creates uneasiness, envy and concern in many people'. Indeed, media and popular reports on the super-rich have focused on extreme wealth, conspicuous lifestyles and the impacts of transnational real estate investments on local housing markets. As we have argued, academics have also played a role in this discursive construction of the super-rich as a distinct and coherent social group.

However, the super-rich make up a diverse group: they generated wealth in very different ways, and they live varying social lives that may take place differently in different cities. Research into these social and spatial variations will help to generate a more nuanced understanding of 'super-richness', and more precise critiques of normative judgements about the super-rich. Notwithstanding this, we have also argued that the super-rich function within broader structural forces that make and sustain their capital, status and networks. Attention to the structural and material production of the super-rich does not remove the spotlight from the super-rich, but it does support going 'beyond merely "exposing" or documenting. . .privileged geographies and valorized transnational lifeworlds' (Pow, 2011, p. 392). It also helps to show that other actors (e.g., governments and transnational corporations), structural forces and intermediaries are implicated in the production of the super-rich.

As Therborn (2000, p. 155) puts it: 'Is the world a system shaping the actors in it and directing their strivings, or is it an arena, where actors who were formed outside act and interact?' The super-rich are important actors in an increasingly networked world of flows and mobility, but they act in a predetermined setting that they cannot fully control. What is required is an analytical stance that emphasizes variations in practices as well as the factors that structure actions of the super-rich in contrast to the overly normative and actor-centric discourses that have shaped the existing literature. In Sassen's (2014, p. 13) words, what 'we are seeing [is not so much] the making. . .of predatory elites but of predatory "formations", a mix of elites and systemic capacities with finance as a key enabler, that push toward acute concentration'. What we need is research that is firmly grounded in such an understanding in order to produce and invigorate transformative scholarship.

NOTES

* The work described in this chapter was substantially supported by a grant from the ESRC/RGC Joint Research Scheme sponsored by the Hong Kong Research Grants Council and the Economic and Social Research Council in the United Kingdom (Project Reference No. ES/K010263/1). We thank participants to the panel on 'The Global Alpha Territory: The Super-Rich and their Place in Contemporary Urbanism' during the City Futures III conference in Paris (18–20 July 2014) for comments on an earlier draft of this chapter. At the time of writing, Sin Yee Koh was Postdoctoral Fellow in the Department of Public Policy at City University of Hong Kong.

1. For example, see edited books such as *Elites Mobilities* (Birtchnell and Caletrío, 2014) and *Geographies of the Super-Rich* (Hay, 2013b).
2. We are not suggesting here that individuals must satisfy one or more of these characteristics in order to be classified as the super-rich. Our claim here is that these are the usual characteristics that have been used in the literature.
3. An obvious example of such broker networks is the niche private wealth management industry that has emerged to service the financial needs of the super-rich (Beaverstock et al., 2011).
4. These include: casinos; coal, palm oil and timber; defence; deposit-taking banking and

investment banking; infrastructure and pipelines; oil, gas, chemicals and other energy; ports and airports; real estate and construction; steel, other metals, mining and commodities; utilities; and telecom services.

5. For an exception, see Nader (2009). Exploring what would happen if the super-rich would be willing to become a driving force in America, organizing and institutionalizing citizen interests, Nader displays a more positive reading of this group.
6. While there are some issues with the Ngram corpus (e.g., representativeness of the corpus, and the socio-cultural and temporal settings influencing how terms are used), this is used here to illustrate the general trends in the usage of the selected terms in English books.
7. And associated website at http://topincomes.parisschoolofeconomics.eu/.
8. For exceptions, see research on class and elites by Mills (1956), Stanworth and Giddens (1974), Sklair (2001) and Savage and Williams (2008).
9. For an exception, see Merrifield (1996) who argues that 'underprivileged' groups are not easily pushed away.

REFERENCES

Alvaredo, F., A.B. Atkinson and T. Piketty et al. (2013), 'The top 1 percent in international and historical perspective', *Journal of Economic Perspectives*, **27** (3), 3–20.
Amin, A. and N.J. Thrift (2002), *Cities: Reimagining the Urban*, Cambridge, UK: Polity.
Atkinson, A.B. and T. Piketty (eds.) (2007), *Top Incomes over the Twentieth Century: A Contrast between Continental European and English-Speaking Countries*, Oxford/New York: Oxford University Press.
Atkinson, R. and S. Blandy (2005), 'Introduction: international perspectives on the new enclavism and the rise of gated communities', *Housing Studies*, **20** (2), 177–86.
Auletta, K. (1982), *The Underclass*, New York: Random House.
Bailey, R.F. (1954), *Guide to Genealogical and Biographical Sources for New York City (Manhattan): 1783–1898*, New York: R.S. Bailey.
Beach, M.Y. (1845), *Wealth and Biography of the Wealthy Citizens of New York City*, 6th edition, New York: The Sun.
Beaverstock, J.V. and J.R. Faulconbridge (2014), 'Wealth segmentation and the mobilities of the super-rich: a conceptual framework', in T. Birtchnell and J. Caletrío (eds), *Elite Mobilities*, Abingdon, UK/New York: Routledge, pp. 40–61.
Beaverstock, J.V., S. Hall and T. Wainwright (2011), 'Servicing the super-rich: new financial elites and the rise of the private wealth management retail ecology', *Regional Studies*, **47** (6), 834–49.
Beaverstock, J.V., S. Hall and T. Wainwright (2013), 'Overseeing the fortunes of the global super-rich: the nature of private wealth management in London's financial district', in I. Hay (ed.), *Geographies of the Super-Rich*, Cheltenham, UK and Northampton, MA, USA: Edward Elgar Publishing, pp. 43–60.
Beaverstock, J.V., P. Hubbard and J. Rennie Short (2004), 'Getting away with it? Exposing the geographies of the super-rich', *Geoforum*, **35** (4), 401–7.
Beddoes, Z.M. (2012), 'For richer, for poorer', *Special Report: World Economy, The Economist*, 13 October, accessed 17 February 2014 at http://www.economist.com/sites/default/files/20121013_world_economy.pdf.
Beresford, P. and W.D. Rubinstein (2007), *The Richest of the Rich: The Wealthiest 250 People in Britain since 1066*, Petersfield, UK: Harriman House.
Birtchnell, T. and J. Caletrío (eds) (2014), *Elite Mobilities*, Abingdon, UK/New York: Routledge.
Blakely, E.J. and M.G. Snyder (1997), *Fortress America: Gated Communities in the United States*, Washington, DC: Brookings Institution Press.
Brenner, N. and N. Theodore (2002), *Spaces of Neoliberalism: Urban Restructuring in North America and Western Europe*, Malden, MA/Oxford: Blackwell.

Capgemini and RBC Wealth Management (2013a), *Insights into Population, Wealth and Investment Behaviors: The Asia-Pacific Wealth Report 2014*, accessed 13 February 2014 at http://www.worldwealthreport.com/apwr.

Capgemini and RBC Wealth Management (2013b), *World Wealth Report 2013*, accessed 13 February 2014 at http://www.capgemini.com/resource-file-access/resource/pdf/wwr_2013_0.pdf/.

Carroll, W.K. (2010), *The Making of a Transnational Capitalist Class: Corporate Power in the Twenty-First Century*, London/New York: Zed Books.

Castells, M. (1972), *La Question Urbaine*, Paris: François Maspero.

Castells, M. (1996), *The Rise of the Network Society. Information Age 1*, Cambridge, MA: Blackwell Publishers.

Cook, A. (2010), 'The expatriate real estate complex: creative destruction and the production of luxury in post-socialist Prague', *International Journal of Urban and Regional Research*, **34** (3), 611–28.

Crawford, M. (1999), 'Blurring the boundaries: public space and private life', in J. Chase, M. Crawford and J. Kaliski (eds), *Everyday Urbanism*, New York: Monacelli Press, pp. 22–35.

Credit Suisse (2013), *Global Wealth Report 2013*, accessed 17 February 2014 at https://publications.credit-suisse.com/tasks/render/file/?fileID=BCDB1364-A105-0560-1332EC9100FF5C83.

Dahl, R.A. (1961), *Who Governs?: Democracy and Power in an American City*, New Haven, CT: Yale University Press.

Davies, W. (2014), 'The tyranny of intermediaries: who writes the rules of our modern capitalism?', *IPPR*, accessed 10 March 2014 at http://www.ippr.org/juncture/171/11905/the-tyranny-of-intermediaries-who-writes-the-rules-of-our-modern-capitalism.

Davis, M. (1992), *Beyond Blade Runner: Urban Control: The Ecology of Fear, Open Magazine Pamphlet Series No 23*, Westfield, NJ: Open Media.

Davis, W. (1985), *The Rich: A Study of the Species*, London: Arrow Books.

Day, P. (2011), 'Fascinating fortunes: our rich list obsession', *BBC News*, 23 April, accessed 22 May 2014 at http://www.bbc.co.uk/news/business-13150500.

Dicken, P. (2011), *Global Shift: Mapping the Changing Contours of the World Economy*, 6th edition, New York: The Guilford Press.

Dolan, K.A. and L. Kroll (2014), 'Inside the 2014 Forbes billionaires list: facts and figures', *Forbes.com*, 3 March, accessed 25 March 2014 at http://www.forbes.com/sites/luisakroll/2014/03/03/inside-the-2014-forbes-billionaires-list-facts-and-figures/.

Domhoff, G.W. (1967), *Who Rules America?* 1st edition, Englewood Cliffs, NJ: Prentice-Hall.

Domhoff, G.W. (2014), *Who Rules America? The Triumph of the Corporate Rich*, 7th edition, New York: McGraw-Hill Education.

Douglass, M., B. Wissink and R. van Kempen (2012), 'Enclave urbanism in China: consequences and interpretations', *Urban Geography*, **33** (2), 167–82.

Economic Intelligence Unit (EIU) and RBC Wealth Management (2012), *Wealth Through the Prism of Culture and Mobility: An Economist Intelligence Unit Report on Behalf of RBC Wealth Management*, accessed 13 February 2014 at http://www.rbcwealthmanagement.com/_assets-custom/pdf/eiu/internationally-mobile-wealthy.pdf.

El-Erian, M.A. (2012), 'New billionaire index reflects competing social forces', *The Huffington Post*, 3 May, accessed 22 May 2014 http://www.huffingtonpost.com/mohamed-a-elerian/new-billionaire-index-ref_b_1321954.html.

Elliott, A. and J. Urry (2010), *Mobile Lives*, London: Routledge.

Elliott, L. (2014), 'Britain's five richest families worth more than poorest 20%', *The Guardian*, 17 March, accessed 17 March 2014 at http://www.theguardian.com/business/2014/mar/17/oxfam-report-scale-britain-growing-financial-inequality.

Featherstone, M. (2014a), 'The rich and the super-rich: mobility, consumption and luxury lifestyles', in N. Mathur (ed.), *Consumer Culture, Modernity and Identity*, New Delhi/Thousand Oaks, CA/London/Singapore: Sage, pp. 3–44.

Featherstone, M. (2014b), 'Super-rich lifestyles', in T. Birtchnell and J. Caletrío (eds), *Elite Mobilities*, Abingdon, UK/New York: Routledge, pp. 99–135.

Frank, R.H. ([1999] 2010), *Luxury Fever: Weighing the Cost of Excess*, Princeton, NJ: Princeton University Press.
Freeland, C. (2012), *Plutocrats: The Rise of the New Global Super-Rich and the Fall of Everyone Else*, New York: Penguin Press.
Gilding, M. (1999), 'Superwealth in Australia: entrepreneurs, accumulation and the capitalist class', *Journal of Sociology*, **35** (2), 169–82.
Gilding, M. (2002), *Secrets of the Super Rich*, Pymble, NSW: HarperCollins.
Gilding, M. (2005), 'Families and fortunes: accumulation, management succession and inheritance in wealthy families', *Journal of Sociology*, **41** (1), 29–45.
Graham, S. and S. Marvin (2001), *Splintering Urbanism: Networked Infrastructures, Technological Mobilities and the Urban Condition*, London/New York: Routledge.
Hamnett, C. (1994), 'Social polarisation in global cities: theory and evidence', *Urban Studies*, **31** (3), 401–24.
Hamnett, C. (1996), 'Why Sassen is wrong: a response to Burgers', *Urban Studies*, **33** (1), 107–10.
Harvey, D. (1973), *Social Justice and the City*, London: Edward Arnold.
Harvey, D. (1989), 'From managerialism to entrepreneurialism: the transformation in urban governance in late capitalism', *Geografiska Annaler. Series B, Human Geography*, **71** (1), 3–17.
Harvey, D. (2004), 'The "new" imperialism: accumulation by dispossession', *Socialist Register*, **40**, 63–87.
Harvey, D. (2010), *The Enigma of Capital: And the Crises of Capitalism*, London: Profile Books.
Haseler, S. (2000), *The Super-Rich: The Unjust New World of Global Capitalism*, Basingstoke, UK/New York: Palgrave Macmillan.
Hay, I. (2013a), 'Establishing geographies of the super-rich: axes for analysis of abundance', in I. Hay (ed.), *Geographies of the Super-Rich*, Cheltenham, UK and Northampton, MA, USA: Edward Elgar Publishing, pp. 1–25.
Hay, I. (ed.) (2013b), *Geographies of the Super-Rich*, Cheltenham, UK and Northampton, MA, USA: Edward Elgar Publishing.
Hay, I. and S. Muller (2012), '"That tiny, stratospheric apex that owns most of the world" – exploring geographies of the super-rich', *Geographical Research*, **50** (1), 75–88.
Hay, I. and S. Muller (2014), 'Questioning generosity in the golden age of philanthropy. Towards critical geographies of super-philanthropy', *Progress in Human Geography*, **38** (5), 635–53.
Hesseldahl, A. (2002), 'The first rich list', *Forbes.com*, 27 September, accessed 20 May 2014 at http://www.forbes.com/2002/09/27/0927richestphotos.html.
Hunter, F. (1953), *Community Power Structure: A Study of Decision Makers*, Chapel Hill, NC: University of North Carolina Press.
Jessop, B. and N.-L. Sum (2000), 'An entrepreneurial city in action: Hong Kong's emerging strategies in and for (inter)urban competition', *Urban Studies*, **37** (12), 2287–313.
Katona, G. and J.B. Lansing (1964), 'The wealth of the wealthy', *The Review of Economics and Statistics*, **46** (1), 1–13.
Keister, L.A. (2014), 'The one percent', *Annual Review of Sociology*, **40** (1), 347–67.
Keister, L.A. and H.Y. Lee (2014), 'The one percent: top incomes and wealth in sociological research', *Social Currents*, **1** (1), 13–24.
Klein, N. (2007), *The Shock Doctrine: The Rise of Disaster Capitalism*, New York: Metropolitan Books/Henry Holt.
Knight Frank (2014), *Global Lifestyle Review: A Comparison of Favourable Tax Locations and Lifestyle Factors*, accessed 13 May 2014 at http://www.knightfrank.com/resources/brochures/2623-kf-lifestyle-report-final.pdf.
Kopczuk, W. and E. Saez (2004), 'Top wealth shares in the United States, 1916–2000: evidence from estate tax returns', *NBER Working Papers*, Cambridge, MA: National Bureau of Economic Research.
Lees, L. (2012), 'Gentrifying the world city', in B. Derudder, M. Hoyle and P.J. Taylor et al.

(eds), *International Handbook of Globalization and World Cities*, Cheltenham, UK and Northampton, MA, USA: Edward Elgar Publishing, pp. 369–77.
Ley, D. (2010), *Millionaire Migrants: Trans-Pacific Life Lines*, Chichester, UK/Malden, MA: Wiley-Blackwell.
Ley, D. and J. Tutchener (2001), 'Immigration, globalisation and house prices in Canada's gateway cities', *Housing Studies*, **16** (2), 199–223.
Ley, D., J. Tutchener and G. Cunningham (2002), 'Immigration, polarization, or gentrification? Accounting for changing house prices and dwelling values in gateway cities', *Urban Geography*, **23** (8), 703–27.
Lundberg, F. (1937), *America's 60 Families*, New York: The Vanguard Press.
Lundberg, F. (1968), *The Rich and the Super-Rich: A Study in the Power of Money Today*, New York: L. Stuart.
Lynd, R.S. and H.M. Lynd (1937), *Middletown in Transition: A Study in Cultural Conflicts*, New York: Harcourt Brace Jovanovich.
McVeigh, T. (2014), 'Inequality "costs Britain £39bn a year"', *The Guardian*, 16 March, accessed 16 March 2014 at http://www.theguardian.com/society/2014/mar/16/inequality-costs-uk-billions.
Merrifield, A. (1996), 'Public space: integration and exclusion in urban life', *City*, **1** (5–6), 57–72.
Merrifield, A. (2013), 'Intervention – whose city? The parasites, of course. . .', *Antipode*, accessed 31 July 2014 at http://antipodefoundation.org/2013/06/18/intervention-whose-city/.
Michel, J.-B., Y.K. Shen and A.P. Aiden et al. (2011), 'Quantitative analysis of culture using millions of digitized books', *Science*, **331** (6014), 176–82.
Mills, C.W. (1956), *The Power Elite*, New York: Oxford University Press.
Murray, C.A. (1990), *The Emerging British Underclass. Choice in Welfare Series No. 2*, London: IEA Health and Welfare Unit.
Nader, R. (2009), *Only the Super-Rich Can Save Us!* New York: Seven Stories Press.
Nathwani, N. (2014), 'On privilege: a leftist critique of the left', *Harvard Political Review*, 29 April, accessed 20 May 2014 at http://harvardpolitics.com/united-states/privilege-leftist-critique-left/.
Neumayer, E. (2004), 'The super-rich in global perspective: a quantitative analysis of the Forbes list of billionaires', *Applied Economics Letters*, **11** (13), 793–6.
Pahl, R.E. (1975), *Whose City? And Further Essays on Urban Society*, 2nd edition, Harmondsworth, UK: Penguin.
Pahl, R.E. (1988), 'Some remarks on informal work, social polarization and the social structure', *International Journal of Urban and Regional Research*, **12** (2), 247–67.
Paris, C. (2011), *Affluence, Mobility, and Second Home Ownership. Housing and Society Series*, London: Routledge.
Paris, C. (2013), 'The homes of the super-rich: multiple residences, hyper-mobility and decoupling of prime residential housing in global cities', in I. Hay (ed.), *Geographies of the Super-Rich*, Cheltenham, UK and Northampton, MA, USA: Edward Elgar Publishing, pp. 94–109.
Pessen, E. (1990), *Riches, Class, and Power: America before the Civil War*, New Brunswick, NJ: Transaction Publishers.
Piketty, T. (2014), *Capital in the Twenty-First Century* [French edition published in 2013 as *Le capital au XXI siècle*, Editions du Seuil], Cambridge, MA/London: Belknap Press of Harvard University Press.
Pinçon, M. and M. Pinçon-Charlot (1998), *Grand Fortunes Dynasties of Wealth in France*, New York: Algora Publishing.
Pow, C.-P. (2011), 'Living it up: super-rich enclave and transnational elite urbanism in Singapore', *Geoforum*, **42** (3), 382–93.
Rodriguez, C. (2014), 'The Young Street: an eccentric millionaire's bold commercial eco-project in Paris', *Forbes.com*, 31 May, accessed 31 July 2014 at http://www.forbes.com/sites/ceciliarodriguez/2014/05/31/an-eccentric-millionaires-bold-commercial-ecoproject-in-paris/.

Rubinstein, W.D. (1981), *Men of Property: The Very Wealthy in Britain since the Industrial Revolution*, London: Croom Helm.
Rubinstein, W.D. (2009), *Who Were the Rich?: A Biographical Directory of British Wealth-Holders, Vol. I: 1809–1839*, London: Social Affairs Unit.
Sassen, S. (1991), *The Global City: New York, London, Tokyo*, Princeton, NJ: Princeton University Press.
Sassen, S. (2014), *Expulsions: Brutality and Complexity in the Global Economy*, Cambridge, MA/London: The Belknap Press of Harvard University Press.
Savage, M. and K. Williams (eds) (2008), *Remembering Elites*, Malden, MA: Blackwell Publishing.
Sayer, A. (2014), *Why We Can't Afford the Rich*, Bristol, UK: University of Bristol Policy Press.
Shaxson, N. (2011), *Treasure Islands: Tax Havens and the Men Who Stole the World*, London: Bodley Head.
Short, J.R. (2004), 'The super-rich and the global city', in *Global Metropolitan: Globalizing Cities in a Capitalist World*, London/New York: Routledge, pp. 109–22.
Short, J.R. (2013), 'Economic wealth and political power in the second Gilded Age', in I. Hay (ed.), *Geographies of the Super-Rich*, Cheltenham, UK and Northampton, MA, USA: Edward Elgar Publishing, pp. 26–42.
Simmons, R. (2011), *Super Rich: A Guide to Having It All*, New York/London: Gotham Books.
Sklair, L. (2001), *The Transnational Capitalist Class*, Malden, MA: Blackwell.
Smith, N. (1984), *Uneven Development: Nature, Capital and the Production of Space*, Oxford, UK: Blackwell.
Smith, N. (1996), *The New Urban Frontier: Gentrification and the Revanchist City*, London/New York: Routledge.
Sorkin, M. (1992), *Variations on a Theme Park: The New American City and the End of Public Space*, New York: Hill and Wang.
Stanworth, P. and A. Giddens (1974), *Elites and Power in British Society*, London/New York: Cambridge University Press.
Stone, C.N. (1989), *Regime Politics: Governing Atlanta 1946–1988*, Lawrence, KS: University Press of Kansas.
The Economist (2014), 'Planet plutocrat', *The Economist*, 15 March, accessed 19 March 2014 at http://www.economist.com/news/international/21599041-countries-where-politically-connected-businessmen-are-most-likely-prosper-planet.
Therborn, G. (2000), 'Globalizations: dimensions, historical waves, regional effects, normative governance', *International Sociology*, **15** (2), 151–79.
Thurlow, C. and A. Jaworski (2012), 'Elite mobilities: the semiotic landscapes of luxury and privilege', *Social Semiotics*, **22** (4), 487–516.
Todd, D. (2014), 'Millionaire migrants: the numbers grow', *The Vancouver Sun*, 6 January, accessed 4 March 2014 at http://blogs.vancouversun.com/2014/01/06/millionaire-migrants-the-numbers-grow/.
Urry, J. (2014a), *Offshoring*, Cambridge, UK: Polity Press.
Urry, J. (2014b), 'The super-rich and offshore worlds', in T. Birtchnell and J. Caletrío (eds), *Elite Mobilities*, Abingdon, UK/New York: Routledge, pp. 226–40.
Useem, M. (1984), *The Inner Circle: Large Corporations and the Rise of Business Political Activity in the U.S. and U.K.*, New York: Oxford University Press.
Van Kempen, E.T. (1994), 'The dual city and the poor: social polarisation, social segregation and life chances', *Urban Studies*, **31** (7), 995–1015.
Veblen, T. ([1899] 1992), *The Theory of the Leisure Class, The Rise of the Network Society. Information Age 1*, New Brunswick, NJ: Transaction Pub.
Volscho, T.W. and N.J. Kelly (2012), 'The rise of the super-rich: power resources, taxes, financial markets, and the dynamics of the top 1 percent, 1949 to 2008', *American Sociological Review*, **77** (5), 679–99.
Wealth-X Institute (2012), *Real Estate and the Asian UHNW Investor*, accessed 13 February 2014 at http://www.wealthx.com/articles/2012/real-estate-and-the-asian-uhnw-investor.

Wealth-X Institute (2013), 'Statistical profile of the Hong Kong UHNWI', accessed 13 February 2014 at http://www.wealthx.com/articles/2013/statistical-profile-of-the-hong-kong-uhnwi/.

Wealth-X Institute and Arton Capital (2014), *A Shrinking World: Global Citizenship for UHNW Individuals: A Special Report*, accessed 30 April 2014 at http://www.wealthx.com/wp-content/uploads/2014/04/Wealth-X-Arton-Capital-A-Shrinking-World-Global-Citizenship-for-UHNW-Individuals.pdf.

Wealth-X Institute and UBS (2013), *World Ultra-Wealth Report*, accessed 17 February 2014 at http://wuwr.wealthx.com/Wealth-X%20and%20UBS%20World%20Ultra%20Wealth%20Report%202013.pdf.

Wilkinson, T.L. (2013), 'Nationality swapping: the latest craze of the world's ultra rich', *Wealth-X.com*, accessed 13 February 2014 at http://www.wealthx.com/articles/2013/nationality-swapping-the-latest-craze-of-the-world%E2%80%99s-ultra-rich/.

Wood, Z. (2014), 'Super rich shift their thrills from luxury goods to costly experiences', *The Guardian*, 30 January, accessed 17 February 2014 at http://www.theguardian.com/business/2014/jan/30/super-rich-shift-experiences-new-status-symbols.

Wooster, R.A. (1967), 'Wealthy Texans', *The Southwestern Historical Quarterly*, **71** (2), 163–80.

Young, I.M. (1990), *Justice and the Politics of Difference*, Princeton, NJ: Princeton University Press.

Zorbaugh, H.W. (1929), *The Gold Coast and the Slum: A Sociological Study of Chicago's Near North Side. University of Chicago Sociological Series*, Chicago, IL: University of Chicago Press.

PART I

WEALTH, SELF AND SOCIETY

3. Historical geographies of wealth: opportunities, institutions and accumulation, c. 1800–1930

*Alastair Owens and David R. Green**

INTRODUCTION

Studies of long-run trends in the distribution of wealth have mainly focused on understanding and seeking to explain patterns of inequality. Since Simon Kuznets (1955) first explored the relationships between inequality and economic growth, a large and important body of research has sought to reconstruct and interpret the historical evidence for an ever-widening range of countries over ever-lengthening periods of time. The most recent addition to the pantheon of studies has been Thomas Piketty's (2014) influential (and, for some, contentious) book, *Capital in the Twenty-First Century* – itself the culmination of a range of efforts to quantify both the extent and the sources of economic inequality. Between Kuznets and Piketty exists a very large number of studies of wealth and inequality, far too numerous to consider in detail here.

The changing composition of wealth, however, is far less understood than its distribution. Indeed, according to Jesper Roine and Daniel Waldenström (2014, p. 69), 'when it comes to historical evidence about wealth composition across the wealth distribution, we know almost nothing'. The importance of composition of wealth is highlighted by Piketty's work in particular. Along with classical economists, he argues that wealth is derived from two sources: capital (including all non-human resources that can be traded on the market and that are therefore able to generate income) and earnings from labour. Piketty points out that over time the rate of return to the former tends to outrun that of the latter, although the trend is not necessarily a smooth one and has, at times, been reversed as a result of economic shocks, war and the taxation of wealth. Nevertheless, taking a long-run view from the early 1800s, from which time estimates of wealth become increasingly accurate, through to the current day, Piketty argues that the differential rate of return to capital versus earnings means that those who own capital resources tend to increase their relative share of wealth faster than those who depend entirely on earnings from labour. Indeed, the income deriving from assets in terms of dividends on share ownership, coupon yields on fixed interest securities, interest on mortgages and bank accounts, rental income and capital gains, is one of the primary sources of inequality, particularly at the upper end of the income distribution. Understanding the composition of income-generating assets, therefore, is important for explaining the processes by which inequality is generated, since different assets can, at different times and in different places, generate varying rates of return. In England, for example, and elsewhere in Europe, the ownership of agricultural land towards the end of the nineteenth century brought in relatively low returns compared to urban land and housing, or share ownership. Finally, the ability to accumulate wealth at a faster rate by one generation is, in turn, passed on through inheritance to the next generation, thereby reproducing and increasing overall levels of inequality over time.

Piketty's arguments also raise important questions about the geography of capital and its relationship to inequality. First there is the question of understanding where the assets that had income-generating potential were located, which entails mapping the shifting geography of economic opportunity from the industrial heartlands of Europe to the frontiers of expansionary states and empires. Second is the question of assessing the extent to which different groups of individuals in varying locations could gain access to these wealth-making opportunities. Over the period covered by this chapter, a range of new institutions, forms of knowledge and expertise emerged that enabled the acquisition of capital and the enhancement or preservation of its income-generating value. These 'institutional frameworks' that underpinned wealth accumulation were increasingly important in mitigating the effects of distance, enabling investors not only to purchase assets that were local to them, but also in providing opportunities for generating wealth from capital that was further afield. In many respects such frameworks drove processes of accumulation. Institutions such as stock exchanges and banks, and technologies from the telegraph to the ticker tape, played a formative role in shaping markets and in bringing wealth-making opportunities that were previously unknown or distant within the reach of investors. In places where these kinds of institutional frameworks were well developed, the possibility of lucrative income generation was there, although proximity alone was not sufficient for this to occur. Access was mediated by power; social class, gender and 'race' were among the many factors that enabled or denied men and women's access to capital through the institutions of the market.

In this chapter we trace the changing composition of personal wealth over time and across space, paying careful attention to the way that institutional developments facilitated this process. We begin around 1800 when, for many European countries, industrialization, urban growth and the development of commercial and territorial interests overseas created new opportunities for accumulation. For world systems analysts, the nineteenth century represented the peak of European – and especially British – global economic influence. Particularly through the extension of forms of colonial rule, European powers at the core of the capitalist world system increasingly dominated and economically exploited those countries that lay at the periphery (see, e.g., Wallerstein, 2011). In terms of the assets that people owned, the emergence of this world system led to a shift away from traditional forms of capital like land to new types of wealth tied to corporate, commercial and financial interests overseas as well as at home. Taking a broad, though selective geographical sweep, our aim is to look at how the ownership of capital by individuals evolved in different settings across the nineteenth and early twentieth centuries. While we emphasize the increasingly global context within which people sought to accrue assets, we argue that the local and regional opportunities for accumulation – shaped by a range of political, economic and cultural imperatives – continued to be significant and can at least in part help to explain patterns of inequality. We first consider some of the institutional and other infrastructures that enabled the amassing and transmitting of different kinds of wealth and assets, noting in particular how they facilitated investment beyond the local scale. The growing sophistication of financial institutions, closer regulation of corporate governance and the ubiquity of financial information and advice (both good and bad) helped to promote a culture of investment that spanned much of Europe, its imperial territories as well as in many of those parts of the world that had broken free of the European imperial embrace. We then explore evidence of the composition of wealth across a range of settings that represent these different locations and that were variously positioned within an uneven capitalist world system. Here we argue that over the course of the nineteenth and early twentieth centuries an increasing number of those with wealth were involved in holding assets that were both geographically but also sectorally diverse, exposing the well-being and security of individuals and families to the vagaries of regional, national and ultimately the world economy. In the 1930s, when our survey ends, this was to prove calamitous for many individuals as they saw the value of shares, securities and other assets ebb away.

INSTITUTIONS AND WEALTH

As Douglass North (1991, p.97) has argued, 'institutions provide the incentive structure of an economy' and in this section we survey the frameworks within which individuals were able to acquire, accumulate and transmit their wealth. As North suggests, these institutional frameworks were deeply implicated in the processes that promoted economic growth and the spread of capitalist enterprise, allowing security over property based on the rule of law as opposed to the arbitrary actions of rulers or despots. In an increasingly complex economy, however, such security extended well beyond the protection of the individual against arbitrary actions and reached into all aspects of economic activity, from contractual arrangements enacted to protect the transfer of property and guarantee rights of ownership, to the legal requirements that ensured the liquidity of the banking system, the stability of the currency and that regulated the integrity of company accounting practices. Extended credit networks across international borders were backed by diplomatic agreements and ultimately military might, whilst the value of specie was also a function of the state's monetary policy. Property rights in capitalist countries, therefore, involved more than just the rights of individual ownership of assets; they also entailed the full involvement of the state in underwriting the institutional frameworks that allowed commerce to function, both within and beyond its borders.

Here we survey some of the key developments that took place over the course of the nineteenth and early twentieth centuries that enhanced property rights in capitalist economies and that underpinned the accumulation of a range of assets. There is not space to offer a comprehensive discussion of institutional developments, so we restrict our consideration to three key areas that help to explain the changing nature and geography of wealth holding and its relationship to inequality: the evolution of the banking and financial system that allowed for the mobilization of capital and funded commercial activity; the gradual extension of company law that reduced the risks of investing in industrial and commercial enterprises; and finally the emergence of new forms of financial information and the systems and technologies that allowed financial knowledge and expertise to move around the world more readily.

Banking and the Financial System

During the course of the nineteenth century, countries throughout Europe and the United States developed a far stronger and more robust set of financial institutions than had been the case in the preceding century. The ability to mobilize significant amounts of capital for large industrial projects, notably the railways and heavy industry, were central to the development of sturdier institutional arrangements that resulted in greater stability in the banking system and relative reductions of risk for individual investors. Key to this process was the emergence of a central lender of last resort, which had some control over monetary policy and note issue, such as the Bank of England, the German Reichsbank and the Banque de France (Tilly, 1998). These institutions, together with the legal frameworks that underpinned their control over the banking system, were significant in reassuring investors that their savings were safe – at least for the most part – and that financial failure was an unlikely event. Panics did erupt, and banks continued to fail, but in the second half of the nineteenth century, at least in most of Europe and the United States, failure of individual banks never seriously threatened to undermine the currency or, indeed, the economy more widely.

In Britain, important banking reforms took place early, significantly enhancing financial stability of the entire banking system. There were no more widespread banking crises after that of 1825–26, although runs on the currency took place and individual banks did, from time to time, fail. From 1844 the Bank Charter Act tied note issue to gold deposits and gave the Bank of England sole authority over banknotes. Private banks in existence at the time retained the right to issue their own banknotes provided that these were secured against deposits. In practice, however, private banknotes dwindled as Bank of England notes became common currency set against the gold standard (Turner, 2014). By mid-century, therefore, the Bank effectively operated as the lender of last resort, underpinning the stability of the entire banking system as well as performing its other roles as lender to the government of the day. Added to this, the growth of joint stock banking from the 1850s and 1860s, allied to the introduction of limited liability for shareholders, encouraged a significant expansion of the banking system. The subsequent amalgamation of many of these separate banks into the five large London clearing banks, with headquarters in the City and with large branch networks distributed across the country, further extended the ability to accumulate deposits and mobilize large amounts of capital that in turn helped to finance economic growth (see Michie, 1992).

In France, similar reforms took place that created the Banque de France as a central bank that also functioned as the lender of last resort and that from 1848 had the sole power to issue banknotes. Similar to Britain, the growth of large joint stock banks in the second half of the century that extended throughout the country helped to provide the large amounts of credit required by industry and commercial development. The same pattern arose in Germany where, after unification, the establishment of the central Reichsbanks in 1876 helped underpin the financial stability of the new nation. The major banks were also important in financing economic development though attitudes to risk varied. In Germany the large corporate banks underwrote much of growth in industry and railways, only turning to small-scale depositors in the 1890s as competition from municipal banks and other institutions geared more towards individual savers grew (Tilly, 1989). The outcome in both countries – mirrored elsewhere in Europe – was that the capital required for economic growth was more readily available through an increasingly secure financial system in which individuals could deposit their savings with ever greater degrees of confidence. Banking crises, as in Britain, were relatively rare, and depositors could with some confidence rest assured that in cases of failure, the central bank would step in as the lender of last resort.

The same could not be said of the United States, where the states' distrust of the federal government hampered attempts to develop a central bank comparable to those established in Europe. The proliferation of state banking, after the charter of the first Bank of the United States had expired in 1836, did nothing to help create a stable financial system and bank failures were therefore far more common there than was the case in Europe. Note issue, too, was chaotic: in 1860 some 1600 or so banks issued their own banknotes. In 1863, to fund the war effort, President Lincoln established the national bank, backed by the federal government, and after Congress started to tax state bank notes in 1865, notes from the national bank became common tender. This helped to improve the efficiency of the banking system, making it possible for individuals to deal in the same currency throughout the country, rather than have to hold separate notes from individual banks (Sylla, 1998).

Limiting Liability, Mobilizing Capital

Whilst the banking network and the stability that central banks imparted to the financial system helped to create more efficient capital markets, there were other equally important reforms that encouraged individuals to invest directly in companies and commercial enterprises. In industrializing Europe at the start of the nineteenth century the traditional way of mobilizing capital, other than through personal borrowing, was via the creation of unincorporated partnerships. But this was a risky venture for investors who may have had little personal knowledge of the firm or its partners, and they could expect little protection from the state when such ventures failed. From the 1840s in England, in response to rising concerns about fraudulent behaviour, legislation was introduced to allow more strictly regulated joint stock enterprises to be established. The Joint Stock Companies Registration and Regulation Act of 1844 created a Registrar of Joint Stock Companies. Partnerships with more than 25 members and transferable shares were required to register and to provide audited accounts on a regular basis. This, in turn, provided investors with better information on which to base their decisions to purchase shares, but it did not protect partners from unlimited liability and therefore becoming accountable for the debts incurred by the company. This, it was argued, both hindered the ability to raise capital and reduced the likelihood that individuals would invest knowing the risk they were taking on in relation to a company's debts. Indeed, considerable concern was expressed that joint stock arrangements merely encouraged unscrupulous directors to persuade passive investors to sink their capital into a venture in which they had little personal interest and perhaps equally little personal concern (Taylor, 2006).

Following a set of business scandals, in which naive and poorly informed investors became liable for companies' debts, pressure mounted to introduce some measures that would limit shareholders' liabilities. In England the move towards limited liability gathered pace in the 1850s, drawing on a wide set of interests ranging from business people keen to extend their ability to attract funds, to social reformers equally keen to protect small, working-class savers from unrestricted risk (see Djelic, 2013 for a discussion of these points). As Donna Loftus (2002, p. 108) has argued, limited liability therefore tied both the middle class and the working class into a common project through shared interests in promoting investment. Limited liability was seen as a way of democratizing investment by reducing the financial risks of failure and thereby encouraging even those with modest means to invest in commercial and industrial enterprises. Indeed, the introduction of limited liability in 1855 and the further consolidation of joint stock companies regulation in 1856 marked the beginning of widespread shareholding in English companies. The latter act, in particular, removed the minimum £10 threshold for shares, extending the possibility of investment in commercial ventures to those of more modest means (Rutterford et al., 2011, p. 3). Further changes took place during the nineteenth century but the essential elements by which the democratization of shareholding was established did not change.

The significance of the legislation in relation to individual patterns of wealth holding was that it opened up the possibility for a far wider group of individuals to invest in shares in the knowledge that the risks they faced were significantly reduced, both because of the better financial regulation of firms as well as the liability they themselves faced as shareholders in joint stock companies. Accompanying these changes to corporate accounting and governance was the fact that share denominations fell in line with the desire to widen the market for investors. By 1900 a £1 nominal value was normal and by the 1920s, shilling shares were commonplace. As a result, as Alec Cairncross (1953, p. 85) argued, gentlemen, solicitors and peers of the realm were joined by 'retailers, professional men, skilled workers and women' as investors in British enterprises at home and abroad. While, by the 1890s, shares had become the single largest item by value in the composition of wealth assessed for death duties in England and Wales (see the discussion below), research by Josephine Maltby et al. (2011) has questioned the view of Cairncross and other contemporary commentators that shareholding had become 'democratized'. At least when it came to owning corporate securities, professional, managerial occupational categories along with 'gentleman' remain dominant in shareholder lists and registers in the first three decades of the twentieth century (Maltby et al., 2011, p. 204).

The changes that occurred in relation to limited liability in England were paralleled in other countries although in some cases similar shifts happened much earlier and took different institutional forms (see Hannah, 2013). In certain instances, such as in Norway, limited liability remained outside state regulation and instead was based largely on private agreements (Hannah, 2013, pp. 8–9). In France, limited partnerships had existed for centuries before more formal procedures were laid down in the 1807 Napoleonic Code to regulate joint stock companies (Hannah, 2013, p. 11). In Prussia the number of individually chartered companies, according to Hannah, was an order of magnitude higher than the few hundred that were formally registered. In the United States, the numbers of corporate companies were many times higher than in other countries, although their average size based on authorized capital was often low relative to those in Britain, Prussia and France and their failure rate was accordingly high.

The English approach to limited liability was also exported to the British Empire in part to promote and protect British investment overseas but also to facilitate trade between the imperial core and periphery (McQueen, 2009). In 'non-settler colonies' such as India it has been argued that while English company law promoted imperial interests, it was not well suited to local forms of business organization and difficult to enforce (Rungta, 1970). In colonies like Australia that were settled in order to pursue British expansionary economic ambitions, company law was more easily implemented, though over time it diverged from an approach that suited British gentlemanly interests at the imperial core more than it did local ones at the periphery (McQueen, 2009). Limited liability, in whatever form it took, actively drew property owners towards the acquisition of shares and other corporate securities. As we note below, this is a trend that we can begin to detect when we examine changing patterns of wealth holding across the nineteenth and early twentieth centuries in a range of locations.

Information: Flows, Technology and Market Integration

Accompanying the development of financial institutions and new forms of corporate regulation in the nineteenth century was a striking increase in the availability of information and expertise related to investment and wealth management. Emanating particularly from financial centres like London, New York and Paris, a large number of financial guides, manuals, periodicals and newspaper columns instructed wealth holders on how best to invest and maintain their assets. Similarly, a growing cast of professionals, from solicitors and brokers, to accountants and financial advisors, served as economic matchmakers, marrying the 'surplus' capital of the wealthy to investment opportunities near and far. Alex Preda (2009, p. 88) argues that over the second half of the nineteenth century 'a visible shift occurs in the vocabulary, literary genres, and the cognitive tools used to represent financial markets, as well as definitions of its objects and actors'. He emphasizes the way that the making, presenting and disseminating of financial advice was increasingly understood as a science and professional practice, grounded in observational technology; a process that served to legitimate certain forms of capitalist investment and wealth accumulation. At the same time, as various literary scholars have argued (see, e.g., Poovey, 2008; Henry and Schmitt, 2009), the nineteenth century gave birth to a flourishing 'culture of investment' where the literary forms and conventions of financial journalism blended and cross-fertilized with fictional genres. This provided a means by which ideas of wealth accumulation and the calculative rhetorics of exchange became deeply embedded within nineteenth-century culture and society, making the 'allure of investment vivid' (Poovey, 2009, p. 40).

Information flows about investment opportunities were also associated with new technologies of communication, notably the spread of the telegraph system that linked the imperial cores and metropoles to their colonial and continental peripheries (Wenzlhuemer, 2012). Cables connecting Britain and the United States were laid across the Atlantic in 1858, and by the mid-1870s the telegraph network stretched from London as far as New Zealand via India and Australia. Telegraph lines spanned the United States from east to west in parallel with the opening of the transcontinental railway and by 1866 the network had consolidated around Western Union's dominance of the market. In Europe by the 1870s, Belgium, the Netherlands, France, Switzerland and southern Germany all had relatively dense telegraph networks, and in the following decades other countries developed their own domestic systems of telegraphy (Wenzlhuemer, 2007). News that had once taken weeks to receive soon took next to no time, annihilating space by time, as David Harvey (1989) termed it, and in the process widening the geographical imagination to incorporate the furthest-flung outposts of the empire (O'Hara, 2010). At the same time, price information was conveyed to the public largely through the daily newspapers, many of which began to be published twice or more a day to accommodate the latest information on the stock markets. In the 1870s, for example, the *Berliner Börsen-Zeitung* was published twice a day: the morning edition contained details at the close of other European stock exchanges, whilst the evening edition contained a review of the Berlin market that day (Baltzer, 2007). At an institutional level, the outcome of these more efficient flows of information was the harmonization of prices of shares and stocks traded in different centres that witness the growing integration of financial markets (Davis and Gallman, 2001; Cassis, 2011, p. 249). As a result of these technological and institutional changes, international trade became more predictable, markets became more integrated and risks became more easily calculable. All served to draw investors into an ever-widening geographical sphere of activity based on accurate and almost instantaneous flows of market information.

As the geography of market opportunities widened, so too did the significance of these new transcontinental forms of communication that provided accurate information about differential rates of return – essential for making rational decisions about investing capital. This expanding geography was reflected in the outward capital flows from the core economies, notably Britain, to peripheral countries. According to current estimates on the eve of World War I, 40 per cent of British overseas investments was in railways, with a further 30 per cent in government or municipal bonds, 10 per cent in the extractive sector and 5 per cent utilities (Bordo, 2000, p. 12). To some extent this reflected growing opportunities brought about by economic expansion into peripheral regions as well as better flows of information about prices. However, it also reflected the growing capital needs of different sectors that also led to the cross-listing of corporate securities on an increasing number of European and North American stock exchanges (Baltzer, 2007, p. 7). In this way it became more likely that investors in one country would become aware of opportunities elsewhere. This was particularly true in global financial centres such as London, where investors typically held more international stocks and shares than did their provincial counterparts.

THE COMPOSITION OF WEALTH

Institutional frameworks like those described above formed an important context to the way that people in different parts of the world accumulated wealth and made decisions about how to invest. To be sure, exactly how these frameworks influenced (or were influenced by) capital accumulation is complex and it would be unwise to make claims about causality. In this section, however, we explore evidence that tells us about what assets people owned and how this varied over time and between different locations.

Sources for Studying the Composition of Wealth

Understanding the composition of wealth rests largely on historical sources generated by the various institutions that sought to regulate its accumulation, transmission and redistribution described above. In particular, taxation records – notably taxes on income and at the point of transfer (usually at death) – have proved most useful to historians in understanding what people owned. However, the evidence is often sporadic and comes with limitations. Income tax data begin to become available from the late 1700s, although their incidence is patchy for many countries until the later 1800s and the information on different sources of income, such as wages, rents and profits from business, varies widely. Similarly, the unit of taxation also varies between countries and also over time: in some countries individuals are the unit of taxation, such as in Italy and Spain, and in others, such as France and the United States, it is the family, and in yet others, the unit of taxation itself changes over time (Roine and Waldenström, 2014, pp. 36–7). Tax avoidance also needs to be taken into consideration when interpreting the evidence, although for much of the nineteenth century the relatively benign rates of taxation provided little incentive to do so.

Most studies of the composition of wealth have used evidence gathered from estate tax (or inheritance tax) returns, supplemented by additional wealth tax returns, household surveys and other listings of wealth. Of these sources, the most consistent is inheritance tax records, although these are not without significant problems of interpretation (see, e.g., Lindert, 1986, pp. 1132–5; Shanahan, 2001, pp. 59–64; Owens and Green, 2012). Measuring wealth using these data, however, is also dependent on how assets were valued, and here significant differences existed between countries and over time, particularly in how real estate was valued in relation to personal wealth (see, e.g., Campion, 1939, pp. 1–11, 39–41; Green and Owens, 2013).

The data have been used in two distinct ways: cross-sectional, by which individuals and families can be compared between countries (and at other spatial scales where data permits) and at different points in time, and across the life course, taking into account varying asset distributions at different ages. Below we focus on cross-sectional studies, taking note of how changes in the composition of wealth reflect changes in the nature of economic activity, notably the shift in some countries from an agrarian to an industrial economy and the growing commercial exploitation of territories by imperial powers.

Taking a cross-sectional approach does present some methodological difficulties, particularly when seeking to understand asset holding in settings where there is rapid economic and social change. Since most of the evidence of wealth holding comes from records associated with death and inheritance, those data tend to have an inherent bias towards capturing the wealth-holding characteristics of older people. Put simply, the evidence of asset ownership gleaned from such documents might not represent the situation for all age groups in a particular setting. An example of the kinds of differences that can result from this bias relates to landownership in Massachusetts. Using census returns, Richard Steckel and Carolyn Moehling estimate that in Massachusetts, throughout the nineteenth century, real estate comprised between about 64 and 77 per cent of taxable wealth for male household heads (Steckel and Moehling, 2001, p. 166). This figure is considerably higher than that which arises when 'end of life' data in the form of estate records are used, partly because of higher wealth thresholds and partly because real estate was less important in the overall portfolios of the relatively wealthy who also tended to be relatively elderly. We must be aware, therefore, that estate evidence can be misleading if used as a surrogate for the wealth-holding population as a whole, providing a skewed understanding of both wealth composition and inequality. Factoring in, and adjusting for, age-related asset ownership requires individual-level data and more complicated data linkage. Not surprisingly, therefore, there are relatively few studies that have been able to explore this dimension and where they have, the sample has inevitably been small. In England and Wales, David Green et al. (2009) have been able to demonstrate the influence of both age and gender on the amount of personal wealth and its composition. In France, Piketty (2014, p. 394) has also been able to identify clear differences in the age–wealth profiles over time, noting the growth of age-related inequality for much of the nineteenth and early twentieth century. Acknowledging these and other limitations of the evidence available for understanding wealth holding, what do we know about asset holding in the nineteenth and early twentieth centuries?

Capital and Assets: Patterns of Wealth Holding, 1800–1930

According to a recent study of sovereign bond spreads, 'the typical investment portfolio of a British gentleman around the turn of the twentieth century was probably more internationally diversified than that of his great grandson living around the turn of the twenty-first century' (Mauro et al., 2006, p. vii). While on the face of it a surprising claim, this view accords with an important body of scholarship examining the relationships between British economic development and imperialism. Via the financial mechanisms of the City of London, this literature demonstrates how British 'gentlemanly capitalists' invested their wealth abroad, thereby playing a critical role in the development of the British overseas interests (Cain and Hopkins, 2001). While the central concern of this literature is not the ownership of assets, it nevertheless raises questions about the way that by the end of the nineteenth century capital accumulation in Britain took place way beyond its shores as the country became the dominant player within the capitalist world system (see also Davis and Huttenback, 1986). But aside from the elite alliance of business magnates and the landed interest that formed this 'gentlemanly capitalist' group, how common was it for others to have financial assets that were overseas based? What was the balance between different kinds of foreign and imperial assets and types of wealth located closer to home? How did the composition of wealth portfolios vary across Britain? And, what about beyond Britain? What was the situation in those countries that were peripheral to the European core that were colonized by European powers or that had secured independence? It is impossible in the space we have here to offer comprehensive answers to these questions, or to capture the complexities of changes in the composition of personal wealth across a wide range of places over a 130-year period. However, while making reference to other locations, we focus our discussion around four examples: England and Wales, Australia, Canada and Brazil. Here, two key trends stand out. The first is a decline in the relative importance of land within wealth portfolios and corresponding rise in other kinds of assets, notably financial ones such as shares and other types of securities. A second, interrelated trend is the widening geographical spread of the assets that people owned, which, like British 'gentlemanly capitalists', frequently took their economic fortunes beyond where they lived.

In the British case, for example, data detailing the ownership of assets as deceased persons' estates were assessed for death duties show a clear rise in share ownership. In the early decades of the nineteenth century shares accounted for around 3 per cent of the value of deceased individuals' total personal estate; by the 1930s the figure had reached almost 40 per cent.[1] This shift not only reflects the broadening of opportunities for investment in shares, at home and abroad – in everything from British railway building schemes to South African diamond mines – it also demonstrates how financial markets were made increasingly intelligible and accessible to would-be investors, and the ways that markets were regulated and risks contained. The label 'shares' obscures a wide range of financial securities in which men and women invested that carried different kinds of risk and, accordingly, different rates of return. One such woman was Hannah Albright, a blind spinster, who died aged 73 at her home in Charlbury, Oxfordshire in 1886, leaving an estate worth £7604. Around this time roughly a third of those who left estates liable for death duties (around 17 per cent of the dying population) owned shares, although a much higher proportion of the very wealthy held them; between 1870 and 1902 some 83 per cent of those with estates valued in excess of £10 000 possessed shares. As someone of more modest means, Albright was relatively unusual in having such a high proportion of her wealth, almost 70 per cent of the total, in the form of shares. Among her portfolio of securities (valued at £5221) were shares in a local gas firm and a nearby canal company; shares in a range of British railway companies, from the Great Western Railway at one end of the country to the Caledonian Railway at the other; as well as holdings overseas, including shares in the Philadelphia and Erie Railroad Company and the Pennsylvania Railroad Company in the United States and in the Montserrat Company in the Caribbean – an agricultural business founded by the famous nineteenth-century British abolitionist and philanthropist Thomas Sturge that grew limes for juicing. Like other investors, Albright not only spread her bets geographically, she also purchased different kinds of investment products, including lower risk debenture stock and preference shares among her railway investments, and potentially more risky but higher yielding ordinary shares elsewhere in her portfolio.

Beyond these financial investments Albright's property portfolio also contained some real estate valued at £1563. Comparable national data on the extent to which individuals owned real estate – land, but also housing and other buildings – are more difficult to obtain as such property was excluded from the death duties until 1853 and, even then, valued in a different way to other kinds of assets. Although housing, particularly that close to expanding towns and cities, remained lucrative, the diminishing importance of real estate as a source of wealth was a long-term trend – at least until much later in the twentieth century. However, in the last three decades of the nineteenth century, estimates suggest that real estate made up around 20 per cent of all wealth passing at death and in this regard Albright's asset portfolio follows the national picture. Roughly one-third of individuals qualifying for death duties left real estate, though, as with shares, the propensity to own land and houses increased with wealth (around 60 per cent of those with estates over £5000 owned real estate).

Data covering the period from 1870 to 1935 (Figure 3.1) show the overall

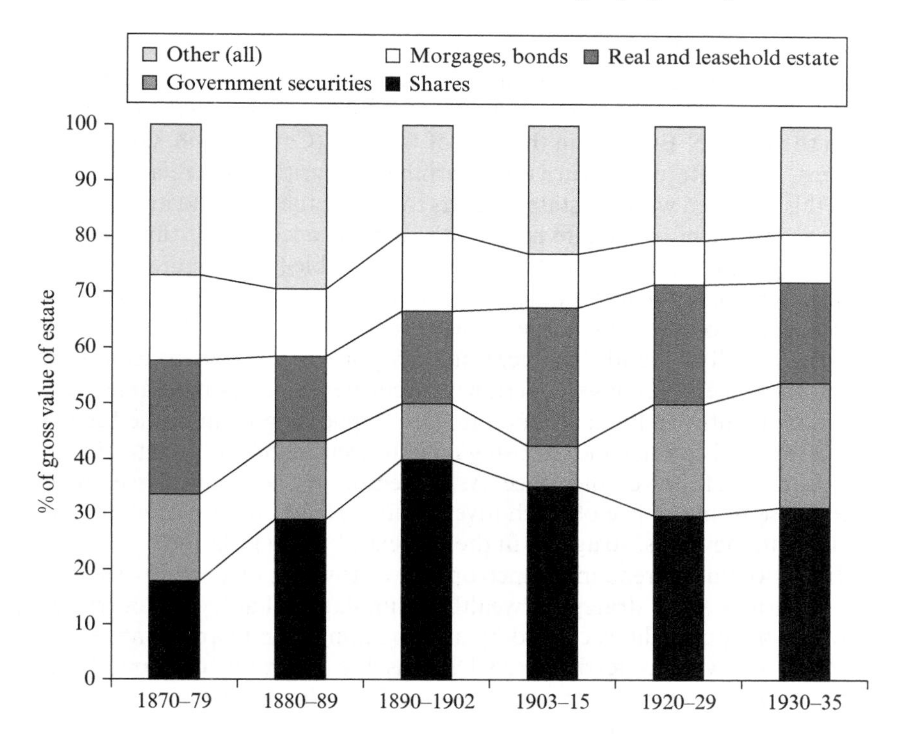

Note: There are no data from 1915 until 1920 due to World War I.

Sources: The National Archives, IR19 Board of Stamps: Legacy Duty Office and Successors: Specimens of Death Duty Account, 1870–1902; IR26 Board of Inland Revenue and Predecessors: Estate Duty Office: Registers of Legacy Duty, Succession Duty and Estate Duty 1870–1903; Annual Reports of the Commissioners of Her Majesty's Inland Revenue, 1903–35.

Figure 3.1 Composition of estates at death, England and Wales, 1870–1935

significance of different kinds of capital and assets among the estates of those who died in England and Wales over this time. Although Hannah Albright did not own any, these data also underline the significance of government securities as an income-generating asset that was held by many British wealth holders. As a state-backed security, 'the funds' as they were known, were widely regarded as a safe investment and were popular amongst widows, spinsters and others searching for a secure income. Evidence suggests, however, that ownership was geographically more concentrated than was the case for other kinds of assets, with investors in and around London more likely to have government securities in their portfolios. This spatial concentration of investors can be explained by the historical evolution of the government securities market that originated with the Bank of England in the City of London (Carter, 1968, Green and Owens, 2003). Rates of return on government securities were relatively low and this, coupled with the state's efforts to reduce the dividend and redeem the national debt in the late nineteenth century, account for their declining significance at that time. However, also notable from Figure 3.1 is the sharp increase in the importance of government securities in the 1920s – a consequence of the mass issue of war bonds after 1914. With rates of return typically around 5 per cent and a vigorous government campaign to promote their purchase, government securities once again became a key asset for wealthy individuals and families, implicating them in the financing of a global conflict and exposing them to charges of war profiteering in the years that followed the peace. As contemporary accounts also emphasize, those trying to live off such investments or, indeed, any fixed income security in the 1920s, struggled in the context of rising inflation.

If the dominant trend in the metropole was towards ownership of shares and securities and a strategy of wealth accumulation that looked outwards towards the possibilities offered by an expanding empire and commercial interests in overseas territories, what was the situation in colonial settings? In British dominions such as South Australia, where the ownership of wealth was more equal than in Europe (Shanahan, 2001, pp. 57–8), land was more important amongst personal wealth portfolios. Given the significance of land to settler schemes, its key role in the commercial exploitation of 'new' territories and its relative abundance compared with Britain where opportunities for its acquisition were more limited, this is perhaps not surprising. In reality, of course, land was not 'abundant'; its appropriation by settlers and the rapid and comprehensive way that indigenous groups were dispossessed of land and other natural resources was a feature of British engagement with many dominion territories. But for newly arriving Europeans, land promised prosperity, security and social mobility, and across the British Empire it was granted freely to settlers by the Crown, offered at subsidized cost, or leased on favourable terms. As Martin Shanahan (2001, p. 73; 2011, pp. 118–19) has shown, real estate accounted for roughly 50 to 60 per cent of the total value of all South Australian estates under £2500 that were submitted for probate between 1905 and 1915 (Figure 3.2). It was inevitably a primary resource in a rural economy that lay distant from international markets and it featured in more than half of all estates brought forward for probate. However, above the £2500 threshold the importance of real estate in terms of value fell, while other assets, notably 'stocks and shares in companies' and 'prop-

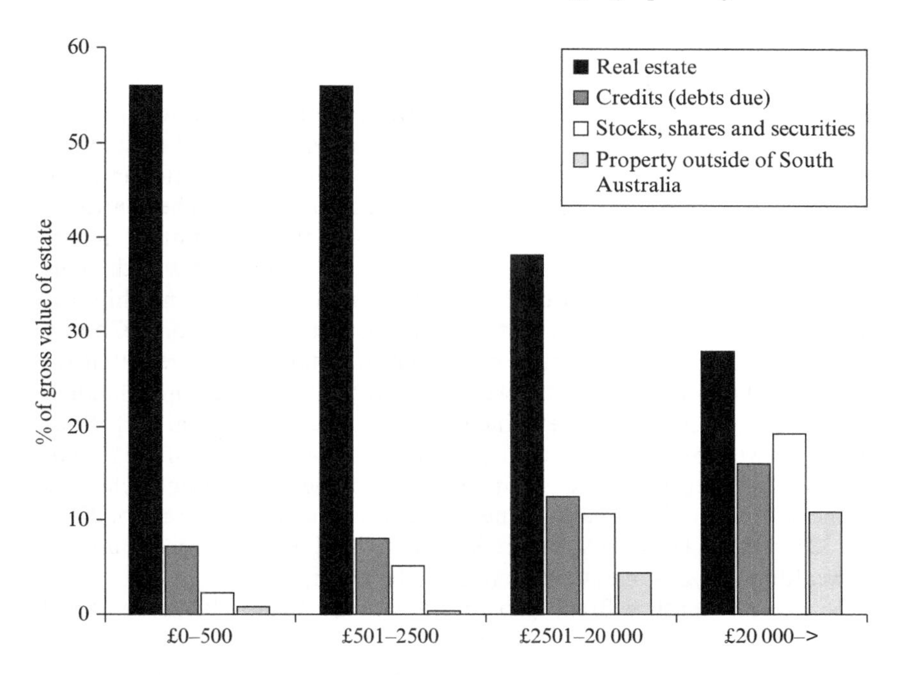

Note: Credits (debts due) amalgamates: mortgages and interest due; bonds, bills, notes, interest due; book and other debts due; stocks, shares and securities amalgamates: ships or shares in shops; treasury bills; government debentures; stocks/shares in banks; stocks/shares in companies; dividends from stock; stock/securities in foreign states.

Source: Shanahan (2001, p. 70).

Figure 3.2 Composition of estates at death in South Australia, 1905–15 (selected assets as percentage of total estate)

erty outside South Australia', grew. More than half of Shanahan's top stratum, those who died with fortunes in excess of £20 000, held wealth outside South Australia (compared with just 4 per cent of those with estates valued at less than £500) suggesting that they were drawn, like their British counterparts, into the financial economy that ranged well beyond the local context. Indeed, Shanahan (2011, p. 119) suggests that in spite of the rural nature of the South Australian economy, a trend away from the ownership of land is detectable there too, 'as other kinds of assets became more attractive'.

The picture was not dissimilar in other British settler colonies. In the Canadian west for example, the dominion government had a land grants policy that provided 160 acres (65 hectares) of land to adult settlers after a period of three years, upon payment of a small fee. Livio di Matteo's (2012, pp. 318–20) analysis of wealth and inequality in Canada between 1870 and 1930 reveals that in the western province of Manitoba, which experienced a settlement boom in the period after 1870, land on average made up between 50 and 60 per cent of the total value of estates and was widely dispersed among the sample of individuals studied. The centrality of the farming economy in this region along with the lack of other investment opportunities resulted in a more egalitarian spread of wealth. This contrasted with Ontario, further to the east, where wealth ownership was more unequal. In the longer-settled rural district of Wentworth County and the urban-industrial centre of Hamilton, estate data reveal that on average only 30 to 35 per cent of wealth portfolios were made up of land; in these locations financial assets (including also business capital and forms of credit) were more important. These findings point to the significance of the links between economic growth and development, resulting in changes in the composition of wealth ownership and inequality (Di Matteo, 2012, pp. 325–30). As Di Matteo concludes, where land was the dominant asset in wealth portfolios, inequality tended to be lower.

Nineteenth-century Brazil provides a contrast to this British colonial world. While still at the periphery of the capitalist world system, after achieving independence from Portugal in the early 1820s, the country experienced profound economic and social change over the middle and later decades of the century. This involved the rapid development of an export economy, driven, in particular, by the coffee trade; major infrastructure construction projects, notably the building of railways; and the creation of a 'modern' financial system. The changes opened up new opportunities for wealth accumulation and were linked with significant social transformations, particularly the decline of slavery and, as in Canada, growing social inequality among the free population. As Zephyr L. Frank (2004) has explored, these shifts had marked impacts on urban areas in Brazil and particularly on the capital, Rio de Janeiro. Using evidence from a sample of estate inventories Frank charts the changing composition of wealth over an 80-year period (Figure 3.3). While urban real estate forms a key component of wealth portfolios across the entire timeframe – unsurprisingly given the rapid demographic expansion of the city, fuelled by European immigration as well as natural increase – the most notable shift is the decline in importance of slaves as an economic asset valued in estate inventories and the rise in the significance of stocks and bonds. In the early to mid-century, Frank shows slave owning to be widespread amongst Rio de Janeiro's middle class (including sometimes amongst families who had themselves secured freedom from slavery). Within entrepreneurial groups African slaves were regarded as a key

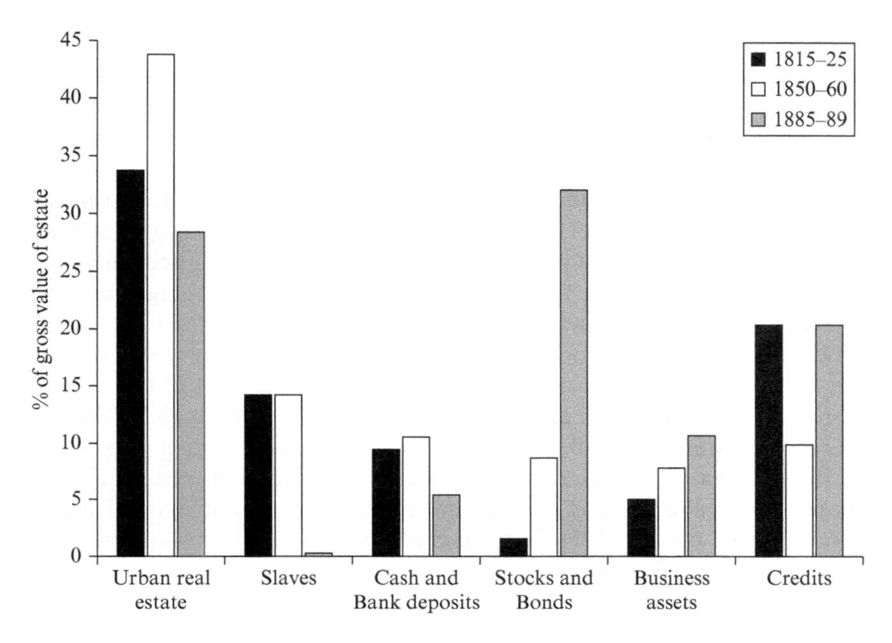

Source: Frank (2004, pp. 57, 73 and 88).

Figure 3.3 Composition of estates at death in Rio de Janeiro, Brazil, 1815–89 (selected assets as percentage of total estate)

economic asset, hired out for wages or put to work in family businesses and sometimes sold on to others at a profit. In the context of limited opportunities for other kinds of investment, Frank suggests that slavery persisted because, at least for a time, it provided an economic foundation for middling social mobility. However, a key shift occurred in the 1850s when the British and Imperial Brazilian authorities effectively closed off the Atlantic slave trade and, with it, a cheap and plentiful supply of labour. Accompanied by the development of new financial institutions including the banking system and a stock market, patterns of wealth holding altered. Ownership of human capital in the form of slaves disappeared, while new financial assets, including railway and government bonds, along with business capital grew significantly (eventually also eclipsing urban real estate). While the proportion of wealth passing at death in the 1880s in the form of stocks and bonds – around 32 per cent of the total – was similar to that in England and Wales, in Rio de Janeiro its ownership was more highly concentrated; the top 20 per cent of wealth holders accounted for 90 per cent of wealth held in stocks and bonds (Frank, 2004, p. 89).

As Frank (2004, 2005) emphasizes, paralleling our own arguments above, the changing nature of wealth portfolios in Rio de Janeiro reflected key institutional developments in the Brazilian setting – in its case, the abolition of slavery, the introduction of a new commercial code and the emergence of a range of complex and sophisticated financial institutions, that promised both security of property and opened up new channels for accumulating wealth, both in Rio de Janeiro and in the developing economy of Brazil as a whole, particularly its southeastern regions. The importance of 'Credits' – credit for goods, notes of credit and cash advances – among wealth portfolios demonstrates the growing sophistication of credit networks within commercial and industrial communities. Though 'informal' this credit system was well embedded and increasingly codified, providing an institutional framework that 'lubricated the wheels of exchange' and stimulated local economic growth through investment and specialization (Frank, 2005, pp. 242, 245). In turn, credit became an important income-generating asset for money-lending individuals and families. Later in the century, the development of Rio de Janeiro's 'modern' banking and financial system tied Brazil into a financial economy that was increasingly global in reach; Rio de Janeiro's wealth-owning class stood alongside British and European investors in sinking their wealth into Brazilian railway securities.

Institutional arrangements also help to explain the differences between other Latin American countries. In Argentina, improved property rights institutions underpinned the growth of personal wealth, albeit to those who supported authoritarian regimes. The annexation and sale of public lands to private owners in Argentina under the Rosas regime, for example, was accompanied by greater rights of ownership, in effect transferring significant amounts of wealth based on rural real estate to urban, middle-class supporters of the regime (Johnson and Frank, 2006, p. 654). In Brazil, the cities were more remote from the countryside and the sale of public lands was of less importance. As a result, the composition of wealth differed in Rio de Janeiro compared to Buenos Aires, notably in the different proportions of real estate held by wealthy individuals in each place. In Rio de Janeiro in the mid-1850s real estate comprised around 45 to 50 per cent of wealth compared to 70 per cent in Buenos Aires (Frank, 2005; Johnson and Frank, 2006, pp. 658–9).

In an economy with easy access to land, real estate is likely to be of far greater importance than other forms of wealth. There were clear differences in the relative importance of real estate between land-rich countries, such as the United States, Australia and Argentina, compared to those that were land poor, such as Britain and other European states. In addition, depending on how easy it was to purchase land in the rural hinterland and the relative value of that land compared to other forms of wealth, there may also have been differences between wealth holders in the metropoles and the periphery.

CONCLUSION

Thomas Piketty's *Capital in the Twenty-First Century* (2014) has reignited widespread interest and debate in the nature of inequalities and their historical development. His claim that differences between income obtained from labour and that generated from capital are central to understanding the emergence of inequalities over the long run, suggests that we need to better understand what these sources of 'unearned' wealth were, and how access to them varied over time, across space and among populations. In this brief survey we have attempted to describe, first, some of the 'institutional' developments that encouraged and enabled different groups of people in different places to acquire different kinds of assets; and, second, to illustrate the range and types of assets that those with sufficient surplus resources to invest tended to favour. If Piketty is correct, our focus on the composition of wealth over the period from 1800 to around 1930 is key to understanding not only the unequal distribution of wealth among populations, but also across different geographical settings and over time, because different kinds of capital proved more or less profitable at different moments in different places. The historical geography of inequality, therefore, is inextricably linked to the historical geography of capital and opportunities for its acquisition and accumulation.

As we have shown here, this is partly a question of what opportunities there existed in a particular location for investing wealth and making money. In British colonial settings, land was an obvious asset and deeply implicated in the economic strategies of settlers and of the colonial powers that sent them there, and was practically given away to those who took the gamble of resettling to the other side of the world. As Livio di Matteo (2012) has argued, this tended to lead to a more egalitarian settler society; land was shared among settlers and investment in other kinds of assets that would allow some wealth holders to 'pull away' from others was more limited. By contrast, in many European countries like Britain, opportunities for investment in land were limited; much of it was tied up under aristocratic ownership or had become less profitable due to cheaper foreign agricultural imports. Here wealth-holding groups turned to other forms of capital, notably shares and financial securities – assets that had the potential to yield greater returns. As this economy in 'paper' assets also spread to Europe's peripheries, so here, as in the countries like Britain, levels of inequality increased. Indeed, understanding the relationships between inequality and asset holding is not simply about mapping opportunities for wealth making that were local to accumulators. The issue is also about the way that some, though not all, social groups and certain, but not all, locations were increasingly able to 'trump' geography, opening up opportunities for money making that were more numerous and more geographically dispersed. The ability to access information about opportunities, to invest wealth through mature and regulated financial institutions that reduced exposure to risk, or to find good advisors and asset managers to seek out the best revenue-generating opportunities on a global scale, was what mattered. So, too, was the backing of the state, be it through diplomatic efforts to promote overseas expansion or through military intervention to protect assets and secure territory.

These processes help to account for patterns of inequality not just between countries but also within them, between those who had the power and resources to access financial expertise and those who had to rely on their own judgement and good fortune; and between those areas, like Rio de Janeiro in the 1880s or London in the 1910s that were hubs in regional, national and global networks of information and those that remained on the periphery. What the estate records – and other surviving evidence of the composition of personal wealth – tell us is how people were placed in relation to these money-making opportunities, infrastructures and information, and how they sought to navigate them within the context of their circumstances. Captured at the point at which assets were being transferred from one generation to the next, they also cast light on the way that wealth and therefore also patterns of inequality were perpetuated over time. Inheritance, as Piketty (2014) also observes, is a further key mechanism in understanding the relationships between capital accumulation and inequality over time. Assets offered income and security at these times of upheaval, smoothing the financial transition as one generation of accumulators gave way to another, by providing stability in old age for surviving partners and an economic springboard for offspring.

As the wealth portfolios of some families grew more varied and geographically far reaching, so over time and across the generations their fortunes, well-being and, ultimately, status, became reliant on the vicissitudes of global economic change. Many of the world's rich watched the value of their financial portfolios tumble following the Wall Street stock market crash of 1929 and the ensuing financial instability of the 1930s. Some never recovered their former wealth, though others, through good fortune or via shrewd advice, rode economic storms such as this without serious loss. To understand how some people got richer, while others lost their fortunes, requires better appreciation of the range of different kinds of wealth that individuals were able to acquire and critically how different groups in different locations were able to access opportunities to acquire assets. Given that acquiring these kinds of assets, according to Piketty, is most often associated with rising levels of inequality, it is even more important that we recognize the significance of historical geographies of accumulation and the institutional processes by which they were underpinned.

NOTES

* This chapter draws upon research undertaken as part of two collaborative projects: 'Women Investors in England and Wales, 1870–1930' funded by the UK's Economic and Social Research Council (Award: Res-000-23-1435); and 'Common Wealth? Wealth-Holding and Investment in Britain and its Settler Colonies, 1850–1914' funded by the British Academy and the Association of Commonwealth Universities. We gratefully acknowledge this support and thank our co-investigators and collaborators – Livio di Matteo, Josephine Maltby, Jim McAloon, Janette Rutterford and Martin Shanahan – and research assistants – Steven Ainscough, Carry van Lieshout, Carien van Mourik and Claire Swan – for their helpful contributions and insights.

1. Since real estate – chiefly land and freehold buildings – was not assessed for duty until part way through the nineteenth century, the figures quoted here refer only to personal property. We discuss land and real estate further below.

REFERENCES

Baltzer, M. (2007), 'European financial market integration in the Gruenderboom and Gruenderkrach: evidence from European cross-listings', *University of Tuebingen Economics Working Paper*, accessed 1 August 2014 at http://dev3.cepr.org/meets/wkcn/1/1617/papers/Baltzer.pdf.

Bordo, M.D. (2000), 'The globalisation of international financial markets: what can history teach us?', paper prepared for the conference on 'International Financial Markets: The Challenge of Globalization', Texas A&M University, 31 March, accessed 1 August 2014 at http://econweb.rutgers.edu/bordo/global.pdf.

Cain, P.J. and A.G. Hopkins (2001), *British Imperialism, 1688–2000*, Harlow, UK: Longman.

Cairncross, A. (1953), *Home and Foreign Investment, 1870–1930*, Cambridge, UK: Cambridge University Press.

Campion, H. (1939), *Public and Private Property in Great Britain*, Oxford, UK: Oxford University Press.

Carter, A. (1968), *The English Public Debt in the Eighteenth Century*, London: The Historical Association.

Cassis, Y. (2011), 'Wealth investment and global finance: international financial centres, 1870–1920', in D.R. Green, A. Owens and J. Rutterford et al. (eds), *Men, Women and Money: Perspectives in Gender, Wealth and Investment 1850–1930*, Oxford, UK: Oxford University Press, pp. 248–64.

Davis, L.E. and R.E. Gallman (eds) (2001), *Evolving Financial Markets and International Capital Flows: Britain, the Americas, and Australia, 1865–1914*, Cambridge, UK: Cambridge University Press.

Davis, L.E. and R.A. Huttenback (1986), *Mammon and the Pursuit of Empire: The Political Economy of British Imperialism, 1860–1912*, Cambridge, UK: Cambridge University Press.
Di Matteo, L. (2012), 'Land and inequality in Canada, 1870–1930', *Scandinavian Economic History Review*, **60** (3), 309–34.
Djelic, M.-L. (2013), 'When limited liability was (still) an issue: mobilization and politics of signification in 19th-century England', *Organization Studies*, **34** (5–6), 1–27.
Frank, Z.P. (2004), *Dutra's World: Wealth and Family in Nineteenth-Century Rio de Janeiro*, Albuquerque, NM: University of New Mexico Press.
Frank, Z.P. (2005), 'Wealth holding in Southeastern Brazil, 1815–60', *Hispanic American Historical Review*, **85** (2), 225–57.
Green, D.R. and A. Owens (2003), 'Gentlewomanly capitalism: spinsters, widows and wealth holding in England and Wales, c. 1800 to 1860', *Economic History Review*, **56** (3), 510–36.
Green, D.R. and A. Owens (2013), 'Geographies of wealth: real estate and personal property ownership in England and Wales, 1870–1902', *Economic History Review*, **66** (3), 848–72.
Green, D.R., A. Owens and J. Rutterford et al. (2009), 'Lives in the balance: age, gender and assets in late nineteenth-century England and Wales', *Continuity and Change*, **24** (2), 307–35.
Hannah, L. (2013), 'The corporate economies of America and Europe 1790–1860', *CIRJE Discussion Papers*, University of Tokyo and London School of Economics.
Harvey, D. (1989), *The Condition of Postmodernity: An Enquiry into the Origins of Cultural Change*, Oxford, UK: Blackwell.
Henry, N. and C. Schmitt (eds) (2009), *Victorian Investments: New Perspectives on Finance and Culture*, Bloomington, IN: Indiana University Press.
Johnson, L.L. and Z.P. Frank (2006), 'Cities and wealth in the South Atlantic: Buenos Aires and Rio de Janeiro before 1860', *Comparative Studies in Society and History*, **48** (3), 634–68.
Kuznets, S. (1955), 'Economic growth and income inequality', *American Economic Review*, **XLV** (1), 1–28.
Lindert, P.H. (1986), 'Unequal English wealth since 1670', *Journal of Political Economy*, **94** (6), 1127–62.
Loftus, D. (2002), 'Capital and community: limited liability and attempts to democratize the market in mid-nineteenth-century England', *Victorian Studies*, **45** (1), 93–120.
Maltby, J., J. Rutterford and D.R. Green et al. (2011), 'The evidence for the "democratization" of share ownership in Great Britain in the early twentieth century', in D.R. Green, A. Owens and J. Rutterford et al. (eds), *Men, Women and Money: Perspectives on Gender, Wealth and Investment 1850–1930*, Oxford, UK: Oxford University Press, pp. 184–206.
Mauro, P., N. Sussman and Y. Yafeh (2006), *Emerging Markets and Financial Globalisation: Sovereign Bond Spreads in 1870–1913 and Today*, Oxford, UK: Oxford University Press.
McQueen, R. (2009), *The Social History of Company Law: Great Britain and the Australian Colonies 1854–1920*, Farnham, UK: Ashgate.
Michie, R. (1992), *The City of London: Continuity and Change Since 1850*, Basingstoke, UK: Macmillan.
North, D.C. (1991), 'Institutions', *Journal of Economic Perspectives*, **5** (1), 97–112.
O'Hara, G. (2010), 'New histories of British imperial communication and the "networked world" of the 19th and early 20th centuries', *History Compass*, **8** (7), 609–25.
Owens, A. and D.R. Green (2012), 'The final reckoning: using death duty records to research wealth holding in nineteenth-century England and Wales', *Archives*, **38** (126), 1–21.
Piketty, T. (2014), *Capital in the Twenty-First Century* [French edition published in 2013 as *Le capital au XXI siècle*, Editions du Seuil], Cambridge, MA/London: Belknap Press of Harvard University Press.
Poovey, M. (2008), *Genres of the Credit Economy: Mediating Value in Eighteenth and Nineteenth-Century Britain*, London/Chicago, IL: University of Chicago Press.
Poovey, M. (2009), 'Writing about finance in Victorian England: disclosure and secrecy in the culture of investment', in N. Henry and C. Schmitt (eds), *Victorian Investments:*

New Perspectives on Finance and Culture, Bloomington, IN: Indiana University Press, pp. 39–57.
Preda, A. (2009), *Framing Finance: The Boundaries of Markets and Modern Capitalism*, London/Chicago, IL: University of Chicago Press.
Roine, J. and D. Waldenström (2014), 'Long run trends in the distribution of income and wealth', *Working Paper No. 2014:5*, Uppsala: Uppsala Centre for Fiscal Studies, Department of Economics, Uppsala Universiteit.
Rungta, R.S. (1970), *The Rise of Business Corporations in India, 1851–1900*, Cambridge, UK: Cambridge University Press.
Rutterford, J., D.R. Green and J. Maltby et al. (2011), 'Who comprised the nation of shareholders? Gender and investment in Great Britain, c. 1870–1935', *Economic History Review*, **64** (1), 157–87.
Shanahan, M.P. (2001), 'Personal wealth in South Australia', *Journal of Interdisciplinary History*, **32** (1), 55–80.
Shanahan, M.P. (2011), 'Colonial sisters and their wealth: the wealth holdings of women in South Australia, 1875–1915', in D.R. Green, A. Owens and J. Rutterford et al. (eds), *Men, Women and Money: Perspectives on Gender, Wealth and Investment 1850–1930*, Oxford, UK: Oxford University Press, pp. 99–124.
Steckel, R.H. and C.M. Moehling (2001), 'Rising inequality: trends in the distribution of wealth in industrializing New England', *Journal of Economic History*, **61** (1), 160–83.
Sylla, R. (1998), 'US securities market and the banking system, 1790 to 1840', *Review of Federal Reserve Bank of St. Louis*, May/June, 83–98, accessed 1 August 2014 at http://research.stlouisfed.org/publications/review/98/05/9805rs.pdf.
Taylor, J. (2006), *Creating Capitalism: Joint-Stock Enterprise in British Politics and Culture, 1800–1870*, London: Royal Historical Society.
Tilly, R.H. (1989), 'Banking institutions in historical and comparative perspective: Germany, Great Britain and the United States in the nineteenth and early twentieth century', *Journal of Institutional and Theoretical Economics*, **145** (1), 189–209.
Tilly, R.H. (1998), 'Universal banking in historical perspective', *Journal of Institutional and Theoretical Economics*, **154** (1), 7–32.
Turner, J. (2014), *Banking in Crisis: The Rise and Fall of British Banking Stability, 1800 to the Present*, Cambridge, UK: Cambridge University Press.
Wallerstein, I. (2011), *The Modern World-System, Vol. IV: Centrist Liberalism Triumphant, 1789–1914*, Berkeley, CA: University of California Press.
Wenzlhuemer, R. (2007), 'The development of telegraphy, 1870–1900: a European perspective on a world history challenge', *History Compass*, **5** (5), 1720–42.
Wenzlhuemer, R. (2012), *Connecting the World: the Telegraph and Globalization*, Cambridge, UK: Cambridge University Press.

4. On plutonomy: economy, power and the wealthy few in the second Gilded Age

Iain Hay

INTRODUCTION

For many years geographers and other social scientists have tended to focus on middle class circumstances and the plight of the poor, attending to the mechanisms by which poverty and economically driven environmental destruction have been produced (e.g., Banerjee and Duflo, 2011; Harvey, 2014), but giving less explicit attention to reproduction of the conditions and circumstances by which the wealthy continue to accumulate their fortunes. It is only more recently that attention has been directed more systematically to those at the apex of affluence (Beaverstock et al., 2004; Hay, 2013; Dorling, 2014; Hay and Muller, 2014). Groundbreaking analysis of historical inequalities in wealth by French economist, Thomas Piketty (2014) and income by English economist Sir Anthony Atkinson (1993, 2003), coupled with the heightening (public) profile of the super-rich (e.g., Frank, 2007; Freeland, 2011, 2012; Di Muzio, 2015), have helped spark interest in not only the lifestyles of the rich and the circumstances under which they live but also the mechanisms that have permitted them to accrue and retain their riches. Provocative but less well-publicized work in these domains has been conducted since the mid-2000s by Ajay Kapur (and colleagues), formerly Chief Global Equity Strategist for Citigroup and now Equity Strategist with Bank of America Merrill Lynch (BoAML). This chapter reviews and elaborates on that decade of work, parts of which have been somewhat inaccessible to public view by virtue of its place in financial advice's grey literature and perhaps also as a consequence of Citigroup's reported gag order on some of its (re)publication.[1] Kapur's interpretations offer interesting insights to the nature of, and prospects for, patterns of wealth inequality.

In October 2005, Ajay Kapur, who at the time performed the role of Chief Global Equity Strategist for Citigroup, suggested that the world is dividing into two key blocs – so-called plutonomies and the rest.[2] In plutonomies, which include several influential economies of the world, notably Australia, Canada, the UK and the USA, economic growth is powered by and consumed by the very wealthy few (Kapur et al., 2005,

p. 8; 2014, p. 1). Kapur's Citigroup team observed that plutonomies have occurred in the past, in sixteenth-century Spain, Holland's seventeenth-century Golden Age,[3] America's Gilded Age of the late nineteenth century and, of course, the 'Roaring Twenties' (Kapur et al., 2005, p. 8). Scholars such as Bartels (2008), Lieberman (2011) and Short (2013) declare that we have now entered what is being termed the 'second Gilded Age', a corollary perhaps of the deeply troubling era of austerity (Peck, 2012; Stanley, 2014). The first Gilded Age was associated with stupendously large fortunes of super-rich tycoons such as John D. Rockefeller, Andrew Carnegie, John Pierpont Morgan and Cornelius Vanderbilt. Not only was it was an era of rapid economic growth and wealth concentration, particularly in the USA, but it was also a period of conspicuous consumption, often expressed in exaggerated aesthetics of old, moneyed European nobility inscribed in the landscape through chateau-like homes such as Cornelius Vanderbilt's 'The Breakers' (Newport, Rhode Island), Otto Hermann Kahn's 'Oheka Castle' (Long Island, NY) and Henry Flagler's vast South Florida mansion, filled with suits of armour and libraries purchased from impoverished European aristocracy. The new, second Gilded Age is a more global phenomenon than its twentieth-century predecessor. It is based on ownership and control of global commodities (e.g., oil, natural gas); control of finances (e.g., Warren Buffett's Berkshire Hathaway); ownership of successful companies (e.g., Google, Oracle, ArcelorMittal, Facebook); and inheritance (e.g., Walton family; Forrest Mars Jr; Liliane Bettencourt [L'Oreal]). Prominent figures in this new Gilded Age include Bill Gates, Carlos Slim Helú, Larry Ellison, Mukesh Ambani, Hong Kong's Kwok family, Amancio Ortega, Charles Koch, Li Ka-shing, and Roman Abramovich.

In three reports for Citigroup clients Ajay Kapur and various colleagues (2005, 2006a, 2006b) set out a thesis that the wealth share of the very rich in plutonomies of this second Gilded Age had become so large that economic activity – and critically therefore investment and wealth-making decisions – in those jurisdictions can no longer be understood through analysis of the behaviour of the average consumer. They opined that the rich are 'so rich that their behaviour overwhelms that of the "average" or median consumer' (Kapur et al., 2006b, p. 8). Kapur and his colleagues speculated further that the forces that drove the late twentieth-century rise in economic polarization will continue for some time (2006b, p.8). These include, for example, 'disruptive technology-driven productivity gains [such as the current technology/biotechnology revolution], creative financial innovation, capitalist-friendly cooperative governments [and tax regimes], immigrants and overseas conquests invigorating wealth creation, the rule of law, and patenting inventions' (Kapur et al., 2005, p. 7).

However, they did also warn of the prospect that there could be a political backlash against the super-rich, a warning Kapur and new colleagues at Bank of America Merrill Lynch repeated much more recently in a 2014 report entitled *Piketty and Plutonomy: The Revenge of Inequality* (Kapur et al., 2014).

While there have been some largely misplaced criticisms of the Citigroup work as an explicitly political treatise – the highest profile of which was Mike Moore's 2009 movie *Capitalism: A Love Story* – the group's reports offered, at their core, challenging big-picture insights to the character of twenty-first-century capitalism as well as to ways in which the bank's retail investors might be able to make more money.[4] In the remainder of this chapter I retrace the central threads of Kapur and his colleagues' work, first describing plutonomy, before linking it briefly to Thomas Piketty's recent treatise (2014) on historical patterns of wealth. I go on to discuss some of the forces that power plutonomy, consider a small number of its economic and geographical implications, and conclude by contemplating plutonomy's future.

KEY CHARACTERISTICS OF PLUTONOMY: ECONOMIC CONTROL AND CONSUMPTION POWER

> So the world is now indeed splitting into a plutonomy and a precariat – in the imagery of the Occupy movement, the 1% and the 99%. Not literal numbers but the right picture. (Chomsky, 2012)

Control of the Economy by the Rich

Plutonomies have two key characteristics that warrant discussion. The first is that the rich control a gigantic and vastly lopsided portion of the economy. For instance, in 2001, the wealthiest 1 per cent of United States households had a net worth that exceeded that of the bottom 95 per cent of households put together (Kapur et al., 2005, p. 10). And as shown in Figure 4.1, by 2012, total household wealth held by the richest 0.1 per cent in the United States stood at 22 per cent, a proportion not seen since the late 1920s (Saez and Zucman, 2014a).[5]

As a consequence, the decisions of the wealthy to reduce savings, for example, have massive effects on other figures such as overall savings rates, current account deficits and consumption levels. In Canada, the wealthiest 86 Canadian resident individuals (and families) hold the same amount of wealth as the poorest 11.4 million Canadians combined.

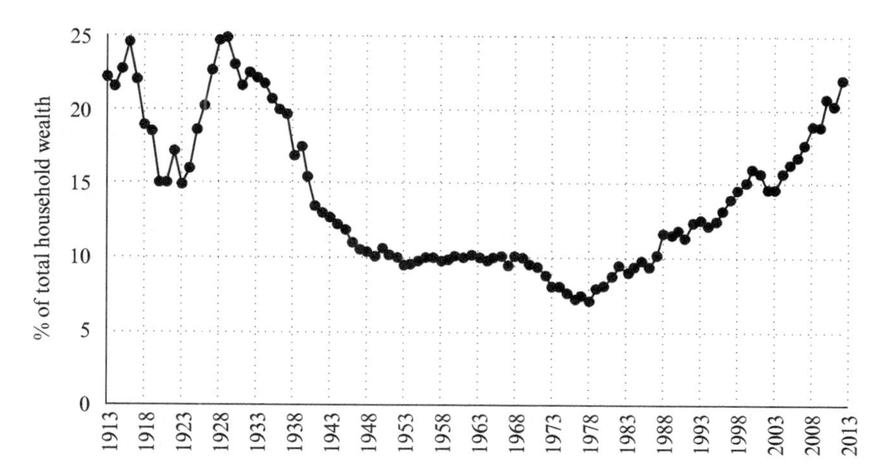

Note: The figure depicts the share of total household wealth held by the 0.1% richest families, as estimated by capitalizing income tax returns. In 2012 the top 0.1% includes about 160000 families with net wealth above $20.6 million.

Source: Saez and Zucman (2014b, Appendix Table B1).

Figure 4.1 Top 0.1% wealth share in the United States, 1913–2012

Expressed in other terms, 0.002 per cent of Canadians have wealth equivalent to that held by 34 per cent of that country's population (Macdonald, 2014, p. 6). Moreover, this is a concentration that has deepened since Canadian wealth data was first collected systematically in 1999. In Australia, the richest 20 per cent of households have a mean net worth of A$2215000 (in 2011–12), accounting for 61 per cent of total household net worth. By contrast, the poorest 20 per cent of households, with average net worth of A$31200, account for a mere 1 per cent of total household net worth (ABS, 2013, p. 18). In India, the wealthiest 5 per cent of households hold 38 per cent of assets while the 'poorest' 60 per cent of the population own just 13 per cent (Institute of Applied Manpower Research, 2011, p. 105). In light of that country's quickening urbanization, the higher levels of inequality in urban areas are disturbing (60 per cent of urban households have 10 per cent of the total value of assets). In places like Chile, Israel, Malaysia, Russia and the Philippines the very wealthy account for a greater share of the economy than their US peers (Kapur et al., 2014, p. 9). Malaysia's first *Human Development Report*, made public in November 2014, exposes the highly concentrated distribution of wealth in that country, observing too that asset inequality is almost double that of income inequality (Muzaffar, 2015, p. 14). Even in China, recent work points to growing wealth inequality. Citing Peking University's Institute of Social Science Survey's *2014 China Welfare Development Report*, Dongxu (2014) notes that the top 1 per cent of households hold one-third of total assets, while the bottom 25 per cent have only 1 per cent (see also Table 4.1). While these figures are not exceptional by global standards, they do mark a dramatic change from the situation as recently as 2000, when China had the second most equal distribution of wealth in the world. Moreover, income inequality in China now ranks among the highest in the world and is much more extreme than in the United States (Gini coefficient of 0.55 in 2012 compared with 0.45 for the USA), a trend that has intensified since China's 1978 economic reforms (Xie and Zhou, 2014). And at a global scale, the wealthiest 66 individuals in the world have a net worth equivalent to the 3.5 billion poorest people (Moreno, 2014). Each of these 66 wealthy individuals has, on average, net worth equivalent to 52 million of their poorer global citizens. The world's richest man, Bill Gates, is 'worth' roughly the same as a mere 156 million of the planet's poorest! In short, although plutonomy's expressions may be more pronounced in some jurisdictions (e.g., Chile, Hong Kong, Israel, Malaysia, the Philippines, Russia, Taiwan, United States) than others (e.g., France, Germany, Japan, Korea), it is not simply a phenomenon of specific jurisdictions; it is a global concern. Moreover, as Saez and Zucman (2014b) note, while the post–Great Depression era witnessed a 'substantial democratization of wealth', from the late 1970s, there has been a substantial reversal, with the share of wealth held by the few wealthiest now approaching levels not seen since the inequitable days of the late 1920s and very early 1930s.

The discussion so far centres solely on wealth, understood to be one's accumulated net assets having monetary value. It is worth making a slight, but important, diversion to consider income distributions (i.e., the money received from work or from returns on investments). Data from the most recent US Federal Reserve Board's Survey of Consumer Finances (SCF)[6] shows that in 2013, American households in the top income band (90–100 percentile of income) had a net worth share of 71.5 per cent (mean net worth of US$3.327 million) and an income share of 57.1 per cent (mean income of US$361 200) (*Federal Reserve Bulletin*, 2014). Moreover, the SCF excludes exceptionally rich households! Tucked away in the study's 'Appendix: survey procedures and statistical measures' (2014, p. 38) is the following note: 'Persons listed by *Forbes* magazine as being among the wealthiest 400 people in the United States are excluded from sampling'. Kapur and his colleagues estimate that the *Forbes* 400 richest families account for about 2.4 per cent of the United States' total net worth (2006b,

Table 4.1 Millionaire households, 2013

	Millionaire Households						Ultra High Net Worth (UHNW) Households (More than US$100 Million in Private Financial Wealth)					
	Number of millionaire households (thousands) 2013			Proportion of millionaire households (%) 2013			Number of UHNW households 2013			Proportion of UHNW households (per 100 000 households) 2013		
1	(1)	USA	7135	(1)	Qatar	17.5	(1)	USA	4754	(1)	Hong Kong	16.8
2	(2)	China	2378	(2)	Switzerland	12.7	(2)	UK	1044	(2)	Switzerland	11.3
3	(3)	Japan	1240	(3)	Singapore	10.0	(4)	China	983	(3)	Austria	9.3
4	(4)	UK	513	(4)	Hong Kong	9.6	(3)	Germany	881	(4)	Norway	8.3
5	(6)	Switzerland	435	(5)	Kuwait	9.0	(5)	Russia	536	(5)	Singapore	7.4
6	(5)	Germany	386	(6)	Bahrain	5.9	(6)	France	472	(6)	Qatar	7.1
7	(7)	Canada	384	(7)	USA	5.9	(7)	Canada	465	(7)	Kuwait	4.5
8	(8)	Taiwan	329	(8)	Israel	4.6	(8)	Hong Kong	417	(9)	New Zealand	4.1
9	(9)	Italy	281	(9)	Taiwan	4.2	(9)	Switzerland	388	(8)	Belgium	4.1
10	(10)	France	274	(10)	Oman	3.7	(10)	Italy	374	(12)	USA	3.9
11	(11)	Hong Kong	238	(11)	Belgium	3.4	(11)	Austria	344	(11)	UK	3.9
12	(12)	Netherlands	221	(12)	UAE	3.3	(12)	Turkey	288	(10)	Israel	3.7
13	(13)	Russia	213	(13)	Saudi Arabia	3.1	(13)	India	284	(16)	Bahrain	3.7
14	(14)	Australia	195	(14)	Netherlands	3.0	(14)	Australia	236	(15)	Canada	3.5
15	(16)	India	175	(15)	Canada	2.9	(16)	Brazil	227	(13)	Ireland	3.4

Note: UAE is United Arab Emirates. Numbers in parentheses are 2012 rankings, determined on the basis of year-end 2013 exchange rates to exclude the effect of currency fluctuations.

Source: The Boston Consulting Group (2014, Exhibit 4).

p. 11). Clearly the addition of figures from the very wealthiest households would skew the SCF data further towards the wealthy.

In some of their recent work, distinguished scholars of wealth and income, Alvaredo et al. (2013, p. 4), report substantial increases in income share held by the top 1 per cent in the United States, with that group's share more than doubling from 9 per cent in 1976 to 20 per cent in 2011. They observe a U-shaped historical trend in income shares going to the top 1 per cent, with high proportions of income accruing to that percentile until the late 1930s, falling through to the late 1970s, and then dramatic upturns in the USA as well as in other 'Anglo Saxon' countries like Australia, Canada and the United Kingdom. Tellingly, they note that because many other high-income countries (e.g., France, Germany, Japan, Sweden) have seen little or no increase in top shares, explanation for concentration cannot reside in factors shared amongst advanced economies. Instead, four main factors are posited to have contributed to heightening income levels amongst the top 1 per cent. The first is changing tax rates that favour high-income earners. Second are changes in the labour market, characterized by differences in bargaining power and greater individualization of remuneration. Third, capital income (e.g., rents, dividends, interest) changes have demonstrated a recovery since the turn of the century and inheritance also appears to be experiencing something of a resurrection. And finally, changing correlations between earned income and capital income have become especially evident in some jurisdictions, including the United States (see Sayer, Chapter 5 in this volume for a discussion of some of these matters). In short, it would appear that the wealthy not only control a disproportionate part of the economy in plutonomies through their wealth, but they are also seizing a disproportionate share of income.

NO AVERAGE CONSUMER

A second key characteristic of plutonomy is that there is no 'average consumer'. 'There are rich consumers, few in number, but disproportionate in the gigantic slice of income and consumption they take. There are the rest, the "non-rich", the multitudinous many, but only accounting for surprisingly small bites of the national pie' (Kapur et al., 2006b, p. 9). For instance, Frank (2011, p. 19) observes that the top 5 per cent of American earners account for 37 per cent of consumer spending, a figure corroborated by Cynamon and Fazzari (2014). The bottom 80 per cent account for only marginally more (39.5 per cent) (Frank, 2010). And the American Affluence Research Center (2014) suggests that it is the wealthiest 1 per cent of US households that account for about 20 per cent of all consumer spending. Matthews (2014) brings the situation to life in his description of a recent US winter shopping period:

> [R]etailers are once again bracing for a miserable holiday shopping season due mostly to the fact that most Americans simply aren't seeing their incomes rise and have learned their lesson about the consequences of augmenting their income with debt. Unless your business caters to the richest of the rich, opportunities for real growth are scarce.

Publicly available data on the spending patterns of the very wealthy remain difficult to obtain, but the occasional insight can be revealing (as Crewe and Martin illustrate in Chapter 16 in this volume on consuming luxury fashion in global cities). For instance, court filings show that Sam Wyly, a former billionaire who made his fortune from Michael's Stores (an arts and crafts retail chain) and Sterling Software, personally spent approximately US$3.75 million per month over the period 2004–14 (Frank, 2014b). Other, sometimes bizarre, examples of expenditure (on 'everyday' items) by the super-rich emerge, including Victoria Beckham's US$33 000 iPhone; Paris Hilton's US$325 000 dog house; and Charlize Theron's US$100 000 clutch bag (CNBC, 2014). And then there is London's annual so-called 'Ramadan Rush' when thousands of wealthy Middle Easterners descend on the city for a month for an orgy of consumption, piloting cars worth up to £900 000 from Harrods to Boodles to purchase £23 000 watches, £12 000 crocodile skin handbags and £39 900 mini Hummers for kids (Hale and Stott, 2014)!

Elsewhere the super-rich are spending at levels vastly disproportionate to their demographic representation. For example, 'wealthy Africans have expensive tastes and a habit of unapologetically flaunting their wealth in ways that would make some Arab Sheikhs look modest' (Spooner, 2014). Examples include Nigeria's US$59 million consumption of French champagne in 2012, trailing only France. Porsche has been unable to keep up with the demand for its cars in Luanda, Angola and set up a thriving dealership in Nairobi, Kenya in 2014 (Juma, 2014; Spooner, 2014). In the past four to five years, Nigerian tycoons Aliko Dangote, Folorunsho Alakija and Femi Otedola each have purchased private jets costing in excess of US$40 million – as well as large private yachts worth tens of millions of dollars and Hyde Park properties (Gbadebo et al., 2013; Mwanza, 2014). And all this from a country where per capita GDP currently stands at US$3000 (Provost, 2014).

The simple result of all of this – from the everyday to the absurd – is the diminishing economic significance of the average consumer: 'We should worry less about what the average consumer – say the 50th percentile – is going to do, when that consumer is (we think) less relevant to the aggregate data than how the wealthy feel and what they are doing. This is simply a case of mathematics, not morality' (Kapur et al., 2006b, p. 11). The very wealthy are simply driving more and more consumer spending and have disproportionate purchasing influence especially in areas such as technology, financial services, travel, motor vehicles, clothing and personal care (Pizzigati, 2011). Kapur and his colleagues go on: 'For the poorest in society, high. . .prices are a problem. *But while they are many in number, they are few in spending power, and their economic influence is just not important enough to offset the economic confidence, well-being and spending of the rich*' (Kapur et al., 2006a, p. 5; emphasis added).

Some of the rationale behind Kapur et al.'s claims about the broader economic and investment significance of the very wealthy can be found in two recent studies, with complementary consequences. First, echoing earlier work by Robert Frank (2007) on the super-rich, Tim Di Muzio (2014, pp. 498–9) argues that the habits of thought and the consumptive practices that characterize differential intraclass competition are focused on distinguishing dominant owners at least as far from members of their peer group as they are from the pecuniary inferiors. Thus, the very rich 'spend considerable sums. . .with the goal of communicating their power and social status to their class peers for greater prestige and advantages' (Di Muzio, 2014, p. 499). Second, Nathan Wilmers (2014) found that wealthy consumers bid up high status or marginally superior services and products in ways that disproportionately reward small differences in producer effort or quality. An outcome is that as the rich grow richer their purchasing decisions may cause wages and prices to polarize even further, as Figure 4.2 suggests. For example, a billionaire in need of a lawyer or accountant or financial advisor may be prepared to pay a substantial premium to secure the (perceived) best advice they can. The most sought-after employees can then bid up their price to levels well in excess of their marginal quality. Those who are not regarded as being of such high quality find themselves on poorer remuneration, working for the less wealthy. Likewise, status consumption drives purchases of other commodities such as wine, watches and motor vehicles[7] (see DePillis, 2015; Frank, 2014a; and Davison, Chapter 17 this volume, for accessible discussions).[8] Thus, as wealth and income inequality grows, so the wealthy account for more and more income and spending! A significant implication of all of this, following Kapur and his colleagues, is that analyses that fail to take account of the impact of plutonomy on spending power are critically flawed.

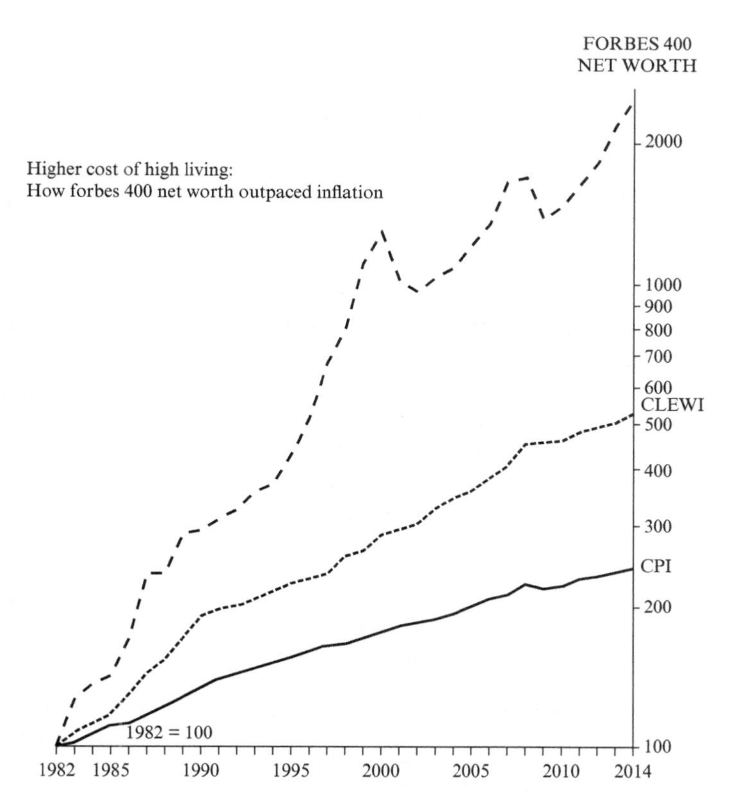

Note: CLEWI = *Forbes* Cost of Living Extremely Well Index, which measures the prices of 40 'superluxe' consumer goods across a number of categories (e.g., Learjet, Dom Perignon champagne, Gucci loafers, Groton school fees, Louis Vuitton travel bag, Olympic-size swimming pool). CPI = consumer price index (USA).

Source: Forbes Media; Bureau of Labor Statistics in Chen and O'Donnell (2014).

Figure 4.2 Cost of Living Extremely Well Index (CLEWI) outpacing inflation, 1982–2014

PIKETTY AND THE PROPAGATION OF PLUTONOMY – WHAT MAKES IT GO?

Roughly nine years after Kapur and colleagues released their work, French economist Thomas Piketty published his remarkable tome, *Capital in the Twenty-First Century*. Piketty (2014) discusses historical circumstances surrounding the tendency towards economic inequality. He provides compelling evidence of a strong long-term trend toward increasingly concentrated income and wealth, concentrations that have also been remarked upon by a diverse array of credible private sector, charitable and international sources (e.g., Credit Suisse, 2014; Oxfam International, 2014; World Economic Forum, 2014). A central argument of Piketty's book is that throughout history the rate of return on private capital (*r*) exceeds the rate of economic growth (*g*). He expresses the relationship in the formula $r > g$. Not surprisingly this implies that the riches of those who hold the greatest wealth will tend to increase at a faster rate than the assets of the less well-off. When a few vocal commentators – notably the *Financial Times* economics editor Chris Giles (2014) – questioned Piketty's analysis, Kapur and his BoAML colleagues (2014, p. 3) entered the fray, offering support for Piketty's work and remarking that BoAML is 'generally comfortable with the thrust of his [Piketty's] analysis'. They went on to reiterate their mid-2000s remarks about the economic significance of the wealthy relative to the 'average' consumer and to deliver a gloomy portent about the prospective intensification of plutonomy:

> When wealth and income are as concentrated as they are, and expected (à la Piketty) to get even more so, examining the 'average' consumer or 'average' investor makes little sense. Examining the fat tail – the behavior of the plutonomists, rather than that of the multitudinous many – is more advantageous to investors. Plutonomists determine and dominate spending and investment decisions and their magnitudes. Any analysis that does not tease out the skewed global income and wealth distribution, but focuses on the average is flawed from the start and is incomplete, **as we step into its deeper extremes**. (Kapur et al., 2014, p. 4; emphasis added)

This final observation about entering plutonomy's deeper extremes is of concern. Recent work by Emmanuel Saez (a frequent collaborator with Piketty) and Gabriel Zucman corroborates that of Piketty. Saez and Zucman (2014b) point to the ways in which wealth is being concentrated increasingly in the hands not just of the top 1 per cent in some countries, but especially in those of the top 0.1 per cent (see also Murray and Peetz, 2014, p. 139):

> Not only is US wealth rising relative to the economy, and is predicted to rise even further, the ownership of this wealth is increasingly in the hands of the top 1%, or more precisely, the top 0.1%. The top 0.1% own about 23% of US wealth, about the same as the bottom 90% put together. In 1978, they owned just a fourth of what the bottom 90% owned. (Kapur et al., 2014, p. 8)

Amidst the rising refrain of depressing distribution data, noted commentator on the wealthy Robert Frank observed presciently in 2011 that, globally, the 'plutonomy is stronger than ever and likely to produce chronic budget deficits, political tensions and more economic volatility for the rich and non-rich alike'. What is more, in their document of 2014 Kapur and his colleagues report that the drivers of the further concentration of wealth, including globalization and capitalism-friendly governments, remain intact. What are those drivers?

As we have already seen from Piketty's (2014) work, a cornerstone of plutonomy can be understood to be the long-term rate of return on private capital (*r*) that exceeds the rate of economic growth (*g*), meaning therefore that wealth tends to accrue to the rich faster than to the poor. Although the forces of wealth divergence generally tend to overwhelm the forces of wealth convergence it does not necessarily follow that uneven distributions of wealth and income will ensue, as decades of successful wealth redistribution policies in many jurisdictions have demonstrated (see, e.g., Reich, 2008 for a discussion of the USA, and Lansley, 2012). However, recent years have brought changed circumstances and policy reforms that have served to focus economic power in the hands of smaller and smaller groups of the wealthy.

Many of the newest pathways to fortunes lie through unsettling, non-linear technology-driven gains in productivity such as the Internet (e.g., Aghdam, 2011) and creative financial innovation (Volscho and Kelly, 2012, p. 694) buttressed by the rule of law (as explored by Dixon and Beaverstock and Hall, Chapters 19 and 20 in this volume respectively). Moreover, as Palley (2007, p. 2) observes, as a process that sees financial markets, financial institutions and financial elites gain greater influence over economic policy and economic outcomes, financialization has transformed economic systems in ways that promote the financial sector relative to the real sector, transfer income from the real sector to the financial sector, and increase income inequality and contribute to wage stagnation. As Beaverstock and Hall note in Chapter 20 in this volume, the super-rich became the beneficiaries of financialization as they were able to accumulate wealth from assets and speculation, play the casino of the international financial system, and become 'active investors' in the global economy.

Strengthening patent protection for inventions also increases economic growth by stimulating expenditures on research and development – which generally makes it more and more difficult for the less wealthy to participate – and heightens income inequality by raising the return on assets (Chu, 2010). According to critical observers (e.g., Stiglitz, 2013; Wikileaks, 2014; Electronic Frontier Foundation, 2015) the secretive Trans-Pacific Partnership Agreement (TPP) – the world's largest economic trade agreement encompassing more than 40 per cent of the world's

GDP – is testament to the growing and problematic privatization of intellectual property and development rights amongst the 12 TPP nations, including Australia, Canada, Chile, Japan, Malaysia, Mexico and the United States. High levels of complexity typically associated with each of these various technological, financial and legal pathways to prosperity mean that they are exploited best by the wealthy and well-educated of the time (Kapur, 2005) and so perpetuate inequalities.

Plutonomy is also propagated through greater global linkages between people and places, which, for example, heighten the range, complexity and potential of profit-making activities to those with the wherewithal to take advantage of them (e.g., Mohiuddin and Su, 2014). Mobility, multiple residency and migration flows characterized by capable, well-resourced emigrants returning to their poorer home country after time studying or working abroad may likewise accentuate wealth and income inequalities. Kapur and his colleagues (2014, p. 23) also suggest that assortative mating is another agent of plutonomy. As the Bennet sisters in Jane Austen's *Pride and Prejudice* were only too aware, pairings on the basis of socioeconomic status (as well as religious beliefs, physical traits and political ideology. . .) may be vital to the preservation of social standing and economic position. But perhaps the key reason for plutonomy's entrenchment is government policy (or purposeful lack of countervailing policy) supportive of capital:

> One of the main reasons you get a plutonomy is you have capital-friendly governments. . . Across the political spectrum it is very tough to destabilize or reverse the nexus of the plutonomists and politicians and policy makers. Normally, there is a very tight correlation between deregulation and a plutonomy. You had periods of deregulation in the 1920s and in the 1980s and those created plutonomies. (Ajay Kapur, cited in Frank, 2011)

Recent widespread commitment to neoliberal policies, including reductions in tax rates for the wealthy (Volscho and Kelly, 2012, p. 693), has fostered the polarization of wealth and income (see, e.g., Muzaffar, 2015). This is a matter to which I return later in the chapter, but first, what are some of the implications of plutonomy?

FROM SAVINGS TO SUSTAINABILITY: SOME IMPLICATIONS OF PLUTONOMY

Kapur et al. (2014) outline a number of implications of plutonomy, focusing specifically on those ten factors that bear on those seeking to profit from contexts in which economic growth is powered and consumed by the wealthy few. Of those implications I wish to focus on an important subset here.

First, plutonomy coupled with asset inflation leads to reduced national savings rates and, conversely, with asset value collapses come heightened savings levels (Kapur et al., 2014, p. 1). At first glance, this might seem counterintuitive. However, during a share market boom, for example, the very wealthy tend to reduce their savings from current income. Their consumption can be funded from dividends and the growing capital appreciation of rising portfolios. During a bust, savings rise to provide a financial buffer. So, for example, BoAML suggest that in the wake of the 2008 global financial crisis (GFC), 'shell-shocked [US] plutonomists (the top 1% accounting for 20% of national income) raised their savings rates from 22.9% to 38.2%. The magnitude of plutonomists' savings dollars. . .accounted for an estimated 60% of the *improvement* in private savings' (Kapur et al., 2014, p. 4; original emphasis). Because, by definition, plutonomists dominate national income and savings pools, their decisions to spend or save overwhelm and override the actions of others. Thus, if 'good times' for plutonomists continue to prevail in the wake of the 2008 GFC, their savings are likely to fall, current account deficits in major economies such as the USA will rise, and emerging markets such as Brazil, Indonesia and China will prosper from rekindled demand for their products and services. As an addendum to this, plutonomy also appears to bring new levels of volatility to the economy. The conjunction of growing dependence on the rich with heightened instability in their income and wealth levels logically implies broader and disturbing levels of economic volatility (Frank, 2011, p. 19).

Second, Kapur et al. (2014) point to the prospect that the spectacular appreciation of luxury real estate markets in plutonomies (e.g., Scatigna et al., 2014; see also Paris, Chapter 12 this volume) may be nearing an end as a result of the winding down of the United States's policy of quantitative easing; heightening anti-corruption campaigns in emerging markets such as China's State Council's draft rules for a property registration system, which would help the government track ownership across provinces and discourage speculation by preventing corrupt officials from building large portfolios (Murray, 2014); and various countries' macro-prudential measures to contain property inflation. These policies include increasing the supply of public lands for development, imposing additional transaction taxes, such as special or additional stamp duties on purchases to discourage speculative short-term real estate trading, or altering loan-to-valuation (LTV) ratios. For instance, since 2009 Singapore has lowered its LTV in some cases from 90 per cent to 40 per cent (Arvai et al., 2014, p. 33).

Third, without policy intervention, emerging markets are 'likely to become entrenched and egregious plutonomies' (Kapur et al., 2014, p. 1) characterized by growing demands for the (imported) consumption preferences of the wealthy as well as for luxury domestic real estate. Unless there are significant changes to disrupt them, plutonomies in developed countries will also continue to deepen (see, e.g., the discussion by Osuoka and Zalik, Chapter 22 in this volume). The various forces that propagate plutonomy and are discussed above, including patent protection, financial deregulation and intensifying global linkages, are said by Kapur and his colleagues (2009, 2014) to be gathering momentum and so, from another perspective, are propelling new global geographies of plutonomy. More than this, Kapur et al. (2009, p. 12) point to the role of immigrants in energizing emerging market plutonomies. In typically colourful prose, they note:

> Check out the brainiacs of Silicon Valley and Wall Street. You will find vast numbers of Chinese, Indians, Israelis, Russians and Eastern Europeans. These are the high dopamine immigrants at the vortex of complexity, pushing technology and structured products, lubricating plutonomy. . . [I]mmigrants constitute only 12% of the US population but have started 52% of Silicon Valley's tech companies, and account for 25% of US patents. They account for half of US PhDs in science and technology. They are now headed back to their home countries, attracted by entrepreneurial opportunities, and repelled by the US visa regime. . . These Chinese, Indians, Eastern Europeans etc. are taking their brilliance home, looking for complexity, ready to engender technological change and wealth waves in their native lands.

Fourth, with growing inequality will come heightened levels of political polarization, instability and geopolitical uncertainty. For instance, in their book, *Polarized America: The Dance of Ideology and Unequal Riches*, McCarty et al. (2006) reveal how political divergence, income inequality and immigration have all increased dramatically in the United States over a 30-year period since the mid-1970s. The increases followed an equally dramatic decline in these indicators over the first 70 years of the twentieth century. Other influential commentators with a similar message include David Rothkopf (2008) and former US Secretary of Labor, Robert B. Reich (2008, 2013). Moreover, 'The rise of more extreme parties in Europe's recent elections reflects greater acceptance of alternative, nationalist ideologies' (Kapur et al., 2014, p. 26). Atop the instability surrounding inequality, for EMs [emerging markets], a world without a dominant hegemon along with the intensification of income and wealth inequality could exacerbate geo-political tensions' (2014, p. 26). As Wilkinson and Pickett (2010) demonstrate so comprehensively in their well-received book, *The Spirit Level*, inequalities in wealth are correlated positively with a broad array of personal and social ills, including reduced life expectancy, poorer mental health, obesity, higher levels of incarceration, poorer educational performance and diminished social mobility and sadly, the greater the gap between rich and poor, the worse the social problems – in both poor and rich countries (Judt, 2010). Moreover, there has been a collapse in intergenerational mobility within plutonomies such as the UK and USA. Children of these countries have, in the early twenty-first century, little hope of improving on the conditions into which they were born (Wilkinson and Pickett, 2010). That 'translates into ill health, missed educational opportunity, and – increasingly – the familiar symptoms of depression: alcoholism, obesity, gambling, and minor criminality' (Judt, 2010 p. 14). To be sure, the wider the wealth gap, the greater the social problems.

There is another deeply significant consequence of plutonomy overlooked by Kapur and his various colleagues but which warrants at least brief mention here. As we move further into the Anthropocene the troubling conjunction of growing economic and political inequalities, a dominant ideology supportive of growth (rather than, say, redistribution) as the solution to almost all social and economic problems, and apparently insatiable, competitive, conspicuous consumption amongst the super-rich, together herald the possibility of more significant ecosystem harm than we have witnessed historically. As Hervé Kempf (2008) observes in his provocative essay, *How the Rich are Destroying the Earth*, politically powerful financial elites – who will be able to isolate themselves from environmental catastrophe longer than the rest of humanity[9] – are not only central to the fabrication of ecological problems, but they also actively and inadvertently obstruct corrective interventions (see Davison, Chapter 17 in this volume for a detailed discussion). And so what is to be done?

UNDOING PLUTONOMY?

> The disposition to admire, and almost to worship, the rich and the powerful, and to despise, or, at least, to neglect, persons of poor and mean condition. . .[is]. . .the great and most universal cause of the corruption of our moral sentiments. (Smith [1759] 2006, I.iii.3, cited in Judt, 2010, p. 23)

As shown above, over 250 years ago Adam Smith remarked on the admiration typically held for the wealthy and contrasting disregard for those less fortunate. Recent patterns of wealth accumulation have occurred remarkably, with little or no public hostility towards, or criticism of, contemporary multimillionaire tycoons, who are hailed as national heroes unless they are found to be out-and-out crooks (Rubinstein, 2004, pp. vii–ix) – and even then some of them continue to be held up as models of success (e.g., Australia's Alan Bond). And notwithstanding the recent fleeting prominence of the Occupy movement, there is just no real sign of credible resistance to, or reform of, the systems and circumstances that have delivered plutonomy even in places such as the UK where its expressions are extreme (Dorling, 2014). Notwithstanding the possibility of war or depression disrupting intensifying patterns of wealth and income inequality (Piketty and Saez, 2003; Dorling, 2014, p. 17), there seems to be little prospect of plutonomy's imminent demise. But, following Kapur et al. (2005, pp. 20–21) it is apparent that plutonomies may be dismantled in three fundamental ways.

First, and least likely, property rights can be revoked. Aside from the fact that this would require a complete overhaul of almost every social relation, the recent, rapid and growing privatization of 'new' forms of property such as intellectual property and genetic property, make this seem so improbable as to be practically impossible.

Second, and a little more moderately, tax and other wealth transfer systems may be revised in ways that, amongst other things, encourage middle class saving and, importantly, require the wealthiest to contribute more to national coffers (see, e.g., Murray and Peetz, 2014, pp. 142–3). As Saez and Zucman (2014a) suggest: 'we need policies that reduce the concentration of wealth, prevent the transformation of self-made wealth into inherited fortunes, and encourage savings among the middle class'. These might include supporting retirement saving (e.g., Australia's superannuation funds; US 401(k) plans; Singapore's new, small tax increases for those earning over S$320 000 per year to fund social spending for the poor and elderly [*The Star*, 14 February 2015, p. 17]) and raising or restructuring dividend, inheritance and capital gains taxes – or indeed imposing such taxes in some jurisdictions. For example, Macdonald (2014, p. 12) observes that aside from municipal property taxes, few Canadian taxes apply to wealth or capital directly. Moreover, as in other jurisdictions such as Australia, capital gains taxes in Canada are levied on only 50 per cent of the original gain. So, notwithstanding various tax dodges, a Canadian (or Australian) who makes $250 000 from the sale of an income-generating asset income will pay less tax than if the money had been generated through job-related income. In some countries (e.g., Austria, Belgium, Belize, Malaysia, New Zealand), there are no capital gains taxes and so profit from the appreciation of assets such as second homes (see Paris, Chapter 12 this volume) goes untaxed.

As pointed out by authors from Hay and Bell (1990) to Delgado (2014) and as Dixon and Beaverstock and Hall remind us in chapters 19 and 20 in this volume, tax reforms would also need to take account of mobile capital's capacities to circumvent jurisdictional regulations. In the foreseeable future this seems very unlikely too. Reforms actually seem to be working in quite the opposite direction, with all-too-evident problems:

> The implication is that in an era of economic globalization – that is, international economic integration and an accompanying increase in capital mobility – governments may find themselves drawn into an internecine 'race to the bottom' in capital taxation, undermining the financing of the welfare state and the provision of public goods generally. Further, taxes must then come to fall unduly on immobile factors, specifically labour, exacerbating labour market rigidities and unemployment. (Delgado, 2014)

Third, the rates at which the rich accumulate wealth can be slowed by changing the balance of power between labour and capital (Kapur et al., 2005, p. 21). Mechanisms to achieve this include domestic labour market regulation to affect minimum wage levels and working hours or protectionist approaches to the import of goods and services. It seems evident from recent events in most plutonomies that none of these is about to occur. There have been few, if any, instances of significant nationalization of assets. Indeed, most of the examples of recent decades are of exactly the opposite – privatization (Judt, 2010). Some of these such as New Zealand's radical privatizations of the late 1980s and early 1990s have been on a mammoth scale. There are few examples of tax rate rises that challenge the rich. And changes to labour market regulation have typically been to the disadvantage of 'developed market labor, especially at the lower end of the food-chain' (Kapur et al., 2005, p. 21). For example, globalization is making it easier for companies to either bring goods in from cheap emerging markets or to move production out to lower-cost countries, as a raft of recent free trade agreements between Australia and some of its Asian trading partners, including China (ChAFTA), Japan (AEPA), Korea (KAFTA) and Malaysia (MAFTA), bear witness. Moreover, some countries have also moved to in-source labour. Overall, globalization offers a labour surplus, keeping a lid on wage inflation.

WHAT NOW?

And so is plutonomy with us for the long term? Is it an inescapable associate of contemporary capitalism? While Piketty (2014) has made it fairly clear that, historically, wealth tends to accrue faster to the wealthy than to the less well off, we need not accept that plutonomy is an inevitable adjunct. The international variations in plutonomy to which I referred earlier (e.g., France, Japan, Sweden compared with Australia, Canada and the USA) are telling geographical indicators of that. As a socially produced system, plutonomy is, of course, subject to socially generated transformation. Plutonomy can continue only when societies support or tolerate vast inequalities in income and wealth. Its annihilation depends on the level of commitment brought politically and socially to the project, a level of commitment from which we seem very far removed (Judt, 2010). For a start, some cultures not only tolerate, but actually have grown to expect, financial inequality, either by a long history of aristocratic distinction and discrepancy, for example, or as an innate consequence of a meritocracy, where rewards are held to reflect innovation, enthusiasm and effort. Moreover, whilst one might expect that in democratic countries, under the 'one person, one vote' principle, economic inequality could be held in check by political equality, there is clear evidence that not only have more of the super-rich reinvented themselves as politicians[10] and used their wealth to circumvent common political stepping stones (Taylor and Harrison, 2008, p. 211) but that 'the rich have been able to use their resources to influence electoral, legislative, and regulatory processes through campaign contributions, lobbying, and revolving door employment of politicians and bureaucrats' (Bonica et al., 2013, p. 105).[11] Echoing Noam Chomsky's (2012) observation that: 'Concentration of wealth yields concentration of political power. And concentration of political power gives rise to legislation that increases and accelerates the cycle', Volscho and Kelly (2012, p. 695) remind us that economic inequality can be self-reinforcing, generating the kind of political inequality that prevents redistributive policies from being applied.

Sadly too, just as reductions in inequality are self-confirming, so an enduring period of inequality can convince us that it 'is a natural condition of life about which we can do little' (Judt, 2010, p. 22). Moreover, in a cruel and ironic associate of this, as inequality rises and the rich get richer, not only do the wealthy turn against redistribution but the voting public may often become more conservative, supporting the very policies that reproduce inequality! (For fascinating discussions of this and related matters, see Bonica et al., 2013 and the conclusion of Dorling's 2014 polemic.) Again, Kapur and his colleagues find the right words and although they refer to American ambitions, the logic applies in other 'aspirational' cultures: 'Perhaps one reason that societies allow plutonomy is because enough of the electorate believe they have a chance of becoming a Pluto-participant. Why kill it off, if you can join it? In a sense, this is the embodiment of the "American Dream"' (Kapur et al., 2005, p. 23).

Of course there does actually have to be some evidence of upward mobility – examples of people who have managed to ascend from a position as member of the precariat to secure a place among the plutonomists. Without really wishing to move this chapter towards its close with tones of conspiracy, there are certainly powerful news and creative industries from HarperCollins[12] to Hollywood, dominated by the wealthy, and long dedicated to nurturing and preserving that dream (see, e.g., Hamill, 2014; and biographical drama *The Pursuit of Happyness*, starring Will Smith and directed by Gabriele Muccino, 2006), despite evidence to suggest that for most it will indeed remain a dream (e.g., OECD, 2010; The Pew Charitable Trusts, 2012).

To conclude, historical and geographical variations in the expression of plutonomy point to the potential for prosperous societies without the egregious inequalities that characterize plutonomy. We may never overcome the historical $r > g$ pattern described by Piketty (2014) and that understandably underpins plutonomy; however, we can certainly take steps to ameliorate plutonomy's worst excesses by, for example, waging 'a non-violent war of attrition on concentrated wealth' (Dorling, 2014, p. 174), changing the balance of power between labour and capital, and revising domestic and international taxation systems. And to do this we must rethink the operation of the state (Judt, 2010, p. 199) both intranationally and internationally. Though not impossible, that certainly will require some doing, for as Judt (2010, p. 202; original emphasis) observes:

> [w]e have freed ourselves from the mid-20th century assumption – never universal but certainly widespread – that the state is likely to be the *best* solution to any given problem. We now need to liberate ourselves from the opposite notion: the state is – by definition and always – the *worst* available option.

Such a fundamental, though clearly not impossible, swing in political thinking is a crucial component of the 'a change in the mood, a. . .shift in what becomes morally acceptable' that Dorling (2014, p. 179) argues is required to overcome the extraordinary inequalities in income and wealth that now confront us.

Won over by the flawed ideology of neoliberalization, distracted by the miscarriage of globalization, and bedazzled by the fortunes of the super-rich, many scholars and members of the public alike have been diverted from important political agendas of economic, social and environmental beneficence and decency. We need now to work concertedly and conscientiously on ways of remoralizing and reimagining the state. Just how are we to make it operate within a globalizing, volatile and unsettling economy in ways that simultaneously minimize problematic inequality, reward individual initiative, satisfy collective needs and ensure sustainable futures?

NOTES

1. A number of bloggers (see, e.g., http://www.cps-news.com/archives/part-one-plutonomy-buying-luxury-explaining-global-imbalances/) have suggested that several of Kapur et al.'s Citigroup memos were removed from their blogs and related sites

following demands to do so from Kilpatrick Townsend and Stockton LLP, a large US-based firm of trademark and copyright lawyers acting for Citigroup. Kilpatrick Townsend and Stockton appear to have issued takedown notices pursuant to Section 512(c) of the US Copyright Act, pointing to Citigroup's 'exclusive ownership of all copyright rights in the U.S. and throughout the world in and to its CIRA [Citi Investment Research and Analysis] reports' (Garris, 2011). A pdf copy of one such letter is available at the Political Gates site (http://politicalgates.blogspot.com.au/2011/12/citigroup-plutonomy-memos-two-bombshell.html).

2. Or as Noam Chomsky (2012) would put it, the precariat – those who live a precarious life at the margins of society.
3. During the Golden Age and at the peak of the associated tulip mania, in March 1637, some single tulip bulbs are reported to have sold for more than ten times the annual income of a skilled craftsman (Dash, 1999).
4. The strategy they recommended was to invest in plutonomy stocks (i.e., shares of companies that produce the goods and services demanded by the super-rich). As one of their own colleagues, Priscilla Luk, is said to have remarked: 'Wow, I can get rich by owning the plutonomy stocks, and then spend my money on those products' (Kapur et al., 2005, p. 25).
5. It is worth pointing out here that while the wealth share of the richest 0.1 per cent increased in the 40 years preceding this, the wealth of those between the top 1 per cent and the top 0.1 per cent actually fell! (Saez and Zucman, 2014a).
6. Although significant, the rise of this new wealth polarization is not worldwide. In places like Scandinavia, France, Germany, the rest of Continental Europe (excluding Italy) and Japan, which Kapur et al. (2006b) characterize as part of the egalitarian bloc, income inequality remains at about the same levels that prevailed in the mid-1970s. It must be stressed of course that wealth does not equate with income.
7. In his work, Ajay Kapur (Kapur et al., 2005) also created a so-called Plutonomy Index, intended to measure the financial performance of those services and products used by the very wealthy (e.g., Beneteau, Bulgari, Porsche, Shangri-La Asia, Shinwa Art Auction, Sotheby's, Tiffany). In the period 1985–2005 that Index is reported to have averaged a return of 17.8 per cent annually, a rate far in excess of the performance of other indices of global shares (Kapur et al., 2005, p. 25). Other indices such as *Forbes* CLEWI (Cost of Living Extremely Well Index) reveal an upward similar trend (Chen and O'Donnell, 2014).
8. In an error echoed by Di Muzio (2014), Kapur and his colleagues mistakenly suggest that many of these toys for the wealthy are so-called Giffen goods, the demand for which increases as price rises (Kapur et al., 2005, p. 9). Certainly the demand for Giffen goods rises as price rises, but this is because of the absence of affordable alternatives (e.g., in time of food shortages, the price of and demand for available commodities may rise, simply because of scarcity). Veblen goods (or ostentatious goods), by contrast, see demand rise with increasing price because people believe the good to be of better quality, higher status and so on.
9. As G.K. Chesterton commented wryly in *The Man Who Was Thursday (A Nightmare)*, 'The poor man really has a stake in the country. The rich man hasn't; he can go away to New Guinea in a yacht' (2009, p. 102).
10. Some of the many examples include Australia's Malcolm Turnbull and Clive Palmer; New Zealand's Prime Minister, John Keys; Italy's Silvio Berlusconi; the former mayor of New York, Michael Bloomberg; and two-time Lebanese Prime Minister, Najib Mikati.
11. A recent report in the *International New York Times* (Forsythe, 2015, pp. 1 and 4) comments on common criticisms that lawmakers are often in the pockets of the wealthy and goes on, somewhat wryly, to observe that in China 'Among the 1,271 richest Chinese. . .a record 203, or more than one in seven, are delegates to the nation's current Parliament or its advisory body. . . [T]hose delegates' combined net worth is $463.8 billion, more than the economic output of Austria' (p. 1).

12. HarperCollins is an operating company of Rupert Murdoch's NewsCorp. NewsCorp also owns Twentieth Century Fox.

REFERENCES

Aghdam, R.F. (2011), 'Dynamics of productivity change in the Australian electricity industry: assessing the impacts of electricity reform', *Energy Policy*, **39** (6), 3281–95.

Alvaredo, F., A.B. Atkinson, T. Piketty and E. Saez (2013), 'The top 1 percent in international and historical perspective', *Journal of Economic Perspectives*, **27** (3), 3–20.

American Affluence Research Center (2014), 'The luxury market', *American Affluence Research Center*, accessed 18 February 2015 at http://affluenceresearch.org/about-us/the-luxury-market/.

Arvai, Z., A.P. Prasad and K. Katayama (2014), 'Macroprudential policy in the GCC countries', *IMF Staff Discussion Note*, March, SDN/14/1.

Atkinson, A.B. (1993), 'What is happening to the distribution of income in the UK?', *Proceedings of the British Academy*, **82**, 317–53.

Atkinson, A.B. (2003), 'Income inequality in OECD countries: data and explanations', *CESifo Economic Studies*, **49** (4), 479–513.

Australian Bureau of Statistics (ABS) (2013), *Household Wealth and Wealth Distribution 2011–12*, Cat. No. 6554.0, accessed 14 August 2015 at http://www.abs.gov.au/ausstats/abs@.nsf/mf/6554.0.

Banerjee, A.V. and E. Duflo (2011), *Poor Economics: A Radical Rethinking of the Way to Fight Global Poverty*, Philadelphia, PA: Perseus Books.

Bartels, L.M. (2008), *Unequal Democracy: The Political Economy of the New Gilded Age*, Princeton, NJ: Princeton University Press.

Beaverstock, J.V., P. Hubbard and J. Short (2004), 'Getting away with it? Exposing the geographies of the super-rich', *Geoforum*, **35** (4), 401–7.

Bonica, A., N. McCarty and K.T. Poole et al. (2013), 'Why hasn't democracy slowed rising inequality?' *Journal of Economic Perspectives*, **27** (3), 103–24.

Chen, L. and C. O'Donnell (2014), 'The Forbes 400 shopping list: living the 1% life is more expensive than ever', *Forbes.com*, 20 October, accessed 19 February 2015 at http://www.forbes.com/sites/liyanchen/2014/09/30/the-forbes-400-shopping-list-living-the-1-life-is-more-expensive-than-ever/.

Chesterton, G.K. (2009), *The Man Who Was Thursday (A Nightmare)*, Rockville, MD: Serenity Publishers.

Chomsky, N. (2012), 'Plutonomy and the precariat: on the history of the U.S. economy in decline', *Truthdig*, 8 May, accessed 6 February 2015 at http://www.truthdig.com/report/item/plutonomy_and_the_precariat_the_history_of_the_us_economy_in_decline_201205.

Chu, A.C. (2010), 'Effects of patent policy on income and consumption inequality in an R&D growth model', *Southern Economic Journal*, **77** (2), 336–50.

CNBC (2014), 'Ridiculously extravagant purchases of the filthy rich', accessed 26 February 2015 at http://www.cnbc.com/id/102054313.

Credit Suisse (2014), *World Wealth Report 2014*, October, accessed 6 February 2015 at https://publications.credit-suisse.com/tasks/render/file/?fileID=60931FDE-A2D2-F568-B041B58C5EA591A4.

Cynamon, B.Z. and S.M. Fazzari (2014), 'Inequality, the Great Recession, and slow recovery', working paper, 24 October, accessed 18 February 2015 at http://dx.doi.org/10.2139/ssrn.2205524.

Dash (1999), *Tulipomania: The Story of the World's Most Coveted Flower and the Extraordinary Passions It Aroused*, New York: Three Rivers Press.

Delgado, F.J. (2014), 'About international tax competition', *Journal of Global Economics*, **2** (4), accessed 12 August 2015 at http://www.esciencecentral.org/journals/about-international-tax-competition-2375-4389.1000124.pdf.

DePillis, L. (2015), 'The inequality snowball effect', *Washington Post*, 25 July, accessed 18 February 2015 at http://www.washingtonpost.com/blogs/wonkblog/wp/2014/07/25/the-inequality-snowball-effect/.

Di Muzio, T. (2014), 'The plutonomy of the 1%: dominant ownership and conspicuous consumption in the New Gilded Age', *Millennium: Journal of International Studies*, **43** (2), 492–510.

Di Muzio, T. (2015), *The 1% and the Rest of Us. A Political Economy of Dominant Ownership*, London: Zed Books.

Dongxu, Z. (2014), 'Top one percent has one third of China's wealth, research shows', *Caixin Online*, 4 August, accessed 23 August 2015 at http://english.caixin.com/2014-08-04/100712733.html.

Dorling, D. (2014), *Inequality and the 1%*, London: Verso.

Electronic Frontier Foundation (2015), 'Trans-Pacific Partnership Agreement', accessed 10 February 2015 at https://www.eff.org/issues/tpp.

Federal Reserve Bulletin (2014), 'Changes in U.S. family finances from 2010 to 2013: evidence from the Survey of Consumer Finances', *Federal Reserve Bulletin*, **100** (4), accessed 6 February 2015 at http://www.federalreserve.gov/pubs/bulletin/2014/pdf/scf14.pdf.

Forsythe, M. (2015), 'A chamber where wealth abounds', *International New York Times*, 3 March, 1, 4.

Frank, R. (2007), *Richi$tan*, New York: Random House.

Frank, R. (2010), 'U.S. economy is increasingly tied to the rich', *The Wall Street Journal*, 5 August, accessed 13 November 2015 at http://blogs.wsj.com/wealth/2010/08/05/us-economy-is-increasingly-tied-to-the-rich/.

Frank, R. (2011), 'The future of our plutonomy: deficits, more booms and busts', *The Wall Street Journal*, 17 February, accessed 6 February 2015 at http://blogs.wsj.com/wealth/2011/02/17/the-future-of-our-plutonomy-deficits-more-booms-and-busts/#.

Frank, R. (2014a), 'How to get rich? Serve the rich consumer', *CNBC*, 30 July, accessed 13 November 2014 at http://www.cnbc.com/id/101879219.

Frank, R. (2014b), 'What billionaires really spend each month', *Inside Wealth*, 23 October, accessed 12 August 2015 at http://www.cnbc.com/id/102115184.

Freeland, C. (2011), 'The rise of the new elite', *The Atlantic*, 4 January, accessed 31 January 2011 at http://www.theatlantic.com/magazine/archive/2011/01/the-rise-of-the-new-global-elite/308343/.

Freeland, C. (2012), *Plutocrats. The Rise of the New Global Rich and the Fall of Everyone Else*, New York: The Penguin Press.

Garris, K. (2011), 'Takedown Notice Pursuant to Section 512(c) of the Copyright Act: NOAPPARENTMOTIVE.ORG', letter to Mr John Schmitt from Kilpatrick Townsend Attorneys at Law, 4 November, accessed 12 March 2015 at http://politicalgates.blogspot.com.au/2011/12/citigroup-plutonomy-memos-two-bombshell.html.

Gbadebo, B., P. Chima and B. Mibzar (2013), 'Nigeria's super rich – how they make, spend their money', *AllAfrica.com*, 19 July, accessed 19 November 2014 at http://allafrica.com/view/group/main/main/id/00025570.html.

Giles, C. (2014), 'Piketty findings undercut by errors', *Financial Times*, 23 May, accessed 6 February 2015 at http://www.ft.com/cms/s/2/e1f343ca-e281-11e3-89fd-00144feabdc0.html#axzz3H1w3og5K.

Hale, B. and L. Stott (2014), 'More blingtastic and vulgar than ever! It's that month when super rich Arabs flock here for an orgy of spending. And this year, they and their cars are taking excess to a new level', *Daily Mail*, 5 August, accessed 3 March 2015 at http://www.dailymail.co.uk/news/article-2717227/More-blingtastic-vulgar-It-s-month-super-rich-Arabs-flock-orgy-spending-And-year-cars-taking-excess-new-level.html.

Hamill, J. (2014), 'The rags to riches story of Britain's thriftiest multi-millionaire', *Forbes.com*, 31 October, accessed 26 February 2015 at http://www.forbes.com/sites/jasperhamill/2014/08/31/the-rags-to-riches-story-of-britains-thriftiest-multi-millionaire/.

Harvey, D. (2014), *Seventeen Contradictions and the End of Capitalism*, London: Profile Books.

Hay, I. (ed.) (2013), *Geographies of the Super-Rich*, Cheltenham, UK and Northampton, MA, USA: Edward Elgar Publishing.
Hay, I. and J.E. Bell (1990), 'Small spaces and big states. Changing state relations in a new global environment', *Tijdschrift voor Economische en Sociale Geographie*, **81** (5), 322–31.
Hay, I. and S. Muller (2014), 'Questioning generosity in the golden age of philanthropy. Towards critical geographies of super-philanthropy', *Progress in Human Geography*, **38** (5), 635–53.
Institute of Applied Manpower Research (2011), *India Human Development Report 2011. Towards Social Inclusion*, New Delhi: Oxford University Press.
Judt, T. (2010), *Ill Fares the Land*, London: Penguin.
Juma, V. (2014), 'Porsche gains 18pc share of Kenya luxury car market in two months', *Business Daily*, 20 July, accessed 23 August 2015 at http://www.businessdailyafrica.com/Corporate-News/Porsche-gains-18pc-share-of-Kenya-luxury-car-market/-/539550/2391058/-/t9wt4f/-/index.html.
Kapur, A., P. Luk and R. Samadhiya (2009), 'Michael Moore, misrepresentation and migrating plutonomies: a secret memo', *Mirae Asset Global Research*, 7 October.
Kapur, A., N. MacLeod and N. Singh (2006a), 'Revisiting plutonomy: the rich getting richer', *Citigroup Industry Note*, 5 March.
Kapur, A., N. MacLeod and T.M. Levkovich et al. (2006b), 'The plutonomy symposium – rising tides lifting yachts', *Citigroup Global Markets*, 29 September.
Kapur, A., R. Samadhiya and U. de Silva (2014), *Piketty and Plutonomy: The Revenge of Inequality*, Bank of America Merrill Lynch Equity Strategy Report, 30 May.
Kapur, A., N. MacLeod and T.M. Levkovich et al. (2005), 'Plutonomy: buying luxury, explaining global imbalances', *Citigroup Global Markets*, 14 October.
Kempf, H. (2008), *How the Rich are Destroying the Earth*, Totnes, UK: Green Books.
Lansley, S. (2012), *The Cost of Inequality*, London: Gibson Square.
Lieberman, R.C. (2011), 'Why the rich are getting richer. American politics and the second Gilded Age', *Foreign Affairs*, **90** (1), 154–8.
Macdonald, D. (2014), *Outrageous Fortune. Documenting Canada's Wealth Gap*, Canadian Centre for Policy Alternatives, accessed 6 February 2015 at http://vipmedia.globalnews.ca/2014/04/canada27s-wealth-gap.pdf.
Matthews, C. (2014), 'Wealth inequality in America: it's worse than you think', *Fortune*, 31 October, accessed 6 February 2015 at http://fortune.com/2014/10/31/inequality-wealth-income-us/.
McCarty, N., K.T. Poole and H. Rosenthal (2006), *Polarized America: The Dance of Ideology and Unequal Riches*, Cambridge, MA: MIT Press.
Mohiuddin, M. and Z. Su (2014), 'Global value chains and the competitiveness of Canadian manufacturing SMEs', *Academy of Taiwan Business Management Review*, **10** (2), 82–92.
Moreno, K. (2014), 'The 67 people as wealthy as the world's poorest 3.5 billion', *Forbes Insights*, 25 March, accessed 6 February 2015 at http://www.forbes.com/sites/forbesinsights/2014/03/25/the-67-people-as-wealthy-as-the-worlds-poorest-3-5-billion/.
Murray, G. and D. Peetz (2014), 'Plutonomy and the one percent', in S.K. Schroeder and L. Chester (eds), *Challenging the Orthodoxy*, Heidelberg: Springer-Verlag, pp. 129–48.
Murray, L. (2014), 'Smart money long gone from China property', *Financial Review*, 18 August, accessed 6 February 2015 at http://www.afr.com/p/special_reports/opportunityasia/smart_money_long_gone_from_china_Oc9avv07ilqMaEQjCZnF4K.
Muzaffar, C. (2015), 'Disparities that threaten Malaysia', *The Star*, 24 February, 14.
Mwanza, K. (2014), 'How Africa's super rich spend their billions', *AFK Insider*, 16 May, accessed 19 November 2014 at http://afkinsider.com/56457/africas-super-rich-spend-billions/.
Organisation for Economic Co-operation and Development (OECD) (2010), *Economic Policy Reforms: Going for Growth 2010*, accessed 26 February 2015 at http://www.oecd.org/eco/growth/economicpolicyreformsgoingforgrowth2010.htm.
Oxfam International (2014), 'Working for the few. Political capture and economic inequality', *Oxfam Briefing Paper No. 178*, 20 January, accessed 23 August 2015 at https://www.oxfam.org/en/research/working-few.

Palley, T.I. (2007), 'Financialization: what it is and why it matters', *Working Paper No. 525*, Annandale-on-Hudson, NY: The Levy Economics Institute.
Peck, J. (2012), 'Austerity urbanism. American cities under extreme economy', *City*, **16** (6), 626–55.
Piketty, T. (2014), *Capital in the Twenty-First Century* [French edition published in 2013 as *Le capital au XXI siècle*, Editions du Seuil], Cambridge, MA/London: Belknap Press of Harvard University Press.
Piketty, T. and E. Saez (2003), 'Income inequality in the United States, 1913–1998', *Quarterly Journal of Economics*, **118** (1), 1–39.
Pizzigati, S. (2011), 'Madison Ave declares "mass affluence" over', *TooMuch*, 30 May, accessed 19 November 2014 at http://toomuchonline.org/madison-ave-declares-mass-affluence-over/.
Provost, C. (2014), 'Nigeria becomes Africa's largest economy – get the data', *The Guardian*, 7 April, accessed 23 August 2015 at http://www.theguardian.com/global-development/datablog/2014/apr/07/nigeria-becomes-africa-largest-economy-get-data.
Reich, R.B. (2008), *Supercapitalism. The Transformation of Business, Democracy, and Everyday Life*, New York: Alfred A. Knopf.
Reich, R.B. (2013), *Inequality for All* [documentary film directed by Jacob Kornbluth], distributed by RADiUS-TWC.
Rothkopf, D. (2008), *Superclass: The Global Power Elite and the World They Are Making*, New York: Farrar, Straus and Giroux.
Rubinstein, W.D. (2004), *The All-Time Australian Rich List*, Sydney: Allen & Unwin.
Saez, E. and G. Zucman (2014a), 'Exploding wealth inequality in the United States', *Washington Center for Equitable Growth*, 20 October, accessed 18 February 2015 at http://equitablegrowth.org/research/exploding-wealth-inequality-united-states/.
Saez, E. and G. Zucman (2014b), 'Wealth inequality in the United States since 1913' working paper, accessed 12 August 2015 at http://gabriel-zucman.eu/uswealth/.
Scatigna, M., R. Szemere and K. Tsatsaronis (2014), 'Residential property price statistics across the globe', *BIS.org* 14 September, accessed 5 February 2015 at http://www.bis.org/publ/qtrpdf/r_qt1409h.htm.
Short, J.R. (2013), 'Economic wealth and political power in the second Gilded Age', in I. Hay (ed.), *Geographies of the Super-Rich*, Cheltenham, UK and Northampton, MA, USA: Edward Elgar Publishing, pp. 26–42.
Smith, A. ([1759] 2006), *Theory of Moral Sentiments*, New York: Dover.
Spooner, S. (2014), 'Rise of Africa's luxury loving super rich – from Nigeria's champagne love affair to South Africa's cosmetics obsession', *Mail and Guardian Africa*, 21 August, accessed 19 November 2014 at http://mgafrica.com/article/2014-08-20-africas-love-of-luxury-super-rich-worth-over-13-trillion-and-they-want-to-spend-it.
Stanley, L. (2014), '"We're reaping what we sowed": everyday crisis narratives and acquiescence to the age of austerity', *New Political Economy*, **19** (6), 895–917, accessed 5 February 2015 at http://www.tandfonline.com/action/showCitFormats?doi=10.1080/13563467.2013.861412.
Stiglitz, J.E. (2013), 'Letter to TPP [Trans-Pacific Agreement] negotiators', 6 December, accessed 10 February 2015 at http://keionline.org/sites/default/files/jstiglitzTPP.pdf.
Taylor, J. and D. Harrison (2008), *The New Elite: Inside the Minds of the Truly Wealthy*, New York: ANACOM.
The Boston Consulting Group (2014), 'Global wealth 2014: riding a wave of growth' (online), accessed 2 November 2015 at https://www.bcgperspectives.com/content/articles/financial_institutions_business_unit_strategy_global_wealth_2014_riding_wave_growth/?chapter=2.
The Pew Charitable Trusts (2012), *Pursuing the American Dream. Economic Mobility Across Generations*. July, accessed 26 February 2015 at http://www.publicradio.org/marketplace-archive/pdf/pew_american_dream.pdf.
The Star (2015), 'S'pore raises taxes on the rich to aid social spending', *The Star*, 24 February, 17.

Volscho, T.W. and N.J. Kelly (2012), 'The rise of the super-rich: power resources, taxes, financial markets, and the dynamics of the top 1 percent, 1949 to 2008', *American Sociological Review*, **77** (5), 679–99.

Wikileaks (2014) 'Updated secret Trans-Pacific Partnership Agreement (TPP) – IP Chapter (second publication)', 16 October, accessed 10 February 2015 at https://wikileaks.org/tpp-ip2/.

Wilkinson, R. and K. Pickett (2010), *The Spirit Level. Why Greater Equality Makes Societies Stronger*, New York: Bloomsbury Press.

Wilmers, N. (2014), 'Can high income consumers increase within-industry wage inequality?' working paper, 25 July, accessed 18 February 2015 at http://dx.doi.org/10.2139/ssrn.2467656.

World Economic Forum (2014), *Outlook on the Global Agenda 2014*, Geneva: World Economic Forum, accessed 6 February 2015 at http://reports.weforum.org/outlook-14/.

Xie, Y. and X. Zhou (2014), 'Income inequality in today's China', *Proceedings of the National Academy of Sciences*, **111** (19), 6928–33.

5. Interrogating the legitimacy of extreme wealth: a moral economic perspective

*Andrew Sayer**

INTRODUCTION

Thanks in particular to the work of Thomas Piketty and Emmanuel Saez, it has become clear that the rich and super-rich have made an extraordinary comeback since the 1970s (Atkinson et al., 2011; Piketty, 2014). This phenomenon is most clear in English-speaking countries in which neoliberalism, and with it, financialization, have been strongest. As their data show, the proportion of national income received by the top 1 per cent and fractions thereof in these countries over the last 100 years follows a U-shaped curve, bottoming out in the post-war boom; in the UK it fell from over 18 per cent in the 1920s to 6 per cent in the 1970s, from which it recovered to over 15 per cent by 2007, followed by a temporary dip and then recovery. Sweden and Norway have shallower U-shaped curves. Italy, Spain, New Zealand and Argentina have also seen a return of the rich, albeit with more fluctuations. China, of course, has seen a dramatic rise in top incomes in recent years. The return of the rich is much more limited in France, Denmark and Japan, which have more L-shaped curves. In Germany, the share of national income taken by the 1 per cent has stayed fairly flat but at a relatively high level since the war. In the Netherlands and Switzerland the income shares of the top 1 per cent have fallen since the post-war boom.[1] Whereas the Great Depression of the interwar period was accompanied and followed by a fall in the proportion of income going to the rich, the recent major recession has seen continued accumulation by the 1 per cent and massive and rapid concentration of wealth at the top in many countries. In 2014 in the UK, the wealth of the richest 1000 people totalled £519 billion, having grown by over 15 per cent since 2012 (1997 figure was £98 billion). The latest figure is enough to fund the entire National Health Service for 4.2 years.[2] Danny Dorling has shown that, over a century (1910–2009), the shares of total income of those within the different fractions of the top 1 per cent in the UK changed together, and rose and fell in the *opposite* direction to the shares of the bottom 99 per cent: when the 1 per cent got more, the 99 per cent got less (Dorling, 2013). But in the UK, as elsewhere, it is primarily those at the top of the 1 per cent who have benefited most.

Both to explain and to evaluate such changes, we need to identify the sources of income received by the rich. Particularly for the latter we need to adopt a moral economic approach to understanding economic activities. By this I mean one that examines and assesses the moral justifications of basic features of economic organization, in particular property relations and what institutions and individuals are allowed and required to do. It treats the economy not merely as a machine that sometimes breaks down, but as a complex set of relationships between people, increasingly stretched around the world, in which they act as producers of goods and services, investors, recipients of various kinds of income, lenders and borrowers, and as taxpayers and consumers (Pettifor, 2006; Alperovitz and Daly, 2008; Graeber, 2011). As a critical approach it goes beyond the usual focus on irrationality and systemic breakdown, to injustice and the moral justifications of taken-for-granted rights and practices. It is not only about how much people in different positions in the economy should get paid for what they do, but whether those positions are legitimate in the first place. Is it right that they're allowed to do what they are doing? One can use the term 'moral economy' also to refer to a kind of economy, as Edward Thompson famously did, but unlike him, though like many more recent users of the term, I regard all economies as moral economies in some respects (Sayer, 2007; Keat, 2010; Wilson, 2014).[3]

A moral economic approach defines economies as systems of provisioning, including but going beyond the cash economy. Economic activity provides use values, whether material (e.g., food) or immaterial (e.g., school lessons, carework), some of which may be commodified, that is, produced for sale. It takes seriously the indisputable fact that we all are social beings, dependent on others as well as ourselves for our survival and well-being. No one is or ever was 'independently rich': financial wealth is just a claim upon wealth in terms of use values produced by others. Unless there are things for sale, money is worthless, and things for sale have to be produced, though not all products need to be produced for sale. Production is prior to exchange.

The concerns of moral economy as an approach include the large differences between what some are able to get and what they contribute, need and deserve. What people should get – a focus of a large literature in political philosophy, including John Rawls's famous *A Theory of Justice* (1971) – is a difficult issue, particularly where it is a matter of what we think people deserve or merit. But what people actually get and do in a capitalist economy has little to do with moral judgement. Particularly in the case of the rich, it can be shown that what they get has more to do with power. I argue that the rich get most of their income by using control of assets like land and money to siphon off wealth that others produce. Much of their income is *unearned*. What is more, over the last 35 years, particularly with the increasing dominance of the economy by finance ('financialization'), the rich have become far richer than before by expanding these sources of unearned income. But while the distinction between earned and unearned income was a staple of classical political economy, socialist thought, and taxation policy, it has fallen out of use in the last 35 years – precisely the period in which unearned income has grown (Mill [1848] 2015; Pettifor, 2006; Hudson, 2011). We need, then, to clarify this distinction. But first it is necessary to disambiguate another key term in our economic vocabulary: 'investment'.

I shall start with these matters of definition and then proceed to provide evidence for my central claim that the extraordinary growth of the wealth of the rich and super-rich stems from the expansion of unearned income.

INVESTMENT

There is a fundamental and deceptive slippage in the use of the word 'investment'. This slippage is central to the legitimation of the rich, and their symbolic domination. Investment is invariably understood to be a good thing, and can provide an appealing cover for a vast range of activities, yet the term is used in two radically different senses: (1) use value/object-oriented definitions focus on what it is that is invested *in*, for example, infrastructure, equipment, or training that enhance what people can do (i.e., it focuses on wealth creation); (2) exchange value/'investor'-oriented definitions focus on the financial gains from any kind of lending, saving, purchase of financial assets or speculation – regardless of whether they contribute to any objective investment (as defined as 1 above), or benefit others. Here the focus is wealth extraction. Henceforth I shall use 'scare' quotes for this second meaning of the term. This is not just an academic distinction: the difference between the two activities is of enormous practical importance for both economic growth and wealth distribution.

The standard move is to elide the distinction and pass off the second as based on the first. Sometimes the two may indeed go together. But it is also perfectly possible for successful investments in the first sense to fail to provide financial benefits to 'investors' in the second sense. The use of my taxes for investing in infrastructure on the other side of the country may benefit others but not me. Conversely, it is equally possible for lucrative 'investments' in the second sense to have neutral or negative effects on productive capacity – through asset stripping, value skimming and rent seeking. Elisions of this vital distinction have become commonplace not so much through a desire to deceive than through ignorance, coupled with the fact that under capitalism individuals have little or no interest in checking whether their 'investments' (as defined as 2 above) have positive, neutral or negative effects on the production of goods and services; to the rentier-'investor', £1 million from rent is no different from £1 million from new productive capacity. Further, for particular 'investors', though not for whole economies, purely extractive kinds of 'investment' such as speculating on asset bubbles can be less risky than objective, wealth-creating investment.

Given the huge difference between these two meanings of the same word, and the contingent relation between the practices to which they refer, we must be on our guard when rich or super-rich individuals justify their wealth by claiming to be 'investors'.

THE EARNED/UNEARNED INCOME DISTINCTION AND ITS APPLICATION

This earned/unearned distinction is fundamental both for understanding economic functions and how distribution is determined, and for evaluating such distributions in ethical terms. Roughly speaking, earned income is what employees and self-employed people get from contributing to the production of goods and services. I do not mean to suggest that the size of their pay reflects what they deserve or contribute, but rather that their pay is at least conditional on providing goods and services that others can use. The relation between what we might think people deserve for their work – however we might want to measure that – and the income they actually get is pretty loose, not least because pay levels are influenced by power and scarcity rather than such considerations. The goods or services they produce and deliver have 'use value' – that is, qualities that make them useful or desirable, such as the nutritious and tasty quality of a meal, or the educational qualities of a maths lesson. Many of these products and services are sold in exchange for money in markets, and so have not only use value but also 'exchange value' and hence are commodities. But many are funded by taxes, providing income for state sector workers, such as police officers and school teachers. Public sector workers, no less than private sector workers, can produce wealth – useful goods and services. These goods and services have costs of production – wages and salaries, the care, education, training and support that enables people to work, and the costs of producing the materials used in the process. Some work has a more indirect relation to the production and distribution of goods and services, but is nevertheless necessary for efficient provisioning. For example, accounting is needed to monitor and manage the use of money and other resources in organizations; an insurance industry is important for providing security; and a legal system and police force are needed to protect people and property.

By contrast, unearned income derives not from providing goods or services that would not otherwise exist and hence incur costs of production, but from controlling an already existing asset, such as land or buildings or spare money, that others lack but need or want, and who can therefore be charged for its use. If the asset already exists, and there are no further costs of production apart from for maintenance, then those who receive income from it do so not because they are in any sense 'deserving' – they have not contributed anything that did not previously exist – but because they *can*. It is a reflection of power deriving from unequal control over existing assets. They use their property not as means of production or 'means of work but as an instrument for the acquisition of gain' (Tawney, 1921, pp. 65–6). This is what Ruskin called 'illth',[4] deriving from what J.A. Hobson later called 'improperty' (Ruskin [1862] 2007; Hobson, 1937). In most cases they have this power of control by virtue of property rights that legally entitle them to use it for extracting payments from others who lack access to such assets. Whereas earned income is work based, unearned income is asset based. In political economy, a person who derives unearned income from ownership of existing assets is known as a rentier.

Insofar as owners can extract money from others and hence be able to buy goods and services without producing goods and services in exchange, then those who are producing goods and services must be producing more than they themselves consume. In other words, they must be producing a surplus. Mere ownership alone produces nothing. The rentier is essentially siphoning off wealth produced by others. In turn, this implies that those who rely on earned income are generally paid less than the value of what they produce.

The owners of such assets are not the only ones who are able to consume without contributing to production of use values, however. Children, the elderly and sick and those unable to work also get a kind of unearned income, whether provided by families or the state. However, these flows of money are generally called 'transfers' rather than 'unearned income' because they are believed – quite reasonably – to be accepted as warranted on the basis of needs: producers are unlikely to object to having to produce a bit more than they themselves consume in order to support them. In practice, just who is deemed entitled to this income varies between societies, but it is generally based on a mixture of recognition of others' needs, love in the case of family members, sense of civic duty and prudent self-interest (in providing a safety net, should things go wrong). The prime justification is usually that the recipients are needy and cannot reasonably be expected to work for an income, or that there are insufficient jobs for them, and that unemployment and ill health can happen to anyone. As always, which transfers are undertaken by the state's taxation and spending depends of course on the politics and balance of power in the country concerned.

But while the transfers that they get are relatively easy to justify, the unearned income of adult, able-bodied rentiers is unwarranted and undeserved. They free-ride on the labour of others. The term 'free-riding' both denotes a particular kind of economic social relation and raises ethical objections to it. This relation between rentiers and producers is indisputably a social relation, one that is exploitative and an important generator of inequality in its own right. It is a relation that has become increasingly complex and globalized, so that British rentiers, for example, siphon off wealth produced by workers not only in Britain but also in many other countries. Let us now examine the main sources of unwarranted, or asset-based unearned income, starting with the classical trio of rent, interest and profit.

The clearest case of the rentier is the landowner. The mere ownership of land or buildings does not make it any more productive, and while it can be claimed that improvements to them should be paid for because they involve production costs and produce enhanced goods, pure rent is not payment for production but merely for access; as Tawney observed, rent is like a private tax on the industry of others (Tawney, 1921). Sources of rent other than land and property, such as intellectual property also provide rentier income – income based on power rather than contribution.

Interest on loans is a charge for the use an existing asset – money – that functions as a claim on the labour and products of others. However, banks do not merely act as intermediaries between savers and borrowers by lending their existing deposits, but create new credit money, simply by writing deposits into the accounts of borrowers; the interest they charge far exceeds any production costs in creating the credit money (Mellor, 2010; Hudson, 2011). Deregulation has allowed them a freer hand in doing this. Lending that fuelled asset price bubbles was a major component of the 2007 credit crunch, and has had an important effect on shifting patterns of economic inequalities, enriching the asset rich at the expense of the asset poor (Engelen et al., 2011). For Marx, it was primarily those who made money out of money –'interest-bearing capital' – who made up 'the class of parasites' (Marx, 1972, p. 545). A moral economic critique of interest situates it in an unequal social relation – usury – in which the lender is typically in a stronger position than the borrower. Although interest payments are taken for granted in contemporary capitalism, outside the formal economy, they are seen as unjust. If we are in financial difficulties and ask a friend for a loan, we would be outraged if they said yes but only on condition that we paid interest and agreed to release our assets as collateral should we not be able to repay. It would be perfectly clear to us that they were taking advantage of our weakness. The injustice of this asymmetry has been the focus of critiques of usury for millennia (Pettifor, 2006; Graeber, 2011).

In classical political economy, profit is the third member of this trio of sources of unearned income. The profits of private (productive) employers derive from their ownership of the means of production and the product, and the dependency of non-owners of means of production on them for employment. Pure capitalists – that is, ones who just own their firms and delegate management to others – are not contributing to wealth creation, but are merely using their power relative to those of propertyless workers to appropriate the difference between costs and the value of what the workers (and managers) produce. Their income is unearned. In these respects, pure capitalists profiting merely from owning businesses producing goods and services look very much like rentiers. But there is also a difference. Even though their income is unearned, it has a 'saving grace': it is at least dependent on supporting productive activity. While, like rentiers, pure capitalists also depend on this surplus for their profit, they are also instrumental in making the generation of that surplus possible, and can only make a profit as long as they do. What is remarkable, though unremarked because it has become normalized with the establishment of capitalist property rights, is that the employees who produce the goods and services have no rights over the revenue that comes from the sale of their output, and no say in what happens to it.

For working capitalists, those who are not only owners but also managers, their income is a mixture of both earned and unearned. In practice, the distinction between employee managers and capitalists and rentiers can be blurred by giving them shares. As organized labour has weakened and delivering shareholder value has become a priority, top managers have been able to use their power to increase their pay in return for maximizing gains for shareholders.

Shares provide another major source of unearned income. Since well over 97 per cent of share transactions are in the secondary market, only in a tiny proportion of cases does the money paid for them go to the company, and thus might be claimed to be a payment for contributing to any objective productive investment. The extraordinary feature of share ownership is not so much limited liability (for losses made by the company) but that it provides a potentially indefinite source of unearned income – an unlimited asset. Both dividends and gains from trading shares are important sources of unearned income. Since growth of demand for shares has exceeded the growth of supply in recent decades, average prices have tended to rise too, creating bubbles, so this source has proved lucrative (Engelen et al., 2011). The development over the last 30 years of the shareholder value movement – a highly successful rentier campaign – coupled with the weakening of trade unions, has made share prices the primary concern of the management of companies.[5] Firms that fail to deliver rising share prices – for example, by ploughing most of their profits into productive investment instead of distributing them as dividends – are disciplined by the market for companies as they become vulnerable to takeover by managements that will deliver shareholder value. Particularly in the light of the power of uncommitted, absentee shareholders, a relevant moral economic question is, what is the justification for this extraordinary difference between the position of workers – dependent and committed – and that of owners who are usually uncommitted and have contributed little or nothing to the work of the organization?

The financial boom and crisis owed much to unearned income from speculation on an ever-growing range of financial assets, and from burgeoning tax avoidance, even as taxes on capital fell. Many in the sector have made their money by arranging major transactions such as mergers and takeovers, involving many millions of pounds, and taking a small percentage commission. It is in their interest to promote as many transactions as possible. In effect, they are standing 'close to a big till' (Ertürk et al., 2007), and it is easy for them to put their hands in it without breaking any laws; the income reflects position rather than contribution. Ertürk et al. (2007) call this 'value skimming' and argue that the financial intermediaries have been a major but hidden beneficiary of financialization, overlooked by those who limit their gaze to bankers or CEOs. High-income but largely anonymous financial intermediaries far outnumber the much-publicized CEOs; Savage and Williams (2008) estimate that in the City of London there are around 15 000 senior intermediaries employed at a principal or partner level in investment banking, hedge funds and other kinds of trading and private equity (as well as those providing support services in law and accounting).

Contemporary mainstream economists are of course likely to object to this account. One of the things financial systems, whether privately or publicly run, are supposed to do is move underused or idle resources, particularly savings, to where they can be used more productively, thereby promoting 'allocational efficiency'. The money made by those who do this is supposedly a reward for enabling this. However, this definition is frequently used as a cover for a quite different version of allocational efficiency that is specific to capitalism, in which the allocation of resources among competing ends is according to where expected rates of financial return are highest. This could be where labour is most exploitable, where consumer incomes are highest, where prospects for extracting rent are best, or where asset inflation is highest (e.g., the latest bubble), or where taxation is lowest. It could be logging in Sumatra, or mining in Peru, or luxury apartments in London, or buying up land in Africa to rent out, or buying shares, bonds or other financial instruments.

SOURCES OF THE WEALTH OF THE RICH

So far I have sketched out the main sources of unearned income involving wealth extraction, but I have yet to show that they are the main source of the wealth of the rich. The relative enrichment over the last 35 years of the top 1 per cent or 0.1 per cent of individuals in the income distribution gives us something to explain but says nothing about the sources of their income. While there is no single source of evidence that covers this comprehensively, a consistent picture emerges from a variety of sources.

It might be thought that one could just define wages and salaries as earned income and subtract them from total income, to get unearned income – receipts of rent, interest, profit, capital gains and speculative gains. But in contemporary economies many of the richest get most of their income in salaries. This makes it appear that rentier-dominated economies are a thing of the past, because now 'the working rich' make up most of the rich and super-rich. But this approach seriously underestimates unearned income as we have defined it, and hence the extent of rentiership. The reason for this is that some people get a salary for performing wealth extraction activities for rentier organizations.[6] So although their income comes in the form of salary and is recorded as such, it is actually rent or interest or profit from mere ownership, and so on, at one remove. Property companies and banks, particularly 'investment' banks and much of the rest of the financial sector are prime examples of organizations most of whose activities fit this description. And as many commentators have noted, it is particularly in these (FIRE – finance, insurance and real estate) sectors that the astonishing growth in incomes, including salaries, at the top has occurred (Engelen et al., 2011; Beaverstock et al., 2013). Further, in the last 30 years, many non-finance companies, from cars to public utilities, have made more money by getting involved in rentier activities in the financial sector than in their traditional sectors, so these increasingly supplement the incomes of the 'working rich' in these sectors too.

Over the years 1979–2012, the top 1 per cent of US households increased their real incomes by 185 per cent, the top 0.1 per cent by 384 per cent and the top 0.01 per cent by 685 per cent, while those of the vast majority of the rest stagnated (Sayer, 2014, p. x).[7] To say the least, it would stretch credulity to claim that this growth of top incomes reflected equivalent increases in their economic contributions. While the share of national income going to workers over the last 35 years has fallen in many countries, labour productivity continued to rise, so 'jobless growth' predominated. According to Tom Palley (2007, p. 10) 'wages of U.S. production and non-supervisory workers (who constitute over 80 per cent of employment) have become detached from productivity growth during the era of financialization'. In the UK the divergence between pay and productivity started later, in the 1990s, and was less dramatic but still marked. Again, only the pay of those at or near the top of the income distribution kept up with economic growth. The weakening of trade unions and employment protection has hit those in the bottom half of the income distribution hardest; the decline in trade union membership correlates with the rising income shares of the top 1 per cent (Oxfam, 2014). In so far as there has been an expansion of employment, it is because a larger proportion of jobs are part-time, and low paid, and there has been a rapid growth of self-employment – more likely to be a sign of precarity than security (Buchanan et al., 2013).

Several studies have revealed similar if not quite so dramatic patterns in a large number of countries:

- In a study of 13 Western capitalist countries, John Peters (2010, p. 93) found that while workers achieved a 4 per cent growth in real wages in the 1970s, growth in real wages from 1980 to 2005 was less than 1 per cent, though those employed in the financial sector did better. In the 1970s and 1980s the wage share of national income in these countries peaked at 78 per cent, but then declined to 63 per cent in 2005 while income from profits, stock dividends, interest and rents rose (see also Perrons, 2012).
- Tali Kristal's 2010 study of labour's share of national income in 16 capitalist democracies arrived at 'an unequivocal conclusion. Since the early 1980s there has been a large and persistent decline in labor's share of national income in most capitalist democracies. The growth of productivity has expanded total income, but in many countries average real wages and fringe benefits have increased more slowly than labor productivity. Meanwhile, income growth has occurred mainly in capitalists' profits, sharply increasing capital's share' (p. 26).[8]
- The United Nations Conference on Trade and Development (UNCTAD) commented: 'In developed countries, the share of

labour income declined, falling by 5 percentage points or more between 1980 and 2006–2007 – just before the global financial crisis – in Australia, Belgium, Finland, France, the Netherlands, Norway, Sweden, the United Kingdom and the United States, and by ten points or more in Austria, Germany, Ireland, New Zealand and Portugal. In several major economies (including France, Germany, Italy and the United States), a significant proportion of the decline in the share of wages had already occurred between 1980 and 1995. This appears to have been linked to a departure from the post-war social consensus, when wage increases closely followed productivity gains' (UNCTAD, 2012, p. 52).

- The International Labour Organization (ILO) found that between the early 1990s and 2007, labour's share of income fell in 51 out of the 73 countries for which data were available. More specifically, labour's share of value added fell by 13 per cent in Latin America, 10 per cent in Asia and the Pacific and 9 per cent in high-income countries. The gap between the top and bottom 10 per cent of waged and salaried workers increased in 70 per cent of countries in this survey (ILO, 2008).
- Both UNCTAD and the ILO show that in most developed countries, especially English-speaking ones, income from capital gains (unearned income) has gone increasingly to the rich.[9]
- Reports from the International Monetary Fund (IMF), the Organisation for Economic Co-operation and Development (OECD) and Oxfam also show a widening of inequalities in most countries, with the rich pulling away from the rest (IMF, 2007, p. 168).[10]

But is this increasing proportion of income going to those at the top a reflection of an increase in their contribution to wealth creation, rather than simply increased wealth extraction? Some argue that recent technological change has favoured higher-paid workers who have the necessary skills, but, as the exhaustive analyses of Thomas Piketty show, if this were the case one would expect wage shares across the top 10 or 20 per cent to have increased, when in fact increases have been heavily concentrated in the top 1 per cent (Piketty, 2014). And when we look at the sectors in which the rich are concentrated – finance and real estate – it is clear that the 'earnings' of many of 'the working rich' come from interest, rent-seeking and value-skimming, albeit paid out as salaries and share options. Particularly between 2000 and 2008, an increasing share of growth was taken as profit by owners and shareholders, contributing further to the divergence of wages and productivity growth (Resolution Foundation, 2012).

Interest payments yield unearned income for lenders, and since lenders tend to be richer than borrowers, interest filters up to the top. In Britain, Hodgson has estimated that only the top 10 per cent get more in interest than they pay out, and within them it is the 1 per cent who get most (Hodgson, 2013).[11] In Germany, the bottom 80 per cent of people in the income distribution pay out far more interest on loans and hidden in the prices of goods than they get back on savings – not surprisingly, for as elsewhere, few of them can afford to save much. Those in the next 10 per cent get roughly as much interest as they pay out, while those in the top 10 per cent get more than they pay out, and within that group the richest of course enjoy the biggest surplus (Kennedy, 2012).

All this is consistent with the argument, made by David Harvey (2007), that there has been a shift from the primary circuit of capital (centred on the production of goods and services) to the secondary circuit of capital in which capital is 'invested' in assets such as land, property, loans, shares and a vast range of financial instruments. This shift, which began in the 1970s, occurred partly because profit margins were being squeezed in the primary sector, due to a combination of increased competition from cheap labour countries and strong labour organization in many sectors in the old industrialized countries. Meanwhile, the growth of private pensions 'invested' in financial products by pension funds and other major harvesters of the savings of the middle classes increased demand for them. But whereas growing demand for everyday products like pizzas and mobile phones prompts increased supply, the same is not true for many key financial assets. Corporations are generally reluctant to issue new shares, because this risks lowering their share prices, and indeed many have spent more on buying back their own shares than on real investment, so increased demand for them has driven up their prices – and with that, top managers' bonuses. Land and property in central locations is also inherently scarce, unless one builds upwards, and property companies have no interest in stopping the inflation of land and property values by building enough to meet demand. Privatization of public utilities also creates lucrative sources of rent from the control of water and energy; as James Meek puts it, our need for these things has been turned into 'a human revenue stream' (Meek, 2012, p. 8). These can in turn be securitized and the income paid out as dividends used as ways for the companies to load up with debt and avoid tax (Allen and Pryke, 2013). The 'working rich' in these companies are the chief beneficiaries of this shift towards rentiership.

'Wealth', in the financial sense, covers many things but is typically defined as total stock of a person's assets minus liabilities. While for most of the 99 per cent their wealth, from homes to clothes, is a stock of use values, for the 1 per cent and particularly those at the top, most of their wealth is primarily an 'investment', used to gain income from ownership of capital assets, via interest, dividends, profits and capital gains. And as Thomas Piketty (2014) shows, in many countries wealth from these sources has been increasing faster than from other sources – primarily wages and salaries – in the neoliberal era. In fact, looking back over a century, the half-century from the 1920s to the 1970s was an anomaly in capitalist history because it was the only time when the ratio of wealth from these sources relative to total income fell significantly. It happened as a result of the growing strength of labour and the destruction of capital, particularly in World War II. Since then the rate of growth of unearned income (not a term Piketty makes much use of) has exceeded the rate of economic growth, resulting in rapid wealth concentration at the top, at the expense of the majority of the population. Provided the rich 'invest' most of their money rather than spend it on consumption, this process will continue, and, as one would expect, the richer they are the higher their marginal propensity to 'invest'. This concentration of wealth means that the children of the wealthy, already advantaged through no effort of their own, stand to inherit huge windfalls – the purest case of unearned income. (Bequests might seem like transfers, as they are gifts, but they differ in that they are not based on need; in fact, given that the children of the rich receive far more than others, they are inversely related to need.) As Piketty notes, this means we are returning to an 'inheritance society' in which there is not only a very high concentration of wealth but large fortunes are passed from generation to generation and social mobility is restricted (Piketty, 2014, p. 351). In France, he estimates that inherited wealth already accounts for a fifth of household monetary resources, and is rising fast (p. 404).[12]

OTHER MORAL ECONOMIC CRITIQUES OF EXTREME WEALTH

Although I have concentrated on the sources of the wealth of the rich, a moral economic critique of their wealth could go beyond this, to include the effects on the distribution of political power, their consumption patterns and environmental considerations. Here I will sketch what these topics might include.

As regards the political aspects, the massive concentration of wealth over the last 35 years has been both cause and consequence of a shift in political power to the rich. In any capitalist economy, those who control the commanding heights of the economy have a structural source of power deriving precisely from this control, which makes governments dependent on them for delivering economic growth. In the neoliberal era this hidden power has been augmented through the weakening of the power of organized labour, as already mentioned. But in addition to the shift in power from labour to capital there has been a shift within capital towards that which dominates the secondary circuit, particularly finance, and with that rentier capitalism. By constructing a narrative of success and modernity, by lobbying, by dominating the media, by buying support through political donations, and infiltrating government and designing policies that suit its interests, this new financial plutocracy has huge influence on political agendas. Its most brilliant coup has been to offload the costs of its financial crisis onto the 99 per cent, and the blame onto the public sector. Its next coup is likely to be pushing through bold new trade treaties that allow big capital to override legislation approved by democratic governments that limit its power. These are the Trans-Pacific Pact (TPP), involving 12 nations round the Pacific Rim,[13] and the Transatlantic Trade and Investment Partnership (TTIP) involving the USA and 28 European Union states. Together they account for 60 per cent of global output. Any state that tries to impose restrictions on capital, say, to protect the environment or protect labour and human rights, regulate finance or keep its public sector public or restrict the growth of corporate property rights is liable to be sued by a closed international court. Few elected politicians have been allowed access to the negotiations or even information about their details. 'Investor–state dispute settlement' mechanisms are to be used to allow big corporations to sue governments before secretive arbitration panels composed of corporate lawyers, bypassing domestic courts and overriding the will of parliaments. This represents a huge victory for plutocracy over democracy (Monbiot, 2013; Wikileaks, 2013). The increased *political* power of capital committed to increasing its already bloated unearned income is truly a force to be reckoned with.

As regards consumption, again it has to be remembered that financial wealth constitutes a potential claim on the goods and services produced as commodities by others. Unless the latter are produced, the wealth is worthless. Thus, a relevant moral economic question might be what justification is there for the work of so many to be controlled by the consumption preferences of so few? One way of thinking about how much concentration of wealth should be allowed is to ask how many people's labour and pay should be dependent on meeting the consumption desires of the rich and super-rich? Carlos Slim Helú (net 'worth' US$87 billion, in 2014), for many years the richest man in the world, owns 90 per cent of the telephone landlines in Mexico, and 80 per cent of the mobile phone networks, an extraordinary politically mediated windfall of economic rent (Thompson, 2006). His income in 2012 was equivalent to the average annual salary of 400 000 Mexicans (Freeland, 2012). If he were to spend it all on consumption one can imagine how large the disproportion would be in production for his benefit. Such inequalities in what people are able to consume creates a massive misallocation of resources, in which the production of exclusive luxuries takes precedence over production for the needs and wants of the vast majority of population, particularly the poor. Increased consumption by the rich can actually make others worse off: in London, purchase of properties as 'investments' by the rich not only allows them both to drive up and free-ride on rising land values, but makes it increasingly difficult for many who work in the capital to live there (see related discussions in this volume by Atkinson, Burrows and Rhodes, Chapter 11, and Paris, Chapter 12).

Finally, as regards the environment, a moral economic approach would question the legitimacy of allowing any to consume more than their share of the Earth's resources (i.e., the world's resources divided by the world's population). To be sure, some may need to consume more than others, for example, people in cold countries need more fuel for heating and insulation than those in hot countries. But in practice, of course, the distribution is not determined by deliberation but by power; why should we assume that the distribution resulting from a free-for-all in which those who control key assets can extract wealth from others must be acceptable? This kind of question is particularly urgent as carbon emissions continue to accumulate, rapidly approaching the point where runaway global warming takes off. Why should anyone be allowed to produce more greenhouse gases from their activities than the Earth's capacity to reabsorb it (already massively exceeded) divided by the world's population? But it is not just that the carbon footprints of rich individuals are so excessive, but that insofar as their wealth depends on continued exponential economic growth, so they can get a return on their 'investments', they have an interest in doing nothing about climate change. Energy companies – amongst the most powerful corporations in the world – are continuing their long-term investment in fossil fuels, and the repayment of debts to the rich depends on economic growth.

CONCLUSION

Much could be added to this explanation of the growth of rentiership, of unearned income based on control of assets, in particular changes in global economic geography. There is also a cultural side to it, as neoliberalism has infiltrated ever deeper into organizational and everyday life, changing behaviours and worldviews. Part of this in turn has been the piecemeal development of narratives celebrating 'the market' and its 'Masters of the Universe' and encouraging the financialization of everyday life (Martin, 2002) – through exhortations to 'release the value' of our homes, become 'entrepreneurial selves', 'leveraging our skillsets'. It also 'moralized' neoliberalism with a discourse of 'wealth creation', 'trickle-down effects', the rich as job creators – specially gifted individuals who we should admire and thank. Ordinary people have been encouraged to see themselves as responsible only for themselves, rather than interdependent, while marketing and managerial discourse strives to create optimism – a 'cruel optimism'[14] – in the face of job insecurity, massive youth unemployment and prohibitive housing costs.

Many academics and other observers have become fascinated with the challenge of understanding the extraordinary complex and arcane practices of financialization – often with a sense of dismay at the way the sector has come to function as master rather than servant of the economy. Many have pointed out how dysfunctional it is to have economies in which the falling share of wealth held by the 99 per cent results in stagnation of demand and hence reduced capacity for paying debt. But the economy is not just a devilishly complex and flawed machine; while it certainly is dysfunctional and contradictory, it is also a set of social relations and there is a danger of missing the injustice of this wealth concentration, insofar as it comes from unearned income, from a massive increase in free-riding. Similarly, debts piled up into the future can only be repaid if the output produced by future generations grows fast enough. Given the size of the debt crisis, this is clearly an impossibility. More importantly, in a world on the brink of runaway global warming, resumed growth in countries with already massively unsustainable carbon footprints is madness – again not only dysfunctional, but unjust in damaging the environment of poorer countries and future generations. Hence the need for a moral economic approach that questions all this and reveals the role of the rich within it.

NOTES

* This chapter is based on arguments developed more fully in my 2014 book, *Why We Can't Afford the Rich*.

1. World Top Incomes Database. See Piketty (2014) for details and analysis. However, these are figures for pre-tax income, and since taxes on the rich have fallen dramatically in the last 30 years, their post-tax income share has risen more.
2. Sources: *Sunday Times* Rich List and *UKpublicspending.co.uk*.
3. Many associate the term 'moral economy' with Edward Thompson's research on regulated markets. Thompson (1971) used the term to refer to an object rather than as a kind of analysis. As such he saw 'the' moral economy as the opposite of the market economy. Unfortunately this historical focus has led many to assume that moral economies ended

with capitalism, and that capitalism is not in any sense a moral economy. I and other more recent commentators argue it is (e.g., Booth, 2004; Keat, 2010). The works of O'Neill (1997), Alperovitz and Daly (2008), Murphy and Nagel (2005), Graeber (2011) and Polanyi (1947) are all consistent with this view.

4. 'Illth' refers to property that either results in a decline in well-being or is put to no use.
5. As Engelen et al. (2011) explain, the shareholder value movement was cause and effect of the rise of large institutional shareholders, and an effect of the weakening of organized labour's ability to act as a stakeholder.
6. Surprisingly, Thomas Piketty misses this point, attributing the rising wealth of 'super-managers' in part to the expanding 'earned income hierarchy' (Piketty, 2014, p. 378).
7. Citing data from Census Bureau and Economic Policy Institute compiled by Colin Gordon. Incomes include transfers. Average top incomes (market income only) from Piketty and Saez's World Top Incomes Database.
8. According to Harry Shutt (2009, p. 124), at the same time returns on 'investments' have been 'higher than in any period of comparable length since the Industrial Revolution', with an estimated 75 per cent of this resulting from appreciation of assets in the USA and Britain, compared with well under 50 per cent on average from 1900 to 1979.
9. The same trends of falling labour shares are reported in IMF (2007, p. 168), though workers in sectors employing a large proportion of graduates increased their shares.
10. See also OECD (2011). I am grateful to Diane Perrons for alerting me to these reports.
11. While this is interesting, it represents national financial systems as closed rather than part of an international system; it also ignores mechanisms of redistribution from poor to rich that do not depend on the creation of interest bearing credit money.
12. Partly because of the limitations of Piketty's (2014) neoclassical framework, he misses many of the moral economic implications of the trends that he tracks. See, for example, his equivocal and evasive assessment of rent and rentiers, in which he seems to imply that mere ownership of productive assets like machinery actually produces something, so that profit is just a payment for that 'contribution' (pp. 422–44). While he calls for redistribution of wealth via a wealth tax (no small thing!) and does refer to rent and rentiers he accepts many of the basic economic institutions that allow unearned income to be extracted and accumulated in the first place.
13. Canada, USA, Mexico, Peru, Chile, New Zealand, Australia, Brunei, Singapore, Malaysia, Vietnam and Japan.
14. The phrase is Lauren Berlant's (2011), though she uses it somewhat differently from how I have used it here.

REFERENCES

Allen, J. and M. Pryke (2013), 'Financializing household water: Thames Water, MEIF, and ring-fenced politics', *Cambridge Journal of Regions, Economy and Society*, **6** (3), 419–39.

Alperovitz, G. and L. Daly (2008), *Unjust Deserts*, New York: The New Press.

Atkinson, A.B., T. Piketty and E. Saez (2011), 'Top incomes in the long run of history', *Journal of Economic Literature*, **49** (1), 3–71.

Beaverstock, J.V., S. Hall and T. Wainwright (2013), 'Servicing the super-rich: new financial elites and the rise of the private wealth management retail ecology', *Regional Studies*, **47** (6), 834–49.

Berlant, L. (2011), *Cruel Optimism*, Durham, NC: Duke University Press.

Booth, W. (2004), 'On the idea of the moral economy', *American Political Science Review*, **88** (3), 653–67.

Buchanan, J., G. Dymski and J. Froud et al. (2013), 'Unsustainable employment portfolios', *Work, Employment and Society*, **27** (3), 396–413.

Dorling, D. (2013), 'Fairness and the changing fortunes of people in Britain', *Journal of the Royal Statistical Society A*, **176** (1), 97–128.

Engelen, E., I. Ertürk and J. Froud et al. (2011), *After the Great Complacence: Financial Crisis and the Politics of Reform*, Oxford, UK: Oxford University Press.
Ertürk, I., J. Froud and A. Leaver et al. (2007), 'Agency: a positional critique', *Economy and Society*, **36** (1), 51–77.
Freeland, C. (2012), *Plutocrats: The Rise of the New Global Super-Rich*, London: Allen Lane.
Graeber, D. (2011), *Debt: The First 5000 Years*, New York: Melville House Publishing.
Harvey, D. (2007), *A Brief History of Neoliberalism*, Oxford, UK: Oxford University Press.
Hobson, J.A. (1937), *Property and Improperty*, London: Gollancz.
Hodgson, G. (2013), 'Banking, finance and income inequality', *PositiveMoney.org*, accessed 21 January 2015 at https://www.positivemoney.org/publications/banking-finance-and-income-inequality/.
Hudson, M. (2011), 'How economic theory came to ignore the role of debt', *Real World Economics Review, No. 57*, 6 September, accessed 15 January 2015 at http://www.paecon.net/PAEReview/issue57/whole57.pdf.
ILO (2008), *World of Work Report 2008: Income Inequalities in the Age of Financial Globalization*, accessed 14 August 2015 at http://www.ilo.org/wcmsp5/groups/public/@dgreports/@dcomm/@publ/documents/publication/wcms_100354.pdf.
IMF (2007), *Spillovers and Cycles in the Global Economy*, accessed 23 January 2015 at www.imf.org/external/pubs/ft/weo/2007/01/pdf/text.pdf.
Keat, R. (2010), 'Every economy is a moral economy', University of Edinburgh, unpublished manuscript, accessed 21 January 2015 at http://www.russellkeat.net.
Kennedy, M. (2012), *Occupy Money*, Gabriola Island, BC: New Society Publishers.
Kristal, T. (2010), 'Good times bad times: postwar labor's share of income in 16 capitalist democracies', *American Sociological Review*, **75** (5), 729–63.
Martin, R. (2002), *The Financialization of Everyday Life*, Philadelphia, PA: Temple University Press.
Marx, K. (1972), *Capital, Volume III*, London: Lawrence and Wishart
Meek, J. (2012), 'Human revenue', *London Review of Books*, **34** (7), 8.
Mellor, M. (2010), *The Future of Money: From Financial Crisis to Public Resource*, London: Pluto Press.
Mill, J.S. ([1848] 2015), *Principles of Political Economy*, CreateSpace Independent Publishing Platform.
Monbiot, G. (2013), 'The lies behind this transatlantic trade deal', *Guardian*, 2 December, accessed 21 January 2015 at http://www.theguardian.com/commentisfree/2013/dec/02/transatlantic-free-trade-deal-regulation-by-lawyers-eu-us.
Murphy, L. and T. Nagel (2005), *The Myth of Ownership*, Oxford, UK: Oxford University Press.
OECD (2011), *Divided We Stand: Why Inequality Keeps Rising*, Paris: OECD.
O'Neill, J. (1997), *The Market*, London: Routledge.
Oxfam (2014), 'Working for the few: political capture and economic inequality', *Oxfam Briefing Paper No. 178*, accessed 23 January 2015 at www.ipu.org/splz-e/unga14/oxfam.pdf.
Palley, T. (2007), 'Financialization: what it is and why it matters', *Working Paper No. 525*, Washington, DC: Levy Economics Institute.
Perrons, D. (2012), '"Global" financial crisis, earnings inequalities and gender: towards a more sustainable model of development', *Comparative Sociology*, **11** (2), 202–26.
Peters, J. (2010), 'The rise of finance and the decline of organised labour in the advanced capitalist countries', *New Political Economy*, **16** (1), 73–99.
Pettifor, A. (2006), *The Coming First World Debt Crisis*, Basingstoke, UK: Palgrave Macmillan.
Piketty, T. (2014), *Capital in the Twenty-First Century* [French edition published 2013 as *Le capital au XXI siècle*, Editions du Seuil], Cambridge, MA: Belknap Press of Harvard University Press.
Polanyi, K. (1947), *The Great Transformation*, New York: Basic Books.
Rawls, J. (1971), *A Theory of Justice*, Cambridge, MA: Harvard University Press.

Resolution Foundation (2012), *Gaining from Growth: The Final Report of the Commission on Living Standards*, accessed 21 January 2015 at http://www.resolutionfoundation.org/wp-content/uploads/2014/08/Gaining_from_growth_-_The_final_report_of_the_Commission_on_Living_Standards.pdf.
Ruskin, J. ([1862] 2007), *Unto This Last*, New York: FQ Classics.
Savage, M. and K. Williams (2008), 'Elites: remembered in capitalism and forgotten in social science', *The Sociological Review*, **56** (S1), 1–24.
Sayer, A. (2007), 'Moral economy as critique', *New Political Economy*, **12** (2), 261–70.
Sayer, A. (2014), *Why We Can't Afford the Rich*, Bristol, UK: Policy Press.
Shutt, H. (2009), *The Trouble With Capitalism*, London: Zed Books.
Tawney, R.H. (1921), *The Acquisitive Society*, London: G. Bell and Sons.
Thompson, E.P. (1971), 'The moral economy of the English crowd in the eighteenth century', *Past & Present*, **50** (1), 76–136.
Thompson, G. (2006), 'Prodded by the left: richest man talks equity', *New York Times*, 3 June, accessed 21 January 2015 at http://www.nytimes.com/2006/06/03/world/americas/03slim.html?pagewanted=1&_r=0&ei=5088&en=87ff5ffac4ee12aa&ex=1306987200&partner=rssnyt&emc=rss.
UNCTAD (2012), *Trade and Development Report 2012*, accessed 23 January 2015 at unctad.org/en/publicationslibrary/tdr2012_en.pdf.
Wikileaks (2013), 'Secret Trans-Pacific Partnership agreement (TPP)', accessed 21 January 2015 at https://wikileaks.org/tpp/pressrelease.html.
Wilson, M. (2014), *Everyday Moral Economies*, Chichester, UK: Wiley.

6. Billionaire philanthropy: 'decaf capitalism'

Ilan Kapoor

INTRODUCTION

My purpose in this chapter is to carry out an ideology critique of the charity work of celebrity billionaire humanitarians, Bill Gates and George Soros. Their charity work, I argue, consists of giving away spectacular sums of wealth along mostly entrepreneurial lines, yet forswearing how such wealth was accumulated in the first place. Indeed, the two billionaires' philanthropy is tied to (disavowed) ruthless business practices, a phenomenon I will refer to as 'decaf capitalism'. My overall argument is that the construction of celebrity corporate philanthropy helps repudiate corporate capitalism's 'dirty' underside, which is to say that celebrity charity helps stabilize and advance the global neoliberal capitalist order.[1]

My focus on Gates and Soros as billionaire humanitarians is meant not to concentrate on their personal motivations or foibles but to help illustrate the broader structural features of our current cultural and political economy. It should not be forgotten that both men are products of the neoliberal deregulation of financial markets of the 1980s and 1990s, accompanied as it was by the emergence of the 'new information economy' and the dot.com boom. This is the period that witnessed notable economic growth in much of the West. But such growth was highly skewed, showing up in the form of a remarkable rise in the ranks of the super-rich on the one hand, and growing social inequality on the other. Thus, while the number of American billionaires during the last two decades has risen 20-fold, the erstwhile economic boom has failed to 'trickle down'. Piketty (2014), for example, has shown how wealth and income inequality in the USA and Europe, particularly over the past 30 years, has increased sharply (cf. Adair, 2010; BBC, 2013). It is precisely such socioeconomic disparity that the global Occupy movement has been trying to highlight; and Piketty's analysis confirms that, since 1980, the share of income going to the 1 per cent in such countries as the USA, UK, Australia and Canada has risen to historically unprecedented levels.

In the midst of such concentrated wealth creation and rising social inequality, Gates and Soros have emerged as celebrities, partly for their status as two of the richest people on earth, and partly for their spectacular generosity, giving away hundreds of millions of dollars to a host of charities. They join the ranks of several other similar corporate billionaire humanitarians such as Ted Turner (media mogul), Warren Buffett (investor), Pierre Omidyar and Jeff Skoll (of eBay fame), Carlos Slim Helú (telecom mogul), Richard Branson (Virgin Group), Hansjörg Wyss (medical devices) and Sergey Brin and Larry Page (Google co-founders). What is peculiar about Gates and Soros (as well as some of their billionaire colleagues) is their appearance of being 'progressive' philanthropists: Slavoj Žižek ironically refers to them as 'liberal communists' (2006, p. 10) because they tend to espouse outwardly left and cosmopolitan causes such as democratic reform and global health. But of course their progressiveness is decidedly liberal in that their view of social responsibility, as we shall see, is circumscribed firmly within a thriving global capitalist order. They are, Žižek continues, 'true citizens of the world. They are good people who worry' (2008a, p. 20).

GIVING: BILLIONAIRE PHILANTHROPY

Bill Gates co-founded Microsoft Corporation in 1975. Under him, the company developed into the largest global software business, dominating personal computer operating systems worldwide. He stepped down as CEO in 2000 (although he remains non-executive chairman), devoting his time to his charitable foundation, the Bill and Melinda Gates Foundation. Apart from being one of the world's richest men (his current net worth is assessed at US$77.2 billion), he is also one of the most 'generous' philanthropists in history, estimated to have given, along with his wife Melinda, close to US$30 billion to date in charity (*Bloomberg Businessweek*, 2010; *Forbes*, 2011, 2014a).

Gates believes in what he calls 'friction-free capitalism' (2006), total transparency in global networking and communication, which is what he thinks the information revolution (pioneered by Microsoft) has made possible. But business and technological innovation, harbingers of a friction-free world, must be integrated creatively with social responsibility according to him. He states that, 'We have to find a way to make the aspects of capitalism that serve wealthier people serve poorer people as well' (2008, p. 9). We can bring self-interest and 'caring for others' together by using 'profit incentives whenever we can' (2008, p. 10). This he calls 'creative capitalism', another term for what others such as Bishop and Green (2008) have termed 'philanthrocapitalism'. It is 'an approach where governments, businesses, and nonprofits work together to stretch the reach of market forces so that more people can make a profit, or gain recognition, doing work that eases the world's inequities' (Gates, 2008, p. 10).

Private enterprise, in this view, allows creativity to flourish in business as much as social programmes and charity work. The implication is that state regulation needs to be kept in check so as not to undermine human creativity. In fact, for Gates, private philanthropy, unencumbered by red tape, can be more successful, sustainable and results oriented than state social programmes (Bishop and Green, 2008, p. 12). But the private sector, and large corporations in particular, must also do their bit, according to Gates. They must view corporate philanthropy (i.e., corporate social responsibility) as making good business sense. And they must give more to fill the gaps left by declining state and non-profit social funding (Bishop and Green, 2008, pp. 13ff.).

The Bill and Melinda Gates Foundation was set up in 1999 (as a merger of two previous foundations). In 2005, the Foundation's endowment stood at US$35 billion, but was further replenished by a US$37 billion pledge (to be disbursed gradually) by Warren Buffett. Today, with an endowment of close to US$40 billion, which is larger than the GDP of several countries, the Foundation is the biggest private charity in the world (Okie, 2006, p. 108; Piller et al., 2007; Gates Foundation, 2014).

The Foundation disburses grants in the order of over US$3 billion annually in three main areas: global health, global development/poverty reduction and US education (Gates Foundation, 2014). Global health programmes take up by far the largest share of grants (about US$14.5 billion disbursed to date), focusing mainly on immunization and vaccine research, HIV/AIDS and maternal and child health. Global development programmes (about US$3.3 billion disbursed to date) centre on financial services for the 'poor' (i.e., microcredit), agricultural development (including funding for a 'Green Revolution' in Africa), and water and sanitation. Finally, US education programmes (about US$6 billion disbursed to date) include support for schools, university scholarships and computer access in public libraries (Gates Foundation, 2011, 2014).

George Soros, like Bill Gates, is also one of the world's richest and most 'generous' entrepreneurs. Trained as an arbitrage trader in London, he moved to New York in the 1960s. There, he set up Soros Fund Management (SFM), which includes his now famous Quantum Fund, the hedge fund that has been the source of his enormous personal wealth. SFM holdings are currently estimated at US$4.2 billion, with the billionaire's own net worth assessed at some US$23 billion (Opalesque, 2009; BBC, 2010; *Forbes*, 2011, 2014b).

Soros is an avid supporter of human rights, while also an opponent of 'market fundamentalism' for its unquestioning promotion of 'free' and unfettered markets (Soros, 2002a, 2002b). He has openly condemned the likes of Vladimir Putin for his anti-democratic rule, while also funding US political organizations to help defeat George W. Bush's 2004 re-election bid. He is thus often seen as a strong critic of neoliberalism and authoritarianism, which may be taken as a Left critique. Yet, influenced by Karl Popper, Soros is in fact a strong defender of liberal democracy, believing, as did Popper, in the frailty of liberal institutions and the need to protect them (Soros, 2002b, p. ix). He is more appropriately a 'liberal reformist', advocating for strong legal frameworks and a modicum of state regulation to make global capitalism more stable and equitable. In a discussion on market capitalism, he writes:

> Of course I am a supporter of the capitalist system. . . I am not advocating the abolishment of the capitalist system but I am arguing that it is imperfect and it has deficiencies. . .we actually do need democracy to make capitalism stable, because we need a social counterweight to offset the excesses of capitalism. (Soros, 2002a, pp. 24, 26; cf. 2002b, pp. 1–7)

This worldview has translated into several notable philanthropic initiatives. In 1991, Soros provided Europe's largest ever educational endowment (€420 million) to create the Central European University (CEU) in Budapest, with the purpose of educating 'a new corps of Central European leaders' (CEU, quoted in Guilhot, 2007, p. 449). During the last two decades, the university has successfully trained a whole generation of economists and policy-makers. As well, in 2010, he donated US$100 million to Human Rights Watch over the next ten years (BBC, 2010).

Aside from these singular initiatives, Soros has established a global network of charitable groups or Open Society Foundations, whose goal is to influence public policy and promote human rights and multi-party elections. With a total annual budget of between US$400 million and US$500 million (Open Society Foundations, 2015), this network has supported a diverse array of local causes in more than 70 countries – from human rights in Zimbabwe and independent media building in Burma, to support for democratic change in Eastern Europe (including the 'Orange Revolution' in Ukraine and the 'Rose Revolution' in Georgia).

TAKING: BILLIONAIRE CAPITAL ACCUMULATION

But in order to give, these billionaires have also taken. And as I will show, Gates's and Soros's engagements to 'save the world' are premised on the prior accumulation of personal fortunes in questionable, if not unsavoury, ways.

Of course, such accumulation has not been as overtly or directly exploitative as that of some of the pioneers of corporate philanthropy such as Andrew Carnegie or John D. Rockefeller, Jr. We have come to associate the latter industrialists with their largesse and support for educational and cultural causes, yet it is all too easy to forget their ruthless corporate tactics. Carnegie's Pittsburgh steel empire, for instance, was built on the suppression of workers' rights, including on the occasion of the notorious 1892 Homestead strike against layoffs and long working shifts, when private security agents were ordered to fire on unarmed workers (Krause, 1992; Žižek, 2006). The same is true for Rockefeller, whose industrial and oil empire was associated with often-brutal labour management tactics, including for example, the 1914 massacre of striking miners in Ludlow, Colorado (Andrews, 2008).

Gates and Soros might have turned from such outwardly harsh methods, yet their business practices are not without serious social consequences, for exploitation under late capitalism has not disappeared; it has simply taken on different, and sometimes veiled, forms.

Gates: Enclosing the Knowledge Commons

Simply put, Bill Gates's enormous wealth derives from the privatization of the knowledge commons. Information/knowledge forms the symbolic substance of our lives – our very individual and collective identity and communications depend upon it – yet by appropriating it, he has made it into tangible property. This privatization of our 'general intellect' is 'a decisive battle which triggered the battle for the "enclosure" of the common domain of software' (Žižek, 2008b, p. 422).

As Stephen Adair (2010) explains, while in our information age, the conditions and relations of production have not disappeared, what has changed is that our world – from commodities to cultural discourses – is increasingly being informatized, transformed into informational bits that can easily be electronically reproduced and exchanged. Theoretically, this should lead to greater abundance for all through the free exchange of digitized knowledge, but such is not the case. Instead, the creative labour that goes into the initial design and display of software has been privatized and highly valorized by the likes of Microsoft: 'When use value can be copied with little or no labor, then valorizing labor. . .may be the only counteracting factor available for capital to prevent prices from falling to a level determined by the commodity's exchange value' (Adair, 2010, p. 251). This means that software companies pay a relatively small initial cost for creative labour, but because digital copying is so easy and cheap, they are able to keep reproduction costs at a minimum, thus profiting from very large economies of scale. They are also able to add a price premium through product branding and logoization, which gives software its cachet and identity. Add to this Microsoft's continuous product updates to keep ahead of technological obsolescence, and you have the makings of massive wealth accumulation.

It is not surprising, then, that the dot.com boom of the 1990s saw the creation of mega-fortunes by such corporations as Netscape, Oracle, Microsoft and Google. The wealthiest people in the world used to be steel magnates and oil barons such as Carnegie and Rockefeller; they are now the sultans of digitized intellect. And they are so successful, not because they have been able to extract traditional surplus value, but because they can use the creative labour of a relatively few employees to extract 'monopoly rents' (Adair, 2010, p. 254).

Critical to the extraction of monopoly rents are intellectual property rights (IPRs) (i.e., royalties, trademarks, copyrights, licensing agreements). It is IPRs that enable the creation of exclusivity and scarcity, allowing Microsoft to collect rents from mostly anyone who wishes to participate in global networking and the intellectual commons. In fact, Microsoft's monopolization of the software industry depends crucially on the strict protection and enforcement of IPRs. Gates has been their ardent defender, supporting patent protection not only for the software sector but also the drug industry by declaring: 'I think if you invent drugs, you should be able to charge for them' (quoted in Rimmer, 2010, p. 328). After all, IPRs insulate his company, permitting it to protect its brand; they help increase his company's market share, squeezing out competitors and potential threats. No wonder, as Adair underlines (2010, p. 250), that we are witnessing today the increasing corporate use of the likes of digital security and signal scrambling (in the face of file sharing and pirating), as well as clamouring for stronger libel and defamation laws (in the face of blogging and paparazzi journalism).

This novel mode of extraction and accumulation engendered by IPRs is often referred to as the 'new enclosure movement'. Rather than being socially managed and available in the public domain, IPRs are mostly monopolized and privatized. Žižek opines that if Gates were allowed a full monopoly over information, 'we would reach the absurd situation in which a private individual would literally own the software texture of our basic network of communication' (2009a, p. 53). Such a problem is not exclusive to the software sector; it is prevalent in many other areas, particularly biotechnology. Thus, farming communities in Northern India may suddenly learn that the basmati rice they have been cultivating and sharing for centuries has been patented by a US biotech corporation (Shiva, 2005, pp. 147–8), while we ourselves may discover that parts of own bodies, 'our genetic components, are already copyrighted, owned by others' (Žižek, 2007).

Microsoft has been able to maintain almost absolute control over the computer software market as a result of its copyright and licensing arrangements. Its operating systems are used by more than 85 per cent of personal computers sold worldwide, with minimal (albeit increasing) competition from Apple and open-source Linux (Bishop and Green, 2008). It also has a history of challenging its competition, squeezing out WordPerfect and Lotus 1-2-3 in the late 1980s, and buying out an average of six companies a year since its creation in 1987, including those responsible for DOS, FrontPage, Internet Explorer, Hotmail, WebTV, and more recently, Skype and Nokia (Stross, 1997). In this regard, it is worth noting that, in the 1990s, Netscape controlled 90 per cent of the global browser market share, but by bundling its browser (Internet Explorer) with its operating system, Microsoft was able to squeeze out Netscape and gain market dominance.

In the late 1990s, the US government took out an antitrust case against Microsoft for engaging in anti-competitive behaviour (in particular against Netscape and Java). After a series of rulings, in 2000, the corporation was found guilty of violating the Sherman Act (i.e., engaging in an illegal monopoly). In 2004, it was handed a record fine of some US$666 million by the European Court for its breach of EU competition law, and in 2008 European antitrust regulators fined it again for a record US$1.3 billion for failing to comply with the 2004 judgement (Castle and Jolly, 2008).

Apart from Microsoft's anti-competitive conduct and extraction of monopoly rent, it is important to underline the significant global structural changes that it has helped bring about in the information industry, most often to the detriment of many in the Third World. True, it may not be fair to pin these changes directly or solely on Gates/Microsoft, but as pioneers and powerful leaders in the global software industry, they are far from blameless. The key issue here is the new global division of labour produced by the growing information revolution, in particular its creation of unevenness and new labour hierarchies and inequalities. Recent research points up, for example, how the outsourcing and offshoring practices of information technology (IT) companies has increased the fragmentation and flexibilization of labour – with mainly women supplying cheap and unprotected labour – while also reinforcing global inequalities by favouring those regions with low-cost, educated workforces and strong IPR protections (urban enclaves in South and Southeast Asia) and ignoring others (parts of the Middle East and Sub-Saharan Africa) (Brenner and Keil, 2005; Carmel and Tjia, 2005; Upadhya and Vasavi, 2008; Gillard et al., 2008).

Such a state of affairs gives new meaning to Gates's much trumpeted ideal of 'friction-free' capitalism. For, it seems, his fantasy of smooth and transparent global networking is made possible only by a grimy underbelly – by outsourcing and offshoring the inconveniences of the IT industry. This is an expression of ideology at its purest, and Žižek offers a scathing critique of it: 'You export the (necessary) dark side of production – disciplined, hierarchical labour, ecological pollution – to. . .[the] Third World. . . The ultimate liberal communist dream is to export the entire working class to invisible Third World sweat shops' (2006, p. 10). The notion of an immaterial and seamless cyberspace is premised on material inertia, on an obstacle-free capitalism, where what Žižek calls the traumatic Real or the social antagonism (i.e., one's own social privilege, the materiality and power of labour, poor working conditions, gender inequality, environmental degradation) is disavowed. The fiction is to make capital movement free and unencumbered, so that software can win over hardware (Žižek, 2008a, p. 17), and the exploitation of labour can be relegated out of sight: 'in the social conditions of late capitalism, the very materiality of cyberspace automatically generates the illusory abstract space of 'friction-free' exchange in which the particularity of the participants' social position is obliterated' (Žižek, 1997, p. 156).

But it's not just that Gates gives with one hand and takes with the other (or, dare I say it, exhibits sleight of hand!); it's also that the very hand he gives with is sullied. Indeed, the Gates Foundation, through which he channels his charity, has been the object of significant criticism in a number of areas. One key problem is that the Foundation tends to fund programmes too narrowly, especially in the health field, where it has now become a major global player. It tends to take a technological approach to health, supporting programmes that are frequently narrowly science-based (e.g., biomedical research, discovery of new medicines) (Cooper, 2008, p. 85; Rimmer, 2010, p. 330). One journalist revealingly describes Gates's zeal for health issues in the following way: 'The way [Gates] talked about wiping out malaria was how he used to talk about wiping out Netscape' (quoted in Cooper, 2008, p. 83). The technical aspects of health end up being too easily separated from larger political economy and social issues (Birn, 2005), so that not enough attention is paid to the more intractable issues of health delivery or broader structures of inequality and patriarchy. Most often, it is not the lack of new drugs, but obstacles such as food insecurity, political repression, or unequal land tenure that affects maternal and child health (Edwards, 2009, p. 41). A notable fallout from its search for quick-fix technological solutions is that, because the Foundation is a leader in the global health field, it diverts energy, resources and commitment away from other important areas such as primary health care, training of health workers, and so on (Edwards, 2009, p. 38).[2]

A second target of criticism is that the Gates Foundation lacks transparency and accountability. Indeed, it is run by three trustees: Bill Gates, Melinda Gates and Warren Buffett, and its main other officers are Gates family members (Bill Gates, Sr) or former Microsoft executives (Eisenberg, 2006). The Foundation is thus directed by a closed, inner circle of family and friends (*The Lancet*, 2009; Rimmer, 2010, p. 330). Its decision-making lacks outside or public scrutiny of any kind, and even excludes representation from programme 'beneficiaries', yet the Foundation plays such an influential global role in health, poverty reduction and education, affecting policy-making, research, funding and thousands of people's lives.

Finally, there is the issue of the Foundation's investments. The problem here is not simply that Gates avoids paying taxes on his charitable donations, as does the Foundation on the investment income from its endowments (Bishop and Green, 2008, p. 11); all charitable donations and Foundations do (in the USA). The more serious difficulty is that the Foundation invests in ways that contradict its own programming goals. Piller et al. (2007) report that it spent US$218 million on polio and measles immunization and research, including for programmes in the Niger Delta. But at same time, they find that the Foundation has invested US$423 million in oil companies such as Royal Dutch Shell, Exxon Mobil, Chevron and Total, companies responsible for serious environmental pollution in the Delta that would not be tolerated in Europe or the USA (Zalik, 2004). In fact, many local Delta leaders blame the oil exploitation in the area for the very health infections the Foundation is combating – malaria from mosquitoes in the stagnant water in oil bore holes, cholera from rivers clogged from oil spills, and lower health immunity among residents (including greater child susceptibility to polio and measles) from breathing in sooty gas flares that contain benzene, mercury and chromium particulates. Piller et al. (2007) estimate in fact that some 41 per cent of the Foundation's assets are invested in companies whose activities run counter to the Foundation's social and health programmes. They state that its portfolio managers are provided with little direction other than to maintain a diversified portfolio. As a consequence, the Foundation has sizeable holdings in major US and Canadian polluters such as Dow Chemical, ConocoPhillips and Telco, and it has forged partnerships with several pharmaceutical companies (e.g., Merck, GlaxoSmithKline), many of which resist the move to generic drugs, pricing medicines beyond the reach of the very patients that the Foundation is trying to help (Bishop and Green, 2008, pp. 64–5; Cooper, 2008, p. 85).

In response to these criticisms, the Foundation announced a full review of its investments, only to later cancel the review, declaring that it stood by its initial policy but would use its shareholder voting rights to influence companies' practices from within (Piller, 2007). Of course, in the absence of public scrutiny of the Foundation, there is no way of verifying if such influence is being exercised, and if so, whether companies are indeed complying.

Soros: Profiting from Destabilization

It is ironic that George Soros talks of socioeconomic stability and the need to 'offset the excesses of capitalism' (2002a, p. 24), given that his hedge funds are precisely a source of instability and excess. As Žižek avers, 'The same Soros who gives millions to fund education has ruined the lives of thousands thanks to his financial speculations' (2006, p. 10).

Soros has been, and remains, a pioneer and leader in the hedge fund business, profiting from neoliberal globalization (of which he has nonetheless been critical), in particular from market deregulation, which saw an explosion of financial leveraging instruments such as his Quantum Fund. The result has been global capital flows and financial markets that today are complex, obscure, highly interdependent and difficult to monitor (trillions of dollars are traded on markets almost every day). Many hedge funds are virtual or even fictitious, and can be moved around at lightning speed. This post-Fordist 'dematerialization' of financial assets has enabled a greater exchange of financial and economic information, but it has also been accompanied, as we shall see below, by the heightened risk, speed and contagion of financial crises. Thus, slight changes in inflation rates or currency values can mean massive capital flight (Aslanbeigui and Summerfield, 2001, p. 10).

The number of hedge funds used to be small in the 1990s, but has increased substantially since, with currently over 8000 of them (Soros's is one of the biggest and most influential), managing about US$2.6 trillion globally (Holmes, 2009, p. 434; Williamson, 2014). Hedge fund investors used to be almost exclusively very wealthy investors, but this too has changed, with the investor pool broadening of late and even many public sector unions investing their pension money (itself the product of declining state support for social security).

Hedge funds have revolutionized the entire approach to risk under late capitalism, particularly through their use of derivatives, which help magnify risk in an effort to make as much money as possible very quickly (Harmes, 1999; Dodd, 2002; Bryan and Rafferty, 2006, p. 62; Caliari, 2007; Holmes, 2009, p. 437). Derivatives have become so complex these days that they use sophisticated computer models that factor in multiple financial indices, future energy prices, or even long-range weather forecasts.

One of the main reasons hedge funds have been so successful is because of the lack of adequate regulation. They are subject to few legal constraints, little fiduciary responsibility, and practically no disclosure requirements, a situation they have taken full advantage of (Holmes, 2009, p. 433). Several operate from offshore centres, and most use tax havens to capitalize on investment gains. All of this erodes governments' tax bases, as well as the ability to monitor transactions or protect against financial crises (Tanzi, 2002).

Since hedge funds are interested only in quick, short-term returns, they frequently harm the long-term interests of people, governments, or companies. Bryan and Rafferty (2006, p. 176) explain, for example, how derivatives tend to erode job security and worsen working conditions by extracting greater labour flexibility and wage competitiveness (i.e., demanding that firms cut costs, increase profits and deliver on share values).

Of course, we may not be able to attribute full or direct responsibility for all this to Soros, but as a major global player in hedge funds, his hands are certainly dirty. In this regard, there are two relatively recent events with which Soros has been notoriously associated, both involving serious financial destabilization. The first is the 1992 European Exchange Rate Mechanism (ERM) crisis, precipitated by Soros, which forced the British government to leave the ERM, costing taxpayers billions of pounds, while profiting Soros to the tune of US$1 billion or more (Harmes, 1999, pp. 16–17). And the second, more serious event was Soros's involvement in the 1997 Asian financial crisis. As Thai property and asset markets began to weaken earlier that year, several hedge funds, including Soros's Quantum Fund, dumped the baht, which created investor panic and rapid capital outflow, spreading to four other countries in the region (South Korea, Indonesia, Malaysia and the Philippines) (Bello et al., 2000, p. 15).[3]

The social impacts of the 1997 Asian crisis were significant and widespread, reversing several decades of socioeconomic gains in the region. Thailand, Indonesia, South Korea, Philippines and Malaysia all witnessed a deterioration in unemployment, poverty, malnutrition and access to education, with women and the poorest sections of society hit the hardest (Bello et al., 2000, p. x; Hill and Chu, 2001, pp. 13–18). Overall, more than 50 million people moved under the poverty line in the region (Aslanbeigui and Summerfield, 2001, p. 9).

A last point on Soros's philanthropic initiatives, for, like those of Bill Gates, they are not without reproach, even as they are paraded as humanitarian. While some of the funding from his Open Society Foundations may be channelled towards organizations that are critical of states' human rights and democratic records, most of Soros's charity is geared towards the political and economic status quo. This is particularly true of his patronage of the Central European University, which, like many other neoliberalizing universities, favours technocratic training and positivist social science methodologies (e.g., econometrics, political modelling) for the education of state administrators. Several university training modules, for instance, are associated with the World Bank, involving seminars on such issues as privatization, corporate governance, or the creation of business schools (Guilhot, 2007, pp. 465–6). Even Soros's backing for 'critical' human rights groups (e.g., Human Rights Watch) is barely a threat to the system; at most, it is reformist – establishing and defending individuals' civil and political rights, but steering clear of the much more politically difficult area of collective socioeconomic rights (i.e., labour, land, or indigenous peoples' rights). Nicolas Guilhot thus writes that Soros's philanthropy is a 'privileged instrument for reinforcing international institutions and producing scientific, professional, social and political infrastructure needed for managing globalization' (2007, p. 464).

CONCLUSION: 'DECAF CAPITALISM'

That Gates and Soros give with one hand, all the while having taken with the other, is an attempt to produce what I will call, drawing on Žižek, 'decaf capitalism'. Indeed, Žižek sees the recent appearance of consumer products that endeavour to remove risk or danger, that try to cleanse their poison or sweeten their astringent, as indicative of the ideological make-up of our age:

> This brings to mind a chocolate laxative available in the US, publicized with the paradoxical injunction: 'Do you have constipation? Eat more of this chocolate!' In other words, eat the very thing that causes constipation in order to be cured of it. This structure of a product counteracting its own essence containing the agent of its own containment is widely visible in today's ideological landscape. (Žižek, 2009b, p. 14)

There are many consumer products that fit Žižek's description – decaffeinated coffee, cream without fat, sugarless beverages, non-alcoholic beer – just as there is a range of contemporary social phenomena that do – phone sex, cybersex and sexting (i.e., sex without sex); green mining and environmentally 'friendly' or 'ethical' oil exploration (i.e., ecological damage without degradation); and war without war, or war without casualties (i.e., distance technological war, on the basis of which we can bomb, say, Gaddafi's bases without involving Western ground troops in Libya, or use armed robots and unmanned drones that may well kill Afghan or Pakistani civilians, as long as no American soldiers' lives are lost) (cf. Žižek, 2004, pp. 507–8). All are ideological attempts to evacuate from reality the dimension of what Žižek calls the 'Real', that is, to purify life, rid it of its inherent dangers and inconveniences.

And so it is with Gates and Soros. They balance out their ruthless profit-making with charity work, thus deploying a sort of 'decaf capitalism' – a capitalism with a human face, a system that exploits but still cares, wreaks social havoc but really worries, institutes a Wild West entrepreneurialism but also a welfare state.[4] Decaf capitalism enables them to rationalize away their monopolistic corporate behaviour and cut-throat financial speculation, or their co-responsibility in labour exploitation and the production of feminized sweatshops. It allows them to continue with business as usual, 'giving back' to counteract the ills of capitalism, all the while becoming the globe's greatest humanitarians.

Today's corporate ethics or environmentalism (e.g., corporate social responsibility, green capitalism) follows the same ideological route. The giant corporations that embrace it also engage in decaf capitalism, on the one hand grabbing as much money as possible, on the other returning a portion of it in the form of charity or green products. They, like Gates and Soros, see no contradiction, or perhaps even relationship, between profit-making and inequality creation, between wealth accumulation and ecological crisis. They so often fail to discern their own complicity in the very 'poverty' or pollution they seek to redress. In fact, they often rationalize the latter through the practice of what Gates calls 'creative capitalism', making business itself the solution to poverty reduction or environmentalism. The attempt once again is to conveniently duck the ills of capitalism, to disavow its production of inequalities, injustices and unevenness.

The important point, for Žižek, is that Gates and Soros, like their corporate colleagues – and for that matter all of us[5] – are ideologically produced, acting both as agents and as pawns to help further the interests of global capitalism. Ideology, in this sense, operates fundamentally at the level of the *unconscious*: capitalism binds us to it libidinally, with the result that we unconsciously *enjoy* it, according to Žižek. Thus, we may well be aware of its limitations or even critical of its excesses (like Soros and Gates), but nonetheless we continue to enjoy it, support it, reproduce it, thereby disavowing what we know to be true (Žižek, 1989, pp. 18, 32–3).[6]

The implication is that it is charity that helps decaffeinate capitalism. It masks and purifies corporate ills, acting as countermeasure to socio-economic exploitation. For Žižek, it not only temporarily redistributes wealth, but also helps avoid war or stem revolution by tempering people's resentment (arising from generalized social inequality):

> [I]t [charity] is the logical concluding point of capitalist circulation, necessary from the strictly economic standpoint, since it allows the capitalist system to postpone its crisis. . . This paradox signals a sad predicament of ours: contemporary capitalism cannot reproduce itself on its own. It needs extra-economic charity to sustain the cycle of social reproduction. (Žižek, 2008b, p. 374; cf. 2008a, pp. 23–4; 2010, p. 240)

If charity is capitalism's necessary decaffeinating agent, allowing the latter to sustain itself while averting rebellion or crisis, it means that Gates and Soros are not philanthropists out of mere personal choice, religious belief, or good Samaritanism, they are businessmen – humanitarians acting (unconsciously perhaps) in the service of capitalism, tranquillizing its worst manifestations, or, to stick with the coffee metaphor, preventing it from overly percolating. They are, so to speak, coffee-pusher philanthropists, keeping people hooked but not wild (or wired), stimulated but not strung out. Their charity work is integral to the logic of capitalism; it helps regulate the system, calming it down when it runs amok. The irony, of course, is that it is the philanthropists' own business activities that help hyperactivate the system in the first place.

What is noteworthy (and implied in the two previous sections) is that the tycoons' decaffeinating philanthropy targets, not systemic problems or institutions, but what Žižek calls 'secondary malfunctions' – narrow science-based health, technocratic policy-making, corrupt and inefficient state institutions, and so forth. 'Precisely because they want to resolve all these secondary malfunctions of the global system, liberal communists [such as Gates and Soros] are the direct embodiment of what is wrong with the system' (Žižek, 2009c, p. 10; cf. 2008a, pp. 23, 37). They end up trying to address only the more outwardly perceptible or 'subjective' violence in the form of poverty, corruption, or individual rights abuse, as opposed to the slower, more torturous, and less immediately tangible structural or 'objective' violence of social inequality, corporate monopoly, dehumanizing working conditions, unequal land tenure, or gender discrimination (Žižek, 2006, p. 10). It is most often these latter broad malfunctions that lead to the former symptomatic subjective violence taken up by the billionaire philanthropists. Thus, pointing to the need for ideology critique to uncover, not the latent meanings of social antagonisms (e.g., poverty), but their disguised meanings (e.g., inequality), Žižek often repeats Bertolt Brecht's famous quote: 'What is the robbing of a bank compared to the founding of a new bank?' (Žižek, 1989, p. 30).

The problem about structural violence though is precisely that it appears abstract, so that Gates and Soros are able both to hide behind and to profit from the facelessness of decaf capitalism. They are able to maintain a certain distance and anonymity from the social impacts of corporate monopoly or ruthless financial speculation, yet at the same time benefit from a system that privileges individual effort, initiative, philanthropy. Such individualization is further magnified by the rise of media hype and celebrity culture. The tendency there is to personalize 'super-successful' businessmen such as Gates and Soros. Žižek notes, for instance, the propensity to appeal to Gates's familiarity as a friend: he is made out to be, not an enigmatic, evil Big Brother, but an ordinary, geeky, nice guy, someone just like us, albeit tremendously talented: 'the notion of a charismatic "business genius" reassert[s] itself in "spontaneous capitalist ideology", attributing the success or failure of a businessman to some mysterious je ne sais quoi which he possesses' (Žižek, 1999, pp. 347, 349). In the process, the power, influence and unsavoury practices of these business leaders are further sanitized (i.e., decaffeinated), naturalizing and familiarizing corporate neoliberalism.

Such decaffeinating predilections are magnified in this instance because the tycoons in question don't simply give (millions of dollars); they give spectacularly (billions of dollars). Mary Phillips (2008) sees such orgiastic and excessive charity as a modern form of 'potlatch', a gift-giving feast with the mediated public display of it as a crying out for status, glory, honour. She quotes Marcel Mauss to reinforce the point: 'The rich man who shows his wealth by spending recklessly is the man who wins prestige' (2008, p. 252). The spectacle of giving, and of giving so much, aims at constructing Gates and Soros as celebrity heroes, providing them with an instantly recognizable brand. For Phillips (as for Žižek, as noted above), such a phenomenon is an attempt by Gates and Soros to ward off their own mortality, but also more importantly, the crisis of capitalism itself: the tycoons' excessive philanthropy helps defuse 'the potential of explosive surplus produced by the US in order to avert [social, environmental] catastrophe' (2008, p. 261).

A final consideration regarding decaf capitalism is its proclivity towards a 'decaf state' (or perhaps a 'decap state' — short for 'decapitated'!). To be sure, the billionaire philanthropists' spectacular giving fits well with the neoliberal gutting of the state: their gifts, like those of the thousands of charitable foundations that have cropped up under neoliberalism, fills a few (among many) of the gaps in state social funding. The problem, however, is that private decisions are being made for public goods (e.g., health care, education, human rights, poverty reduction) (see, for a discussion, Hay and Muller, 2014). Elites decide, according to their own priorities, prejudices, or idiosyncrasies, what causes matter, how much to spend on them, and in what manner. Enlightened benevolence and individual heroics thus replace collective will, with the (decaf/decap) state sidelined into adulation and gratefulness.

The related issue here is the lack of political legitimacy and accountability: state-funded programmes have at least a modicum of public oversight and recall; their deregulation and privatization means they now answer only to a clique of private individuals. Not only are we left with the corporate world deciding what 'poor' or marginalized communities need, but we must also trust in corporate 'voluntary' self-regulation (e.g., accountability or certification codes that are part of corporate social responsibility). Yet, isn't something amiss when private organizations such as the Gates Foundation have annual budgets greater than that of the World Health Organization and can more or less dictate policy on issues such as HIV/AIDS or malaria immunization?

The flip side of the decaffeination of the state, of course, is not just that it cannot step up, but also that so often it will not. Private philanthropy appears to have sanctioned governments (in both the Global North and South) to abrogate their social responsibilities, letting them off the hook. The state can thus shirk its duties towards marginalized communities, human rights, or health, because the likes of Gates and Soros are there to fill in. It can ignore the lack of adequate regulation of big corporations or hedge funds, even though this might negatively affect jobs, consumers, business competition, or old-age pensions. The post-political landscape of decaf capitalism is one in which magnanimous elites spearhead both social programming and rabid entrepreneurialism without account, while the state is content to sit back and even applaud, equally without account.

NOTES

1. I am drawing here on Žižek's notion of ideology, which maintains that there is a founding inconsistency or gap (i.e., 'the Real') in our structures of signification: reality is forever ruptured by gaps, contradictions or antagonisms. Ideology (or an ideological fantasy) then is that which attempts to obscure the Real, to cover over these gaps, deadlocks, imperfections (Žižek, 1989, p. 45; Kapoor, 2013).
2. Note as well that the Gates Foundation has 'given away' free software to US libraries and schools, but this 'generosity' can be characterized more as a marketing tool (for Microsoft) than philanthropy as it helps build a clientele.
3. The Asian financial crisis is famous for prompting then Malaysian Prime Minister Mahathir to make anti-Semitic remarks about Soros, accusing him of causing the crisis. For his part, Soros admits being involved in the crisis, but not of precipitating it (1999, pp. 208–9).
4. The latest twist is the call by philanthropists such as Warren Buffett to raise taxes on the mega-rich in the USA, and an appeal by Bill Gates for Western political leaders to adopt a financial transaction tax to be used as a new source for foreign aid (Lobe, 2011). Once

again, this is an ideological attempt to rationalize capitalism: what Buffett and Gates are effectively saying is, 'let us continue to engage in ruthless financial speculation, let capitalism wreak havoc, but let us help sweeten the pot after the fact with state welfare and foreign aid'. The only (possibly) positive dimension here is the endorsement of a state tax, which would at least be publicly overseen (although in Gates's proposal, it would for all intents and purposes be handed over to the increasingly privatized foreign aid industry).

5. While we are all ideologically complicit in global capitalism, all of us contributing to its reproduction, Soros and Gates are undoubtedly more powerfully positioned in this ideological nexus by dint of their socioeconomic power, privilege and wealth.
6. Ideology critique, for Žižek (1989), is then about coming to terms with these unconscious libidinal attachments.

REFERENCES

Adair, S. (2010), 'The commodification of information and social inequality', *Critical Sociology*, **36** (2), 243–63.

Andrews, T. (2008), *Killing for Coal: America's Deadliest Labor War*, Cambridge, MA: Harvard University Press.

Aslanbeigui, N. and G. Summerfield (2001), 'Risk, gender, and development in the 21st century', *International Journal of Politics, Culture and Society*, **15** (1), 7–26.

BBC (2010), 'Financier Soros to donate $100m to Human Rights Watch', *BBC News*, 7 September, accessed 20 March 2010 at www.bbc.co.uk/news/world-us-canada-11218868.

BBC (2013), 'US income inequality at record high', *BBC News*, 10 September, accessed 30 September 2013 at www.bbc.co.uk/news/world-us-canada-24039202.

Bello, W., N. Bullard and K. Malhotra (2000), 'Introduction' and 'Notes on the ascendancy and regulation of speculative capital', in W. Bello, N. Bullard and K. Malhotra (eds), *Global Finance: New Thinking on Regulating Speculative Capital Markets*, London: Zed.

Birn, A.-E. (2005), 'Gates's grandest challenge: transcending technology as public health ideology', *The Lancet*, 11 March, accessed 14 August 2015 at http://www.thelancet.com/pdfs/journals/lancet/PIIS0140673605664793.pdf.

Bishop, M. and M. Green (2008), *Philanthrocapitalism: How the Rich Can Save the World*, New York: Bloomsbury Press.

Bloomberg Businessweek (2010), 'The 50 most generous philanthropists', accessed 14 August 2015 at http://www.bloomberg.com/ss/07/11/1115_philanthropy/index_01.htm.

Brenner, N. and R. Keil (2005), *The Global Cities Reader*, London/New York: Routledge.

Bryan, D. and M. Rafferty (2006), *Capitalism with Derivatives: A Political Economy of Financial Derivatives, Capital and Class*, Basingstoke, UK: Palgrave Macmillan.

Caliari, A. (2007), 'Regulation of hedge funds: why is it a social security issue?', *Center of Concern*, accessed 14 August 2015 at https://www.coc.org/node/6296.

Carmel, E. and P. Tjia (2005), *Offshoring Information Technology*, Cambridge, UK: Cambridge University Press.

Castle, S. and D. Jolly (2008), 'Europe fines Microsoft $1.3 billion', *New York Times*, 28 February, accessed 12 March 2010 at www.nytimes.com/2008/02/28/business/worldbusiness/28msoft.html.

Cooper, A. (2008), *Celebrity Diplomacy*, Boulder, CO: Paradigm Publishers.

Dodd, R. (2002), 'The role of derivatives in the East Asian financial crisis', in J. Eatwell and L. Taylor (eds), *International Capital Markets: Systems in Transitions*, New York: Oxford University Press.

Edwards, M. (2009), 'Gates, Google, and the ending of global poverty: philanthrocapitalism and international development', *Brown Journal of World Affairs*, **15** (2), 35–42.

Eisenberg, P. (2006), 'The Gates-Buffett merger isn't good for philanthropy', *Chronicle of Philanthropy*, 20 July.

Forbes (2011), 'The world's billionaires', *Forbes.com*, accessed 24 March 2013 at www.forbes.com/wealth/billionaires.

Forbes (2014a), '#1 Bill Gates', *Forbes.com*, accessed 25 April 2014 at www.forbes.com/profile/bill-gates/.

Forbes (2014b), '#29 George Soros', *Forbes.com*, accessed 25 April 2014 at www.forbes.com/profile/george-soros/.

Gates, B. (2006), 'Beyond business intelligence: delivering a comprehensive approach to enterprise information management', Microsoft Corporation Executive e-mail, accessed 15 July 2010 at www.microsoft.com/mscorp/execmail/2006/05-17eim.mspx.

Gates, B. (2008), 'A new approach to capitalism', in M. Kinsley (ed.), *Creative Capitalism*, New York: Simon & Schuster.

Gates Foundation (Bill and Melinda Gates Foundation) (2011), 'Programs and partnerships', accessed 6 June 2012 at www.gatesfoundation.org/programs/Pages/overview.aspx.

Gates Foundation (Bill and Melinda Gates Foundation) (2014), 'Foundation fact sheet', accessed 24 March 2014 at www.gatesfoundation.org/Who-We-Are/General-Information/Foundation-Factsheet.

Gillard, H., D. Howcroft and N. Mitev et al. (2008), '"Missing women": gender, ICTs, and the shaping of the global economy', *Information Technology for Development*, **14** (4), 262–79.

Guilhot, N. (2007), 'Reforming the world: George Soros, global capitalism and the philanthropic management of the social sciences', *Critical Sociology*, **33** (3), 447–77.

Harmes, A. (1999), 'Hedge funds as a weapon of state? Financial and monetary power in an era of liberalized finance', *YCISS Occasional Paper No. 57*, York University, Canada.

Hay, I. and S. Muller (2014), 'Questioning generosity in the Golden Age of philanthropy. Towards critical geographies of super-philanthropy', *Progress in Human Geography*, **38** (5), 635–53.

Hill, H. and Y.-P. Chu (2001), 'An overview of the key issues', in *The Social Impact of the Asian Financial Crisis*, Cheltenham, UK and Northampton, MA, USA: Edward Elgar Publishing.

Holmes, C. (2009), 'Seeking alpha or creating beta? Charting the rise of hedge fund-based financial ecosystems', *New Political Economy*, **14** (4), 431–50.

Kapoor, I. (2013), *Celebrity Humanitarianism: The Ideology of Global Charity*, London/New York: Routledge.

Krause, P. (1992), *The Battle for Homestead, 1880–1892*, Pittsburgh, PA: University of Pittsburgh Press.

Lobe, J. (2011), 'Bill Gates to support "Robin Hood" tax', *Al-Jazeera*, 24 September, accessed 2 September 2013 at www.aljazeera.com/indepth/features/2011/09/2011924125427182350.html.

Okie, S. (2006), 'Global health – the Gates-Buffett effect', *The New England Journal of Medicine*, **355** (11), 1084–8.

Opalesque (2009), 'Investing – Soros Fund reports $4.2 bln holdings', 17 August, accessed 12 March 2010 at www.opalesque.com/54134/soros%20fund/Investing_reports246.html.

Open Society Foundations (2015), 'About us', accessed 14 August 2015 at https://www.opensocietyfoundations.org/about.

Phillips, M. (2008), 'Tycoon philanthropy: prestige and the annihilation of excess', in D. Crowther and N. Capaldi (eds), *The Ashgate Research Companion to Corporate Social Responsibility*, Aldershot: Ashgate.

Piketty, T. (2014), *Capital in the Twenty-First Century* [French edition published 2013 as *Le capital au XXI siècle*, Editions du Seuil], Cambridge, MA: Belknap Press of Harvard University Press.

Piller, C. (2007), 'Gates Foundation to keep its investment approach', *Los Angeles Times*, 14 January, accessed 14 August 2015 at http://www.latimes.com/business/la-na-gates14jan14-story.html.

Piller, C., E. Sanders and R. Dixon (2007), 'Gates Foundation money works at cross purposes', *Los Angeles Times*, 7 January.

Rimmer, M. (2010), 'The Lazarus effect: the (RED), campaign and creative capitalism', in T. Pogge, M. Rimmer and K. Rubenstein (eds), *Incentives for Global Public Health*, Cambridge, UK: Cambridge University Press.
Shiva, V. (2005), *Earth Democracy*, London: Zed Books.
Soros, G. (1999), *The Crisis of Global Capitalism: Open Society Endangered*, 2nd edition, New York: PublicAffairs.
Soros, G. (2002a), 'Against market fundamentalism: "the capitalist threat" reconsidered', in L. Zsolnai (ed.), *Ethics and the Future of Capitalism*, London: Transaction Publishers.
Soros, G. (2002b), *George Soros on Globalization*, New York: PublicAffairs.
Stross, R. (1997), *The Microsoft Way*, Reading, MA: Addison-Wesley.
Tanzi, V. (2002), 'Globalization and the future of social protection', *Scottish Journal of Political Economy*, **49** (1), 116–27.
The Lancet (2009), 'Editorial: what has the Gates Foundation done for global health?', **373** (9675), 9 May.
Upadhya, C. and A. Vasavi (2008), 'Outposts of the global information economy: work and workers in India's outsourcing industry', in *In an Outpost of the Global Economy: Work and Workers in India's Information Technology Industry*, London/New York: Routledge.
Williamson, C. (2014), 'HFR: hedge fund assets grow 16.7% year record $2.6 trillion in 2013', *Pensions & Investments*, accessed 26 March 2014 at www.pionline.com/article/20140121/ONLINE/140129967/hfr-hedge-fund-assets-grow-167-to-record-26-trillion-in-2013.
Zalik, A. (2004), 'The Niger Delta: petro-violence and partnership development', *Review of African Political Economy*, **101** (4), 401–24.
Žižek, S. (1989), *The Sublime Object of Ideology*, London: Verso.
Žižek, S. (1997), *The Plague of Fantasies*, London: Verso.
Žižek, S. (1999), *The Ticklish Subject: The Absent Centre of Political Ontology*, London: Verso.
Žižek, S. (2004), 'From politics to biopolitics. . .and back', *The South Atlantic Quarterly*, **103** (2/3), 501–21.
Žižek, S. (2006), 'Nobody has to be vile', *London Review of Books*, **28** (7), 10.
Žižek, S. (2007), 'Censorship today: violence, or ecology as new opium for the masses', accessed 15 February 2011 at www.lacan.com/zizecology1.htm.
Žižek, S. (2008a), *On Violence: Six Sideways Reflections*, New York: Picador.
Žižek, S. (2008b), *In Defense of Lost Causes*, London: Verso.
Žižek, S. (2009a), 'How to begin from the beginning', *New Left Review*, **57** (6–7), 43–55.
Žižek, S. (2009b), 'Brunhilde's act', *International Journal of Žižek Studies*, Special Issue, **4**, accessed 20 February 2011 at zizekstudies.org/index.php/ijzs/article/viewFile/294/362.
Žižek, S. (2009c), *First as Tragedy, Then as Farce*, London: Verso.
Žižek, S. (2010), *Living in the End Times*, London: Verso.

7. Making money and making a self: the moral career of entrepreneurs

*Paul G. Schervish**

INTRODUCTION

The purpose of this chapter is to take an initial step in constructing a social psychology of the entrepreneur by examining the confluence of financial world-building and moral self-construction connected to the process of entrepreneurship. In effect, being an entrepreneur is a moral career, a joint venture of making money and making a self. The participation of individuals in the historically given conditions of entrepreneurial wealth generation creates a specific type of agent and not just a specific type of social organization.

I base my analysis on findings generated from intensive interviews with 49 entrepreneurs drawn from a subsample of a study of 130 individuals with, in 2014 dollars, a median yearly income greater than US$1 million and a median net worth in excess of US$16.5 million.[1]

I first set an empirical and theoretical context for the chapter. Second, I set out the three economic rules governing every successful practice of entrepreneurship. I call this the productive secret of money to emphasize that entrepreneurs studiously come to unveil and honour the rules as the necessary conditions for success. Third, I discuss the experiences, realizations, and practices that comprise the journey to becoming an entrepreneur. Fourth, I explain how entrepreneurs execute the productive secret by tailoring their own strategic, financial and spiritual secrets of business development and moral self-construction.

EMPIRICAL AND THEORETICAL CONTEXT

Other chapters in this volume bring us up to date on historical, statistical, theoretical and policy-oriented research about the wealthy and the world of wealth. I mention the Ewing Marion Kauffman Foundation (2015) that supports (e.g., Renz et al., 2014) and publishes research from various sources, including 1323 articles on entrepreneurs and 696 articles on entrepreneurship.

I set the context for what follows in two ways. The first is empirical; the second is theoretical. First, the vast majority of today's top-tier wealth holders are or were entrepreneurs. It is making money as an entrepreneur that propels individuals into the top tiers of wealth. Recent analysis by John Havens (2015) at the Center on Wealth and Philanthropy, based on the 2013 Federal Reserve Survey of Consumer Finances [SCF] (National Opinion Research Center, 2013), indicates that the majority of the top 1 per cent of wealth holders are entrepreneurs, even with the SCF excluding members of the *Forbes* 400 in order to prevent their being identified in the data. In 2013 the wealthiest 1 per cent of US households were those whose net worth was US$7.880 million or greater.

Among the top 1 per cent, 76.6 per cent owned one or more unincorporated or closely held business in whole or in part, amounting to 43 per cent of the aggregate net worth of these business owners. Among this 76.6 per cent, 88.3 per cent took an active management role in one or more businesses that they owned. The close relationship between wealth and entrepreneurship is also revealed by looking at the top 0.1 per cent of households with net worth of US$30 million and above: Havens's analysis indicates that fully 81.7 per cent of this group owns one or more unincorporated or closely held business in whole or in part, amounting to 51.9 per cent of the aggregate net worth of these business owners. Among this 81.7 per cent, 82.5 per cent currently take an active management role.

Internationally, not only has there been a dramatic upswing in the number of ultra high net worth individuals (UHNWIs), but many developed nations have a higher per capita ratio of top-tier wealthy than the USA. According to the 2015 edition of *The Wealth Report* (Knight Frank, 2015), in 2014 there were 172850 individuals around the world with net worth of US$30 million or more. This seems low since the 2013 Federal Reserve's SCF indicates that there are 125998 households in the USA alone with net worth of at least US$30 million. This difference is due in part to the fact that the SCF measures households, in which it may take the net worth of two or more individuals to reach US$30 million. Regardless, *The Wealth Report* anticipates that over the coming decade there will be a 34 per cent increase in individuals with net worth at or greater than US$30 million, with staggering growth in many developing nations. If nearly 80 per cent of the top 1 per cent and nearly 82 per cent of the top 0.1 per cent in the USA are business owners, we may assume that this at least is true in established developed nations, more true in countries that have joined the developed world in the past 30 years, and even more true for the rise of the UHNW population in those less developed nations with favourable conditions for above-average economic growth. For instance Vietnam's UHNWIs are forecast to grow 159 per cent in the next decade.

The second context of the chapter draws attention to how foundational treatises on the institutional aspects of capitalism all address the social-psychological dispositions that form personalities. Marx and Engels's (1848) appraisal of entrepreneurs is well known. Capitalists are harbingers of the progressive bourgeois revolution but with the emergence of wage labour, become exploitative and produce the contradictory forces leading to the demise of capitalism. In addition to such political-economic analysis, Marx and Engels's historical materialism includes much about the social-psychological consequences of a mode of production. In *A Critique of the German Ideology* (1846), they write that the:

> mode of production must not be considered simply as being the production of the physical existence of the individuals. Rather it is a definite form of activity of these individuals, a definite form of expressing their life, a definite mode of life on their part. As individuals express their life, so they are.

In *The Theory of Moral Sentiments*, Adam Smith (1759) speaks of employers as the key to the modern wealth of nations; organizationally they seek their own interest in a way that leads to the rising standard of living of others. Those on the path to riches may at first be tempted to make accumulation of opulence their goal. However, self-consciousness propels their enterprising endeavours. Their inner spectator teaches that gathering wealth is an intermediate end whose ultimate end is to position them to acquire approbation and deflect disapprobation. The only true advantages 'we expect from that great purpose of human life that we call "bettering our condition". . .are being noticed, attended to, regarded with sympathy, acceptance, and approval'. The rich person is pleased 'that he has made himself worthy of their [his fellow creatures'] favourable regards'. This leads to a 'consciousness of merit, i.e. the consciousness of deserving to be rewarded' (Smith, 1759, p. 28).

For Weber the orientations and purposes of entrepreneurs, wrought first by the Protestant ethic and then the spirit of capitalism, brought about the institution of capitalism. Every aspect of Weber's historiography has been challenged. But still standing is his notion that capitalists were motivated by the value-rational (*wertrational*) form of social action – a 'belief in the value for its own sake of some ethical, aesthetic, religious, or other form of behavior' (1978, pp. 24–5). Weber explicates his thesis that making money derives from the value-rational worldly inner calling of Protestants (cf. 1978, p. 541) to strive for expanded accumulation. In terms of this chapter, Weber suggests, in a way Smith might agree, that *making a self* is the source of a personality fashioned for *making money*. And Marx is clear that every mode of production is not just a 'physical' reality but also for its inhabitants a 'definite form of expressing their life' (Marx and Engels, 1846) that defines their being.

THE PRODUCTIVE SECRET: ALIGNMENT TO THE RULES OF ENTREPRENEURSHIP

Entrepreneurs must become aligned to specific socially given rules surrounding how money is invested in a business to earn profits. Drawing on Bourdieu (1984) I refer to these general rules of the market and capital accumulation that frame wealth acquisition as a 'structured field of possibles', the external, objective conditions for successful entrepreneurial activity. I call these rules the 'productive secret of money' because entrepreneurs must unearth and embrace these rules as cognitive insight and emotional intuition.

The Rule of Market Imbalance: Finding Where 'There is a Need and the Supply is Zero'

The first rule of money that prospective entrepreneurs must obey is that successful generation of above-average rates of return depends on their locating an objective imbalance of supply and demand in a particular product market. 'We tried, in each case, to solve problems that needed solving', says Detroiter Seth Arvin,[2] explaining how his chemical company grew from a basement enterprise to a multimillion-dollar public corporation. Discipline, hard work, risk, and some more or less stringent control over other people's work may well become crucial at some point for abundantly harvesting the fruits of entrepreneurship in the form of wealth. But none of these virtues automatically leads to success, argues respondent Brendan Dwyer. The starting point of any venture, he insists, is locating a potential market vacuum or unfilled need where 'there is a need and the supply is zero'.

The Rule of Ideas to Close Market Gaps: 'From Such Ideas Fortunes are Fashioned'

Dwyer also enunciates the second rule: to derive the ideas and muster the resources to resolve this imbalance between need and supply. It is a matter of grasping *ideas* – discovering the insight about just how to solve the gap between demand and supply. 'It is from such ideas', he contends, 'that fortunes are fashioned'. There was a void 'for a certain type of specialty insurance', he explains, and when 'we exposed the market place to that idea. . .the market place responded'.

Sometimes the idea is encountered through happenstance or suggested by someone else. But it is not just an idea that gets recognized, it is the idea as *opportunity*. Commenting about his decision to begin his book distribution company, Dwyer remarks, 'This was not my idea, but I recognized it'. Such ability to discern and act upon opportunities in good ideas requires a certain enterprising sensitivity that many entrepreneurs proudly extol as one of their distinctive personal attributes. Says Vincent Pierce, founder of a series of successful electronics ventures, 'I felt I could see more clearly than others the frontiers of technology and that was clearly the best opportunity'.

The Rule of Affecting the Rate of Return: 'Above Average Returns Without Above Average Risks'

In addition to identifying a specific product market where demand exceeds supply, the entrepreneur must uncover and embrace a second objective rule of money. This second rule of money distinguishes the entrepreneur from other types of investors such as venture capitalists, futures traders, or long-term bond investors. In investment theory, the concept of expected value or probable return on an investment summarizes the outcome of the complex relationship between the amount, duration, level of risk, and rate of return of an investment. Generally speaking, higher expected rates of return are associated with higher risk. The distinctive aspect of entrepreneurship, says respondent Dwyer, is that 'above average returns without above average risks can best be obtained by adding your own intellectual capital to your money in such a way that the total return will be greater than the sum of the returns on each'. The unique characteristic of entrepreneurship as an investment strategy is its ability to offer – indeed, require – the active engagement of the investor in producing the return on investment. It is not surprising, then, that entrepreneurs so committed to actively managing their incipient investments should point to specific inspirations, hunches, commitments, sacrifices, and breaks as the key to their success, rather than to their adherence to abstractly formulated rules of money.

The Dialectic of Virtue and Fortune

For instance, the first rule about locating a market imbalance is often formulated 'as being in the right place at the right time' or as having received a lucky break or fortuitous lead. The second rule of efficacious creation of one's own rate of return tends to be enunciated as various practical 'keys to success'. Respondents attribute their prosperity to hard work, product design, quality control, proper treatment of customers and employees, and other business practices by which they distinguish their distinctive contribution. I now turn to the accounts offered by the respondents about how they founded and run their businesses.

Entrepreneurs must follow the rules, roles, and relationships of the market and capital accumulation, and must heed a corresponding personal discipline. The entrepreneurial process of self-construction and world building parallels the identity formation process by which the inherited struggle to overcome their servitude to the rules of inheritance in an effort to make money serve their desires (Schervish and Herman, 1988). The entrepreneurial process takes place, like the inheritor process, always within the larger framework of an individual's life history and often entails one or more liminal (Turner, 1969) or transitional periods of tension, hard times, questioning and separation. But if the inherited process is characterized by the use of virtue to gain ascendancy over wealth, the entrepreneurial process is marked by the application of virtue to derive wealth.

BECOMING AN ENTREPRENEUR

Phase 1: Great Expectations

The first stage of the entrepreneurial journey revolves around the development of the aspiration for financial success. It generally extends from the period of youth through the acquisition of a first job, and sometimes even without a first job. It is a time to internalize the desire for financial independence and to recognize that working for someone else as an employee is the major impediment to fulfilling that desire. Later phases of the process find the entrepreneur applying insight to action. But most important in the first phase is gathering insight: recognition of ambiguity, contact with the entrepreneurial spirit, and a prefigurative vision of themselves as entrepreneurs. The 'early years' represent the initial inculcation of *virtue* and the first efforts to overcome the hand dealt by fate or *fortune*. The neophyte begins the long process of internalizing a drive for success, the habit of hard work, the confidence to take risks, and an orientation of delayed gratification.

Images of humble beginnings

Some younger entrepreneurs have enjoyed affluent backgrounds or, like Donald Trump, have inherited a smaller business that they parlay into larger enterprises. But the entrepreneurial biography generally begins with an account of humble beginnings if not actual poverty (compare this with the discussion by Schulz and Hay, Chapter 8 in this volume). Even when there is some evidence of a secure and even affluent financial background, the respondents emphasize how, as Jesuit John LaFarge (1954) entitles his autobiography, 'the manner is ordinary'. Allison Arbour, who runs a major Midwestern US advertising firm, says that her father's position as director of purchasing for a small manufacturing company only made her family 'middle class. . .if we were that far up the ladder. We always had food, clothing, and all the good things'.

While those of modest financial background stress their relatively humble beginnings, those who actually endured poverty in their youth highlight it more emphatically. Such respondents speak of economic hardship in their childhood homes; a first-hand knowledge of the perils of financial insecurity suffered by family and friends; and the experience of seeing a family business go under or a parent – usually the father – lose a job, leave home, or die. Respondents introduce such hardship passages not just to fill in the early years of their biographies but also to emphasize their active virtue in overcoming obstacles on the road to where they are today. Ethan Wright voices this contrast between origins and destiny as knowing that as a kid 'we were quite poor' but 'we always had a clean house and. . .enough food to eat'.

A saga of parental inadequacy and family mobility in search of better opportunities precipitated the inspired drive for economic achievement and material display of Roger Ulam, currently riding the crest of a billion-dollar business. After his father died at the age of 29 and 'I was four going on five' with a mother unable to support the family, Ulam spent the next 13 years shuttling between boarding schools and relatives.

The mobilizing sting of poverty or financial insecurity can derive as well from early working life rather than childhood, recounts William Erwin, a medical supply wholesaler from New England. He grew up in a 'comfortable' lifestyle, but is 'still carrying psychological scars from having been poor' in his early work life because 'I was a father. . .having a child every year for seven years. Kind of being locked into a financial situation that was very mediocre at best. . . We were always on a kind of a roller coaster'.

The self-made generate the legitimating notion of 'deserving rich' at the outset of their accounts and weave this theme throughout their narratives (compare with problematic meritocracy discussed in Schulz and Hay, Chapter 8 this volume). Indeed, the story of financial achievement begins for most of our respondents with various vignettes of misfortune or, at best, with recollections about the even-handedness of fortune. To a person, respondents insist that they were not particularly privileged or spoiled in their youths. They tell of their own interventions to better their lot. Their activity to make more of their starting point in life is something they tend to be most proud of. To explain their achievement as deserved, our respondents devote narrative time to recounting challenges and hardships of fortune and recount how they exercised virtue, discipline and effort to make more of the hand dealt to them.

Frontier virtues

The respondents, living under the cultural umbrella of Calvinism and the Enlightenment, adhere to a belief in the relatively unlimited potency of *character* for shaping fortune. The early years arm respondents with the capacity to combat the vagaries of fortune. They teach the efficacious practice of virtue, what Aristotle concludes is the habit of doing good, or what one respondent, Dale Jayson, defines negatively as 'developing the habits of doing the things that non-successful people aren't doing'.

These entrepreneurs invariably emphasize how they have inculcated frontier virtues such as hard work, the ethic of investment, simplicity of lifestyle, thrift, care for others, and delayed gratification. Lamenting that his own kids are 'never going to know the atmosphere of Smalltown, USA', Russell Spencer extols his own 'disciplined upbringing' that included the family motto that 'you've got to go to college', the childhood injunction against going out on school nights and staying out late, and the expectation that he would work every summer.

If school learning is discounted, the object lessons of adolescence are acclaimed as major formative experiences. Spencer always remembered his father's gentle but firm admonition to him at age ten that 'people don't like to be called Polacks, Dagos, Wops, or Niggers or anything like that'. 'Smalltown USA is interesting', he adds, in how it shapes your values because 'you know the bank president and you know the milkman by first name. And the cop knows you and he knows your dad. And there's a sense of "you'd better not screw up" because it's going to get back to him pretty quickly'.

Strength of character

Underlying the array of *specific* virtues constituting the moral character of the entrepreneur is something even more crucial. This is the appropriation of the 'active' virtue of continuous self-improvement through disciplined training, what Machiavelli calls *virtu*, that embodied power or force that holds fortune 'in check' ([1514] 1961, p. 130). To build a business becomes not just a way to earn a living or become financially secure, but also a daily moral test. For respondent Roger Ulam, the disciplined development of a moral self began, appropriately enough, during his stint in the Marine Corps where, as he puts it, 'I saved most of the money I made. I didn't go out on liberty and raise hell like the other guys did. . . I spent a lot of time in the library reading and studying. I came up with my five priorities which are spiritual, social, mental, physical, and financial – in that order'.

The rigorous process of what Dale Jayson calls 'self-actualization' boldly exemplifies such self-generation of virtuous character. Jayson's dramatic narrative captures especially well the development of self as an active repository of efficacious power. Disillusioned that white team members were getting played ahead of him in high school, Jayson learned from his father, 'Son, you got to be twice as good or maybe three times'. This lesson, proffered by 'a guy with no formal education', showed him 'where the boundary is' between success and failure. It firmly embedded 'the philosophy that whatever it takes, you do it'. 'Strive for perfection', admonishes Jayson. 'The closer you come to being perfect, the more secure you are'.

The prefiguration

The interviews do not provide the kind of information that would enable us to compile an exhaustive list of the determinants of a successful entrepreneurial career. However, several key elements can be identified. For instance, one frequent element of an entrepreneur's background is some kind of direct encounter with the workings of entrepreneurship. This often occurs in childhood through the activity of a family member, but it sometimes comes through their own youthful entrepreneurial initiatives. As William Erwin intimates in his interview, 'My dad was a commissioned salesman selling paper products – a very, very avid reader – and helped me with having kind of an entrepreneurial spirit'. As a youth, recounts Ethan Wright, he witnessed his father move in and out of a couple of businesses, but Wright really cut his teeth by working for his uncle who looked upon him 'as the son he didn't have'. Ulam, who grew up without such familial role models, secretly adopted McDonald's founder Ray Kroc as a remote surrogate mentor.

These models implant the belief that agents need not simply find a place in the world but can wilfully found the world itself. Part of that personal self-formation that accompanies entrepreneurship is the creation of a strong individuality defined by our respondents as the anticipation that they can mould the world to their interests. This is the great expectation – not some naive hope for some unknown beneficiary to provide an inheritance, but the purposeful self-directed quest to form a worldly domain of principality commensurate to their expanded individuality.

Such a demanding sense of self derives not only from contact with enterprising parents or surrogate mentors. It also arises from internalizing the entrepreneurial path, often at an early age. A remarkably consistent finding is that virtually every one of our entrepreneurs report, without our prompting, some youthful great expectation to become successful in their own business. 'When I left Cincinnati I wanted to be a great football player, and I wanted to be a millionaire', says interviewee Raymond Wendt, putting it as directly as possible. Prospects for financial principality become focused in a prefigurative entrepreneurial identity often accompanied by an experiment of rudimentary entrepreneurship.

Respondents consistently recall a youthful projection of themselves into a financially secure future, 'to sightsee it out', as Jayson phrases it. They envision themselves as thriving entrepreneurs or professionals, establish this as their life's goal, and set out to attain their dream. 'I was a teenage tycoon in my head', proclaims Ulam who engaged in numerous money-making schemes going back to high school. Electronics entrepreneur Pierce says, 'As early as my junior year in college I had the vague but distinct idea that some day I would like to start a company'. Eva Radkey testifies to the abiding power of the vision by citing how an earlier dream to produce a speciality pastry became actualized only as she floundered for purpose amidst a devastating mid-life crisis. And like a moth drawn to the flame, William Erwin just could not resist the call to entrepreneurship: 'It wasn't so much that I had to do it; but something inside of me wouldn't let me not do it'.

Phase 2: Breaking Away – The Limits of Working for Others

None of the transitions between phases can be cleanly demarcated, especially as the incipient entrepreneur makes the transition from employee to self-employed. In this stage the unsettled employee undergoes the tension, uncertainty and self-testing associated with the exercise of virtue. Such virtue is required to train would-be entrepreneurs in the skills and discipline needed to risk striking out on their own and to scan astutely for a lucrative opening.

Liminality: 'without a life-preserver in the shape of a salary'

Armed with the push of childhood economic insecurity, the pull of their prefigurative expectation, and the virtue of discipline and training, our respondents enter the world of business with individuality in search of a principality. It is no exaggeration to say that the search becomes a moral quest eventuating in a virtual redefinition, not just of themselves as creative economic actors, but also of their whole perception of the productive capacity of money. The first step in this self-evolution is the transition from the relatively secure status as an employee within an enterprise of someone else's making to the more precarious status as an entrepreneur within an enterprise of one's own making. 'The businessman pure and simple', commends Andrew Carnegie, 'plunges into and tosses upon the waves of human affairs without a life-preserver in the shape of a salary; he risks all' (Rischin, 1965, pp. 4–5).

Some respondents make their initial foray into 'the waves of human affairs' directly as entrepreneurs, often in some family business, but many follow the more common path of trying to fulfil their aspirations first by taking jobs that draw on their training and interests. With only a few exceptions, respondents received college degrees and many pursued graduate work. They obtain favourable employment placements with potential for long-term careers and financial advancement.

Despite such auspicious prospects, respondents uniformly come to question whether alignment to the rules of money that govern them as employees can fulfil their heightened expectations for financial security or personal independence. This period of liminality begins with entrance into the dual process of disaffection from current conditions of employment and attraction to the alternatives of 'buying one's own time', as William Erwin puts it.

Dissatisfaction with a particular job is often given as the precipitating event, even though something more positive is almost always going on in the transition from employee to entrepreneur. For instance, Russell Spencer says, 'I got out of college and I ended up in the bank, which to me then sounded very appealing. And I don't think I really understood that there was no money in it'. Many other respondents also begin their disaffiliation by focusing on the monetary considerations. They complain that being employed too stringently limits the upper boundary of income and wealth they can aspire to. But other aspects of self-employment become cherished as well. 'By and large', explains Erwin, 'there were certain benefits that legitimately accrue to an owner that are not available to someone whose income is totally [derived from] working for someone else. . .[such as] car expenses and things of that nature' but these are 'nothing of a grand nature', he continues, in comparison to the fact that 'your time was your own; you were building something'.

The fact is that prospective entrepreneurs simply *outgrow* what even the best employment position has to offer. Growing dissatisfaction with an employment position becomes translated not into moving to another job but into moving outside of the employment relationship altogether. The key to the transition from employee to entrepreneur, then, is not any specific job dissatisfaction but the stark recognition that their expectations for financial security, autonomy and personal happiness cannot be satisfied if they sell their time to someone else rather than buy it themselves. Harbouring great expectations certainly makes our prospective entrepreneurs less content with any employment constraints, but it is their already strongly developed individuality that enables them to consider that working for others in any capacity is the foremost barrier to surmount. The liminal transition from employee to entrepreneur is thus a new way of acting. But even more fundamentally, it is a new self-understanding, a new way of being.

Liminality: the search to quell the 'restless hope'

Those who must take leave of their status as an employee as well as those who move directly into entrepreneurship must enter an interlude of virtuous search for the appropriate entry into entrepreneurship. This search becomes a vocational quest, especially for the neophytes who must track down their initial exposure to entrepreneurship in contrast to those who need only select among the opportunities already within their purview. This quest to find the proper inroad is a relentless search. Aspiring entrepreneurs struggle for cognitive insight into how best to position themselves to fulfil their aspirations, to quell what Matthew Josephson, the renowned chronicler of the 'Robber Barons', calls their 'restless hope' (1934, p. 33). After terminating a successful but ultimately unsatisfying tour of duty as a 'peddler' in his uncle's business, Ethan Wright tells how he began to 'figure out what I really wanted to do with myself'. He was sure of only one thing: 'a drummer keeps beating [in my head]: "financial independence, financial independence"'.

The prelude to the productive secret

The search to fulfil the drive for individual autonomy eventuates in entrepreneurs learning the productive secrets surrounding a specific market segment: the location of a market niche where demand outstrips supply, an effective idea to close that gap, and the necessity of committing their own efforts to obtain that high level of return on investment that distinguishes entrepreneurship from other forms of investment.

Viewing the world in a peculiarly distinctive way is prior even to the determination of a market niche. Before learning a business, the entrepreneur learns a way of scrutinizing the world. 'A quality of iron enters the soul', Josephson says; the future entrepreneur 'acquires a philosophy suited to opportunities' (1934, p. 35). 'It takes a certain kind of desire', explains Wright. 'There are people who will look at land and say there's a wonderful place to grow roses or to have cattle roam, but [the real entrepreneur] will only look at it as though it were dollar bills: "land I should have bought, land worth this but it could be worth that"'. In Marxist terms, this is the shift from perceiving goods and services as use values to perceiving them as exchange values, that is, as commodities valued for their market capacity rather than for the value of satisfaction they bring.

They are produced and sold not according to the logic of fulfilling personal enjoyment, but according to the logic of expanded accumulation.

Learning to approach the world for its exchange rather than for its use value can rightly be called the productive secret of the productive secret. It is the grasping of a new truth, the attainment of maturity by the capitalist mind. It is the inner voice, says Wright, associated with 'the way I've been trained and [how] my mind works'. It says 'if you do it this way you'll make money'. Market analysis is not just an interesting insight, it is also the recognition of an opportunity for action. Chicago hotel supplier Benjamin Ellman says, 'The difference between successful people and very successful people, I've always felt, is. . .a unique ability to take advantage of opportunities. Everybody gets opportunities but most people don't recognize them as opportunities'.

The break

Not only does this second phase itself vary in intensity and duration. It also varies in how many of its substages different respondents go through. The major factor extending or truncating the second phase is how prominently the testing period of liminality just discussed figures in the business biography of the entrepreneurs. That is, how challenging are the initial roadblocks that those on the march from employee to employer must skirt on the road to success. The greater the impediments to be overcome in accumulating either human or financial capital, the more we hear a story of liminality and virtue, and the more the early stages of Phase 2 are emphasized. The more our respondents enjoy the benefits of assistance and leads, the more they experience a relatively smooth non-liminal transition, the more they recount a story of fortune and breaks surrounding the last stage of Phase 2. In virtually every instance, the retrospective accounts cite – and usually emphasize – the benefit of a break at the onset or at some crucial turning point of the entrepreneurial career.

Fortune is viewed as most generous by those who have had 'greatness thrust upon them' – as Malvolio proclaims in Shakespeare's *Twelfth Night* – even though they have neither weaved nor toiled to earn it. Relatives, friends, or individuals encountered by chance, provide a relatively smooth transition from a general aspiration for economic independence to a concrete apprenticeship in the business world. These informal mentors offer unsolicited partnerships, investment opportunities, ownership positions in small or fledgling businesses, or just plain good advice. Ellman, for instance, stumbled upon his prospects when, in the early 1950s as a budding manufacturer of hotel supplies, he felicitously heeded the advice of a chance acquaintance he met on the road. This stranger suggested that he get back in his car and visit the founder of a chain of highway establishments known as 'motels'. It turned out to be the beginning of a long friendship. A deal was struck that day to provide speciality furniture for every motel to be opened in that chain.

Phase 3: Making it and Making a Self

Having come to align themselves to the rules of entrepreneurship, the increasingly endowed entrepreneurs now begin to align entrepreneurship to themselves. As they 'make it' financially, a peculiar empowerment of the wealthy begins to take hold. Instead of the rules of the world being represented in the life of entrepreneurs, the will of the entrepreneurs becomes represented in the specific institutional shape of their businesses and in the personal shape of their biographies. As they come to 'make it' we find that entrepreneurs are making their selves and making the world in their image. Such ability to manufacture the environment for oneself and others is the deific power of creation – a capacity so broadly and purposefully exercised from this point on to warrant for our respondents the designation of 'monarch' and their domain that of 'principality'.

FOUR SECRETS FOR PERSONALLY TAILORING THE PRODUCTIVE SECRET

Strategic secrets: learning the ropes

Once at the helm of an enterprise, the entrepreneur is no longer just a worker or investor but what Carnegie terms a 'merchant' or 'maker'. But being a maker means more than 'to make some something tangible and sell it', as Carnegie defines it (Rischin, 1965, p. 4). It means also to make one's environment as well as to fashion oneself into an empowered being. Entrepreneurs must locate and work to their advantage some imbalance between supply and demand. Within this requirement, however, 'making it' becomes a highly personalized endeavour.

With remarkable consistency, our respondents proudly recount the strategic secrets of their success. These are the set of specific investment, labour, production, and marketing strategies derived from their budding individuality and to which they attribute their success. Our respondents withhold no secrets about their formula for their success and are ready – even anxious – to reveal the trade secrets of their entrepreneurial and managerial achievement.

What eventually becomes a defining characteristic of the wealthy – control over their destiny – begins as a series of strategic lessons to be learned about how to conduct a business and comport themselves so as to retain a competitive edge and, just as important, feel positive self-regard about their accomplishments. Ralph Pellegrino is the owner of a retail hardware chain who gets involved in every detail of the business and says, 'The thing I enjoy most is being on the floor with the customers. . . I'll go up and put my arms around them, and I do love people. I could pick up a thousand things that were wrong, the feel you have, mostly because it's customer-oriented like that'.

In addition to being customer oriented, the strategic secrets of success that we hear about include treating employees with respect, providing good working conditions and benefits, producing high-quality products, working hard, and, as David Stephanov counsels, simply being tougher than anyone else. 'There's four ways to get wealth', he suggests, 'You inherit it, you work for it, you borrow it, or you steal it'. He assures us, however, that the only way he got wealth 'was to work for it'.

While Stephanov is tougher than others, Rebecca Jacobs is quicker and more engaged. The Detroit-based importer attributes her success to four strategic secrets that she strings together in her account: (1) work in a way that is principled and self-engaging; (2) hire and associate with the best professionals; (3) resist any temptation to indulge in self-pity for being a woman in a tough business environment; and (4) have the creative mentality of a winner. Jacobs avows, 'I set my own goals. I'm a dreamer, I think that's one of the big assets that I have. I just know I'm going to win'.

Although Allison Arbour later 'did some time' – as her father put it – at a major university, this prominent entrepreneur who became the first woman president of her city's major business organization simply 'walked out of college' never to earn a degree. She learned from being defrauded by two early partners: having 'to hustle big' under the pressure of commission work, diving right into new projects, and working alongside the 'old war horses' in her industry. Her key to success has been her readiness 'to get my hands dirty' and to say 'let me try that'. She says 'I had to touch people in the business. . .I guess, to understand it. Maybe, if I had enough educational background to comprehend things, I wouldn't have had to work so hard to learn. But everything I learned, I learned by doing'.

Business success and the financial secret of money: 'the cream on top of the cake'

Establishing a financial principality coincides with crafting an individuality. Just as entrepreneurs pursue their expectations by materializing their individuality in the form of a business, the wealth earned by that business feeds their individuality. By learning how to translate business success into malleable liquid assets and then applying those resources to the fulfilment of their interests, they uncover the financial secret of money. In addition to learning how to invest money to build a principality and individuality in the world of production, they learn how to build a principality and individuality in the realm of consumption.

Respondents who have learned the financial secret of money recognize that their empowerment in areas other than business depends upon learning how – and when – to disinvest time and money from their businesses in order to foster other goals. They reveal at least an implicit knowledge of this financial secret in how they translate business success into an even bigger financial success. They talk about 'knowing when to get out by selling the firm', 'going public', 'and 'liquidating' all or part of their business assets. What makes them successful is learning the productive secret of money. What makes them consciously empowered as 'wealthy' or 'financially secure' in a broader sphere of life is learning that just as money can be put into a business, it can also be taken out to serve other desires and interests. Discovering that they are in fact wealthy, that indeed they have *disposable* income, is an important prerequisite for the wealthy moving from being disposed over by their money to disposing over it.

Only as his business got 'going nicely', and he re-emphasized his insatiable drive to save and invest, recalls Ethan Wright, did the drum beat of 'financial independence' become 'quieted down'. He returned to his musical interests and spent time doing things with his family because he 'never knew what it was like not to work Saturdays'. Then, he recounts 'a couple of dramatic changes started taking place'. First, he realized that his business had been a success, that he had become an accomplished practitioner capable of maintaining that success. And second, he recognized that he could consider himself wealthy if he were able to transform his fixed capital into liquid assets.

It was precisely 'when we decided to take the company public', explains Arvin that he first realized that he was wealthy. Up to this point, 'I did have money, but it wasn't a lot of money', the chemical entrepreneur recollects. 'We still lived in a modest house and our kids had no idea that we had any money'. When the company went public, 'my wife bought me a Rolls-Royce because she owned twenty per cent of the company'. From that point on, things changed. 'After I got the Rolls-Royce people began to treat me with respect and I acquired some prestige. I guess maybe that's when I realized that money was power'. When the Park Company approached him for the sale, Arvin wondered 'what would this do for us?' His answer is typical: it would 'give our children security, things to do, what they want to do; give us a chance to be philanthropic, to do the things I wanted to do for the community'.

Eva Radkey learned the financial secret as two New York investors purchased her speciality pastry chain of 18 outlets. At first, she resisted the buy-out but after six months of consideration she agreed to the offer, with two conditions: that she continue a relationship to the company 'doing the promotion and the PR' and that the company 'stay with my name' on the logo. Like Arvin she discovered that she now had discretionary income. She put some into real estate but most she distributes generously to her children. 'I'd rather help my children while we're still alive, not when I'm six feet under'. When asked whether she considers herself wealthy, however, Radkey responds by redefining wealth as health. 'Yes [I am wealthy], as long as I have my health. You can have all the money in the world but if you don't have your health and happiness. . .'.

Like many others who have reached financial security, Jacobs repeats the refrain that 'money does not buy happiness', health, or 'being a nice person'. Voicing a contemporary version of Smith's (1759) notion that wealth first appears as attraction to possessions but gravitates to deeper functions, Jacobs expounds on what money *can* buy. 'I love the money. . .I love all the things it supports. Money gives peace of mind. It gives you security. You know you're going to eat and you know you're going to be able to do all those things'. Her poor childhood in Detroit, she clarifies, did not make her unhappy. 'I have wonderful memories of singing and going to the beach and going to Belle Isle and sleeping outside because it was too hot in our house and we all got together'. Still, she is pleased that 'money can. . .make your life easier for you' and for family members. For instance, she was able to fulfil her dying father's lifelong wish to visit Israel and to provide him with six weeks of the best of care at the Mayo Clinic. Financial success breeds a reconsideration of interests. Like Jacobs, for whom wealth 'also made me realize how short life is and how important family is', many begin to reach for a new conception of themselves and their financial vocation.

Phase 4: Renewed Quest for Principality and Individuality – The Kalpataru Tree and the Spiritual Secret of Using Money for Non-material Purposes

In the fourth phase the empowered entrepreneurs take up the quest to discover and carry out a deeper set of interests. This, we shall see, does not mean they abandon their business and investment strategies altogether; only that the purposes to which they apply their empowered individuality become broadened into a principality based on a fuller range of religious, humanistic, political, or social goals. Locating these non-material interests and dedicating personal and financial resources to realize them is to learn and apply the spiritual secret of money.

The spiritual secret of money

The spiritual secret of money is the deeper hidden ability of money to liberate entrepreneurs from the demands of the productive discipline of money. Once wealth is achieved, it is possible to reverse the causal nexus between subject and object. Learning the financial secret moves entrepreneurs from 'being consumed' by money to consuming it in accord with their desires. Commitment to the spiritual secret takes things still further. It is not everyone who learns or seeks to learn the spiritual secret of money. Those who do, however, move from attending to the quantity of their interests to the quality of their wants, including self-development and philanthropy.

Two dynamics converge in the lives of those for whom enterprise has brought financial security and personal confidence. The first, as we have already seen, is the incontrovertible fact that those empowered by wealth can buy what they want – at least in the material realm. The second is that entrepreneurial success often induces new and different wants. This is neither automatic nor easy for the wealthy. As New York interior decorator Carol Layton explains, the temptations of wealth are legion. 'I love having money', she remarks. 'It makes it possible for me to do whatever I want to do'. Having attained financial security, however, raises the question about what she wants to do with her resources. 'First you have children and you want to help make them financially secure. And two, if you're like we are, I have a need for ongoing things. I want the business to go on. I want to build something that doesn't die when I die'. But at this point in her life she begins to explore a philanthropic perspective: 'I get more active in philanthropic organizations', whereas she could not even think of this in the process of accumulating her fortune. 'You get bogged down in keeping the business going'. After leaving Harvard without writing her thesis, Layton then 'went into something as crass as money making'. Gradually she came to realize that she 'had to do something' by way of a social 'contribution'. To do so, she explains, required her to 'divorce the money a little bit from how people feel about things' in a materialistic culture.

Looking only to what 'I want to do' is no way to live, counsels Boston builder, Charles Dore. 'There's so many people that are just into themselves and can't seem to get out of themselves, can't seem to see the broad picture', explains Dore. 'People don't want to see. They draw the curtain. You start to tell them about Ethiopia, or poverty and – and these people basically are good but they tend to pull the blind. How the hell do you get that blind up?'

Houston Smith (1958) summarizes the Hindu spirituality of riches in the story of the Kalpataru tree. This is the wishing tree that freely gratifies any expressed desire. It is impossible to say whether the moral economy of the Kalpataru tree will move any particular individual toward virtue or hedonism. What is clear, however, is that the intent of the tree's aphorism is to counsel us to reflect on the quality rather than the quantity of what we want.

Wall Street investor Ron Markewicz reveals the alluring invitation of the Kalpataru tree. Until now he has been an explicit proponent of free enterprise, in business and in life. In both, he says, people go it alone, with 'merit the only criterion'. He now grasps 'the sort of abstract notion that if you have had good fortune smile upon you, then you should sort of smile back'. Life 'balances on some cosmic scale', he explains. 'If you have had a lot of good fortune come your way, somewhere there are scales that are adjusting. When things get too far out of balance with what you've got versus your seeding, trying to put something back, then somehow there isn't any harmony'.

Renewed liminality: 'my problem was that I had made it'

The encounter with the Kalpataru tree entails a renewed liminality as the entrepreneur takes up a quest to chart a post-prosperity agenda. For Ethan Wright, coming to realize that his business had reached the point where it could offer him financial security 'was a cruncher'. 'It was like a seizure. . .I could do anything I wanted to do. But I didn't know what the hell to do. So the first thing I did was I hired myself a psychoanalyst'. As if standing before the Kalpataru tree Wright was no longer in search of empowerment or capacity but in search of direction. 'I had arrived at what the goal was, which was financial independence. . . So I had to find something to do. And what you end up doing, I think, what I discovered is you go to those things that truly interested you'.

Many entrepreneurs uncover the spiritual secret of money early on even as they build and consume their fortunes. For instance, Brendan Dwyer manages to formulate his religious version of the spiritual secret of wealth in the midst of his most prosperous business undertaking. 'My role is to attempt to harmonize my [business] activity with the roles that the Creator would like to see born', explains Dwyer, citing the norm of inner peace as the touchstone of the spiritual secret, 'and when I'm in conflict with that, then I'm going to have a certain tension and disunity'.

Still, most entrepreneurs mirror the course followed by Dallas insurance magnate Henry Nielsen. Only after he made his millions did he begin to reflect on where he should go from there. 'That's one of the things that I've got to make a decision on', he says in reference to making a shift to a philanthropic career. Even though he is 'probably going to put half a million or a million in a trust this year' he doesn't 'have the slightest idea' of what he wants to do with his charity. His liminal transition from business priorities to philanthropic concerns requires not only that he 'get somebody in to run the store, which I'm working on', it also requires that he learn how to 'play out' his involvement 'on the social scene'.

Economic vocation: 'to take on any kind of a project, you've got to have some wealth'

Nielsen is on the mark in recognizing the importance of wealth for exercising agency in initiating ventures. 'To take on any kind of a project, you've got to have some wealth', he advises. 'So you got to permit people to make wealth. You have to have it to do some good. If you want to go down and. . .do a new YMCA project, you don't go to the people that need the YMCA to do it'.

Rebecca Jacobs admits that feeling guilty about having so much money is part of 'why I give a lot to charity'. But her posture is more complex, drawing more on positive impulses of spirituality than guilt:

> Well, I guess it goes back to being lucky, really. I feel that I'm very lucky that I have so much. And sometimes I kind of feel bad when I think about our country, the way [the poor] live. I go out to baseball games now in Detroit, and have to go in a bad section [of town]. I don't say, God I feel guilty that I have so much and they don't. I always just think that I'm quite lucky and I always ask myself is there a little bit more that you could do to help others.

Jacobs's economic vocation revolves around locating her wealth in the contexts of a vertical relation to God, a horizontal tie to other people and a personal quest for self-fulfilment. 'I believe in God. There is a God that watches over all of us. Whatever religion you are'. She recognizes that she can have what she wants but has managed to include the welfare of others within the horizon of what brings her satisfaction. 'And I think that, in order that I can make a contribution and have a happy life, that what we have to do is help others. And that makes you happier. Maybe that's selfish, but it does, it makes me happier'.

Despite the testimony of Jacobs, Singer, Dwyer, and numerous others, there is of course no automatic or inexorable positive relation between wealth and spiritual depth. Wealthy entrepreneurs neither ask nor answer the question of deeper spiritual existence more frequently or better than anyone else. What is true, however, is that the path to a spirituality of money for the wealthy passes by the Kalpataru tree because, more than anyone else, they are capable of a financial choice. I find that among entrepreneurs the fullest spiritual development or individuality is reported by those who become secure in their economic achievement, transform their interests into deeper wants, and begin to devote themselves and their wealth to self-discerned humanistic or religious goals.

CONCLUSION

Although I write about the path to becoming a prosperous entrepreneur, with proper adaptation my analysis also pertains to the world's thriving poor who benefit from micro-loans and investment bonds. It also applies to hands-on philanthropists bringing innovations in products and processes to the world, community activists engaged in social movements, and non-profit and governmental officials who use their resources to close the gap between what others need and what they can supply. There is no question that taken together these initiatives have the potential to also create individuals with a means and a motive for accruing finances, and with the personal aspiration for the development of self, family, and community.

Embedded in my descriptive findings are three theoretical implications about the nature of entrepreneurship, ideology and agency. First, entrepreneurship is a career and a moral career connecting enterprise and existence. As entrepreneurs learn the productive, strategic, financial and spiritual secrets of money, they undergo a series of transpositions in consciousness regarding their desires, capacities and obligations. Second, the normative discourse through which entrepreneurs recount their stories reveals the nature of ideology to be an entire exposition of causal logic rather than simply an array of ideas. Entrepreneurs portray their stories as the unfolding of a disciplined struggle in which they apply virtue to overcome obstacles and to make more of their lives than what was bequeathed them at the outset. Obeying virtue makes them both moral and financially successful. Third, and most important, is that the fundamental class trait of the wealthy is to be not just agents but *hyperagents*.[3] The underlying commonality tying together all the critical versions about the wealthy is their capacity to shape the institutional environment within which they and others exist. Critical research on the role of the rich regards the vast number of wealthy and ultra-wealthy either, with Marxist theory, as exploitative owners of the mode of production; or, with elite theory, as clandestine architects of directions for the nation's culture, selection of political candidates, legislative decisions, and government regulations; or with consumption theory, as perverting societal values by their control of what is sold, bought and fancied. Common to each of these approaches is that the wealthy have the capacity to shape the world around them through the exercise of temporal, spatial and psychological empowerment (Schervish et al., 1994). Whether through building and decorating their homes, securing their investments, educating their children, carrying out philanthropic agendas, planning exotic vacations, purchasing political access, or owning the means of production, the distinctive attribute of the wealthy is their ability to conform their environment to their wills.

I view agency theory (Giddens, 1984; Sewell, 1992; Emirbayer and Mische, 1998), as arguing that agency is the ability to exercise strength of character within the constraints of fortune. As such, the wealthy generally, and entrepreneurs in particular, can be viewed as agents par excellence. In the dialectic of socialization and social construction, the wealthy are masters of social construction. What agents can accomplish only through social, political, or philanthropic movements, hyperagents can achieve relatively single-handedly. While agency entails finding the best place for oneself within existing rules and resources, 'hyperagency' entails the establishment of the rules and resources by which one chooses to live and within which others must live. Hyperagency is the unifying concept that captures ownership of the mode of production, domination by elites, and the production of consumption. Hyperagency is the class trait of all those with exceptional financial capacity; and is more than anything else – to borrow from Marx – the law and the prophets of wealth.

NOTES

* I wish to thank Andrew Herman my former graduate student for his invaluable assistance on the original research that led to this chapter. I am more immediately grateful to John Havens for providing the special analysis included in the empirical context and to Lisa Kaloostian for her insightful and competent editorial help in preparing this chapter. I am most thankful to Thomas B. Murphy for setting me on the course of my investigations on wealth and philanthropy, for his intellectual contributions, and his founding and long-term financial investment in our work.

1. The interviews were conducted as part of the Study on Wealth and Philanthropy sponsored by the T.B. Murphy Charitable Trust at Boston College from 1985 through 1988. The sample was distributed over ten metropolitan areas of the United States. Respondents were contacted through a branching technique whereby initial respondents referred us to wealthy friends and associates. Since this research, we have interviewed or surveyed over 400 more wealth holders and given the dramatic era of growth during which the original information was gathered makes it relevant to today's ascendency of entrepreneurs and entrepreneurship.
2. All names and, where necessary, places and types of business activity have been changed to preserve the anonymity of the respondents.
3. I am indebted to Andrew Herman for the formulation of this term.

REFERENCES

Bourdieu, P. (1984), *Distinction: A Social Critique of the Judgement of Taste*, trans. R. Nice, Cambridge, MA: Harvard University Press.

Emirbayer, M. and A. Mische (1998), 'What is agency?' *American Journal of Sociology*, **103** (4), 962–1023.

Ewing Marion Kauffman Foundation (2015), website, accessed 7 April 2015 at http://www.kauffman.org/.

Giddens, A. (1984), *The Constitution of Society*, Berkeley, CA: University of California Press.
Havens, J.J. (2015), 'Memorandum: net worth and entrepreneurs', Boston, MA: Center on Wealth and Philanthropy, Boston College.
Josephson, M. (1934), *The Robber Barons: The Great American Capitalists, 1861–1901*, New York: Harcourt, Brace and World.
Knight Frank (2015), *The Wealth Report*, accessed 7 April 2015 at http://www.knightfrank.com/wealthreport.
LaFarge, J. (1954), *The Manner is Ordinary*, New York: Harcourt Brace.
Machiavelli, N. ([1514] 1961), *The Prince*, trans. G. Bull, Baltimore, MD: Penguin.
Marx, K. and F. Engels (1846), *A Critique of the German Ideology* [online], Marx/Engels Internet Archive, accessed 6 April 2015 at https://www.marxists.org/archive/marx/works/download/Marx_The_German_Ideology.pdf.
Marx, K. and F. Engels (1848), *The Manifesto of the Communist Party*, accessed 7 April 2015 at https://www.marxists.org/archive/marx/works/download/pdf/Manifesto.pdf.
National Opinion Research Center (2013), *Survey of Consumer Finances*, Chicago, IL: University of Chicago.
Renz, D.O., M.L. Taylor and R. Strom (eds) (2014), *Handbook of Research on Entrepreneurs' Engagement in Philanthropy*, Cheltenham, UK and Northampton, MA, USA: Edward Elgar Publishing.
Rischin, M. (ed.) (1965), *The American Gospel of Wealth*, Chicago, IL: Quadrangle Books.
Schervish, P.G. and A. Herman (1988), *Empowerment and Beneficence: Strategies of Living and Giving Among the Wealthy. Final Report: The Study on Wealth and Philanthropy*, Boston, MA: Center on Wealth and Philanthropy, Boston College.
Schervish, P.G., P.E. Coutsoukis and E. Lewis (1994), *Gospels of Wealth: How the Rich Portray Their Lives*, Westport, CT: Praeger.
Sewell, W.H. Jr. (1992), 'A theory of structure: duality, agency, and transformation', *American Journal of Sociology*, **98** (1), 1–29.
Smith, A. (1759), *The Theory of Moral Sentiments*, accessed 7 April 2015 at http://www.earlymoderntexts.com/assets/pdfs/smith1759.pdf.
Smith, H. (1958), *The Religions of Man*, New York: Harper and Row.
Turner, V.W. (1969), *The Ritual Process: Structure and Anti-Structure*, Chicago, IL: Aldine.
Weber, M. (1978), *Economy and Society*, Berkeley and Los Angeles, CA: University of California Press.

8. Taking up Caletrío's challenge: silence and the construction of wealth eliteness in Jamie Johnson's documentary film *Born Rich*

Sam Schulz and Iain Hay

INTRODUCTION

In a review published in 2012, Javier Caletrío raised the question of how, in an era typified by a radical widening of the social and economic gap, academic concern with inequality remained so firmly fixed upon the poor. Echoing sentiments raised previously by Beaverstock et al. (2004), Caletrío argued that the very rich remain protected by a 'veil of silence' so effective as to ensure their 'invisibility and impunity' (p. 136). He called upon social science researchers to find new ways of examining social systems that seem, sometimes exclusively, to operate in favour of those who are already privileged. This chapter joins emerging efforts (e.g., chapters 2 and 5 by Koh, Wissink and Forrest, and Sayer in this volume) to take up that challenge.

For race theorists, turning the analytic gaze back upon privilege is not new: they have long valued the need to avert critical attention 'from the racial object to the racial subject' (Morrison, 1992, p.90) so as to reorient the sociological and geographical focus from 'victims' of racism and common sense assumptions of 'race' as synonymous with non-white people, to the prioritization of whiteness as an area of critical endeavour (Back and Solomos, 2000, pp.21–2). A key finding in this field is that mechanisms of privilege remain largely unseen and unexamined by those who benefit most from them, and this in turn advances a notion of 'everyday privilege', a phenomenon that is protected and reproduced through its routine denial (McIntosh, 2002).

In this chapter we make a similar theoretical inversion by turning our analytic gaze from an exploration of poverty and disadvantage to the privileges associated with super-wealth. Savage and Williams (2008) suggest there has been an absence of 'elite studies' and they ascribe this lacuna to the dominance of quantitative analyses that focus primarily on statistically significant groups. But it is also possible that the flagrant nature of super-rich privilege has driven a research gap: who needs to point out that the richest 200 (or fewer) individuals in the world having the combined wealth of roughly half the world's population is unfair?[1] Super-wealth of this magnitude seems so extreme as to be an anomaly.

WHY STUDY THE SUPER-RICH?

To better understand the nexus between privilege and disadvantage, we aim to give super-wealth deserved sociological attention. Along with Caletrío, several writers (Hay and Muller, 2012; Koh et al., Chapter 2 this volume) highlight the strategic worth of doing so. In Piketty's (2014) view, a focus on super-rich privilege necessitates a shift of focus from exploitation to accumulation as the central dynamic of capitalism. This is similar to the ways in which whiteness theorists have explored racialized disadvantage through deconstructing and unveiling unearned racial privilege. Such privilege is not necessarily the result of calculated manoeuvres but arises, in part, from everyday structuring and maintenance of racialized social relations. With respect to capitalist social arrangements, Savage (2014, p. 600) explains: '[A] focus on accumulation recognizes that class relationships are not zero sum games, that all agents, differentially positioned within society, develop sensible (in their own terms) strategies to secure and advance their position'. What, then, are 'sensible' (and therefore natural) ways of being for the super-rich? And if we acknowledge the interlocking nature of social hierarchies, what do naturalized expressions of super-rich power mean for the rest of us?

From a post-structuralist standpoint, power courses throughout society and all subjects have access to it (Blood, 2005, p. 48). Yet individuals occupying super-rich subjectivities have far greater access, and hence scope, to shape the world we share. In this sense, power is organized and circulates through ruling class lives, and today's super-rich have access to increasing concentrations of power.[2] The roots of this situation run deep and it is beyond the scope of this chapter to theorize a nuanced backdrop.[3] Suffice to say the super-rich have historically been enculturated to 'secure new territories or conditions favourable to their enterprises; to gain personal advancement [and] to make the world the kind of place *they* want to live in' (Donaldson and Poynting, 2007, p. 9; original emphasis). This is not to suggest that self-interest is exclusive to super-rich subjectivities, but rather that 'super-rich' self-interest carries with it far greater consequences.

It should also be noted, however, that the characteristics of eliteness to which Donaldson and Poynting allude move beyond a logic of individualism. Elite subjectivity in their view is socially produced, and the tremendous access to social power enjoyed by elites is 'catastrophic' insofar as it has far-reaching effects. For instance, today's wealthy elite control 'far more of the earth's scarce resources than any other class' (2007, p. 240), their everyday lifestyles are both 'hyperconsumptive' (Paris, 2013, p. 97) and 'hypermobile' (Budd, Chapter 15 this volume; Elliott and Urry, 2010), their field of action has consequently expanded (Wedel, 2009, as cited in Caletrío, 2012, p. 140), and where they choose to alight, the super-rich typically affect the physical and social landscape 'with scant regard' for anybody else (Hay, 2013, p. 10).

Put simply, the relative ability of super-rich people to affect society is greatly magnified; they consume more, control more, their problems and proclivities influence more, and while everyone has agency to promote degrees of social change, the super-rich simply have 'more'. For a growing number of researchers, the super-rich therefore constitute a valuable starting point when it comes to exploring new configurations of power and social inequality (Caletrío, 2012, pp. 136–7).

GETTING TO THE SUPER-RICH

Despite the value of studying elites, an obstacle for most researchers is access to them (Aguiar and Schneider, 2012; Hay and Muller, 2012). The very rich 'easily avoid too much scrutiny' and have mechanisms at their disposal that allow them to regulate outside access (Higley, 1995, p. 2). They are in fact 'surrounded by layers of social insulation' (Donaldson and Poynting, 2007, p. 66) that create 'a bubble of privilege' (Maxwell and Aggleton, 2010). For Wedel (2009), elites are thus elusive subjects and this presents methodological challenges for academics who, like us, lack the cultural capital or resources that might otherwise open doors.

While 'distance' may be favoured in some approaches to social research, too much distance can also prove problematic – as Probyn warns, validity and its usefulness must always be tested 'on our own pulses' (1993, as cited in Donaldson and Poynting, 2007, p. 14).[4] Defining elites and negotiating vast distances from them are, therefore, key challenges when it comes to studying super-rich lives.

For our purposes we draw from Beaverstock et al. (2004, p. 402) who, in terms of definition, note that while prosperity through inheritance was once a hallmark of super-wealth, a new class has arisen whose affluence is generated primarily through strategic investments. The participants in Johnson & Johnson heir Jamie Johnson's maiden film, *Born Rich*, which forms the basis of this chapter, draw from across these points of origin, with some being the successors of 'old money' and others beneficiaries of this new wave of strategic capital. All, however, share in a number of largely unearned concessions including that work for the super-rich is essentially optional and that they benefit collectively from a culture 'which celebrates and affirms their rites of accumulation' (Donaldson and Poynting, 2007, pp. 10, 167).

Whether their net worth puts them in the millionaire, multimillionaire, or even billionaire, bracket, in this chapter we thus view super-rich individuals as members of a class (however heterogeneous or globally dispersed, yet always particularized by local context) who share a number of excessive privileges, and a 'cohesive community with its own forms of. . .self-perpetuation' (De Camargo, 1981, in Donaldson and Poynting, 2007, p. 11), which we seek to examine by interrogating those super-rich individuals' 'sense-making'.

To circumvent issues of access, we borrow from Donaldson and Poynting's (2007) 'found' life history. Life history methodology holds that an individual life may be used as a resource for analysing society (Frankenberg, 1993). To speak of a life is to speak of the relations that produce it. Life history commonly involves lengthy, one-on-one interviews with an individual subject (Jackson and Russell, 2010) and one or more interviews as a basis for analysis. But, faced with the same dilemmas of access as us, Donaldson and Poynting (2007, p. 24) turned to autobiographies and biographies of ruling class men as their primary research materials, and attainment of data saturation to generate a collective portrait of ruling class masculinity. In other words, they regarded an explicit focus on the similarities and themes to emerge from the materials as common to the class itself.

This approach to class analysis (and, subsequently, to the social and spatial impacts of ruling class practices) takes us beyond rigid arguments over class boundaries. It embraces a view of class as 'lived reality' (Donaldson and Poynting, 2007, p. 10), and holds the individual subject as socially produced (Rose, 1998). In other words, individuals are both shaped by and capable of shaping social relations, yet incapable of stepping outside the relations that produce them.[5] This deconstructive standpoint turns on the notion that 'individuals' lives are the place in which societal changes are played out' and thus 'the actions of individuals make up the history of which they are part' (Donaldson and Poynting, 2007, p. 12). Life history can therefore help to illuminate socio-structural relations within which individuals are located, which shape the social logics that sustain ruling classes.

In exploring lives of the really rich, we wish therefore to use elite subjects' own perspectives as entry points for analysis. We seek insights to the worlds of the super-rich through learning how being 'born rich' is applied to social logics that underpin certain patterns of practice, while closing others down. This, then, is a simplified version of the found life history that co-opts Johnson's documentary rendering of the lives of some of his closest friends. And like the sociologically aware 'class traitors' whose writing deepened Donaldson and Poynting's understanding of 'filthy rich blokes' (2007, pp. 16, 12), Johnson's insider status affords us access to lives and lavish homes from which we would otherwise be excluded. Our interest, however, is not any empirical 'Truth', but the truths that Johnson's subjects deploy to make sense of themselves and others. To our thinking, this promises rich insights to the rationales and practices that shape and give meaning to (wealth) eliteness.

ANALYSING VISUAL TEXTS

The key text underpinning our analysis is Jamie Johnson's documentary film *Born Rich* (2003), set in the United States. *Born Rich* comprises interviews with 11 ultra-wealthy young people covering topics as broad as growing up, schooling, attitudes toward wealth, interactions with outsiders, friendships, relationships with parents, dispositions toward work, leisure practices, and beliefs about dating and marriage – lines of inquiry that might otherwise inform a life history interview. Owing to the insider/outsider (or autoethnographic) standpoint adopted by Johnson, he too appears as an interview participant, and his personal reflections and conversations with his reluctant father play out as a backdrop to the documentary.[6]

To this text we apply a mode of visual discourse analysis to explore what the world looks like for these rich young people and how power runs through their accounts. At the basis of such an approach is an understanding of discourse as more than just conversation: 'it refers to all the ways in which we communicate with one another, to that vast network of signs, symbols, and practices through which we make our world(s) meaningful to ourselves and others' (Gregory, 1994, p. 11). Discourses are neither fixed nor singular (Foucault, 1981) but overlap in competition for control of subjectivity (Ashcroft et al., 2000, p. 184). Discourse therefore provides individuals with multiple means of constructing the world, and our aim is to illuminate the hegemonic discourses underpinning the participants' narratives in sections of Johnson's film.

Consequently, we are mindful of the field of visibility that characterizes elites 'by what kind of light it illuminates and defines certain objects and with what shadows and darkness it obscures and hides others' (Dean, 1999, p. 41). In other words, eliteness is produced and reproduced in discourse and 'within the rules of a discourse, it makes sense to say only certain things' (St Pierre, 2000, p. 485). Within sets of power relations such as those that keep elite power and subjectivity in place, the concept of discourse thus illustrates 'how language gathers itself together according to socially constructed rules and regularities that allow certain statements to be made and not others' (2000, p. 483). And as Johnson demonstrates, a key defining feature of eliteness is the tenet *not* to talk about money.

Documentaries like Johnson's also represent a niche; they are critical attempts to illuminate aspects of society that might otherwise remain hidden. And while a new wave of documentaries and 'reality' television (e.g., BBC2's 2015 *The Super Rich and Us*; Bravo television's 2010–15 *The Real Housewives of Beverly Hills*; CNBC's 2014 *Secret Lives of the Super-Rich*) has since saturated Western media with exposés of the lives of the rich and famous, *Born Rich* is an early, earnest exploration of wealthy lives by one of the ruling class's own. Albeit limited, our analysis contributes to Johnson's work by locating his film within a broader context of contemporary theorizing on eliteness.

Through immersion in the text and through identification and transcription of key themes and passages (see Rose, 2012) – categories that are delimited, in our case, by the story the filmmaker wishes to tell – we are able to consider how silence operates as a complex cultural practice that naturalizes super-rich power; how being super-rich is conceptualized, justified or problematized by super-rich individuals; and how a range of practices and structural relations work conjointly to reproduce ruling class status.

SILENCE AS CULTURAL PRACTICE

The 11 participants in *Born Rich* include Ivanka Trump (real estate heiress), Georgina Bloomberg (media heiress), S.I. Newhouse IV (publishing heir), Luke Weil (gaming industry heir), Cody Franchetti (textile heir), Stephanie Ercklentz (finance heiress), Josiah Hornblower (Vanderbilt/Whitney heir), Carlo von Zeitschel (European royalty), Christina Floyd (professional sports heiress), Juliet Hartford (A&P supermarket heiress), and Jamie Johnson himself (filmmaker and Johnson & Johnson heir). While the inherited wealth of each of the participants at the time of filming is not divulged, we know from comments made by some that, even in today's figures, their wealth sets them apart: Newhouse remarks, 'it would be a very low estimate for me to say 20 billion'.

From the outset, Johnson establishes the importance of silence to the film's narrative, a phenomenon that circumscribes the lives of all his ultra-wealthy contemporaries: 'The thing is nobody wants to talk about money. It's like this big taboo always lurking under the surface. That's why I want to ask other kids who were born rich about that one subject everybody knows is not polite to talk about'.[7] The opening scenes comprise a montage of Jamie's 21st birthday celebration, an auspicious event when he will inherit 'more money than most people could earn and spend in a lifetime' (Jamie Johnson). Against this backdrop, numerous visual cues invoke a symbolic imaginary of super-wealth: fine champagne, sumptuous interiors, family portraits exemplifying a long history of affluence, jewels, tuxedos, and, for the well-heeled young women in attendance, 1920s' flapper couture redolent of an era of self-indulgence. For Rose (2012, p. 163), the imaginary represents 'the field of interrelations between subject and other people or objects', but as we soon come to see, the relational construction of super-rich subjectivity is about much more than the presence of the aforementioned 'objects' of wealth; codes around silence are equally significant. Although Johnson gives voice to the recognized cultural code that it is 'impolite' to discuss money – a theme that emerges time and again, particularly with reference to Jamie's father who 'never talked about his inherited wealth', leaving a young Jamie confused 'about a secret [he] wasn't supposed to know' – a more expansive exploration of silence is also pursued, initially by way of the relationships we see playing out between the super-rich party-goers and those from lower classes. While the former are seen indulging in these opening scenes, a headwaiter cautions his charges behind the scenes of the extravagant affair: 'This is a party for one of the Johnson & Johnson family members. . . You may not know which of these people are Johnson & Johnson. . .so treat everyone as if they are'.

Uniformed waiters nod dutifully, acknowledging symbolically that in super-rich circles servants especially are trained to venerate the elite, to be compliant, submissive and essentially 'invisible' (Romero, 1992, as cited in Donaldson and Poynting, 2007, p. 64). Given the socially and spatially segregated experiences of most super-rich young people – separated from the bulk of humanity by where they live, school or holiday, by the exclusive nature of super-rich modes of transport, and by virtue of extended contact with nannies or other hired help (often in the absence of their super-wealthy parents) – exchanges with servants function as silent and taken-for-granted cultural practices that reinforce 'a thorough and early appreciation. . .of their own ontological superiority' (Donaldson and Poynting, 2007, p. 43). This sense of superiority is made patent in the film when Luke Weil, gaming industry heir, exclaims boldly: 'Did you ever have someone piss you off?. . .I can just say, f*** you, I'm from New

York. My family can buy your family, piss off. And this is petty, and this is weak, and this is very underhanded, but it's so easy'.

The remaining interview participants maintain a comparatively composed veneer, revealing in less brazen if still pretentious terms, the extent of the power they share. Nevertheless, the initial scenes from *Born Rich* set the tone for Johnson's exploration of eliteness as, in part, the bestowal of unearned privileges that are often so natural as to be invisible to ruling class individuals themselves. Such privileges give shape and meaning to elite lives, and are reinforced by beliefs and practices to which broader society contributes. Reflecting on his own experiences and from a standpoint that contrasts markedly from those of his interviewees, Johnson explains:

> I live in a country that *everyone* wants to believe is a meritocracy. We want to think that everyone earns what they have. . . I know my family gives away millions of dollars each year to charity, but how does that exactly level the playing field? I mean, what did I do to earn the kind of money I'll own at midnight tonight? All I did was inherit it.

We can gather from this statement that the maintenance and reproduction of eliteness is not the purview of rich individuals alone but is embedded in broader social relations. Elite power is a 'naturalized' part of growing up that is validated through the deference of the few outsiders who spend extended periods with the very rich – hired help – and super-rich power is also grounded in structural relations and regulated by discourses that organize widely circulated codes concerning merit, politeness and charity.[8] Eliteness is thus the product of a complex interplay of relations that Johnson endeavours to examine.

But doing so is dangerous, and we learn this early in the film when family attorney, Peter Skolnik, offers Jamie sobering legal advice:

> I can't in good conscience recommend to the kinds of people you want to talk to about the kinds of things you want to talk about that they participate. . . I mean you're dealing with families that have always made it gospel that you don't talk about your money. And now you want to put them in a film?. . . I'm not surprised that you're having trouble.

To examine elite codes of silence is, potentially, to activate the systems that enforce and reinforce them. Social institutions, such as the law, and hired professionals, such as attorneys, are mobilized (and mobilize themselves) within discourses of eliteness to protect and conceal the super-rich, subjects who in turn become agents of elite power. Luke Weil provides clear illustration of the rationality arising from these dynamics. He does this first by stating his reluctance to be in the film and the concomitant means by which silence is policed by fellow elites:

> I would worry about my parents seeing it and I would worry about some of my friends seeing it, you know? Some of them would think that my involvement in it is inappropriate. Just the notion of talking about one's wealth. . .describing a bunch of people's wealth, it's tacky.

And later, unhappy with his filmic representation, Weil exercises his capacity to sue for defamation, an action to which Skolnik remarks: 'The irony here is that Luke can clearly afford to deal with his problems by suing'. In this sense, elite power is not only preserved and policed by elite individuals, but by the people and institutions called upon to assist them, and by the beliefs (concerning charity or politeness, for instance) in which broader society shares.

We might add, however, that being super-rich does not equate to being silent about one's wealth entirely – certain carefully regulated displays of wealth are in fact to be celebrated:

> The point is not that the super-elite are reluctant to display their wealth – that is, after all, at least part of the purpose of yachts, couture, vast homes, and high-profile big-buck philanthropy. But when the discussion shifts from celebratory to analytical, the super-elite get nervous. (Freeland, 2012, p. xii)

Hence, characterizing these opening scenes is a combination of messages about silence. First, is the obvious and well-known code within elite culture that it is 'not polite' to talk about money – a code that conjoins 'silence' with 'manners/good taste'. But silence also operates more intricately to produce elite subjectivity in highly regulated ways. Silence manifests itself in taken-for-granted practices to which elites are exposed from an early age, such as the behaviour of servants toward them. Expectations concerning how they ought to be treated by the outside world are deeply ingrained, making the elite, in F. Scott Fitzgerald's (1926) terms, different from you and me. Moreover (formal) social institutions, such as the law, are mobilized to reinforce these widely observed silences, and to ensure that elite wealth remains celebrated, but not interrogated.

In turn, elite families have the resources and cause to exercise considerable legal might in their own interests. This feedback loop (between legal and political sway, and the capital that enables it) is in Freeland's (2012, p. xvi) terms both cause and consequence of the rise of super-elite power. Individuals growing up in a super-rich primary habitus are groomed to be servants of a system that protects them, as S.I. Newhouse (media heir) reveals when he suggests that with super-wealth comes an urge to 'escape from the mold', from which there is 'no escape'. This indicates a form of 'elite governmentality'; or the integrated means by which agents of the system, informed by disciplined rationales, are cultivated to maintain a seemingly 'natural' social order by enforcing (yet rarely questioning) these same cultural codes: 'like some sort of secret handshake, these codes form [class] bonds that are difficult for some dissenting [elites] to see until they themselves experience being surveilled' (McLaren et al., 2000, p. 111).[9] By contravening silence as a fundamental 'elite' code, Johnson positions himself as a 'dissenting elite', a move that begins to part the veils of elite silence.

BEING SUPER-RICH

Of course, not all of the participants in *Born Rich* understand or embody eliteness in identical ways. They represent different historical entry points to super-wealth, and silence works differently in their narratives. These orientations toward wealth are largely inherited (i.e., passed down along family lines and refracted through parental behaviour), and other than Josiah Hornblower (Vanderbilt descendant and Whitney heir), who struggles notably with his position, it is really only Johnson who subverts the foundations of ruling class subjectivity when openly questioning what he did to 'earn' his unearned privilege.

In contrast to Johnson, several of the participants openly naturalize social hierarchy, negating the need to vindicate inordinate wealth. Cody Franchetti (textile industry heir) provides a patent example: 'classes have always existed in humankind. . . Not to be supercilious and look down from an older culture, but really, you [North America] are discovering this just now'. The camera pans across a well-appointed sitting room adorned with accoutrements typically associated with wealth. Finally, it rests on Franchetti who, playing the piano, explains:

> I personally do not believe at all in any moral obligation for anyone. But I've seen that people with money that don't do anything are unhappy. . . While the poor are limited by their poverty it is the duty of the rich to cultivate themselves in their idle time. And if they use their own riches for this they can reach the highest elevation. . . Guilt is for old women and nuns.

Franchetti spends his time engaged in cultural pursuits to avoid an idle malaise. He adopts a sense of *noblesse oblige*, or 'the notion that there are responsibilities and obligations associated with this privilege' (Donaldson and Poynting, 2007, p. 42), which in his case takes the form of an individual 'project of the self' (Foucault, 1988; Giddens, 1991, p. 200). Illustrating means by which this project might play out – for instance, we see Franchetti selecting the right telephone for the right occasion, reading philosophy, and presenting a collection of slim-line (and hence, we are told, superior) encyclopaedias, all examples of his refinement – Franchetti draws close attention to his finely cut suit, a particular penchant given his historical links to the industry:

> The lapels, you see, they're high. There is nothing worse than when you see these jackets with these lapels that go like this [pulls one side of his jacket down]. See, they're low riding. [President] Clinton wears this kind of thing, looks like a restaurant owner. It's so vulgar. This is so [demonstrates his high lapels]. . .has an aristocratic thing to it.

A desire for expensive, hand-crafted suits (or indeed any expensive luxury items) conveys the subtle ways in which economic and cultural capital reinforce one another, for without the capital and situated knowledge to invest in such items, individuals cannot employ their taste, symbols, materials and ideas strategically within the *socius* to advance their 'positional identity' (Drummond, 2010, p. 376). Economic and symbolic investments of this nature thus invoke a broader system of relations that enable elites to 'display their membership of a global and very privileged club' by wearing the uniform of affluence at the expense of those who cannot (Donaldson and Poynting, 2007, p. 240).

The disposition toward super-wealth that Franchetti propounds commonly involves drawing on various social markers to situate oneself within hierarchical relations (Maxwell and Aggleton, 2010, p. 8). But what remains unspoken within this logic are the ways in which elite acts of self-cultivation are never disconnected from the contexts that give them meaning. Rather, they rely upon the construction of less-refined, 'vulgar', lower-classed Others – (perhaps remarkably in Franchetti's case, President Clinton) – as well as the routine elision of social inequality with the rhetoric of taste in the dialectical construction of elite subjectivity. In this sense, they rely upon and reproduce social hierarchy.

S.I. Newhouse (media heir) and Luke Weil (aforementioned) both share in the degree of wealth enjoyed by Franchetti. However, we soon learn that these men have been raised with different standpoints on 'being' super-rich. Newhouse explains that despite being heir to billions of dollars, his inheritance is not a given: '[I]t doesn't work that way in my family. You've gotta *earn* your way. If you don't go to school *at least somewhat*, if you don't work for the family and put in your time, you don't get shit'. Not dissimilarly, Weil remarks: 'My father told me from a very early age that I could do whatever I wanted. . ."As long as you *do something*, you're going to be set up"'. However, 'doing something', it transpires, is a flexible condition and particularly so for Weil, who goes on to explain:

> I attended, in the entirety of my first academic year, I think less than eight academic obligations total, and that includes tests and exams. . .And I'd get these

> really evasive letters from deans saying, 'though you were technically below the number of credits required for a continuing education at Brown University we do feel that with continued academic progress you will da da da da da'. I'm just like, Jesus; you guys can't throw me out, can you! [Laughs] This is ridiculous.

Not only does this excerpt illustrate how super-rich individuals, like Weil, learn early of their imperviousness to the consequences of their own behaviour, but also how elite institutions, including universities, reinforce these indirect lessons. For the children of the very rich, education operates differently from the ways it does for the less well-off. Whereas those from lower classes may view education as a means of social mobility, the super-rich already top the social ladder. Consequently, elite schools and universities function not only as vehicles for super-rich families to maintain their class position; they also draw from associations with the super-rich to uphold their own status. Weil goes on to make that (inter) dependence clear:

> I drove up [to the college] with the chairman of the board of trustees from there; he was my dad's best friend at the time. The second day of school the headmaster. . .came up to me and personally introduced himself. Things like that you tend to notice.

These kinds of symbiotic relations are borne out in the literature. For instance, Connell et al. (1982, p. 51) write that elite parents are 'richer and more powerful than all teachers' who they consequently tend to view as their 'paid functionaries'. Ostrander (1984) argues that attendance at an elite school is one of the many means by which the upper class 'create and maintain the exclusivity of their way of life and their social interactions' (as cited in Donaldson and Poynting, 2007, p. 86). Moreover, '[s]uch networks not only make useful connections possible, indeed inevitable, but by excluding children who do not fit, they very practically ensure the right marriages and the consolidation and continuity of the networks themselves' (Donaldson and Poynting, 2007, p. 118), a point we will return to later.

While super-rich subjects such as Weil may co-opt a rhetoric of 'doing something' to justify inordinate wealth, in reality their privileged status is already fortified by formal institutions, whether they do 'anything at all'. Institutions such as schools thus preserve elite power by creating borders that exclude much more than they include. And in this way, privilege and disadvantage are linked.

But an alternative standpoint to 'doing something' to justify one's wealth is 'doing very much'. The discourse of meritocracy – or the idea that super-wealth may be justified through hard work, intellect or merit – emerges in several of Johnson's interviews. Georgina Bloomberg, media heiress and successor to a net worth of around US$18 billion, explains that 'having the last name Bloomberg sucks', and that '[her] struggle is to have people see the person behind that name'. Thus in her struggle to prove herself (and so embody a particular orientation to super-rich subjectivity), Bloomberg competes in equestrian sports.

As the camera pans past groups of exceptionally well-dressed spectators, we follow Bloomberg through the Palm Beach Polo Equestrian Club – a stunning milieu bespeaking a history of concentrated wealth. Ultimately, we are granted entry to her new barn and introduced to several of her horses, including 'Nadia', bought recently in Europe. It is against this framing that Bloomberg states with self-assurance, 'I don't really care who accepts what I do or how I act, I'm doing what I love', and perhaps to her credit, she went on to qualify as a show jumper in the 2008 Olympics.

Ivanka Trump (real estate heiress), Christina Floyd (professional sports heiress) and Stephanie Ercklentz (finance heiress) also invoke discourses of meritocracy when enunciating their standpoints on wealth, in each case referring to their wealthy fathers. Ercklentz explains that her German-born father 'put himself through college' in order to then 'study at Harvard Law School'. Owing to his subsequent wealth, Ercklentz herself was: 'raised in the lap of luxury in New York, totally lucky, going to private schools my entire life and kind of having whatever I wanted but always knowing that if you lose it all, all you have is this [points to her head]'. Floyd also positions her wealthy father as something of a 'self-made man', explaining to Johnson:

> My dad is an amazing person. . . Fortunately for him he was good enough to turn pro and make it on his own with golf and not have to have a college degree. He sort of went from nothing and made it into everything, which is very admirable.

Likewise, Ivanka explains how proud she is to be a Trump and of everything her father has achieved. Recollecting a pivotal moment in her young life that crystallized her position on super-wealth, Trump explains:

> I remember once my father and I were walking down 5th Avenue and there was a homeless person sitting right outside of Trump Tower. . . And I remember my father pointing to him and saying, 'That guy has eight billion dollars more than me' because he was in such extreme debt at that point. . . And I think I only thought about it like a year or two ago and I found it interesting. It makes me all the more proud of my parents that they got through that.

Discourses of meritocracy advance the notion of a level playing field and erase the power of difference (Warren, 2003, p. 56). Trump feeds into this logic when implying that her father (Donald), whose own father had an estate worth more than US$150 million, shares in the same pool of resources as someone who is homeless. Trump effectively secures a moral pretence through recourse to family 'pride' while whitewashing the positional benefits from which her father has drawn to build (and rebuild) his empire. Such benefits, while often invisible to their prime beneficiaries, include one's position within race, class and gender relations, as well as the networks (or social and cultural capital) afforded by an elite education.[10]

Ercklentz overlooks the same relations when suggesting that, if her fortune slipped away, all she would have is her mind – a perspective that fails to account for the 'cash value'[11] attached to her exclusive social networks. Evidence of these networks is on display as a backdrop to Ercklentz's narrative, which we see portrayed in snapshots of a charmed New York life: shopping in exclusive stores, attending black tie events, mingling with her socialite mother, both women draped in fur. Such imagery suggests, symbolically, that contrary to her earlier comments, being super-rich is about much more than the cultivation of mind. Rather, the successful embodiment of moneyed eliteness involves development of a certain style: rubbing shoulders with the right people, having access to the right social spaces, knowing what to wear (and how to wear it), and exhibiting a relaxed confidence that Khan (2011, pp. 77–113) describes in terms of 'ease' – traits and attributes that are developed in elite settings to which few have access.

As an illustration, Ercklentz explains that while working briefly at the investment banking giant Merrill Lynch – an opportunity that itself conveys influence – she was praised, not for numerical acumen, but instead for her ability to mix with affluent clientele:

> That was my best skill. Like I got handwritten notes from CEOs of companies saying I was great, I was fabulous, if I ever needed anything to call them. But I didn't have the drive. . . I was like, my friends are at Downtown Cipriani [a luxury restaurant] right now, it's 10 o'clock at night, they're having bellinis and I'm cranking out numbers that will never even get *looked* at. And that's where I drew the line.[12]

With respect to elite attributes and the reproduction of broader systems of social inequality, Khan contends that 'what seems natural is made, but access to that making is strictly limited' (2011, p. 9). Consequently, when merit is awarded for socially constituted distinctions – for example, handwritten notes from CEOs – differences in life outcomes 'appear a product of who people are rather than a product of the conditions of their making' (2011, p.9).

Floyd's claim that her father 'made everything out of nothing' constitutes a comparable sleight of hand by overlooking the fact that his own father owned a golf driving range – a detail indicating not only some wealth, but exposure from a young age to an exclusive sport long associated with wealthy white men (see Donaldson and Poynting, 2007, p. 196). And Bloomberg also performs a discursive manoeuvre that obscures her positional advantage by failing to credit the considerable investments (both financial and symbolic) underpinning her sporting success, and the historical association of horse riding with royalty, wealth and the monopolization of space: as Englade (1997, cited in McManus, 2013, p. 155) once mischievously remarked, 'horse ownership is the only currency that distinguishes the truly wealthy from the rest of us'.

For all these women, pride, self-assurance, admiration and merit function as veils of silence screening the foundations of their unearned privilege. Each advances the logic that all it takes is hard work to rise to the top of the classed, raced and gendered social hierarchy. A corollary of course is that those in less advantaged positions (i.e., the homeless person in Trump's vignette or those requiring 'welfare') have simply not 'worked hard enough' (see, e.g., Stanford and Taylor, 2013). It is this way, Khan (2011, p. 15) suggests, that privileged young people in elite environments learn to emphasize hard work and talent when rationalizing their good fortune or high educational scores. They learn, like these wealthy women featured in *Born Rich*, to naturalize the class benefits of their position and invest in a mode of 'rhetorical silence' – a manoeuvre that justifies unearned privilege through recourse to pride or merit.[13]

Despite ostensibly diverse standpoints on 'being' super-rich, Johnson's participants all justify inordinate wealth by valorizing individual projects of the self – whether it be through taking part in competitive, highbrow pursuits; 'doing something' to earn one's inheritance; working hard; or developing 'cultivation'. Their narratives feed into the same epistemological foundations, which reproduce elite power. In turn, these dynamics give rise to a range of dispositions and social practices that are accepted in elite circles. In Trump's case, pride over her family's hard-earned wealth permits an impulse to appropriate space for her own benefit – an impulse that is reflected in the scenes in which she is repeatedly set: gazing out over the New York City skyline. Moreover, Trump's desire to seize and shape space is framed as 'natural' (and is thus beyond critique) when she goes on to suggest that real estate development is, simply, in her 'blood': 'I love looking at the New York skyline and being able to figure out what I'm going to add to that. . . I've basically always wanted to go into real estate development. It's in the blood I guess. [Giggles]'

REPRODUCING SUPER-RICH STATUS

At the same time as being introduced to super-rich young people engaged in modes of rhetorical silence, we are also made privy to exclusive social spaces that help to naturalize their hold on power. Among these is the iconic South Hamptons – an expensive beach resort on Long Island, New York, well known for its exclusivity – which features as one of several key milieus in Johnson's film.[14] Here we follow a bubbly Floyd fresh from sunbathing on a driving tour of her favourite spots. Floyd stops intermittently to indicate places of interest, explaining along the way that if she gets her timing right she can catch her parents' helicopter in and out of New York City, or if she wants to go shopping, she can simply use their charge card.

These features of Floyd's everyday life resonate with comments made by several of the other participants: Hornblower remembers growing up thinking that everybody had a family museum; Johnson thought that 'all kids' lived on vast properties with carriages and horses; Weil recalls being taken to and from school in limousines; and Trump remembers how 'different' outsiders treated her owing to her parents' wealth. These everyday, yet extraordinary, features of the participants' lives help to illustrate how growing up rich makes the super-rich different from you and me and how privilege makes their social relations 'vastly different to those experienced by other people' (Donaldson and Poynting, 2007, p. 241). Such topographies of being screen the super-wealthy from the social problems to which their hyper-consumptive and self-interested behaviours contribute. As a case in point, Donaldson and Poynting argue:

> [t]he multiple homes this class owns take up vast amounts of space and consume unimaginable amounts of money, energy and others' labour just to keep running. Crowded as the world is, this class never has to suffer from such discomfort [with] one rich man [occupying] easily one thousand times more space than any ordinary person. (2007, pp. 240–41)

But as Floyd escorts us around South Hampton problems like crowding or debt quickly fall away to be replaced by a pristine cultural landscape comprising vast waterfront estates, sweeping lawns, stately homes and exclusive clubs – in Pow's terms, the kind of 'purified spaces customized for elite social reproduction' (2013, p. 62). Stopping in front of one such exclusive establishment, Floyd remarks: 'This is extremely, extremely the WASPiest of WASPs[15] [giggles], but it's beautiful, as you can see. All the courts are grass; you have to wear all whites to even be here'.

The camera moves past the establishment's facade (replete with 'members only' signage) to reveal rows of neatly manicured courts populated with 'all-white' players wearing all white attire. Against this crisp white setting our attention is drawn to a single black player, his 'difference' highlighting the deeply racialized social geography of South Hampton. Johnson asks: 'Is he a member?' To which Floyd replies:

> Well he's probably a pro. . .I think the people at the club would probably not be that excited if someone came in with a black person. Especially someone – not that I have any problem with it, *please* – but especially being that my parents for example are relatively new to the club or to all of this in general. So like, we're already sort of on 'watch', you know? [Giggles]

Here we see not only the way in which eliteness is particularized in time and place – and how in this place its contours are so evidently white – but also how recipients of elite privilege learn to develop innocent strategies for preserving their unearned advantage whilst remaining members of associations that trade on social inequality. Despite the patently racialized character of the club to which Floyd has recently been granted access, she nevertheless manages to distance herself from any semblance of racism by claiming that personally she has 'no problem' with black people. This comment rests on a view of racism as an independent and overt act, 'limited to willed, concerted activity' (Frankenberg, 1993, p. 47). In her adoption of such an overtly individualistic viewpoint, Floyd fails to comprehend racism as a system silently shaping her own opportunities and sense of self, and thus 'intimately and organically' linked to her daily experiences (1993, p. 6).

As our tour of South Hampton proceeds we soon learn that is not just country clubs in this immaculate part of the world that position their privileged clientele as (unwitting) contributors to broader systems of social inequality. Ercklentz remarks: 'When you go out in South Hampton you *definitely* will go to Conscience Point'. The camera cuts to a close-up of Britt West, manager of Conscience Point nightclub, who explains to Johnson:

> I will come in at about 7 o'clock and do the table seating for the night: what party? How important are they? Where are they going to be seated in the club? Obviously [we will allow] attractive women because that looks good for us. We [also] want to see Cristal champagne sitting on the table because there are other people that are in the crowd that will want to ultimately emulate that. . . We have a two bottle minimum per table. Our cheapest bottle is $250 and champagne prices go all the way up to a magnum of Cristal [which] would be $900.

The exclusiveness of Conscience Point is assured, not only by market mechanisms (i.e., the high cost of drinks), but by strict admission rules. Thus, exclusive clubs trade on elite power (and in so doing advance elite status), while functioning as spaces where class is organized in routine and practised ways. In this sense, being elite is less about 'conscious' or deliberate acts on the part of elite individuals than it is about the routine ways in which their lives are structured by the spaces they inhabit: 'it is about how they exist, whom they meet, what they say, [and] what they can do' (Donaldson and Poynting, 2007, p. 165).

Exclusive establishments function as mechanisms that circumscribe elites' lives by regulating their options for socialization. West goes on to explain: 'they all know each other and that is their dating pool. . .it's pretty much limited to that circle'. Yet despite the very obvious and overt ways that elite schools, clubs, establishments and resorts all regulate elites' interactions through systematic methods of exclusion, when asked about her own dating life, Ercklentz reflects:

> I've never actually dated outside of my social background, never. It's really weird. Um, I've never really thought about it, it's just like your compatibility, somebody on your same wavelength. Like. . .[if I married someone of a lower class] he'd have to understand the fact that I love going shopping and spending all this money.

Hornblower adds, equally perplexed:

> I think it'd probably be weird to bring somebody [of a different class] back to your *homes*, to your summer house or maybe to introduce them to your parents. . . My parents they're so, you know, they live in this world where they might be like, 'so, where do you summer?' Or something like that? (Emphasis added)

In Ercklentz's narrative we thus see how a veil of silence (or an 'elite blind spot') enables her to naturalize dating within her circle by reducing it to a matter of 'wavelength'. In this formulation, class plays no role in insulating super-rich lives, and the mechanisms that work to exclude 'outsiders' are bracketed from view. The broader implication here is that those with the greatest power to affect social relations are enculturated to see the world in highly limited ways.

CONCLUSION

The aim of this chapter has not been to provide a critique of Johnson's insider observations of the lives of his super-rich friends, but to employ *Born Rich* as a vehicle for extending a sociology of ruling class subjectivity. Taking the lead from Caletrío, we have endeavoured to explore how silence works to consolidate this class's hold on power. The significance of this project lies not only in the fact that super-rich power is concentrating rapidly, but that super-rich privilege is, in many ways, reliant upon the disadvantage of others.

A key theme running through Johnson's film is silence: wealthy elites are taught not to talk about money. This time-honoured code operates behind a facade of 'good taste' while stifling opportunities for (self-) critical reflection on self-interest, accumulation or inequality. Failure to interrogate unearned privilege in turn enables super-wealthy subjects to draw unproblematically upon discourses of meritocracy; discourses that elide the many benefits of their wealth beneath veils of 'hard work', 'talent' or 'intellect'. In so doing, discourses of meritocracy excise the cultivation of valuable cultural traits, such as 'ease' or the ability to socialize with other elites, from the privileged social contexts to which few have access. These dynamics fortify elite power while concentrating the disadvantage of excluded others.

Wider society contributes to these relations. The deference shown toward elites by hired help and service providers, and the ways in which consumer culture commends markers of super-rich privilege (i.e., fine suits, yachts, luxury cars), reinforce the positional superiority of the wealthy. Institutions such as the law safeguard codes of super-rich silence. And social geographies of the elite structure their everyday lives through mechanisms of exclusion that make 'being' super-rich, in more ways than not, simply a matter of living.

The interplay of these relations forms the bedrock ontology of eliteness: a disposition shaped by an unrealistic sense of superiority and a blinkered view of the world from which the super-rich are socially and spatially separated. One consequence of these relations is that elites are effectively disconnected from the lives of others they so greatly affect. While being super-rich can be validated in diverse ways, including through recourse to a 'natural' sense of *noblesse oblige*, through doing 'something', or indeed through working hard (a perspective that is arguably the most insidious in that it positions 'meritorious' elites as existing beyond critique), the standpoints taken by Johnson's participants (and tragically rehearsed so often by others seeking to rationalize the position of the wealthy) feed into the same epistemological foundations, the self-justified beliefs that sustain elite power.

Super-rich subjectivities change with the contexts that produce them and are therefore unavoidably connected to the socio-spatial worlds we share. While our chapter reveals those relationships on mere fragments of the rapidly changing global terrain of super-wealth, we believe it does go some way to addressing Caletrío's (2012) challenge, shining light where darkness has historically endured. And it is only through parting the veils that conceal super-rich power that the foundations of poverty and disadvantage can be more properly exposed.

NOTES

1. See Donaldson and Poynting (2007, p. 9). Also see Elliott and Pilkington (2015) for updated statistics, which estimate that, by 2016, and on current trends, 1 per cent of the world's population will own more wealth than 99 per cent.
2. By way of example, a recent report in the *International New York Times* (Forsythe, 2015, pp. 1 and 4) comments on common criticisms that lawmakers are often in the pockets of the wealthy and goes on, somewhat wryly, to observe that in China 'Among the 1271 richest Chinese. . .a record 203, or more than one in seven, are delegates to the nation's current Parliament or its advisory body. . . [T]hose delegates' combined net worth is $463.8 billion, more than the economic output of Austria' (p. 1).
3. Writers such as Freeland (2012) and Khan (2011) go some way toward doing so.
4. See Donaldson and Poynting (2007, p. 14) for a detailed discussion of the ethics of empathy, distance and standpoint in life history work of this nature. We adopt a similar 'post-structuralist' stance aimed neither at psychologizing nor critiquing the objects of our study, but at identifying the ways in which power circulates through super-rich lives. In this sense, we are not so much interested in individual identities or personalities, but in the means by which subjects, marked by wealth, are positioned within and negotiate social relations.
5. This ontological standpoint underpins the concept of deconstructed subjectivity. Rather than an unfettered agent existing independently of structure, such as in humanist ontological conceptions, here we utilize Foucault's subject, who is neither entirely autonomous nor enslaved, but mediates the process of becoming within discourse (Matza, 1966, in McGrew, 2011, p. 245; Foucault, 1979; Sawicki, 1994, pp. 103–4).
6. Johnson's second film, *The One Percent* (2006), provides a fitting follow-up to *Born Rich* by exploring the politics of wealth distribution in the United States and the ways in which disproportionate shares of that nation's wealth are put in the hands of a small percentage of its population.
7. Unless specified otherwise, all direct quotations in this chapter are drawn from *Born Rich* (2003).
8. For useful discussions of some of the structural processes that sustain super-rich power, see Judt (2010), Volscho and Kelly (2012) and Bonica et al. (2013).
9. Foucauldian conceptions of governmentality – or how subjects come to engage in self-regulating practices that constitute them as 'particular types of subjects' (Markula and Pringle, 2006, as cited in Ayo, 2012, p. 100) – are useful here for understanding how eliteness is both entrenched and socially engineered.
10. For instance, Trump attended private Fordham University (The Jesuit University of New York) for two years before relocating to the Wharton School of the University of Pennsylvania (a private Ivy League university).
11. In a similar vein, Lipsitz (1998) discusses the way that 'being white' within the context of race-structured societies has a cash value insofar as accounting for advantages that come to individuals by virtue of their racialized identity.
12. It is of course possible that knowing of her family finance connections, some of those CEOs saw developing links with her and her moneyed family as means of advancing their own careers and social networks.
13. Crenshaw (1997) discusses rhetorical silence in the context of whiteness studies, where 'being white' in societies that are mapped by whiteness enables white individuals to overlook the foundations of their racial privilege. Here, we use the concept similarly to convey how the subjects in *Born Rich* avoid interrogating the foundations of their privilege in routine, everyday ways, for instance through recourse to pride or merit.
14. For some discussions of the Hamptons and eliteness, see, for examples, Hsu (2009) and Mears (2014).
15. White Anglo-Saxon Protestants.

REFERENCES

Aguiar, L.L.M. and C.J. Schneider (2012), *Researching Amongst Elites: Challenges and Opportunities in Studying Up*, Farnham, UK: Ashgate.
Ashcroft, B., G. Griffiths and H. Tiffin (2000), *Post-Colonial Studies: The Key Concepts*, London: Routledge.
Ayo, N. (2012), 'Understanding health promotion in a neoliberal climate and the making of health conscious citizens', *Critical Public Health*, **22** (1), 99–105.
Back, L. and J. Solomos (eds) (2000), *Theories of Race and Racism: A Reader*, London/New York: Routledge.
Beaverstock, J., P. Hubbard and J.R. Short (2004), 'Getting away with it? Exposing the geographies of the super-rich', *Geoforum*, **35** (4), 401–7.
Blood, S.K. (2005), *Body Work: The Social Construction of Women's Body Image*, London/New York: Routledge.
Bonica, A., N. McCarty and K.T. Poole et al. (2013), 'Why hasn't democracy slowed rising inequality?' *Journal of Economic Perspectives*, **27** (3), 103–24.
Caletrío, J. (2012), 'Global elites, privilege and mobilities in post-organized capitalism', *Theory Culture Society*, **29** (2), 135–49.
Connell, R.W., D.J. Ashenden and S. Kessler et al. (1982), *Making the Difference: Schools, Families and Social Division*, Sydney: Allen and Unwin.
Crenshaw, C. (1997), 'Resisting whiteness' rhetorical silence', *Western Journal of Communication*, **61** (3), 253–78.
Dean, M. (1999), *Governmentality: Power and Rule in Modern Society*, London: Sage.
De Camargo, A. (1981), 'The actor and the system: trajectory of the Brazilian political elites', in D. Bertaux (ed.), *Biography and Society*, London: Sage.
Donaldson, M. and S. Poynting (2007), *Ruling Class Men: Money, Sex, Power*, Bern: Peter Lang.
Drummond, M. (2010), 'The natural', *Men and Masculinities*, **12** (3), 374–89.
Elliott, A. and J. Urry (2010), *Mobile Lives*, London: Routledge.
Elliott, L. and E. Pilkington (2015), 'New Oxfam report says half of global wealth held by the 1%', *The Guardian*, 19 January, accessed 20 January 2015 at http://www.theguardian.com/business/2015/jan/19/global-wealth-oxfam-inequality-davos-economic-summit-switzerland.
Englade, K. (1997), *Hot Blood*, London: Macmillan.
Fitzgerald, F.S. (1926), 'The rich boy', *Redbook Magazine*, January and February.
Forsythe, M. (2015), 'A chamber where wealth abounds', *International New York Times*, 3 March, 1, 4.
Foucault, M. (1979), 'What is an author?', trans. J. Harari, in D. Lodge (ed.), *Modern Criticism and Theory: A Reader*, London/New York: Longman, pp. 141–60.
Foucault, M. (1981), 'The order of discourse', in R. Young (ed.), *Untying the Text: A Post-Structural Anthology*, Boston, MA: Routledge and Kegan Paul, pp. 48–78.
Foucault, M. (1988), 'Technologies of the self', in L.H. Martin, H. Gutman and P.H.Hutton (eds), *Technologies of the Self*, Amherst, MA: University of Massachusetts Press, pp. 16–49.
Frankenberg, R. (1993), *White Women, Race Matters: The Social Construction of Whiteness*, Minneapolis, MN: University of Minneapolis Press.
Freeland, C. (2012), *Plutocrats: The Rise of the New Global Super-Rich and the Fall of Everyone Else*, New York: Penguin Books.
Giddens, A. (1991), *Modernity and Self-Identity: Self and Society in the Late Modern Age*, Stanford, CA: Stanford University Press.
Gregory, D. (1994), *Geographical Imaginations*, Oxford, UK: Blackwell.
Hay, I. (2013), 'Establishing geographies of the super-rich: axes for analysis', in I. Hay (ed.), *Geographies of the Super-Rich*, Cheltenham, UK and Northampton, MA, USA: Edward Elgar Publishing, pp. 1–25.

Hay, I. and S. Muller (2012), 'That tiny, stratospheric apex that owns most of the world – exploring geographies of the super-rich', *Geographical Research*, **50** (1), 75–88.
Higley, S.R. (1995), *Privilege, Power and Place: The Geography of the American Upper Class*, Lanham, MD: Rowman and Littlefield Publishers.
Hsu, H. (2009), 'The end of white America', *The Atlantic*, January/February, accessed 11 March 2015 at http://www.theatlantic.com/magazine/archive/2009/01/the-end-of-white-america/307208/.
Jackson, P. and P. Russell (2010), 'Life history interviewing', in D. DeLyser, S. Herbert and S. Aitken et al. (eds), *The SAGE Handbook of Qualitative Geography*, London: Sage, pp. 172–93.
Johnson, J. (dir.) (2003), *Born Rich* [documentary], New York: HBO.
Johnson, J. (dir.) (2006), *The One Percent* [documentary], New York: HBO.
Judt, T. (2010), *Ill Fares the Land*, London: Penguin.
Khan, S. (2011), *Privilege: The Making of an Adolescent Elite at St. Paul's School*, Princeton, NJ/Oxford, UK: Princeton University Press.
Lipsitz, G. (1998), *The Possessive Investment in Whiteness: How White People Profit from Identity Politics*, Philadelphia, PA: Temple University Press.
Markula, P. and R. Pringle (2006), *Foucault, Sport and Exercise: Power, Knowledge and Transforming the Self*, New York: Routledge.
Matza, D. (1966), 'The disreputable poor', in R. Bendix and S.M. Lipsest (eds), *Class, Status, and Power: Social Stratification in Comparative Perspective*, New York: Free Press.
Maxwell, C. and P. Aggleton (2010), 'The bubble of privilege. Young, privately educated women talk about social class', *British Journal of Sociology of Education*, **31** (1), 3–15.
McGrew, K. (2011), 'A review of class-based theories of student resistance in education', *Review of Educational Research*, **81** (2), 234–66.
McIntosh, P. (2002), 'White privilege: unpacking the invisible knapsack', in P.S. Rotherberg (ed.), *White Privilege: Essential Readings on the Other Side of Racism*, New York: Worth, pp. 97–102.
McLaren, P., Z. Leonardo and R. Allen (2000), 'Epistemologies of whiteness: transforming and transgressing pedagogic knowledge', in R. Mahalingham and C. McCarthy (eds), *Multicultural Curriculum: New Directions for Social Theory, Practice and Policy*, New York: Routledge, pp. 108–26.
McManus, P. (2013), 'The sport of kings, queens, sheiks and the super-rich: thoroughbred breeding and racing as leisure for the super-rich', in I. Hay (ed.), *Geographies of the Super-Rich*, Cheltenham, UK and Northampton, MA, USA: Edward Elgar Publishing, pp. 155–70.
Mears, A. (2014), 'How the other half parties: distinction and display among the jet set', paper presented at Council for European Studies Conference, 14 March.
Morrison, T. (1992), *Playing in the Dark: Whiteness and the Literary Imagination*, Cambridge, MA: Harvard University Press.
Ostrander, S. (1984), *Women of the Upper Class*, Philadelphia, PA: Temple University Press.
Paris, C. (2013), 'The homes of the super-rich: multiple residences, hypermobility and decoupling of prime residential housing in global cities', in I. Hay (ed.), *Geographies of the Super-Rich*, Cheltenham, UK and Northampton, MA, USA: Edward Elgar Publishing, pp. 94–109.
Piketty, T. (2014), *Capital in the Twenty-First Century* [French edition published in 2013 as *Le capital au XXI siècle*, Editions du Seuil], Cambridge, MA/London: Belknap Press of Harvard University Press.
Pow, C.-P. (2013), '"The world needs a second Switzerland": onshoring Singapore as the liveable city for the super-rich', in I. Hay (ed.), *Geographies of the Super-Rich*, Cheltenham, UK and Northampton, MA, USA: Edward Elgar Publishing, pp. 61–76.
Probyn, E. (1993), *Sexing the Self. Gendered Positions in Cultural Studies*, London: Routledge.
Romero, M. (1992), *Maid in the U.S.A.*, New York: Routledge.

Rose, G. (2012), *Visual Methodologies: An Introduction to Researching with Visual Materials*, London: Sage.
Rose, N. (1998), *Inventing Our Selves*, Cambridge, UK: Cambridge University Press.
Savage, M. (2014), 'Piketty's challenge for sociology', *The British Journal of Sociology*, **65** (4), 591–606.
Savage, M. and K. Williams (eds) (2008), *Remembering Elites*, Oxford, UK: Blackwell.
Sawicki, J. (1994), 'Foucault and feminism: a critical reappraisal', in M. Kelly (ed.), *Critique and Power: Recasting the Foucault/Habermas Debate*, Cambridge, MA: MIT Press, pp. 347–64.
Stanford, S. and S. Taylor (2013), 'Welfare dependence of enforced deprivation? A critical examination of white neoliberal welfare and risk', *Australian Social Work*, **66** (4), 476–94.
St Pierre, E.A. (2000), 'Poststructural feminism in education: an overview', *Qualitative Studies in Education*, **13** (5), 477–515.
Volscho, T.W. and N.J. Kelly (2012), 'The rise of the super-rich: power resources, taxes, financial markets, and the dynamics of the top 1 percent, 1949 to 2008', *American Sociological Review*, **77** (5), 679–99.
Warren, J.T. (2003), *Performing Purity: Whiteness, Pedagogy and the Reconstitution of Power*, New York: Peter Lang Publishing.
Wedel, J.R. (2009), *Shadow Elites: How the World's New Power Brokers Undermine Democracy, Government, and the Free Market*, New York: Basic Books.

9. 'One time I'ma show you how to get rich!'[1] Rap music, wealth and the rise of the hip-hop mogul

Allan Watson

INTRODUCTION

In little more than two decades, rap music, and its associated hip-hop culture, has gone from underground phenomenon to a multimillion-dollar business. It has gained a ubiquitous presence and lasting foothold in the market economy, achieved transnational acceptance, and co-opted the mainstream music industry to generate huge sums of money. In 2000, it was estimated that rap music generated US$1.8 billion in sales in the United States alone (Kun, 2002), and in 2006 US$1.3 billion, equivalent to 11.4 per cent of the entire US music market (Miller-Young, 2008). But hip-hop culture also extends beyond music and dance; it is not merely a form of entertainment but has become a lifestyle, helping to break down the separation between pop culture and daily life (Jenkins, 2011), influencing fashion and consumption of a wide range of lifestyle goods. As Miller-Young (2008) argues, hip-hop has become the main form of legibility for African American culture, and one that has diasporic and global effects.

As a well-established industry with considerable economic power, hip-hop has created significant economic opportunities for recording artists, producers and entrepreneurs (Cox Edmondson, 2008). This has especially been the case for African American men, for whom it has provided an opportunity for employment and wealth creation on a scale that has not been seen in any other industry (Murray, 2004). It is the only industry in which their connection with the difficult conditions of the American ghetto, rather than being a disadvantage, has become an advantage; as Basu and Werbner suggest, 'black hip hop capitalists, whether big or small, have gained footholds in the cultural industries, not in spite of, but because of their potent black street aesthetics' (2001, p. 253). Rappers, so often identified solely with 'the street' would, through ethnic entrepreneurship, become executives, and would make rap music central to the changing business practices and aesthetics of the contemporary music industry (Negus, 1999).

Through corporatization and commodification, hip-hop culture has become an extremely efficient device for extracting profit from consumers (Holmes Smith, 2003). In the case of a small number of ethnic entrepreneurs, their ability to tap into the business in an entrepreneurial and innovative manner has brought about the opportunity to create substantial personal wealth. These hip-hop 'moguls' have identified and adopted management solutions to commodify, package and symbolically manage a politically and socially volatile minority underclass' expressive culture (Holmes Smith, 2003) in a way that has sold not only to the young African American population, but much more widely, including to white suburban middle classes:

> They are truly remarkable: multimillion-dollar empires with global reach, selling a huge variety of products to some of the most desirable demographics (young, high income, acquisitive, interested in luxury products). These empires own brands that are rapidly approaching icon status, most developed in just the last decade or two. (Oliver and Leffel, 2006, p. 2)

As key agents in taking hip-hop to the mainstream, and creating a lifestyle that transcends race and ethnicity (Cox Edmondson, 2008), they have enjoyed a payday that no other music entertainers of colour have ever seen in any genre of music (Jenkins, 2011).

It is these hip-hop moguls that this chapter explores. The aims of the chapter are two-fold. First, the chapter seeks to situate the economic success of the hip-hop moguls in the context of the development and nature of the hip-hop business. In particular, the chapter describes the entrepreneurial dynamic that has developed in the hip-hop industry, and the ethnic entrepreneurship from which this dynamic has developed. The chapter provides brief case studies of the two most successful male moguls – Sean 'Diddy' Combs and Shawn 'Jay-Z' Carter – which emphasize how these individuals have successfully fused commercial artistic success with entrepreneurial achievement (Cox Edmondson, 2008) to generate personal net worths in excess of US$500 million.

Second, the chapter considers the moguls' relentless quest for wealth with reference to the social meaning of wealth in hip-hop culture. Here, to understand the spectacular displays of wealth that have become the aesthetic of choice (Murray, 2004) for these moguls, the chapter situates the material excesses that dominate hip-hop culture within the context of the battles for status and campaigns of respect that have become part of the 'street code' for the hustler in the ghetto, as well as considering the gender issues arising from this. Further, with reference to this street code, the chapter considers the tension that exists for moguls between 'making it' – achieving economic success – and 'selling out' – losing their connection with the street and thus their subcultural values. This is a central and controversial issue within hip-hop culture (Sköld and Rehn, 2007), and one that bears heavily on the politics of the music and defines the hip-hop mogul as a political figure.

THE HIP-HOP BUSINESS AND AFRICAN AMERICAN ENTREPRENEURSHIP

Initially, rap music was not a genre of music that seemed to correspond to the normal commercial criteria of the major music corporations. This was especially the case given that it was not seen to appeal to their majority white middle-class consumer market. Many in the corporate world wrote rap off as a faddish street trend, and even those who did recognize its potential commercial value were nervous of public reactions to its often violent and misogynistic lyrics and imagery. Corporations were slow therefore to recognize the lucrative nature of the hip-hop business. Indeed, as Johnson (2008) notes, for the better part of the first decade of its commercial existence, the music was largely ignored by corporate music interests.

The failure of major record companies to capitalize on the hip-hop phenomenon left a window of opportunity and a power vacuum into which entrepreneurial African Americans could move, exploiting new venture opportunities (Cox Edmondson, 2008). Throughout the 1980s and 1990s, African American entrepreneurs made, produced and sold the hip-hop music that was in demand by a growing consumer base, often setting up independent record labels and positioning themselves as the Chief Executive Officers (CEOs). As Basu and Werbner (2001) note, with few prior capital and symbolic resources, African Americans created, and continue to create, a music industry based around networks of entrepreneurs. These entrepreneurs, as Negus suggests, have created rap as a 'self-conscious business activity as well as a cultural form and aesthetic practice' (1999, p. 489).

In part due to the growing recognition that hip-hop not only appealed to African American working class youth, but to a much wider demographic, by the mid-1990s the major record companies had come to dominate the hip-hop music industry. The companies were either directly producing the most popular forms of hip-hop, or entering into lucrative distribution deals with the most successful of the independent labels. There are many examples of independent record labels set up by African American musicians and producers, but few have had enough proven or potential commercial success to attract major corporate interest. Those that have include Russell Simmons's Def Jam, Dr. Dre's Death Row Records,

Jermaine Dupri's So-So-Def Recordings, Sean 'Diddy' Combs' Bad Boy Records and Jay-Z's Roc-A-Fella Records.

Throughout the 1990s, forms of ethnic entrepreneurship – especially setting up independent record labels and production companies such as in the examples highlighted above – tended to be the only way for young African American businessmen and women to enter the music industry as professionals, due to their systematic exclusion from positions of power within corporations (Garofalo, 1994). As Negus (1999) noted, rap producers and musicians had not been co-opted into the boardroom in ways comparable to the type of osmosis that has occurred in other genres such as rock, as part of a deliberate attempt to maintain a distance between the corporate world and the genre culture of rap. However, since the turn of the century, in an attempt to compete in the hip-hop genre with successful African American entrepreneurs running their own record labels, a number of the major record companies have recruited other African American entrepreneurs to executive positions overseeing their hip-hop divisions (Cox Edmondson, 2008).

Perhaps the most prominent example is Antonio 'L.A.' Reid, who with Kenneth 'Babyface' Edmonds founded LaFace Records in 1989 through a joint venture with Arista Records, and would go on to become president and CEO of Arista Records between 2000 and 2004. When Sony and BMG merged in 2004 to create Sony Music Entertainment, Reid became the Chairman and CEO of The Island Def Jam Music Group in February 2004. In the same year, Reid would install another African American entrepreneur, the hip-hop mogul Jay Z, as president of Def Jam recordings. In July 2011, Reid became the Chairman and CEO of Epic Records, owned by Sony Music Entertainment. In a relatively short time, the influence of African Americans has spread across many levels of the music industry (Cox Edmondson, 2008).

As Cox Edmondson (2008) suggests, the appointment of these ethnic entrepreneurs to executive positions, within major record companies, often by acquiring the small firms that they operate, is a strategy aimed at eliminating them as competitors. But it is also about being able to tap into the extremely lucrative hip-hop market, something that had previously been problematic for large corporations run by white middle-class executives who did not understand the aesthetics and narratives of the genre. In employing African American entrepreneurs, the corporations look to draw on those individuals' cultural capital, which as Basu and Werbner (2001) state, is not necessarily financial or educational, but rather takes the form of social networks and familiarity with the aesthetic and musical nuance of the genre. These are not things that large companies can emulate:

> [T]he aesthetics and narratives of rap music are so deeply embedded in the daily lives of people, from rapping to each other at house parties or open-mics, to using and appreciating the style, clothing and linguistic innovations of black English, to knowing people who have contacts with other rappers, producers and so forth, to preclude easy appropriation. (Basu and Werbner, 2001, p. 250)

There exist many African American cultural workers and entrepreneurs who are less visible than the CEOs in the music business, but who are nevertheless drawing on cultural knowledge to become employees, freelancers or ethnic entrepreneurs in the music industry or in new creative fields. Rap music is the chief creation of a much wider cultural field of hip-hop that can be defined through the language, attitude, style and fashion that have become manifest in music, film, television networks, magazines and other popular media (Crossley, 2005). It is a cultural field in which the opportunities for employment and entrepreneurialism are great, creating opportunities for 'scores of black stylists, clothing designers, graphic artists, Web designers, producers, publicists, writers, editors, and A&R reps, and make-up artists, among many others' (Murray, 2004, p. 6). Basu and Werbner (2001) for example, describe how in Los Angeles there is a wide variety of rap-related businesses ranging from hip-hop journalists and stylists to small management production companies, with turnovers ranging from US$30 000 to US$300 000 annually. In short, as a cultural form, hip-hop has provided one of very few significant employment and wealth-generating opportunities for African Americans, and has employed them in unprecedented numbers.

Aside from the corporate commercialization of hip-hop, perhaps the most important development in enabling African American musicians to make money has been control of their intellectual property (in contrast to intellectual property ownership in the software industry, discussed by Kapoor in Chapter 6 of this volume). Although African American musicians have worked in the music industry for many years and have contributed significantly to global popular culture, traditionally they were paid very little for their services, usually being hired on a temporary basis. They typically did not own the intellectual property for their material; rather, this was owned by white-run record labels, producers and publishers, who could make millions of dollars from their work. The current generation of hip-hop entrepreneurs recognized that, through ownership of their own material, they could generate significant incomes for themselves, challenging the 'exclusionary forces of racism' that Basu and Werbner (2001, p. 243) argue have prevented African American artists taking command of their intellectual property.

THE HIP-HOP MOGULS

The mainstream hip-hop that followed the corporatization of the genre in the mid-1990s would prove to be a huge opportunity for many hip-hop artists and entrepreneurs, and there are now numerous African American hip-hop millionaires (Peterson, 2006). The so-called hip-hop 'moguls' represent a small number of ethnic entrepreneurs who have risen from the ranks of the many hip-hop microenterprises to accumulate significant personal wealth, to the tune of hundreds of millions of dollars. They are generally artists who have established successful companies that own the rights to their production, as well as the production of other hip-hop and R&B artists. Yet they also have business interests that have extended well beyond music. Drawing on the economic power of hip-hop and their own celebrity status, they have developed business interests in fashion, sport, alcohol and a wide range of other consumer products that appeal to the young consumers of the hip-hop generation. Some have themselves become iconic brands – cultural icons and representative symbols of globalized hip-hop culture.

Amongst the most well-known are Russell Simmons, Percy Miller, Sean 'Diddy' Combs, Shawn 'Jay-Z' Carter and Andre 'Dr. Dre' Young – each of whom has a current estimated net worth in excess of US$300 million. Estimates of net worth in 2014 suggest that there are currently 15 hip-hop entrepreneurs with a net worth in excess of US$100 million (Table 9.1), while two – Andre 'Dr. Dre' Young and Sean 'Diddy' Combs – have seen their net worth continue to increase towards the $1 billion mark.

There are a number of notable points regarding this list. First, all of the above artists/producers are African American, with the exception of Eminem, which is demonstrative of the fact that while hip-hop has commercial appeal to a broad population, the actual production of hip-hop remains dominated by African American artists. Second, all of the above artists are based in the United States, which not only represents the world's largest domestic market for hip-hop, but also the most globally influential producer of music in the hip-hop genre. While the spread of hip-hop culture across the globe has benefited these moguls, there have also been opportunities for artists based outside the United States to make their fortune; South Korean rapper/pop star Psy, for example, has an estimated net worth in the region of US$8 million to US$10 million on the back of his hit single 'Gangnam Style' (McIntyre, 2014). Hip-hop continues to penetrate into new markets, for example recently in the Middle East. In these new markets, entrepreneurial rappers are beginning to emerge. Take, for example, Dubai-based DJ Bliss, a producer, radio host, TV presenter, founder of the Bliss

Table 9.1 Hip-hop artists and producers with an estimated net worth exceeding US$100 million in 2014

Name	Estimated Net Worth, 2014 (Millions, US$)
Andre 'Dr. Dre' Young	780
Sean 'Diddy' Combs	700
Shawn 'Jay-Z' Carter	560
Percy Robert 'Master P' Miller	350
Russell Wendell Simmons	325
Curtis '50 Cent' Jackson	270
Usher	180
Ronald 'Slim' Williams	170
Bryan 'Birdman' Williams	170
Eminem	170
Dwayne 'Lil Wayne' Carter	140
Ice Cube	140
Calvin Cordozar 'Snoop Dogg' Broadus, Jr	135
Kanye West	130
LL Cool J	100

Source: Data accessed 28 October 2014 at http://www.celebritynetworth.com.

entertainment company, and the face of Beats By Dre in the Emirates (Tennent, 2013).

However, most notably missing from the above list are female hip-hop artists and producers, none of whom have yet reached a personal net worth in excess of US$100 million. This does not mean, however, that female hip-hop artists have not amassed significant personal fortunes; indeed a number have an estimated personal net worth of, or in excess of $50 million dollars, most notably Queen Latifah (Table 9.2). The term 'hip-hop mogul' then, although most commonly applied to male entrepreneurs, cannot be applied to men alone. I return to the issue of gender later in the chapter.

What follows are profiles for two of the most financially successful of the male hip-hop moguls: Sean 'Diddy' Combs and Shawn 'Jay-Z' Carter. The purpose of these profiles is to emphasize how these individuals have fruitfully fused commercial artistic success with entrepreneurial achievement, developed significant marketing power and established new forms of partnerships and deals with established firms.

Table 9.2 *Estimated net worth of leading female hip-hop artists and producers*

Name	Estimated Net Worth (US$ Millions)
Queen Latifah	60
Nicki Minaj	50
Missy Elliot	50
Lil' Kim	18
Mary J Blige	10
MC Lyte	8

Source: Data accessed 28 October 2014 at http://www.celebritynetworth.com.

Sean 'Diddy' Combs

With a net worth estimated at US$580 million in 2013, climbing to US$700 million in 2014, Sean Combs is the wealthiest of the hip-hop moguls. Combs has successfully transformed himself into one of the recognizable brands in hip-hop, making the *Forbes* 'Richest 40 Under 40' list in 2002 before he was 30 years old (Cox Edmondson, 2008). According to the *Forbes* 'Hip Hop Cash Kings' list, Combs's earnings in 2013 were estimated to be US$50 million, and in 2012 US$45 million. Combs began his career as many in the music industry do: as an intern (at Uptown Records in New York City), performing mundane tasks such as fetching coffee and delivering tapes. Combs would, however, go on to become one of Uptown's top executives. The company was subsequently acquired by MCA of the Universal Music Group in 1990, and Combs would go on to become chairman and CEO of his own company, the Bad Boy World Entertainment Group.

A record label, management firm and production company, Bad Boy Group began in 1993 and very quickly entered a joint venture with Arista Records that included a US$15 million deal for distribution rights and netted Combs an annual salary of US$700 000 (Oliver and Leffel, 2006). Combs signed a raft of successful artists including the infamous rapper The Notorious B.I.G., while his in-house production team, known as the 'Hitmen', produced music for a series of famous rap and R&B artists. From 1994 to 1997, the label reportedly achieved US$100 million in sales (Oliver and Leffel, 2006). In 2002 the joint venture ended and Bad Boy Records would subsequently switch back and forth between the Universal Music and Warner Music Groups. Currently the group has a roster of 12 artists. A highly commercially successful rapper himself, Combs released his first single in 1997 on his own label, and his first album in the same year made him US$7.64 million. His next two albums would make him US$2.66 million. Furthermore, as a leading producer of rap and R&B music, he has earned a portion of the royalties on the sales of albums by many leading artists, something that has formed a significant part of his net worth.

Combs's business interests, however, extend well beyond music. In 1998 he formed his own clothing line, 'Sean John', which has reportedly seen retail sales of US$525 million annually. Other business ventures have included a movie production company (including ventures into acting), a fragrance, two restaurants, as well as various brand endorsements. In 2004, Combs was ranked by *Inc*. magazine as one of the top ten celebrity entrepreneurs, as well as being named by *Time* and CNN on their list of 25 'Global Business Influentials'. In October 2007, the French vodka brand Ciroc offered Combs a 50 per cent share of the profits to help to develop and market the brand as brand ambassador and an equal share owner (see Palan and Mangraviti in Chapter 21 of this volume for a discussion of company purchases of individual goodwill). According to *Forbes*, sales rocketed from 120000 cases in 2007 to 1 million in 2011 (O'Malley Greenburg, 2011a), and in 2014 *Forbes* reported that this deal entitles Combs to eight-figure annual pay-outs and a nine-figure windfall if the brand is ever sold (O'Malley Greenburg, 2014).

Combs has also developed a number of interests in television. In 2002, he took over MTV's *Making the Band* franchise, which led to an exclusive deal with MTV for future television projects, and he went on to oversee a number of TV shows as executive producer. On 21 October 2013 Combs launched Revolt TV, a music channel aimed at 18- to 34-year-olds. The channel went live to around 22 million Comcast subscribers and 12 million Time Warner cable TV customers (Block, 2013). In early 2014, Combs reportedly made a US$200 million bid for Fuse, a TV music channel that reaches 73 million subscribers (Gensler, 2014).

Shawn 'Jay-Z' Carter

The net worth of rapper Shawn 'Jay-Z' Carter is comparable to that of Sean Combs, estimated at US$500 million in 2013 and US$520 million in 2014. Unlike Combs, Carter began his career not on the business side of the industry but on the performing side, as a young artist making cameo appearances on several albums and on stage in the late 1980s and early 1990s. Unable to find a major record label that would give him a record deal, Carter created the Roc-A-Fella Records independent record label in 1995 along with Damon Dash and Kareem Biggs. Jay-Z released his debut album *Reasonable Doubt* on Roc-A-Fella in 1996, which included an appearance by famed rapper The Notorious B.I.G., and for which Carter owned the rights. The album eventually reached platinum status[2] in the United States, and from there he would go on to become one of the world's most commercially successful rap artists. In 1996, Carter and Dash received offers from Sony and Def Jam recordings for a stake in Roc-A-Fella Records, with a 33 per cent stake eventually being sold to Def Jam for US$1.5 million. With 67 per cent of the profits going to Carter and Dash, they would accrue millions of dollars from Carter's future music sales. In 1997, Jay-Z released his follow-up album *In My Lifetime, Vol. 1*. Its executive producer was Sean Combs; it was distributed by Def Jam and it would also reach platinum status. His 1998 release *Vol. 2. . . Hard Knock Life* would be even more successful, certified five times platinum in the United States and selling over 5 million copies. To date he has released 12 solo albums plus a number of collaborative albums, most of which have sold 3 million copies or more, swelling Carter's personal wealth.

Carter would continue to profit not only as a recording artist but also as a businessman. In 2004, Def Jam bought the remaining shares in Roc-A-Fella for US$10 million, and later that year Carter became the president of Def Jam recordings as part of a three-year deal reportedly worth US$8 million to US$10 million per year, plus ownership of his own masters (O'Malley Greenburg, 2011b). In 2008, Carter and concert promoter Live Nation announced a US$150 million 360-degree deal[3] that consolidated all of Carter's music-related ventures – including recording, touring and management of other artists – under Live Nation. This included an advance of US$10 million for the next album; US$25 million upfront plus an additional US$20 million for certain publishing rights; and US$50 million to cover the costs of starting a new record and talent agency called Roc Nation. In addition, a touring advance of US$25 million was written into the deal, and touring would subsequently see Carter earn around US$1 million per show – US$25 million in 2009 alone (O'Malley Greenburg, 2011b). In 2013, as part of a broader deal between mobile phone manufacturer Samsung and Roc Nation, estimated at US$20 million, Samsung paid US$5 million for a total of 1 million copies of Carter's latest album, *Magna Carta Holy Grail*, to be given away to the first million owners of their devices who claimed them, a full 72 hours before its official release (Forde, 2013).

Aside from music, another of Carter's main business interests has been in fashion. In 1999, along with Damon Dash, Carter founded the urban clothing brand Rocawear, which within 18 months produced US$80 million in revenues, in part through cross-promotion of their products within Carter's music. In 2003, Carter's involvement in the fashion business continued with the launch of Reebok's S. Carter sneaker, with each box containing a CD preview of Carter's forthcoming album, the *Black Album*. Then in 2006 Carter took sole control of the Rocawear brand at the cost of US$22 million, before selling the rights to the Rocawear brand to the Iconix Brand Group for US$204 million in March 2007. However, Carter entered an agreement with Iconix to endorse, promote and manage Rocawear, and to identify and buy new brands, for a reported US$5 million per year, ensuring a continued income stream (O'Malley Greenburg, 2011b).

Alongside music and fashion, sport has become one of Carter's largest business interests, and he paid a reported US$4.5 million to become part-owner of the Brooklyn Nets (National Basketball Association) team, and bought a stake in their new home, the Barclay's Center (an indoor arena in Brooklyn, New York), which he later sold for US$1.5 million. Further, in April 2013 ESPN reported that Carter would be launching a sport management group, Roc Nation Sports, which would act as an agency for leading sports stars. Other business interests are varied, including the 40/40 night club chain, real estate, beauty products and fragrances. In 2002, Carter, along with business partners Damon Dash and Kareem Burke, took over the US distribution rights of the Armadale Vodka brand from the Scottish firm William Grant and Sons, resulting in another cross-marketing vehicle that could be promoted in Carter's songs.

THE SOCIAL MEANING OF WEALTH IN HIP-HOP CULTURE

To understand fully the rise of the hip-hop moguls and their drive to accumulate wealth, it is necessary to comprehend the social meaning behind these practices of wealth accumulation. As Negus (1999) notes, the activities involved in producing popular music should be thought of as 'discursive practices, which are interpreted and understood in different ways and given various meanings in specific social situations' (1999, p. 491). In the case of hip-hop music, the social meaning of wealth can only fully be understood if one views it as part of a social structure emerging from a difficult economic reality, namely, the difficult social conditions in housing projects in North America: 'In essence, for better or for worse, rap music is one modern response to the social and economic ailments of the collective African American community, which include joblessness, disempowerment and poverty' (Crossley, 2005, p. 504). Specifically, hip-hop culture and rap music emerged from socially invisible working class and poor African American youths in the South Bronx, New York (Peterson,

2006). It represented a form of musical expression for their feelings about inner city lives that lacked economic security and were often dangerous and violent (Cox Edmondson, 2008). As Johnson (2008) notes, the isolation of poor African American communities, the criminalization of African American men and the militarization of policing to control them, and post-industrial transformations in the economy, were all occurring as the hip-hop aesthetic was developing. Thus these social conditions can be considered the performance conditions from which hip-hop culture and rap music in particular have emerged. Indeed, they are regularly detailed in rap lyrics and the narratives of hip-hop culture (Peterson, 2006).

Sköld and Rehn (2007) argue that, given that rap developed in impoverished urban areas, it is fairly self-explanatory why tales of success would start to flourish in the discourse. Indeed, the 'rags to riches' stories of hustlers-turned-rappers have become central to the discourse of hip-hop culture:

> [O]ne must view individual wealth and the consumptive practices that accompany it as a legitimate outcome of a strenuous striving to succeed, and representative therefore of an unexpected (almost divinely ordained) social mobility that arose against the grain of conventional wisdom. . . Material goods present life-and-death stakes for the mogul's version of utopia because they represent socioeconomic benchmarks of achievement that blacks have been told will forever elude their grasp. (Holmes Smith, 2003, pp. 80–83)

However, such a romantic view of wealth creation can act to hide the complex relationship that exists between wealth, the acquisition of material possessions, the maintenance of 'respect' and the concept of manhood in hip-hop culture (Watts, 1997). It is a culture in which, Rose argues, there is 'a never ending battle for status, prestige, and group adoration, always in formation, always contested, and never fully achieved' (1994, p. 36). Wealth has long been central to this battle for status. One only has to look to the numerous feuds that have occurred between rap artists, involving attacks on a competing rapper's economic status through lyrical diatribes of 'violence, bling and consumption fantasy' (Peterson, 2006, p. 901), to understand how material possessions become part of a campaign of respect (see Watts, 1997).

On the street, hustlers look to enhance their reputation and self-image through self-presentation; to present themselves as a threat to other hustlers, as someone not to be messed with. As such, ostentatious displays of wealth act as a mode of self-recognition (Miller-Young, 2008); they are to do with proving a hustler's place in the world (Rehn and Sköld, 2005), gaining respect and demonstrating manhood. These spectacular displays require conspicuous and spectacular consumption in order to assemble expensive items for show, items that are key in enhancing reputation. What arises is a competitive, 'fierce materialism' (Crossley, 2005, p. 504), in which items are treated as trophies and treasures. These treasures not only demonstrate escape from poverty, but also demonstrate and create self-worth and self-esteem:

> [T]he Hustler is blinded by the glare of materialism, constitutes his social identity in accordance with the street code, and actualizes his self-worth as an objectified street agent by objectifying others; that is, by blinding others with the oppressive glare of the materialistic presence. (Watts, 1997, p. 54)

Over the past couple of decades, a generation of hip-hop musicians, producers and consumers has become almost synonymous with materialism (De Genova, 1995; Crossley, 2005), and a whole language has developed in hip-hop culture to describe the hyper-consumption of luxury goods – perhaps the most well known being the use of 'bling-bling' to describe expensive, oversize jewellery. Conspicuous consumption forms an important part of rap music, both lyrically and visually. For Rehn and Sköld (2005) the 'bling-bling attitude' has come to be *the* dominating attitude in rap music and hip-hop culture. These narrative strategies in hip-hop culture 'enter into a pact with American cultural outlets and are selectively enhanced so that urban (and suburban) youth can share in an artist's attempt to "live large" by replicating and consuming the imagery' (Watts, 1997, p. 43). By taking part in spectacular consumption, they can 'participate in the transposition of poverty into profit' (1997, p. 51).

Gender, Sexualization and Wealth

It is not, however, only material objects that are considered to be trophies within these spectacular displays; so too are women. While issues around gender are not new to hip-hop – as Hunter notes, through 30-plus years of development hip-hop's 'troubling gender relations have never left the spotlight' (2011, p. 17) – this most recent incarnation of hip-hop has created new questions regarding both the voice of women and their representation and sexualization. In the early stages of hip-hop's global commercialization in the 1990s, there was a proliferation of female rap artists. Some, such as Mary J Blige and Lil' Kim were 'introduced' by established male artists, while others such as Queen Latifah, Lauren Hill, MC Lyte and Missy Elliot launched themselves (Phillips et al., 2005). During this period, female artists would have significant success in the sales charts. Yet, as rap has become increasingly mainstream and homogenized, this chart success has evaporated; this, Hunter (2011) argues, is demonstrative of the fact that whiter audiences have less interest in the voice of African American women rappers.

Yet as female rappers have become more invisible to rap audiences, women have become a more visible – and highly sexualized – part of this new consumer-driven hip-hop that is centred on consumption, ownership and displays of masculinity. Women video dancers, for example, have become more and more common, with scantily clad women used to 'enhance the heteronormative masculinity of the male rappers in the video' (Hunter, 2011, p. 18). In this way, women have become part of conspicuous displays of power and wealth, being used by male hip-hop artists to authenticate their hyper-masculinity, entrepreneurial empowerment, sexual prowess and power over women and other men (Miller-Young, 2008). This rise in the sexualization of women in hip-hop and of soft-core sexual performance can be seen to have developed concurrently with the rise of strip-club culture, and is something that seems to have deepened even further with the emerging new genre form of hard-core hip-hop pornography (see Miller-Young, 2008).

It becomes increasingly difficult for female rappers to find an audience for their voices as they attempt to position themselves either against or within this new, highly sexualized context: as Hunter summarizes, 'black women dancers sell; black women rappers don't' (2011, p. 30). As Phillips et al. (2005) highlight, men outnumber women in the artistic arena, and in the corporate arena, of hip-hop. Further, they argue that as hip-hop has expanded beyond music into video production, clothing and lifestyle products, 'the processes impeding women's participation and power-sharing have only become more widespread' (2005, p. 254). Therefore, while as noted in the previous section some women rappers have made personal fortunes in hip-hop, they are relatively small in number and cannot match the earnings of many male artists. Hence the term 'mogul' seems to be almost exclusively used in the popular press to refer to male hip-hop artists, exacerbated by the masculinist biases already in place in the domains of advertising and news reporting.

MOGULS: GETTING RICH AND 'SELLING OUT'?

What makes the hip-hop mogul significant, Holmes Smith (2003) argues, is the way in which, through their celebrity, they appeal to the power of socially competitive consumption as a vehicle for personal fulfilment. The mogul, he suggests, 'inspires more downtrodden constituents to "buy-in" to the emerging paradigm of accessible luxury and social status and in the process assumes an influential role as social mediator' (2003,

p. 71). The influence of these individuals as mediators and as role models is particularly strong when one considers the fact that the entertainment industry in the late twentieth century and early twenty-first century has been one of the few options of hope for upward mobility for African Americans (Rehn and Sköld, 2005). Considered as living proof that the poverty of ghetto life can be overcome, the mogul's glamorous lifestyle serves as symbolic proxy for 'the more mundane strivings of those with whom the mogul shares an apparent racial or ethnic affiliation' (Holmes Smith, 2003, p. 71), and embodies the dream of the poor from the ghetto. Drawing on their own 'rags-to-riches' stories of success against the odds – against drugs, violence and poverty – the opportunistic hip-hop mogul has recognized that the utopian aspirations of his downtrodden constituents can be bought and sold, 'and the higher the price the better' (2003, p. 82).

The question then becomes one of how a hip-hop mogul who creates music that appeals to a wider audience to gain huge wealth, but who as a result loses the connection with the street code that informed his music in the first instance, can be seen to retain their position as social mediator and role model for a working class and poor African American population. Often termed 'selling out', the difficulties of achieving economic success whilst not losing subcultural values – 'making it' and at the same time 'keeping it real' in a way that signifies that one is not removed from the community – continues to be a central and controversial issue within hip-hop culture (Sköld and Rehn, 2007):

> The real contradiction of the music and culture resides in its worldwide acceptance and homogenization (sanitation) by the culture industry. Hip-hop artists now appear in clothing and soda commercials, hawking everything from jeans to junk food. They appear in sit-coms, game shows, and Hollywood films. All of this exists while the music drastically attempts to maintain its authentic hard-core street credibility. Herein resides the fundamental conflict. (Murray, 2004, p. 19)

As Basu and Werbner (2001) argue, making money is not seen as a vice in itself, but when accompanied by this type of 'selling out' – which dilutes the musical centrality of the African American aesthetic/sensibility – then problems of authenticity arise. Indeed, Shawn 'Jay-Z' Carter in particular has regularly been criticized by the music press for selling out, that is to say producing music intended to sell to a mass audience to create personal wealth, becoming distant from the street and the 'true' aesthetic of rap music. A tension thus exists between 'street dictates' and the market strategy of these moguls and music companies; as Holmes Smith notes, 'if the mogul cannot claim to understand and be able to tap the volatile energies of the street he will cease to exist as a viable figure of commercial and cultural enterprise' (2003, p. 85).

The moguls, however, have cleverly managed to turn this tension into a symbiosis by creating themselves as particular personas, in which they become an individual who is at once something special and separate, and part of the community and shared social context of the ghetto. Central to these personas are the conspicuous displays of wealth. Given the centrality of competition to the rap aesthetic, moguls aim to situate themselves as winners of a game, nullifying their rivals through the 'oppressive glare' of their 'materialistic presence' (Watts, 1997, p. 54). This has given rise in hip-hop to narratives around the 'player' – someone playing the 'game' of becoming rich and successful; and 'player-hater' – a term used to criticize people who are jealous of, or who do not respect, successful people. These materialistic battles between rival hip-hop artists remain one of the key ways for hip-hop artists to legitimize their roots in hip-hop culture and assert their continued connection with the street code. At the same time, these battles have also proven to be an effective vehicle for enticing listeners and boosting sales (Rehn and Sköld, 2005), leading to accusations that some of the more highly publicized battles have been publicity stunts (see Peterson, 2006).

Displays of wealth are not the only way in which rappers deal with the tensions between business and the street. Sköld and Rehn (2007) note the way in which rappers are prone to talk about their innovative ways of getting a business started; being able to find and exploit new business empires, they argue, are character traits that seemingly outrank even their ostentatious spending habits. These narrations regarding the various business endeavours in which rappers are engaged are told and retold in rap lyrics and in hip-hop magazines. As Negus (1999) notes, many rap consumer magazines make frequent reference to business management and 'career planning', presenting articles in which rap moguls discuss their commercial strategies and business plans explicitly. Indeed, as Sköld and Rehn (2007) suggest, one might get the impression that stories regarding business and commercial success are more important to hip-hop magazine readers than the music itself. Not only does this imply a strategy of self-actualization through business (Rehn and Sköld, 2005), but talking about business in this way is also a method by which the moguls have looked to build the legitimacy of their connection with the street. They do so through making the connection between business acumen and 'street smart', as a way of emphasizing that the hustling skills learnt on the streets have played a key role in the way they 'handle their business':

> Hear any rappers talk about industry success, business acumen, and entrepreneurial skillsets, and they'll most probably follow it up by telling you that they

> learned their ways on the street, just hustlin' or engaging in an outright criminal lifestyle, in short, by leading their lives in hard-core urban reality. (Sköld and Rehn, 2007, p. 60)

Take the mogul Shawn 'Jay-Z' Carter again, for example, who it was earlier noted has been widely criticized for 'selling out' to gain commercial sales: Carter often emphasizes his background in the drugs trade as a way of legitimizing the link between making money in the music business and hustling on the street. Where rappers profess an upbringing in poverty in their lyrics, but have in reality have a more privileged background, they open themselves up to being discredited by press and audiences. The Canadian rapper Drake, for example, grew up in an affluent neighbourhood in central Toronto, landed a role on a hit TV series at age 15, and then set out on a lucrative career in the hip-hop music industry. He has gleaned much criticism for continuously emphasizing his relationships with poverty through his music, including the release of a track titled 'Started from the Bottom'. But the success that Drake, and other rappers from more privileged backgrounds have had – such as Kanye West, whose mother was an English professor at Clark Atlanta University, and father a photojournalist – are demonstrative of the fact that in the commercial era of hip-hop, the realities of an artist's background matter less than the hip-hop persona they portray through their music.

The Politics of Wealth

It is also important to recognize that the tension between authenticity and 'selling out' in rap music is a tension that bears heavily on the politics of the music. When rap music emerged from the streets of Brooklyn, lyrics typically were concerned with the lived realities of urban poverty. It was through rap music that a street-level politics of struggle and survival began to surface, giving the marginalized a voice that was raw, real and authentic (Jenkins, 2011):

> The conceptual metaphors found in rap music are symbolic of more than just the psychological and physical nature of the hip-hop experience. They are emblems of the social inadequacies that still exist in the United States and of the economic and institutional disadvantages that many African Americans still face. (Crossley, 2005, pp. 509–10)

Rap music, then, could be situated within a strong tradition of African American activism. However, as rap became more corporatized and commercialized and rappers became wealthier, the street politics would give way to the bling-bling culture of spectacular consumption and displays of wealth. As Watkins (2004) notes, the linkage of corporate strategies and marketing techniques with the expressive cultures of African American youth undeniably alters the trajectory of hip-hop. This has been seen as a threatening counter-movement to the politically aware origins of hip-hop music, and as an indication of the decline of African American politics (Rehn and Sköld, 2005). Peterson (2006), for example, argues that for many young people, hip-hop not only drives consumerism, but also creates socio-political apathy.

How do we consider, then, the figure that has been at the centre of this commercialization of hip-hop and benefited most from it financially – the hip-hop mogul – with regard to the politics of the music? For Holmes Smith (2003), while the mogul may lament the plight of the masses, they do not sacrifice their own quest for the American good life on their behalf. In this regard, the hip-hop mogul is not only seen as 'selling out', but is also seen as agent in the decline of African American street-level politics. However, to consider the mogul in this way would be to ignore that bling-bling culture is in itself highly political. It is political in the sense that it represents the ways in which African Americans have successfully made inroads into the music industry, and have exploited the industry on their own terms to redistribute wealth and create African American capital in a way that has not have been achieved by African Americans in any other area of the economy. Thus, the wealthy hip-hop mogul, rather than an agent in the weakening of African American politics, becomes the key agent of its redefinition:

> For Jay-Z the bling-bling thus becomes an instrument for doing something else; a way of redrawing the industry map, of moving borders and upsetting governing orders within the music industry. A means with deterritorializing powers which has enabled him as a rapper and former crack-dealer from the streets of Brooklyn to succeed in entering the industry, upsetting it from the inside. (Rehn and Sköld, 2005, p. 27)

In this way, bling-bling becomes both an economic language and political movement (Rehn and Sköld, 2005) that has replaced politicized defiance with an economic one (Murray, 2004).

CONCLUSIONS

Whether it is considered 'selling out' or just good business, given that the main consumers of hip-hop music are primarily white suburban males (Samuels, 1995), the main successes of (predominantly male, African American) hip-hop moguls resides in their ability to symbolically reproduce the street code and commodify it 'in the form of an easy-to-open package of hip' (Watts, 1997, p. 51). The 'hood' has become a commodified virtual reality, which entrepreneurial rap artists have packaged and capitalized upon (Sköld and Rehn, 2007), and the huge personal wealth of the moguls has been derived from their ability to turn 'a potent African American urban cultural idiom into a palatable mainstream product' (Basu and Werbner, 2001, p. 256).

In no small part through the actions of an aggressive corporate sphere, the hip-hop industry today is both hugely profitable and highly corporatized. The entrepreneurial moguls, rather than resisting corporate moves into hip-hop, have in fact been able and willing to further aid in the commodification of the music, which has resulted in great financial reward for themselves (Johnson, 2008). They have successfully fused commercial artistic success with entrepreneurial achievement, developed significant marketing power, and established new forms of partnerships and deals with established firms (Cox Edmondson, 2008). Through their street-inspired lyrics, spectacular displays of wealth and business acumen, we see how the two spheres of art and economy/business have collapsed into one another to become one and the same (Rehn and Sköld, 2005). As such, the rise of the hip-hop moguls can be seen as legitimizing a set of exchange relations between public culture, rap music and the rap industrial complex (Watts, 1997).

Yet, at the same time, the rise of the hip-hop moguls is representative of how African Americans have not only created a music industry around a culture of entrepreneurship (Basu and Werbner, 2001),but have also retained control of hip-hop music and culture in the face of the major music corporations. As Basu and Werbner suggest, their entrepreneurship involves 'not stoicism and frugality, but an appreciation of conspicuous consumption, enjoyment, outlaw narratives and subversive styles' (2001, p. 247). As such, they have recoded the transgressive roots of rap music and hip-hop culture within the context of capitalist consumption (Murray, 2004).

NOTES

1. The title for this chapter is from the song 'Get Rich' by Tyga, from the album *Hotel California*, released on 9 April 2013 under Young Money Entertainment, Cash Money Records.
2. The number of sales required for an album to reach platinum status generally depends upon the population of the territory in question. For example, in the United States an album is awarded platinum status after selling 1 000 000 units, while in the UK it is only 300 000 units.

3. A '360-degree deal' is a business relationship between a recording artist and a record company in which the company provides financial support for the artist (e.g., direct advances, funds for marketing, promotion and touring) while the artist gives the company a percentage of all of their income, including from sources other than recorded music, such as live performance and merchandise.

REFERENCES

Basu, D. and P. Werbner (2001), 'Bootstrap capitalism and the culture industries: a critique of invidious comparisons in the study of ethnic entrepreneurship', *Ethnic and Racial Studies*, **24** (2), 236–62.

Block, A. (2013), 'Sean "Diddy" Combs betting big on Revolt TV', *Billboard*, 21 October, accessed 8 April 2014 at http://www.billboard.com/biz/articles/news/tv-film/5763143/sean-diddy-combs-betting-big-on-revolt-tv.

Cox Edmondson, V. (2008), 'A preliminary review of competitive reactions in the hip-hop music industry', *Management Research News*, **31** (9), 637–49.

Crossley, S. (2005), 'Metaphorical conceptions in hip-hop music', *African American Review*, **39** (4), 501–12.

De Genova, N. (1995), 'Gangster rap and nihilism in Black America: some questions of life and death', *Social Text No. 43*, 89–132.

Forde, E. (2013), 'Jay-Z's Samsung deal signals a musical future where the rich get richer', *The Guardian*, 4 July, accessed 30 May 2013 at http://www.theguardian.com/music/musicblog/2013/jul/04/jay-z-samsung-music-future.

Garofalo, R. (1994), 'Culture versus commerce: the marketing of black popular music', *Public Culture*, **7** (1), 275–88.

Gensler, A. (2014), 'Sean "Diddy" Combs makes $200 million bid for Fuse', *Billboard*, 10 March, accessed 8 April 2014 at http://www.billboard.com/biz/articles/news/tv-film/5930372/sean-diddy-combs-makes-200-million-bid-for-fuse-report.

Holmes Smith, C. (2003), '"I don't like to dream about getting paid": representations of social mobility and the emergence of the hip-hop mogul', *Social Text*, **21** (4), 69–97.

Hunter, M. (2011), 'Shake it, baby, shake it: consumption and the new gender relation in hip-hop', *Sociological Perspectives*, **54** (1), 15–36.

Jenkins, T.S. (2011), 'A beautiful mind: black male intellectual identity and hip-hop culture', *Journal of Black Studies*, **42** (8), 1231–51.

Johnson, C.K. (2008), 'Danceable capitalism: hip-hop's link to corporate space', *The Journal of Pan African Studies*, **2** (4), 80–92.

Kun, J. (2002), 'Two turntables and a social movement: writing hip-hop at century's end', *American Literary History*, **14** (3), 580–92.

McIntyre, H. (2014), 'At 2 billion views, "Gangnam Style" has made Psy a very rich man', *Forbes.com*, 16 July, accessed 29 October 2014 at http://www.forbes.com/sites/hughmcintyre/2014/06/16/at-2-billion-views-gangnam-style-has-made-psy-a-very-rich-man/.

Miller-Young, M. (2008), 'Hip-hop honeys and da hustlaz: black sexualities in the new hip-hop pornography', *Meridians: Feminism, Race, Transnationalism*, **8** (1), 261–92.

Murray, D.C. (2004), 'Hip-hop art: notes on race as spectacle', *Art Journal*, **63** (2), 4–19.

Negus, K. (1999), 'The music business and rap: between the street and the executive suite', *Cultural Studies*, **13** (3), 488–508.

Oliver, R. and T. Leffel (2006), *Hip-Hop, Inc. Success Strategies of the Rap Moguls*, New York: Thunder's Mouth Press.

O'Malley Greenburg, Z. (2011a), 'Why Diddy will be hip-hop's first billionaire', accessed 8 April 2014 at http://www.forbes.com/sites/zackomalleygreenburg/2011/03/16/why-diddy-will-be-hip-hops-first-billionaire.

O'Malley Greenburg, Z. (2011b), *Empire State of Mind: How Jay-Z went from Street Corner to Corner Office*, London: Penguin Books.

O'Malley Greenburg, Z. (2014), 'Diddy's net worth: $700 million in 2014', Forbes.com, 18 April, accessed 28 April 2014 at http://www.forbes.com/sites/zackomalleygreenburg/2014/04/18/diddys-net-worth-700-million-in-2014/.

Peterson, J. (2006), '"Dead presence": money and mortal themes in hip hop culture', *Callaloo*, **29** (3), 895–909.

Phillips, L., K. Reddick-Morgan and D.P. Stephens (2005), 'Oppositional consciousness within an oppositional realm: the case of feminism on rap and hip-hop', *The Journal of African American History*, **90** (3), 253–77.

Rehn, A. and D. Sköld (2005), '"I love the dough": rap lyrics as a minor economic literature', *Culture and Organization*, **11** (1), 17–31.

Rose, T. (1994), *'Black Noise': Rap Music and Black Culture in Contemporary America*, Hanover, NH: Wesleyan University Press.

Samuels, D. (1995), 'The rap on rap: the "black music" that isn't either', in A. Sexton (ed.), *Rap on Rap: Straight-Up Talk on Hip-Hop Culture*, New York: Delta.

Sköld, D. and A. Rehn (2007), 'Makin' it, by keeping it real: street talk, rap music, and the forgotten entrepreneurship from the "hood"', *Groups & Organization Management*, **32** (1), 50–78.

Tennent, J. (2013), 'No one is paying attention to Dubai's mega-rich rappers', World Hip Hop Market, 18 November, accessed 29 November 2014 at http://worldhiphopmarket.com/no-one-is-paying-attention-to-dubais-mega-rich-rappers/.

Watkins, C.S. (2004), 'Black youth and ironies of capitalism', in M.A. Neal and M. Forman (eds), *That's the Joint: The Hip-Hop Studies Reader*, New York: Routledge, pp. 557–78.

Watts, E.K. (1997), 'An exploration of spectacular consumption: gangsta rap as cultural commodity', *Communication Studies*, **48** (1), 42–58.

10. Biographies of illicit super-wealth

Tim Hall

INTRODUCTION

As well as holding great material wealth, the super-rich have immense normative power across a range of cultural discourses. They shape desires and resultant spending patterns across a wide economic and cultural spectrum (Frank, 1999). This chapter is concerned, in part, with the normative power of super-wealth, and its consequences, and explores it through the lens of that subset of the super-wealthy whose money has been derived from illicit economic activities. Illicit economic activities here refer to either those activities that are located within illegal markets, for example, illegal narcotics markets, or activities that are illegal but take place within legal markets, such as financial fraud. In addition we might also recognize and include the actions of corrupt public officials. This chapter aims to say something about the illicitly super-wealthy, of whom the super-rich literature has said very little to date, and also about the normative power that comes with wealth, again, something that the salient literatures have only explored in a limited sense thus far. It aims, therefore, to add a little range and nuance to the multidisciplinary literatures of the super-rich. Implicit in this chapter are a number of questions regarding the literature of the super-rich. First it wonders what these literatures have said about illicit wealth. It also considers the extent to which we can apply observations about the super-rich derived from its extant multidisciplinary literatures to those members of this elite who derive their wealth by illicit or outright criminal means. Further, it argues that much literature about the super-rich says little about the diversities within this admittedly tiny, but by no means homogeneous, group. By looking specifically at the illicitly super-wealthy the chapter considers the extent to which exploring variety within the super-rich might enhance its literatures.

Despite their relatively small number there is significant diversity amongst the illicitly super-rich. At the risk of simplification and of overstating their categorical distinction, we might recognize three categories of the illicitly super-rich. These are criminal entrepreneurs, financial fraudsters and corrupt public officials. Criminal entrepreneurs are those who operate primarily within markets that are formally illegal, notwithstanding various qualifications about the ambiguities around the status of many markets and of commodities as they pass through a diversity of regulatory spaces (Abraham and van Schendel, 2005; Hall, 2013; Hudson, 2014). Having said that, once they have established any sort of success their criminal identity is likely to become blurred as they seek to colonize legitimate markets for the purposes of money laundering, diversification or risk aversion. Both Pablo Escobar and Dawood Ibrahim, discussed later in this chapter, are examples of criminal entrepreneurs. Financial fraudsters refer to those who operate within legal markets, and who, initially at least, are typically recognized as legitimate actors within these markets. Their purpose is to defraud these markets for personal gain, protection or professional position. Often their initial intentions are not criminal but rather their motivations for crime within these markets can be genuine attempts to pursue profit or recover losses from bad investment decisions and trades. Nick Leeson, the 'rogue trader' whose unchecked high-risk investments brought down Barings Bank in 1995 would be a case in point here (Leeson, 1999). Bernard Madoff, discussed below, is an example of a financial fraudster. Corrupt public officials are those operating with legitimate, public economic and political arenas who abuse their power for private gain. This chapter will discuss examples of the first of these two categories of the illicit super-rich.

The super-criminal is an enduring figure within popular discourses. Central to the construction of this archetype is the accumulation of incredible wealth (and/or power). The Cold War was a particularly fertile period for the imagination of a series of ciphers that stood for the West's geopolitical fears of the East. James Bond famously fought Dr Julius No, Auric Goldfinger, Emilio Largo and Ernst Stavro Blofeld amongst others, while the secret agents, Napoleon Solo and Illya Kuryakin, the men from U.N.C.L.E., crossed swords with several operatives from the shadowy, and unfortunately named, criminal organization T.H.R.U.S.H. More recently the association of crime = power = wealth has recurred in the popular US television drama *Breaking Bad* (2008–13), where the mild-mannered high school chemistry teacher, Walter White, turned to crystal meth production to provide for his family financially after a cancer diagnosis. In the process he accumulated millions of dollars, which he stored, neatly piled, in his lock-up garage whilst his wife tried in vain to launder it through the family's car wash business. Further, much of the normative power of gangster iconography, derived from the recycling of mythologies located in films such as *The Godfather* (1972), *Scarface* (1983) and *Goodfellas* (1990), so pervasively deployed within contemporary rap music culture, for example (see Chapter 9 by Allan Watson in this volume), derives from its supposed associations with conspicuous consumption, wealth and forms of hyper-masculinity (Hobbs and Antonopoulos, 2013).

Crime = power = (super-) wealth is a pervasive, intuitively convincing and enduring cultural discourse. However, it is one more rooted in our fears, or our questionable aspirations, than in any empirical reality. There are super-rich criminals for sure, but they are few in number. Their ranks are boosted in our minds by the tendency of the media to inflate the standing and achievements of mid-ranking players from the criminal underworld, mythologies those individuals are often only too keen to buy into (Hobbs, 2013; Hobbs and Antonopoulos, 2013). So, whilst super-rich criminals may be tiny in number, a crucial observation that underpins research into the super-rich generally is their 'hyperagency' (Hay and Muller, 2012, p. 81), their ability to shape localities, regions and transnational economic systems. An agenda that has underpinned the advancement of research into the super-rich in recent years has been predicated on their being powerful agents in the world. The hyperagency of super-rich criminals is a less clearly empirically observable concept. Whilst individual 'Mr Bigs' have significant impacts on their regions and upon specific transnational flows, it cannot be plausibly argued that the global criminal economy is more fundamentally shaped by a small cabal of super-criminals than the licit neoliberal global economy is bent to serve the needs of a small global elite of the super-wealthy. However, what is apparent is the ability of a class of super-rich criminal whose wealth derives from illegal activities within the financial markets of global cities to take advantage of the changes to the regulation of these markets that have been a characteristic of the trajectory of neoliberal global capitalism until the late 2000s.

But first a methodological note. It is very difficult, if not impossible, to define which criminals might be regarded as super-rich. Criminals will only be successful if they are either able to hide their wealth or evade the law. Neither of these makes for easy access for the researcher. However, the elusiveness of the super-rich subject is not limited to those whose wealth was obtained through illicit or illegal means. Leaving aside the whole question of the ambiguities between licit and illicit wealth generation, others have noted that the super-rich generally are a very difficult population to conduct research with (Hay and Muller, 2012, p. 83). The three main subjects of this chapter, Pablo Escobar, Dawood Ibrahim and Bernard Madoff, are chosen partly for the unambiguity of their membership of the super-rich club. Inevitably, though, the material upon which this account is based is secondary and partial.

THE GLOBAL ILLICIT ECONOMY

In aggregate terms, the generation and movement of illicit money, goods and services by organized criminal groups are significant globally (Hall, 2013). Although there is debate about the extent of the global organized crime economy (Madsen, 2009, pp. 81–2), and accurate estimates are impossible, there is widespread agreement that organized crime forms a significant proportion of global economic activity. Quantitatively most estimates put the activities within the global organized crime economy as constituting roughly 15–20 per cent of total global GDP (Galeotti, 2005a, p. 1; Glenny, 2008; Madsen, 2009). Although some commentators have argued that we have seen some containment of organized crime since the mid-1990s (Moynagh and Worsley, 2008, p. 177) the prevailing view is one of continuous growth in the trafficking of goods of various kinds and the associated provision of illicit services by organized criminal groups (Madsen, 2009, p. 25; see also Bhattacharyya, 2005; Aas, 2007, pp. 25, 125). These flows are dominated by money generated by the production and distribution of narcotics, which have been estimated as constituting between 50 and 70 per cent of the economic activities of organized criminal groups (Galeotti, 2005a, p. 1; Glenny, 2008, p. 262). This trade alone, fuelled by great public demand and failed international prohibition regimes, represents approximately 8 per cent of world trade, suggesting a global industry of roughly the same size as the textile industry (Moynagh and Worsley, 2008, p. 176).

The trafficking of people is now well established as a major activity of organized criminal groups (Aguilar-Millan et al., 2008, p. 8; Madsen, 2009; Westmarland, 2010, p. 124). Some commentators have suggested that people trafficking represents the fastest growing source of income for organized criminal groups, trades that are underpinned by combinations of expanding global inequalities, regional social and political unrest, particularly, but not exclusively, within regions of the Global South, the prevalence of cultural traditions of migration, fostering and the remittance of money in some regions, the hardening of many international borders and, once again, great demand within civil society and the licit economy in the Global North for illegal labour and sexual services (Passas, 2001; Bhattacharyya, 2005, pp. 159, 174; Manzo, 2005; Madsen, 2009, p. 43; Westmarland, 2010). Widely circulating estimates suggest that roughly 4 million people are trafficked globally per year, with, for example, between 400 000 and 500 000 women trafficked into the EU annually since the late 1990s. Global revenues from people trafficking stood at roughly US$10 billion per year in the mid-2000s (Galeotti, 2005a, p. 3; Wright, 2006, p. 98; Aguilar-Millan et al., 2008, p. 8; Madsen, 2009, p. 45) although it should be noted that some estimates put these figures considerably higher. Bhattacharyya, in moving beyond the numerical significance of people trafficking, has argued that the sexual and economic exploitation of people in this way is integral to '"the sewing together of the world"' (2005, p. 169) through contemporary globalization, with trafficked individuals acting as vital reserves of flexible labour and forms of sexual capital important to the economic development of certain places such as urban Thailand (2005, pp. 171, 177).

Other significant, and apparently growing sources of revenue for organized criminal groups include the trafficking of licit goods to take advantage of tax differentials or to avoid tax or duty (Madsen, 2009), most notably cigarettes, and the trade in counterfeit goods. Cigarette smuggling, for example, has been identified as an activity of geopolitical, as well as economic significance. Production of counterfeit goods has also been identified as a major revenue stream for organized criminal groups. This is now a trade that accounts for approximately 7 per cent of total world trade, two-thirds of which originate from production sites in China (Phillips, 2005; Glenny, 2008; Saviano, 2008). Organized criminal groups are also responsible for the trafficking of a host of other goods, as well as the provision and control of illicit services such as gambling and prostitution as well as cybercrime, crimes against the environment, robbery, extortion and corruption (Albanese, 2005; Aas, 2007; Madsen, 2009; Neal, 2010).

Organized crime is underpinned by the globally significant illicit capital flows associated with the process of money laundering. This is a process whereby organized criminal groups distance their monies from their criminal origins. Money laundering and other illicit flows both blur the boundaries between the criminal and licit economies and also, it has been argued, play important roles in the sustenance of the global financial system more widely (Hampton and Levi, 1999; Brown and Cloke, 2007, p. 319; Deneault, 2007). Estimating the global extent of laundered finance is inherently difficult. Estimates that are available vary but suggest figures of between 2 and 5 per cent of global GDP (Levi, 2002, p. 879; Reuter and Truman, 2004). The illicit mobilities associated with this finance are not merely 'paper' crimes or ones only involving the reclassification of capital on computer screens. Rather, they have been recognized as having a number of significant material impacts. They have been identified as potentially undermining the stability of currencies and national economies – for example, in the case of the 2001 crash of the South African rand (Nordstrom, 2007, p. 168).

Offshore financial centres, for example, have developed as key elements of the wider architecture of neoliberal, global capitalism and central regulatory spaces employed by corporations seeking advantageous, yet legal, tax arrangements (Roberts, 1995; Hampton and Levi, 1999; Deneault, 2007). In total, offshore financial centres contain some 4000 licensed offshore banks (Madsen, 2009, p. 114) that service significant proportions of global GDP, with estimates suggesting perhaps up to one-fifth is held in offshore funds (Dick, 2009, p. 98). Further, there is evidence that roughly 40 per cent of world trade in goods and 65 per cent in hard currency is serviced through offshore financial centres (Nordstrom, 2007, pp. 65, 172). Although small in number and tiny in spatial extent, these centres exert significant influence across global financial systems and act as key nodes in these networks. Illicit and licit finance then mingle in the same material and virtual spaces, evidence of an opacity inherent within the global economy that many accounts fail to fully acknowledge (Brown and Cloke, 2007). Offshore financial centres show that the instruments and institutions developed to aid contemporary neoliberal globalization have been appropriated and employed with alacrity, and in empirically similar ways to licit corporations, by criminal entrepreneurs of various stripes who are faced with all the advantages of opacity offered by offshore financial centres and only very weakly enforced and largely ineffective anti-money laundering legislation (Wright, 2006; Weinstein, 2008, p. 29; Madsen, 2009, p. 112; Sharman, 2011). This suggests that illicit financial flows, of which laundered money constitutes a significant element (Hampton and Levi, 1999; Deneault, 2007; Madsen, 2009), are very much part of the economic mainstream and that tackling them requires more than just catching rogues and criminals but rather looking more fundamentally at reducing or eliminating the opacity of the wider global financial system.

Financial fraud of various scales is a recurrent element of financial markets. There are many examples of major financial fraud netting, in some cases, many millions of pounds, dollars or euros to the perpetrators. Often executed with great complexity these frauds come in many forms. It is impossible to estimate the global extent of fraudulently obtained capital at any one time. However, a cursory glance at the scale of individual cases suggests the potential significance of financial fraud as a source of illicit super-wealth. The death of the media proprietor Robert Maxwell in 1991 revealed that he had fraudulently appropriated hundreds of millions of pounds from the pension funds of the companies he had run (Mackay, 1992). The businessman Asil Nadir was jailed for ten years in August 2012 for the theft of £29 million from the British company Polly Peck in the late 1980s (Neville, 2012). In 2012 the Texan financier Allen Stanford was found guilty of perpetrating fraud worth US$7 billion and was subsequently jailed for 110 years (Dart, 2012). The case of Bernard Madoff, which dwarves all of the above, is discussed below. Systematic estimates of the extent of fraudulently obtained capital would offer valuable additions to the academic literatures of organized criminality and super-wealth.

THE BIOGRAPHICAL APPROACH

Biographical literatures have been relatively sparsely deployed as a source and method within the multidisciplinary literatures of organized crime (though see Hobbs, 2004, p. 413). This is so for a number of reasons, most notably the inaccessibility of actors operating within this terrain. Few reliable or comprehensive biographical accounts of organized criminals exist, although financial fraud is more thoroughly documented within the international financial media. Those accounts that do exist are often tainted by sensationalism, or voyeuristic or egotistical exaggeration. Where much is known about the lives and activities of organized criminals, typically historical actors such as Al Capone or the Kray twins, there is a tendency for these figures to extend a normative reach over the more messy, mundane and plural terrain of organized crime economies (Hobbs, 2013). Much of the mythologization of organized crime stems from the selection of aspects of such well-known crime figures, their physical or sartorial appearance, apparent notions of honour or territorial allegiances, for example, and their extension and amplification through the cultural lenses of the mass media and entertainment industries. Rather, scholars of organized crime have pursued more systematic or structural aspects of the topic than local or biographical elements. For this reason they have tended to favour accounts such as the regional case study where individual actors feature relatively little within the narrative. Further, there is a limit to what the biographical can tell us about the wider terrain of organized crime given the exceptionalness of genuinely super-rich crime lords. Influential scholars of organized crime have done much to undermine the mythologization associated with this personalized rendition of organized crime. Rather, ethnographically informed accounts have tended to reveal worlds of hard graft, little glamour, save for some localized kudos, and limited financial reward (Hobbs, 2013).

What then do we gain from exploring the biographies of super-rich criminals given the qualifications outlined above and the prevalence of alternative approaches within the literature? Most basically, in some instances, such actors are undoubtedly important agents within the developmental trajectories of regions where organized crime is deeply embedded and, in a few instances, is the dominant element within regional economies such as parts of Afghanistan whose economy is closely locked into the production of opium (Goodhand, 2009), Northern Mexico, which is a key transit zone for drug trafficking into the US market (Vulliamy, 2010) and even British Columbia in Canada, which has seen a major growth in cannabis production following recession in some of its more traditional industries (Glenny, 2008, pp. 245–55). Here, to ignore the roles of very powerful individual criminals produces somewhat colourless and denuded accounts. The practices of financial fraudsters are also significant in the ways that the reactions to their exposure has shaped regulation of the markets they operate within and the nature, structures and practices of regulating bodies (Chung et al., 2009). The biographies of those operating fraudulently within legal financial markets are relatively well documented through autobiographical (Leeson, 1999) and biographical narratives (Elkind and McLean, 2004; Arvedlund, 2009) and through the international financial press (see Sender, 2008, for example). These are potentially valuable sources of empirical detail of the scandals and practices they document. In addition, however, if appropriately contextualized and interpreted they may point to historical continuities or structural weaknesses of markets and their regulatory regimes. These sources, deployed appropriately offer peopled readings of these realms, highlighting the roles of agency within their reproduction or transformation. Further, by being exceptional in terms of their being rare does not necessarily mean that they are atypical or exceptional in other ways. Their biographies may articulate aspects of the practices of organized criminal economies that are of more systemic significance. In a world where reliable accounts of the micro-scale practices operating within this economic realm are rare, such accounts should not be dismissed. In addition, individuals such as the ones discussed in this chapter often occupy ambiguous positions straddling, in multiple ways, the licit and illicit economic, political and social worlds of their regions. This ambiguity, the elusiveness of any categorical distinction between licit and illicitness, has been cited as key to understanding the wider significance of organized crime as an economic and political phenomenon (Hall, 2013). The biographies of individuals who appear to be operating across this ontologically elusive distinction are potentially revealing of the ways in which the articulations between these realms are practised. Finally, these accounts potentially speak to the extant literatures of the super-rich (Gilding, 1999; Beaverstock et al., 2004; Petras, 2008; Hay, 2013). To date the insights derived from these literatures have emerged largely from examining the practices of those operating licitly, avowedly at least, within licit economic realms. Currently we know little about how these insights play across to those engaged in extensive criminal entrepreneurialism. Biographies of super-rich criminals then are troublesome in many ways. However, within the multidisciplinary literatures of the super-rich and of organized crime, they offer potential insights to complement those derived from other approaches.

It is worth emphasizing though two more notes of qualification here. First, the story of illicit super-wealth that emerges through biographical accounts is contingent very much on the subjects selected. This chapter focuses in part on the cases of two major criminal entrepreneurs who have emerged from relatively impoverished, turbulent local and regional settings in the Global South. There are many other stories to be told about illicit wealth though, many of them emerging much closer to the heart of the Western financial system. The inclusion of the case of Bernard Madoff here is an attempt to extend the narrative of the chapter. Given the political renditions of organized crime that have emerged in the West (see Woodiwiss and Hobbs, 2009; Hobbs and Antonopoulos, 2013) particularly those that construct it as an ethnically specific, external threat, it is important to recognize the contingencies inherent in such accounts. Finally, there is a danger of imagining organized crime through the biographies of super-rich criminal entrepreneurs overstates somewhat the structural coherences of their operations (Kenney, 2007). Whilst the individuals discussed here, particularly their brand and reputations for violence, are important they should not be considered criminal masterminds controlling vast empires as a puppeteer controls a puppet. Rather, the operations of criminal enterprises are more diffuse affairs.

THAT TINY STRATOSPHERIC APEX WHO OWN MOST OF OUR IMAGINATION

The subjects of this discussion, Pablo Escobar (born Medellin, Colombia 1949, died Medellin 1993), Dawood Ibrahim Kaskar (born Bombay, 1955) and Bernard Madoff (born 1938, Queens, New York City) have been selected because they undoubtedly qualify as super-rich and have also been the subjects of some of the few extended discussions in the academic and serious journalistic literatures (Castells, 2000; Glenny, 2008). First it is worth dealing with their wealth. Unsurprisingly there are no reliable estimates of their absolute wealth. Both Escobar and Ibrahim appear regularly on lists of the world's wealthiest gangsters. However, such lists tend to be frivolous exercises where almost any gangster with a name will make an appearance at some point. However, in the cases of Escobar and Ibrahim there is substance to back up these assertions. In the early 1990s Dawood Ibrahim's annual turnover was estimated at a quarter of a billion dollars by the Indian Police, and this was from his Indian operations alone (Glenny, 2008, p. 156). Pablo Escobar's wealth was such that, at the height of his operations in the 1980s, he was able to offer to pay off Colombia's foreign debt as a route to legitimacy within the Colombian state (Castells, 2000, p. 205). It would appear that this was a realistic proposition rather than mere braggadocio. Bernard Madoff is a former investment advisor and former non-executive chairman of the NASDAQ stock market. Madoff's career in finance came to an end in 2008 when he was arrested and subsequently convicted in 2009 of 11 federal felonies relating to a massive, possibly the largest ever, Ponzi scheme. For this he received a sentence of 150 years in prison. The size of the scheme, which had been running since at least the 1980s, has been estimated at US$65 billion (Chung et al., 2009) although more sceptical assessments put it in the region of US$10–17 billion (Hays et al., 2009). Madoff had been apparently investing money from a large number of clients from hedge funds to small investors, hospitals and charities and producing steady returns over many years. In reality these returns were being financed by money from new investors. The revelation of Madoff's crimes, their longevity and extent and the failure of US financial authorities to detect them severely shook investor confidence in US and international financial markets and led to calls for significant regulatory reform (Chung et al, 2009).

Rather than rehashing details of their biographies, this discussion reinterprets details from these three lives, drawn primarily from Castells and Glenny's accounts and those of the Madoff scandal in the global financial media, in the light of concerns pursued by scholars of the super-rich. In doing so it aims to add some empirical range to these discussions whilst highlighting the need, potentially, for greater conceptual nuance. The discussion is presented as a series of themes of relevance to the super-rich literature, illustrated through these cases.

Case 1: Super-rich criminal entrepreneurs tend to be, and largely remain, deeply embedded in their home regions

'Weightless' hypermobility has been observed as a characteristic of the super-rich. However, there is some debate in the literature about the relative mobility or fixity of the super-rich. Hay and Muller (2012), for example, have questioned Beaverstock et al.'s (2004) ideas about the weightless and mobility of the super-rich, suggesting instead that many super-rich are rooted in place. The biographies examined in this chapter also suggest that rootedness in and to place is characteristic of the criminally super-rich. Such actors often emerge out of the turbulent social, economic and political conditions of their home regions and, despite the often considerable geographical reach of the networks they are connected with, tend to retain strong ties to these regions. The question of the extent to which criminal entrepreneurs, of all types, ever really transcend their home regions is a valid one. It is not uncommon for criminal entrepreneurs, even major or super-rich players, to remain resident in these regions, drawing on the support of local gang members and sometimes the local populations and political community. Super-rich criminal entrepreneurs certainly share the desire of the super-rich generally for privacy and security; however, in their case they tend to seek it within highly secure, sometimes opulent, compounds, located within the very regions from which they built their networks and perhaps the regions where their reputations retain the greatest currency. Criminal reputations then are not necessarily entirely transferable commodities (Varese, 2011), which might explain the more constrained mobilities of this group of the super-rich. Further, criminal mobilities of all kinds are further constrained by practical issues such as regulatory regimes and extradition arrangements. The practicalities of hypermobile citizenship can undoubtedly be negotiated, but they appear more problematic for the criminally super-rich than for the super-rich generally (see below).

This regional embeddedness is demonstrated well by both Pablo Escobar and Dawood Ibrahim. Escobar's entry into crime came through his efficient trafficking of stolen gravestones. A lucrative trade, this equipped him with the expertise, network logistics and connections that he was later able to apply to the smuggling of cocaine when that market opportunity arose in the 1980s (Castells, 2000, p. 203). Escobar remained deeply embedded within the Antioquia region of Colombia and specifically its capital city of Medellin. The area from the 1970s was deeply in recession following the collapse of its textile industry. The influx of cash from the exportation of cocaine allowed Escobar to invest in the region and build considerable status and support. At one point he was elected to the Colombian Congress before being removed at American insistence (2000, p. 204). Drawing on the case of Escobar, admittedly exceptional but not entirely atypical, Castells has spoken of this regional embeddedness as a key characteristic of major narcotics entrepreneurs:

> The attachment of drug traffickers to their country, and regions of origin, goes beyond strategic calculation. They were/are deeply rooted in their cultures, traditions, and regional societies. Not only have they shared their wealth with their cities, and invested a significant amount (but not most) of their fortune in their country, but they have also revived local cultures, rebuilt rural life, strongly affirmed their religious feelings, and their beliefs in local saints and miracles, supported musical folklore (and were rewarded with laudatory songs from Colombian bards), made Colombian football teams (traditionally poor) the pride of the nation, and revitalized the dormant economies and social scenes of Medellin and Cali – until bombs and machine guns disturbed their joy. The funeral of Pablo Escobar was a homage to him by the city, and particularly by the poor of the city: many considered him to have been their

> benefactor. Thousands gathered, chanting slogans against the government, praying, singing, crying and saluting. (2000, p. 205)

Dawood Ibrahim's origins were similarly located within the turbulent popular politics and street life of his home city Bombay, later renamed Mumbai, in India. Ibrahim, the son of a police officer, grew up in a poor neighbourhood in central Bombay and became involved in petty street crime at a young age. The power base for his step up into more organized criminality came with his, along with his elder brother, taking control of the Young Party, a Muslim political group, in the 1970s, which they subsequently remodelled into a criminal outfit (Glenny, 2008, pp. 148–9). Ibrahim, like other organized criminal entrepreneurs in India at the time, took advantage of the demand for alcohol, consumer goods and gold, difficult and expensive to obtain under India's post-war protectionist economic regime, to gain slices of the extensive smuggling operations around Bombay (Glenny, 2008; Weinstein, 2008). Despite fleeing to Dubai in 1984 to escape arrest, Ibrahim remained closely connected to and directly involved in Bombay/Mumbai's underworld, overseeing the partial unification of the city's various organized crime groups in the late 1980s and drawing financial tribute from those units who traded under his brand (Glenny, 2008). Later, Ibrahim's organization, D Company, shifted into drug trafficking following the liberalization of the Indian economy in the early 1990s that undercut the traditional revenue streams of many Indian organized crime operations (Weinstein, 2008). Later, after alleged involvement in the 1993 Bombay bombings and other terrorist activities, Ibrahim relocated to Karachi in Pakistan where he remains, protected, it has been widely reported, by the Pakistani intelligence services (Glenny, 2008). More recently, Ibrahim and D Company have been linked to corruption and match fixing in cricket, a massive betting market in India (Hawkins, 2012). Although Ibrahim left Bombay/Mumbai in 1984 and oversaw transnational criminal operations, it is questionable whether he truly achieved the weightlessness that some have attributed to members of the super-rich and transcended his regional origins and embeddedness.

Identification with a place or region, particularly if it is the home region, is potentially of significance within the realm of financial fraud. Bernard Madoff's connection to New York and specifically its suburban Jewish communities appears to have been important within his subsequent criminal activities. As well as being a member of the financial and commercial establishment in New York for many years, Madoff deployed his agency across some of the city's socio-political contexts such as country clubs and charity boards (Chung et al., 2009). As well as part of an effort to construct an image of exclusivity and trust these settings were important sources of investment for Madoff:

> Bernard Madoff rose to prominence on Wall Street by raising money from people who in many cases were largely like himself – New Yorkers, often based in the Long Island suburbs, with ties to the city's financial and business community. Mr Madoff, 70, was a member of the club, if there is such a thing in New York. (Sender, 2008, p. 20)

Whether Madoff viewed this milieu in entirely instrumental terms is not known but what is clear is that they facilitated his fraud directly through maintaining contacts with potential investors and indirectly in that they helped Madoff to be seen as a respectable, trustworthy individual with an apparent affinity for more than just wealth. This insider status may have also played a part in the financial authorities' failures to detect Madoff's fraud for many years despite a number of warning signs (Plender, 2008). These observations suggest the potentials of more territorialized readings of financial fraud than have tended to prevail to date.

Case 2: The international migratory mobilities of criminal entrepreneurs operating within illegal markets tend to be shaped by geographical, cultural, linguistic, trading and regulatory factors

Despite being regionally embedded, super-rich criminal entrepreneurs are far from immobile (Varese, 2011). Their operations are typically networked and transnational and they display a propensity for international migration. At times they move because they are driven out of their regions of origin by fear of arrest and prosecution or regional turbulence; at times they are drawn by more welcoming regulatory regimes elsewhere; often they move due to a combination of these factors. Undoubtedly the tight bonds between neighbourhood and gangs that characterized organized crime in earlier phases of its development are loosening. For example, discussing the case of Russian criminal organizations in a global era, Galeotti (2005b, p. 58) can confidently claim 'the old organising principles of ethnic ties or place of origin are breaking down. Many gang or network names relate to the place where they first formed or where their original leaders came from, but these often have little bearing on their present location or operation'. However, this is not the same as suggesting these groups enjoy unfettered hypermobility.

Dawood Ibrahim's migration to Dubai in 1984 demonstrates the ways in which these migrations can be shaped and constrained. Dubai is relatively proximate to Bombay/Mumbai. Despite being some 2000 kilometres (1200 miles) apart across the Arabian Sea, the two ports are closely networked through a history of trading, much of it extra-state smuggling.

The gold smuggling operations that D Company were involved in prior to the liberalization of the Indian economy were a modern manifestation of these histories of trade (Glenny, 2008, p. 152). It was these links, and Dubai's welcoming attitude to foreign wealth of all kinds, that determined Ibrahim's relocation there.

Such somewhat fettered international migration has been observed more broadly amongst criminal entrepreneurs. Russian criminal entrepreneurs, for example, have shown a particularly high propensity to migrate internationally. Galeotti (2005b, p. 58) argues that in the last two decades 'the Russian *mafiya* moved quickly to establish itself as an international phenomenon'. This is partly a consequence of the turbulent conditions within Russia threatening the stability of criminal operations, and partly a consequence of Vladimir Putin's response to the anarchic conditions that had prevailed in Russia since the 1990s, including a hard-line attitude towards organized crime within the country (2005b, p. 55). This forced Russian criminal entrepreneurs to seek international opportunities and havens for themselves and their monies. These criminal entrepreneurs were drawn to a range of locations with links of various kinds to Russia. These destinations included countries with cultural links to Russia through significant Russian populations and political links through Warsaw Pact membership or their being former Soviet states. In the United States, whilst the initial entry points included the long-standing Russian community in Brighton Beach in Brooklyn, New York, Russian *mafiya* operations have since diffused to other American cities (2005b, p. 61). Within Southern Europe there has been significant Russian criminal migration to Israel, which has a Russian-speaking population totalling more than 1 million, and to Cyprus, which, whilst having no historical links to Russia, offers an appealing climate, a convenient geographical location and relaxed banking regulations, convenient for the laundering of criminal monies (Siegal, 2003; Galeotti, 2005b).

This is confirmed by Frederico Varese's (2011) extensive empirical research into the problematic international migrations of organized criminal groups. Varese's arguments challenge the cliché, born out of countless alien conspiracy thesis renditions of organized crime, that organized criminal groups migrate easily across space and successfully colonize foreign territories. Rather, Varese's research, based on analysis of the migrations of the southern Italian crime group the 'Ndrangheta to Piedmont and Veneto in northern Italy, Russian Mafia groups to Rome and Budapest, the historical migrations of the Sicilian Mafia to New York City and foreign Triads in China, reveals that the transplantation of organized criminal groups beyond their home territories is 'fraught with difficulties' (2011, p. 4). He finds that some crime groups are able to transplant successfully but this is far from inevitable. He argues that:

> outside their home region, mafiosi would struggle to corrupt the police and collect reliable information. . .it might be taxing to make victims believe that the person standing in front of them belongs to a menacing foreign Mafia. A reputation for violence depends on long-term relations, cemented within independent networks of kinship, friendship, and ethnicity. It is next to impossible to reproduce them in a foreign land. (2011, p.4)

Super-rich criminal entrepreneurs, and indeed those of more modest rank, are mobile internationally then. However, given the more restricted nature of these migrations, the enclaves occupied by these actors tend to be those favoured by others of their ethnic or national background or those renowned international havens for organized criminals. There is undoubtedly market logic at play within these migratory decisions but they do not seem entirely reducible to it. It is highly unlikely that super-rich criminal entrepreneurs are not involved in the luxurious 'bubble neighbourhoods' (Hay and Muller, 2012, p. 79) beloved of the super-rich, particularly those of high property value located in global cities. However, such involvement has not been the subject of serious scholarship to date. Although speculative, it is likely that any involvement of criminal entrepreneurs within these exclusive spaces will follow the migration of their licitly wealthy fellow nationals, is an emergent trend, and is somewhat peripheral to the more general international spatialities of these criminal groups.

Case 3: Super-rich criminal entrepreneurs profit from the exploitation of economic, political and/or regulatory vulnerabilities in their regions of origin

Organized criminality, whilst being a ubiquitous feature of economic systems of all kinds, varies considerably in its extent in different regions (Van Dijk, 2007; Hall, 2010a). Although it is difficult to identify a set of location factors that underpin regional concentrations of organized crime, economic and political vulnerabilities associated with state collapse or regional transition, difficulties in the extension of state sovereignty and rule of law across territories, and severe economic retrenchment within the licit economy appear to be particularly significant in many cases (Hall, 2010a, p. 8). It is within such apparently deeply unpromising economic and political circumstances that the generation of illicit super-wealth is sometimes possible. Indeed Hall (2010b, p. 844) has argued 'it is frequently the very conditions that condemn these spaces to the periphery of the new international division of labour that make them such attractive sites for criminal economic activity'. While some have called for redrawings of global maps of economic dynamism (Nordstrom, 2011, p. 13) the literatures of super-wealth generation might also be extended to acknowledge the potentials of such sites.

Both Pablo Escobar and Dawood Ibrahim benefited from the effects of challenging conditions in their home regions. In Escobar's case the severe economic recession in Medellin's traditional employment sectors, most notably the textile industry, in the 1970s opened up the possibility for dynamic criminal economies to emerge within what was previously a very entrepreneurial milieu (Castells, 2000, p. 205). Clearly, it is not inevitable that extensive transnational organized criminal activity will follow the deindustrialization of regions or that super-rich criminal entrepreneurs will emerge from the rubble of the industrial economy. Such things remain the exception. Rather, it appears that the combination of factors and circumstances is key. Within his account of the emergence of Medellin as a centre of global cocaine production in the 1980s and 1990s, Castells identified a range of other factors that were also important to these outcomes. These included extant, underutilized transnational drug trafficking connections and locally rooted, relatively educated, criminal entrepreneurs who were plugged into these networks (2000, p. 205). Indeed, in the case of Colombia and cocaine, chance seemed to have played a significant role.

In Dawood Ibrahim's case, whilst it was initially the opportunities opened up for criminal entrepreneurs by India's protectionist economic policy that ran from the end of World War II until the early 1990s, it was the economic turmoil following a protracted labour dispute in Bombay's textile industry in 1985 and the mass joblessness that followed that entrenched organized crime within the city's economic sphere and that Ibrahim exploited (Glenny, 2008, p. 151). The economic hardships suffered by many following the dispute forced some to turn to organized criminal groups who stepped in to provide the work that the state and the legitimate economy were unable to. The problem of surplus labour in the cities of the Global South providing an economic resource that organized criminal groups are able to tap into has been recognized more widely within the literature (Daniels, 2004; Moynagh and Worsley, 2008, p. 181).

In the case of financial fraud generally and the activities of Bernard Madoff in particular, the vulnerabilities that allowed his criminal practices to continue for many years related to the regulation of financial investment and trading. Undoubtedly this was a key factor in the longevity of Madoff's fraudulent practices. Madoff's interactions with regulators over the course of his fraud apparently did nothing to affect his criminal activities:

> Mr Madoff's broking operation was regularly inspected by Finra, the industry regulator, but it never looked at the investment business he ran on the side.

> Nearly 20 years into his scam, Mr Madoff was finally required to register as an investment adviser, but the SEC (U.S. Securities and Exchange Commission) division that regulates that side of the securities business is seriously under-staffed and never inspected his books. Of 11 300 investment advisers, only about 10 per cent are examined every three years – by about 400 staff. (Chung et al., 2009, p. 9)

The direct response of regulators has been structural change, internal organization and attention to the specific regulatory loopholes that Madoff exploited for many years (2009, p. 9).

Case 4: Illegal markets and those who command them have paradoxical impacts across the economic, political and cultural spheres of their regions of origin

The impacts of organized criminality on regions, especially those lacking alternative development avenues, have been the subject of some debate (Hall, 2013, p. 371). Some have argued that these impacts are inevitably negative, noting the high levels of violence typically associated with well-developed organized crime economies and their negative impacts on legitimate economic activities (Van Dijk, 2007, pp. 51–3; Glenny, 2008, p. 291). Others though have noted the influx of money and investment associated particularly with the export of narcotics and other illicit commodities, and their contributions to infrastructural development and stability in some regions that the state finds hard to reach (Bagley, 2005; Bhattacharyya, 2005, p. 118; Nordstrom, 2007; Lee, 2008, p. 345; Goodhand, 2009). It is worth trying to reflect on the role of the super-rich criminal entrepreneur in these ways though, rather than just of organized criminal activity generally. Here the abilities of super-rich criminal entrepreneurs to affect regions would seem to depend upon the interplay of a number of qualities that, whilst not the exclusive preserve of super-rich criminals, they seem to command to greater degrees than their more mid-ranking colleagues.

Super-rich criminal entrepreneurs possess a form of hyperagency (Hay and Muller, 2012, p. 81) that can be deployed in ways that produce significant regional, material and cultural impacts of the kinds discussed in the case of Pablo Escobar through his cultivation of the marginalized populations of Medellin (Castells, 2000, p. 205). Criminal entrepreneurs might be adept at exploiting conditions of apparent chaos but they soon tire of them if they threaten to undermine the efficiency or continuation of their enterprises. Craving monopolistic conditions, popular support and some form of stability, criminal entrepreneurs with sufficient resources (money, muscle and strategic insight) typically take steps to achieve this. Criminal economies are fluid and dynamic entities, inherently so, so to do this as efficiently as Escobar appeared to, to the extent that he was effectively labelled, internally and externally, as a geopolitical threat, is extremely rare. Only a tiny number of super-rich and powerful criminal entrepreneurs such as Escobar in Colombia have achieved this level of agency, despite what sensationalist media might tell us. Typically the 'organized crime empire as alternative state' outcome is the product of general conditions of organized criminal activity rather than the result of a single individual's deployment of their hyperagency. Indeed, some have argued that the charismatic crime boss/cartel model of the Colombian cocaine industry overstates the agency of these individuals and their organizations (Kenney, 2007, p. 234). Undoubtedly Escobar, since his death, has been the subject of an extensive process of mythologization, as is typical of many historical figures from organized crime (Hobbs, 2013), which makes the task of determining their hyperagency and its impacts all the more difficult.

The hyperagency of major criminal entrepreneurs is, crucially, realized in symbolic as well as material ways. As well as exuding an aura of benevolence, organized criminal groups must also exude a plausible aura of threat. Violence and the threat of violence are vital actual and symbolic commodities within criminal economies, which are deployed widely as regulatory mechanisms (Hobbs, 2013). In the small number of cases where this can be deployed on an industrial scale they become a key aspect within criminal hyperagencies. Their impacts on regions can be devastating and their roles in producing criminal cults significant. In both Escobar's and Ibrahim's cases a significant number of their violent resources were deployed against the state in ways that were equally oppositional/terrorist as much as they were regulatory/criminal. However, the oppositional and regulatory roles of these acts are not mutually exclusive by any means. Escobar engaged in an extensive assault on the Colombian state in an attempt to end the extradition treaty that Colombia had signed with the United States in the 1980s (Castells, 2000, p. 205). Ibrahim is remembered for his role in the 1993 Bombay bombings in which 350 people were killed and in excess of 1200 injured. He has been linked to a number of other terrorist incidents as well. The bombings were a response to the widespread rioting in the city the previous year that had been directed towards its Muslim populations. Glenny (2008, p. 166) has located the bombings within the proxy war that unfolded between the Indian and Pakistani secret services in the 1990s.

Where the symbolic and material resources that can be deployed across regions can be embodied within particularly powerful individuals such as Escobar and Ibrahim, they coalesce into that important yet elusive commodity 'reputation'. Reputation is an associative, if somewhat geographically contingent, commodity, perhaps most akin to the notion of brand within the legitimate economy, which is carried by the lower-level operatives and units operating on behalf of powerful criminal entrepreneurs. Whilst, like threat, it is a ubiquitous quality of organized criminal economies, it is likely to be particularly powerful where it can be traced, in the perceptions of those affected, to the personality of recognizable criminal entrepreneurs. Here we see the ways in which gangsters of all stripes and standings are able to shape the normative spheres of localities, territories, regions and even nations. Regions denuded of opportunities within the legitimate economic realm and where statehood is weakly articulated appear particularly vulnerable. It is here that troublesome wealth, and in these contexts it is a relative concept, emerges as a powerful social and cultural, as well as economic force shaping relations and aspirations amongst marginalized populations. Speaking of the social prominence of gangsters (thieves) in the former Soviet republic of Georgia, for example, Slade argues that:

> [al]though the thieves' 'law' may not have instrumental value for the majority of the population, it still represents a certain ethical stance and an alternative culture that can prove attractive in a country suffering from an acute problem of 'alienated statehood' in which people feel no affinity or allegiance to the state. The ethos of the bandit can manifest itself in very negative ways all through society, spreading a climate of fear and distrust throughout social relations. It can be popularized through film and media channels that can at times seem to be in love with the danger and sexiness of the Mafia. . . This control might be indirect; the image of the thief (some of whom are now extremely rich) may simply come across as appealing in a country ravaged by economic disaster and war. (2007, p. 178)

In such contexts, should super-rich criminal entrepreneurs, or those perceived as super-rich by impoverished populations, emerge, their normative power should not be underestimated. Perhaps then, globally as well as regionally, the normative power of the super-rich criminal entrepreneur outweighs their material impacts.

Reputations in financial fraud tend to be constructed in slightly different ways to those of criminal entrepreneurs. That is not to say they are not as important a commodity. Rather than threat it is reassurance that is important here. People who are being defrauded need to have been convinced that the investments they are lured into making are legitimate, safe and/or lucrative. In Madoff's case, as was noted earlier, his reputation as a member of New York's financial and cultural establishment, 'the club' (Sender, 2008, p. 20), was important in sustaining the illusion that he was a legitimate investor of his client's money and that this investment was safe. Reputation here, as in the case of criminal entrepreneurs, appears to be territorially grounded to some extent. There was certainly a distinctive geography to many of the investors who were defrauded by Madoff. This was primarily defined around economic and cultural characteristics and connections to Madoff and their attendant geographies within the greater New York City area and its subsequent national diaspora spaces. Again, this points to the potentials of more embodied, territorialized readings of financial systems generally and super-fraud particularly.

CONCLUSION

The case of the organized criminal and illicit wealth contrasts somewhat to that of studies of the super-rich in general. There we have a situation of a tiny elite, of whom very little is known, who have immense power, and '"that owns most of the world"' (Hay and Muller, 2012, p. 75). Here we have a tiny criminal elite of whom a huge amount is either known or imagined, albeit much of this subject to mythologization and whose normative power within the cultural discourse of criminality far outweighs their power in reality. However, although they are to some extent presented in the media and popular discourse in caricatured terms they should not be dismissed as trivial or unimportant. Rather, they play important roles within wider renditions of organized crime, including within official discourses, that portray a messy and elusive phenomenon (Hobbs, 2013) as malleable and susceptible to official responses. But in fact, those official responses are typically inadequate to address the problems of organized crime and illicit wealth generation whether they emerge from the operation of criminal entrepreneurship, financial fraud or corruption. In addition to this, in a small number of cases, members of the criminal elite are important players regionally, and perhaps occasionally, nationally and in some cases within the operations of particular legal or illegal markets. The academic literatures of the super-rich have said little about illicitly obtained wealth to date, for reasons that are entirely understandable. Considering this small subset of the super-rich highlights, a little, the limits of salient discussions. However, there remains much to be said about these individuals and their economic, social, political and normative influences.

REFERENCES

Aas, K.F. (2007), *Globalization and Crime*, London: Sage.

Abraham, I. and W. van Schendel (2005), 'Introduction: the making of illicitness', in W. van Schendel and I. Abraham (eds), *Illicit Flows and Criminal Things: States, Borders and the Other Side of Globalization*, Bloomington, IN: Indiana University Press.

Aguilar-Millan, S., J. Foltz and J. Jackson et al. (2008), 'The globalization of crime',

The Futurist, **42** (6), accessed 16 August 2015 at http://www.researchgate.net/publication/216807865_The_Globalization_Of_Crime.
Albanese, J. (2005), 'North American organised crime', in M. Galeotti (ed.), *Global Crime Today: The Changing Face of Organised Crime*, London: Routledge, pp. 8–18.
Arvedlund, E.E. (2009), *Too Good to be True: The Rise and Fall of Bernie Madoff*, Harmondsworth, UK: Penguin.
Bagley, B. (2005), 'Globalisation and Latin American and Caribbean organised crime', in M. Galeotti (ed.), *Global Crime Today: The Changing Face of Organised Crime*, London: Routledge, pp. 32–53.
Beaverstock, J.V., P. Hubbard and J.R. Short (2004), 'Getting away with it? Exposing the geographies of the super-rich', *Geoforum*, **35** (4), 401–7.
Bhattacharyya, G. (2005), *Traffick: The Illicit Movement of People and Things*, London: Pluto Press.
Brown, E. and J. Cloke (2007), 'Shadow Europe: alternative European financial geographies', *Growth and Change*, **38** (2), 304–27.
Castells, M. (2000), *End of Millennium*, 2nd edition, Oxford, UK: Blackwell.
Chung, J., S. Daneshkhu and B. Masters et al. (2009),'A bitter dividend', *Financial Times*, 24 June, 9.
Daniels, P.W. (2004), 'Urban challenges: the formal and informal economies in mega-cities', *Cities*, **21** (6), 501–11.
Dart, T. (2012), 'Cricket's court jester guilty of $7bn fraud', *The Times*, 7 March, 3.
Deneault, A. (2007), 'Tax havens and criminology', *Global Crime*, **8** (3), 260–70.
Dick, H. (2009), 'The shadow economy: markets, crime and the state', in E. Wilson (ed.), *Government of the Shadows: Parapolitics and Criminal Sovereignty*, London: Pluto Press, pp. 97–116.
Elkind, P. and B. McLean (2004), *The Smartest Guys in the Room: The Amazing Rise and Scandalous Fall of Enron*, 2nd edition, Harmondsworth, UK: Penguin.
Frank, R.H. (1999), *Luxury Fever: Money and Happiness in an Era of Excess*, Princeton, NJ: Princeton University Press.
Galeotti, M. (2005a), 'Introduction: global crime today', in M. Galeotti (ed.), *Global Crime Today: The Changing Face of Organised Crime*, London: Routledge, pp. 1–7.
Galeotti, M. (2005b), 'The Russian "Mafiya": consolidation and globalisation', in M. Galeotti (ed.), *Global Crime Today: The Changing Face of Organised Crime*, London: Routledge, pp. 54–69.
Gilding, M. (1999), 'Superwealth in Australia: entrepreneurs, accumulation and the capitalist class', *Journal of Sociology*, **35** (2), 169–82.
Glenny, M. (2008), *McMafia: Crime Without Frontiers*, London: Bodley Head.
Goodhand, J. (2009), 'Bandits, borderlands and opium wars: Afghan state-building viewed from the margins', *Markets for Peace? DIIS Working Paper No. 26*, Copenhagen: DIIS.
Hall, T. (2010a), 'Where the money is: the geographies of organised crime', *Geography*, **95** (1), 4–13.
Hall, T. (2010b), 'Economic geography and organised crime: a critical review', *Geoforum*, **41** (6), 841–5.
Hall, T. (2013), 'Geographies of the illicit: globalization and organized crime', *Progress in Human Geography*, **37** (3), 366–85.
Hampton, M. and M. Levi (1999), 'Fast spinning into oblivion? Recent developments in money-laundering policies and offshore finance centres', *Third World Quarterly*, **20** (3), 645–56.
Hawkins, E. (2012), *Bookie, Gambler, Fixer, Spy: A Journey to the Heart of Cricket's Underworld*, London: Bloomsbury.
Hay, I. (ed.) (2013), *Geographies of the Super-Rich*, Cheltenham, UK and Northampton, MA, UA: Edward Elgar Publishing.
Hay, I. and S. Muller (2012), '"That tiny, stratospheric apex that owns most of the world" – exploring geographies of the super-rich', *Geographical Research*, **50** (1), 75–88.
Hays, T., L. Neumeister and S. Shlomo (2009), 'Extent of Madoff fraud now estimated at

far below $50bn', *Haaretz*, 8 March, accessed 24 October 2014 at http://www.haaretz.com/news/extent-of-mad off-fraud-now-estimated-at-far-below-50b-1.271672.
Hobbs, D. (2004), 'The nature and representation of organised crime in the United Kingdom', in C. Fijnaut and L. Paoli (eds), *Organised Crime in Europe: Concepts, Patterns and Control Policies in the European Union and Beyond*, Dordrecht: Springer.
Hobbs, D. (2013), *Lush Life: Constructing Organized Crime in the UK*, Oxford, UK: Oxford University Press.
Hobbs, D. and G.A. Antonopoulos (2013), '"Endemic to the species": ordering the "other" via organised crime', *Global Crime*, **14** (1), 27–51.
Hudson, A. (2000), 'Offshoreness, globalization and sovereignty: a postmodern geo-political economy?', *Transactions of the Institute of British Geographers NS*, **25** (3), 269–83.
Hudson. R. (2014), 'Thinking through the relationships between legal and illegal activities and economies: spaces, flows and pathways', *Journal of Economic Geography*, **14** (4), 775–95.
Kenney, M. (2007), 'The architecture of drug trafficking: network forms of organisation in the Columbian cocaine trade', *Global Crime*, **8** (3), 233–59.
Lee, R. (2008), 'The Triborder–terrorism nexus', *Global Crime*, **9** (4), 332–47.
Leeson, N. (1999), *Rogue Trader*, 2nd edition, London: Sphere.
Levi, M. (2002), 'The organization of serious crimes', in M. Maguire, R. Morgan and R.Reiner (eds), *The Oxford Handbook of Criminology*, Oxford, UK: Oxford University Press, pp. 878–913.
Mackay, A. (1992), 'City urged to pay up for Mirror fraud', *The Times*, 18 July, 2.
Madsen, F. (2009), *Transnational Organized Crime*, New York: Routledge.
Manzo, K. (2005), 'Exploiting West Africa's children: trafficking, slavery and uneven development', *Area*, **37** (4), 393–401.
Moynagh, M. and R. Worsley (2008), *Going Global: Key Questions for the 21st Century*, London: A & C Black.
Neal, S. (2010), 'Cybercrime, transgression and virtual environments', in J. Muncie, D. Talbot and R. Walters (eds), *Crime: Local and Global*, Cullompton: Willan, pp. 71–104.
Neville, S. (2012), 'Nadir jailed for 10 years for theft Polly Peck theft', *The Guardian*, 23 August, accessed 18 August 2015 at http://www.theguardian.com/business/2012/aug/23/asil-nadir-10-years-jail-polly-peck.
Nordstrom, C. (2007), *Global Outlaws: Crime, Money, and Power in the Contemporary World*, Berkeley, CA: University of California Press.
Nordstrom, C. (2011), 'Extra-legality in the middle', *Middle East Review*, **41** (Winter), 10–13.
Passas, N. (2001), 'Globalization and transnational crime: effects of criminogenic asymmetries', in P. Williams and D. Vlassis (eds), *Combating Transnational Crime: Concepts, Activities and Responses*, London: Frank Cass, pp. 22–56.
Petras, J. (2008), 'Global ruling class: billionaires and how they "make it"', *Journal of Contemporary Asia*, **38** (2), 319–29.
Phillips, T. (2005), *Knockoff: The Deadly Trade in Counterfeit Goods*, London: Kogan Page.
Plender, J. (2008), 'Sting in the tale told by history's biggest swindlers', *Financial Times*, 19 December, 7.
Reuter, P. and E.M. Truman (2004), *Chasing Dirty Money: The Fight Against Money Laundering*, Washington, DC: Institute for International Economics.
Roberts, S. (1995), 'Small place, big money: the Cayman Islands and the international financial system', *Economic Geography*, **71** (3), 237–56.
Saviano, R. (2008), *Gomorrah: Italy's Other Mafia*, London: Macmillan.
Sender, H. (2008), 'Member of the "club" who won the allegiance of colleagues', *Financial Times*, 13 December, 20.
Sharman, J.C. (2011), *The Money Laundry: Regulating Criminal Finance in the Global Economy*, Ithaca, NY: Cornell University Press.
Siegal, D. (2003), 'The transnational Russian mafia', in D. Siegel, H. van de Bunt and D. Zaitch (eds), *Global Organized Crime: Trends and Developments*, Dordrecht: Kluwer Academic Publishers, pp. 51–61.

Slade, G. (2007), 'The threat of the thief: who has normative influence in Georgian society?' *Global Crime*, **8** (2), 172–9.
Van Dijk, J. (2007), 'Mafia markers: assessing organized crime and its impact upon societies', *Trends in Organized Crime*, **10** (1), 39–56.
Varese, F. (2011), *Mafias on the Move: How Organized Crime Conquers New Territories*, Oxford, UK/Princeton, NJ: Princeton University Press.
Vulliamy, E. (2010), *Amexica: War Along the Borderline*, London: Bodley Head.
Weinstein, L. (2008), 'Mumbai's development mafias: globalization, organized crime and land development', *International Journal of Urban and Regional Research*, **32** (1), 22–39.
Westmarland, L. (2010), 'Gender abuse and people trafficking', in J. Muncie, D. Talbot and R. Walters (eds), *Crime: Local and Global*, Cullompton, UK: Willan, pp. 105–36.
Woodiwiss, M. and D. Hobbs (2009), 'Organized crime and the Atlantic challenge: moral panics and the rhetoric of organized crime policing in America and Britain', *British Journal of Criminology*, **49** (1), 106–25.
Wright, A. (2006), *Organised Crime*, Uffculme, UK: Willan.

PART II

LIVING WEALTHY

11. Capital city? London's housing markets and the 'super-rich'

Rowland Atkinson, Roger Burrows and David Rhodes

INTRODUCTION

In recent years there has been significant popular interest in the spectacle, impact and increasingly conspicuous presence of the 'super-rich' within London's population and the glittering symbolic landscape that has accompanied them. Despite more than half a decade of austerity and recession – following the global financial crisis of 2008 – life at the very top has, it seems, been very good for the very wealthy and the professions that service their needs and whims (Hay and Muller, 2012). A recent report (Capgemini and RBC Wealth Management, 2015), for example, estimates that there were some 14.6 million high net worth individuals (HNWIs) – each with US$1 million or more of investable assets – distributed around the globe in 2014. Yet in 2008 there had been just 8.6 million, suggesting that this population has grown by almost 70 per cent in just six years. Of this group, more than half a million (550 000) now reside in the UK, with the great bulk of them, perhaps unsurprisingly, living in London. The annual 'rich lists' produced by *The Sunday Times* are also helpful in identifying the individuals and families who possess huge amounts of wealth. The analysis for 2014 (*Sunday Times Magazine*, 2014) included for the first time a supplementary 'super-rich' list and suggested that, as of 2014, there were 104 individuals with wealth of more than £1 billion resident in the UK, worth a total of £301 billion (approximately US$500 billion). This means that the UK now has more pound sterling billionaires per capita than any other country in the world and that London is now far and away the city with the greatest number – some 72 (compared to Moscow with 48 and New York with 43).

The reaction to the colossal wealth of many of London's inhabitants (rather than the frequent focus of income disparities), especially as it relates to some profound changes in many local housing markets at a time of otherwise growing levels of austerity, has split into two broad positions. Some view the massive injections of capital, especially into bricks and mortar (and increasingly glass and steel), as an enviable badge of confidence and a clear fillip to the wider city economy (notably this is a position taken by the city's own mayor, Boris Johnson). Others, however, view the same evidence as a sign of a spatial condensing of wealth that signals deeper and more worrisome transformations within London's already prohibitively expensive housing markets with the potential to undermine the city's vibrant and diverse culture (Haywood, 2012). Even despite such injections of capital investment London remains an economy distinct from that of its hinterland given the excesses of its prices.

For those who see problems there is also the sense that we are seeing an acceleration of processes of housing-class dislocations that have a long history in the city. This battle for belonging, between those struggling to find a place in the city (whether as owner or tenant) and those who can so easily purchase houses as investors or landlords, now forms a major element of the social politics of the city. Stories of vacant homes with staggering price tags, lying wholly or temporarily empty, have further fuelled the impression that it is 'money looking for a home' in the city, rather than Londoners, that acts as the major driver of these changes. The slowly growing conflict around housing resources and the symbolic character of the city has come to include middle-income groups and the extremely wealthy. In the recent past the fractious debates of gentrification (Butler, 2003; Slater, 2006) and displacement processes in London (Atkinson, 2000; Slater, 2009) were pitched around contests for space by the low-paid and the traditional working classes against new middle class occupiers of more rundown and cheaper areas of the city. Yet the dynamic at the heart of these shifts remains familiar – an intensification of gross wealth inequalities globally and nationally.

The shift upward in the class positions now focused around competition for London's housing resources appears to signal something profound. Despite critiques from right of centre commentators the debates provoked by Thomas Piketty's (2014) monumental study, *Capital in the Twenty-First Century*, suggests that much contemporary empirical social science – including urban studies – has occurred during a very specific 'historical blip' in which levels of social inequality, for the most part at least, were lessening. For Piketty, when the rate of return on capital (very broadly defined) is greater than the rate of economic growth, then economic inequality inevitably increases as income derived from capital outperforms income derived from other sources, such as salaries and wages (for a critical discussion see Andrew Sayer, Chapter 5 in this volume). For most of the history of capitalism this has indeed been the case, except for a brief period in the middle of the twentieth century – between about 1930 and the late 1970s when economic growth rates were greater than returns on capital and labour movements and social democracy generated municipal provisions and various social entitlements (most of which are now back under threat). The knocks that capital experienced in this period now look to be due to rather unusual circumstances: two World Wars, the Great Depression, the establishment of redistributive welfare states, the growth of the negotiating power of trade unions and a few decades of rapid economic growth. However, since the 1980s the long-term rates of return to capital have again become greater than income growth. So the concentration of wealth drawn to invest in London's housing markets may suggest a fundamental secular shift in which capital dynamics are returning to those that pertained prior to the 1930s. In such a context the most affluent people in the world are disproportionately the offspring of the 'super-rich' (with a notable increase in their ranks during the 'Big Bang' deregulation of the London financial centre and global grabs of state assets by what have become new cadres of billionaires); *ceteris paribus*, the rich and their descendants will get richer and, even if economic growth is sustained, concentrations of wealth and ever-greater levels of social inequality will continue apace (Burrows, 2013).

These are big questions that we will not be able to fully adjudicate upon here; how any new dynamics of global wealth accumulation becomes manifest in spatial relations and the built environment is clearly no small matter. However, such questions frame what is at stake in discussions of the reasons for the massive changes in the fortunes of London's contemporary housing markets – money is drawn to spaces appearing to provide a greater return than is available through other forms of investment where people have the resources to make such investments. This is not to argue, of course, that any such changes in wealth dynamics are purely 'economic'. We now know enough about the history of neoliberalism to recognize that it is, fundamentally, a political project that requires a particular form of statecraft (Mirowski, 2013). To understand what is happening to housing markets in London, for example, at least three contextual political factors must be understood. First, the current political class within the UK is largely compliant with the interests of global capital, distanced from the poverty of housing conditions for the many but also from the anxieties of the local 'well off' who often now feel displaced by the new waves of the global uber-wealthy (Atkinson, 2015). Second, this same political class is engaged in the gross retrenchment of the old welfare system, which is now being used as a legislative vehicle to assist in clearing social space for those on high incomes, as tenants and low-income households are priced out of spaces previously reserved for them by the state (Fenton et al., 2013). Third, and most importantly, an unregulated private housing market – where investment and exchange values are grossly foregrounded over consumption and use values – is disadvantaging those who were already struggling to survive its excesses (Harvey, 2014).

Though all of these transformations and contestations form the backdrop to our wider investigations, here we focus on the broad patterns of change and newfound concentrations of the very wealthy in London and its environs. In this brief contribution we chart the changing character of London's central housing markets (particularly the boroughs of Kensington and Chelsea, Westminster and Camden) and the wider social and cultural changes that have appeared as a result of this growing concentration of wealth and inequality in the city. Of course London has long been one of several, pre-eminent centres of global wealth and power (Sassen, 2010), yet a number of factors have combined in recent years to propel a new and more forceful wave of investment and location decisions that suggest, at least as far as the extremely wealthy are concerned, the city is now positioned at the apex of global rankings of wealth (if not always liveability, as some recent assessments have indicated). Yet this shift is so notable because it has appeared at a time of a major economic recession accompanied by swingeing cuts to local public services and welfare provision. In a sign of the kernel of vitality that capitalism retains for its most privileged exponents (Harvey, 2005), the crisis of the system and its costs to the bulk of the population has not been matched by a threat to the congealing of further power and money at the top of income and wealth spectrums. The relative insulation of the very rich from the crisis has provoked widespread social anger and a growing concern with the analysis of inequality (Wilkinson and Pickett, 2010) and the fracturing of urban society more generally (Graham and Marvin, 2001; Dorling, 2013).

These conditions form the background to what we report on here. In our recent work we have sought to chart the spatial contexts and consequences generated by the ways in which the structure of the economic system and its political exponents have produced the concrete and observable phenomenon of massive advances in wealth for those at the top, and its realization through investment and residence in key neighbourhoods within London. The city has become a pre-eminent location for capital accumulation and wealth storage, particularly in prestige areas (the prime and super-prime markets), yet it has also become the playground of the extremely wealthy who find in these spaces the array of support systems and services that ease their lives and assist in generating feelings of relative safety at street level (Atkinson, 2015).

The underlying politics of this work is the paradox of *growing* capital abundance for the already wealthy – *pace* Piketty – at a time when the city's socially and spatially marginal have seen their conditions attacked in the name of austerity programmes through welfare spending caps, bedroom taxes and state-led gentrification and clearance of the last filaments of affordable and public housing in more expensive areas. In this context, mapping and understanding the shifts in capital investment (stories of 'ghost' neighbourhoods where owners choose only to buy but not to live such as The Bishops Avenue or areas within London's West End and Belgravia – see also Chris Paris, Chapter 12 in this volume) and the wider impact that wealth has produced for the neighbourhoods they touch, and beyond, is important because we need to make these connections in order to explain the full workings of the city and the kind of social injustices that continue to mark it. The city appears to have sold its poorer residents down the river (almost literally as many are displaced to properties beyond the capital) (Fenton, 2011) while opening its arms, metaphorically, to those who have benefited from rounds of primitive and advanced forms of capital accumulation under the global expansion of neoliberalism and civic instability in other global regions. It seems to us right to ask how capital and the wealthy find uses for the city that tell us of the operation of capitalism and the kind of urbanism opening up in cities like London.

These concerns form the bedrock of our research in the 'Alpha Territories' of central London, the areas in which the bulk of the super-wealthy now live, redressing the long-standing critique that the social sciences have tended to ignore or neglect these phenomena due to the costs and difficulties associated with gaining data or access (Burrows, 2016). We draw on commercial socio-demographic profiling systems generated by Experian, census data and over a hundred interviews gathered by our team with the wealthy and their real estate and service intermediaries that generate the life support system for their city lives (Pinçon and Pinçon-Charlot, 1999). The structure of our contribution is as follows. First the broad changes that have occurred in the property markets of central London over the past decade are detailed. We then move to describe the kinds of housing and social stress outside these 'Alpha Territories' before discussing in more detail the kinds of vacant/investment and present/mobility activities of the super-rich and the kinds of physical restructuring of the city that have been generated by the desires of the very wealthy. We conclude by saying something about the kind of nascent social politics that appears to be emerging as a result of these tensions.

A WORLD OF EXCESS? THE LOCATION AND HOUSING MARKET IMPACTS OF THE EXTREMELY WEALTHY

There are any number of different ways that we might begin to map out the neighbourhood locations across London where the very wealthy predominate and where we should thus focus our analysis. Elsewhere (Burrows, 2013) we have argued that, although far from perfect, it is data originating from 'commercial sociology' (Burrows and Gane, 2006) rather than from the state or the academy that currently provides us with the most nuanced sense of the geographies of the global super-rich. ACORN, MOSAIC and a number of other geodemographic systems are able to geocode every residential address in the UK based upon a huge amount of spatially referenced data sourced from myriad commercial and official sources. The MOSAIC system, owned by Experian, for example, holds over 400 separate items of data on almost 49 million adults in the UK. These adults can then be 'associated' with individual addresses. There is not always a one-to-one correspondence between a person and an address, however, as some people are 'associated' with more than one – students, people with two or more homes, those with multiple residences across the globe (Paris, 2013) and so on. The nature of the 'association' may also vary – it could be voter registration, responsibility for payment of utility bills, a mobile phone or some other account. However, the data do broadly allow each address in the UK to be classified in relation to the types of adults 'associated' with it. The MOSAIC system classifies each address into one of 67 different 'types'. It geocodes the very wealthy together under the auspices of the 'Alpha Territory' of which there are considered to be four distinct types: 'Global Power Brokers'; 'Voices of Authority'; 'Business Class'; and 'Serious Money'. Although such labels may not always be to the taste of social scientific sensibilities the veracity of the statistical clusters upon which they are based have often been found to correspond extremely well with more ethnographic descriptions of the neighbourhoods they seek to describe (Parker et al., 2007).

The 550 000 HNWIs resident in the UK, identified in the latest *World Wealth Report* (Capgemini and RBC Wealth Management, 2015), are highly likely to be 'associated' with addresses located within this overall 'Alpha Territory' in the MOSAIC schema; however, they are most likely to be concentrated in areas dominated by 'Global Power Brokers'. Across the UK as a whole we can locate some 1 759 984 adults (in 2010) associated with addresses in the 'Alpha Territory' but within this, only 144 553 identified as 'Global Power Brokers'. If we are interested in the geodemographics of the 'super-rich', addresses associated with these 144 553 might be thought of as offering us the most intense concentrations, whilst a focus on the 'Alpha Territory' as a whole will provide us with a broader indication of where such people are most likely to reside. The Alpha Territory (AT) as a whole is described in the MOSAIC documentation as groups of people with substantial wealth who live in the most sought-after neighbourhoods in the UK. Within this: 'Global Power Brokers' (GPBs) are described as wealthy and ambitious 'high flyers' living predominantly in the very best urban flats; 'Voices of Authority' (VoA) are described as influential 'thought leaders' living predominantly in comfortable and spacious city homes; members of the 'Business Class' (BC) are described as business leaders, often approaching retirement and living in large family homes in the most prestigious residential suburbs; and, finally, 'Serious Money' (SM) is described as families with considerable wealth living predominantly in large, exclusive detached houses with large amounts of disposable income. In addition to these very basic descriptions of the distinctiveness of each of these wealthy geodemographic types MOSAIC provides a plethora of other measures that distinguishes them from each other. Each of these geodemographic (ideal) types thus describe the specificities of particular socioeconomic, cultural, generational, political and perhaps even affective territorial fields. Thus rather than having to rely upon measures based upon the spatial overlaying of a range of more or less appropriate variables – average house prices, average incomes, property types, demographics and so on – a geodemographic approach allows us to conceptualize and measure territories as possessing particular amalgams of a wide range of measures that, in combination, often possess a range of emergent properties not otherwise readily decipherable; they could then be thought of as attempts to operationalize the granular socio-spatial 'compounds' that result from the variable historical and political interplay of affective, cultural, economic, environmental and social life.

The data suggest that (as of 2010) there are some 641 777 adults living in the ATs in London (36.5 per cent of the total for the UK). Of these 137 727 are classified as GPBs (95.3 per cent of the total in the UK); 278 825 as VoA (47.1 per cent); 157 540 as BC (21.7 per cent); and 67 685 as SM (22.8 per cent). Not surprisingly, the distribution of this population within Greater London is anything but even. The most affluent group – the GBPs – occupy a very circumscribed geography focused on the west and southwest of central London, with a smaller outpost in the northwest. At an even more detailed level of analysis we can begin to closely examine the numbers and concentrations of this, the wealthiest and most London-centric, AT type. Table 11.1 shows the postcode districts in central London where the greatest numbers can be found; these are the very core territories of the London 'super-rich'. Well over 10 000 adults can be found

Table 11.1 London postcode districts with the largest number of adults classified as 'Global Power Brokers'

		Number of Global Power Brokers	Concentration (%)
1	London SW1 Belgravia	14018	29.6
2	London SW3 Chelsea	13112	56.6
3	London NW3 Hampstead	12029	26.9
4	London W8 Kensington	11568	58.0
5	London W2 Paddington	9493	20.7
6	London SW7 South Kensington	9066	50.9
7	London W11 Notting Hill	8916	30.9
8	London W1 West End	6806	26.5
9	London NW8 St John's Wood	6555	22.7
10	London W14 West Kensington	4637	14.7

Source: Authors' own compilation.

in each of Belgravia (14018), Chelsea (13112), Hampstead (12029) and Kensington (11568). However, the greatest concentrations can be found in Kensington (58.0 per cent of the population), Chelsea (56.6 per cent) and South Kensington (50.9 per cent). As we will see below, this geodemographic operationalization of the spatial distribution of the 'super-rich' corresponds very closely with a set of contiguous areas that can be identified as part of the 'super-prime' housing market: an archipelago of gilded ghettoes and territories in London's heart.

THE POLITICS OF WEALTH AND THE SCALE OF HOUSING CHANGE IN LONDON

In a city of around 8 million and a metropolitan area with some 13 million people the numbers of people involved here may seem like a drop in the ocean, even when we add the estimated 281000 millionaire HNWIs, yet the character and landscape of the city, the tone of its political debates and the nature of its economy have been more deeply captured and re-crafted by the super-rich than numbers alone might indicate (Freeland, 2012). These changes began to appear in earnest in the 1980s, accelerated from around the start of the new millennium and, curiously, have only been amplified further under the austerity conditions of the financial crisis. Three factors can be identified in explaining this growth of indigenous and imported wealth and its concentration in London.

First, new waves of accumulation among financial (Froud and Williams, 2007) and political elites have generated historically unprecedented concentrations of wealth in the new classes of owners and bureaucrats who feel that global regional instabilities make the UK a first-rate asset class for investment in property or financial vehicles operated by its financial heart in the City (for additional discussion see chapters 10 and 18 by Hall and Short respectively in this volume). The perception of London as a centre of tax and finance stability has helped to channel vast global supplies of capital looking for a safe haven (Shaxson, 2012) and good prospects for future growth. Exchange rates favourable to foreign investors have fuelled a buying frenzy among the rich and a wider class of property investors who may never even have set foot in the assets that they snap up 'off plan'.

Second, and related, London's property market is seen as one of the safest bets of almost any asset class, often viewed as outperforming gold, the stock market or any other investment vehicle. As we discuss below, London's housing prices have seen remarkable growth in the 'prime' and 'super-prime' areas as well as the tiny 'ultra-prime' housing market that has developed to take advantage of this.

Third, the interests of a coalition of the city's governors (political, real estate and finance elites) have conjoined and now seek to prop up their legitimacy by offering a narrative of these changes that puts great emphasis on the need for unfettered controls on property sales, capital and corporate investment. This somewhat unholy alignment of interests has generated a powerful discourse, regularly relayed in the media, that trumpets the gains to the city generated by a cosmopolitan class of international owners who might otherwise be scared off were any controls, taxes or tariffs to be put in place. This alliance is increasingly in conflict with a new constituency that ranges from those on no/low incomes to those on much higher salaries who have seen their own, and often their children's, housing opportunities dismantled by the exorbitant market realities now set in train. This disparate group have similar interests, consolidated by the new force of conditions and impacts generated by the very wealthy. Set against a logic of endless expansion, massive stresses on transport infrastructure and exclusionary costs to entry of housing, the working classes of the city appear to tilt impotently at the excesses and theft of the city by capital at a time of intense housing need and stress by the absolute bulk of the population.

The combined result of these changes has placed London's housing even more emphatically in the class of a speculative unit, rather than a space of use and inhabitation (Harvey, 2014; see also Sayer, Chapter 5 this volume). The conjunction of these forces has produced a city system at odds with the need to produce a viable social and municipal realm for those on less than 'stellar' incomes and resources (we touch on these wider conditions later). It is clear that London has seen newfound wealth finding a home in the world's capital, servants and factotums often in tow. This supply of wealthy individuals is seemingly limitless, generated by unstable regional contexts from which capital has flown, spurred by the gains of primitive accumulation in countries like Russia, corrupt and dirty money washing through property systems as a safe and respectable bet or simply from those 'clean', or otherwise, who see London as the pre-eminent location to live and/or invest. This distinction is an important one given the frequent media focus on the 'buy to leave' nature of much investment – large homes lying vacant as the city's workers of scant or modest means struggle to find their own place. Stories of tax loopholes, light property tax and ill-gotten gains (Sayer, 2014) have furthered the sense of the 'cuckoo-like' qualities of capital now forcing out the ambitions of wider sections of the population.

What then is the actual scale of these changes in the London property market? An analysis of Land Registry data suggests that between 2011 and 2013 there were 16 715 properties that sold for £1 million or more, 683 that sold for over £5 million and 119 that sold for over £10 million. The same data suggest that there were some 10 000 transactions of properties worth £1.35 million or above. If one maps these 10 000 transactions (10 186 in actuality due to some transactions having exactly the same value) it is possible to get an indication of the major recipient spaces of global property capital that have occurred in recent years. This is shown in Figure 11.1 and highlights the concentration of transactions in the central, southwestern and outer southeast and northwest parts of London. This map thus not only broadly corresponds with that revealed by the spatial distribution of the 'Alpha Territory' using our geodemographic measure, it also correlates with predominant understandings of the main areas of historically wealthy neighbourhoods, with more wealth condensing in established areas that have cultural cachet and prestige and which often figure prominently in the cognitive maps of foreign buyers. Areas like Knightsbridge and key outlets, like Harrods, are central to the demands of potential buyers whose direct knowledge of the city may be limited and to whom agents often steer clients as AAA rated property investments.

The internationalization of the London residential property market that we focus on here has been notable for many decades; this feature of the market has been amplified to the extent that around 70 per cent of all new-build homes in the super-prime areas of central London are now estimated to go to wealthy foreigners. It is perhaps this feature that in large part has driven rampant house price inflation in the central areas

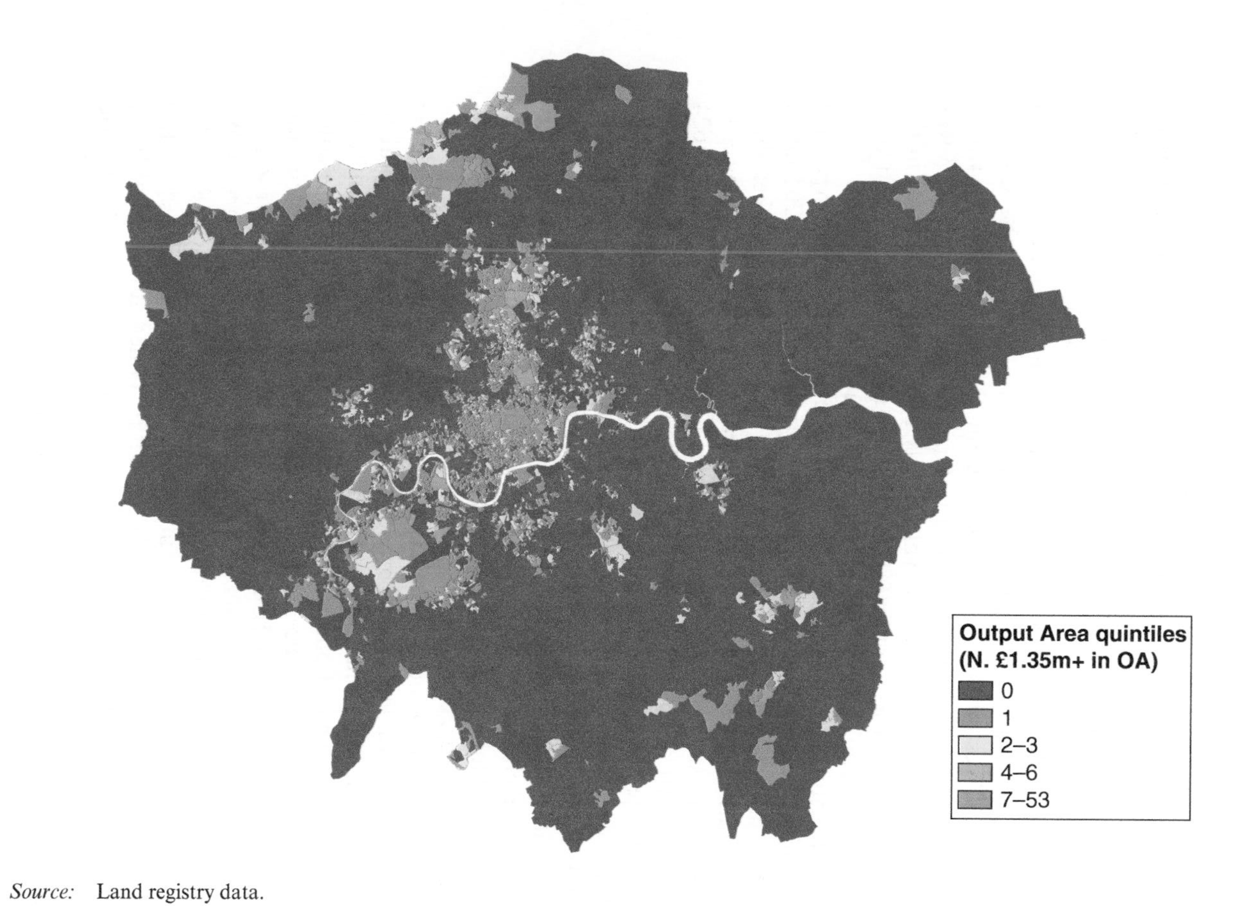

Source: Land registry data.

Figure 11.1 Top 10 000 house sales in London between 2011 and 2013, £1.35 million+

of London. Notably, prices had risen by nearly two-thirds (61 per cent) in prime central London in the preceding three years (Savills, 2012). Our own analysis of land registry data (often considered problematic in its relative capture of the most expensive transactions) suggests that over the last three years there has been a 12 per cent increase in the median price of all property between 2011 and 2013 for all London transactions. However, for areas where the 'Alpha Territory' predominates we find that its subset of 'Global Power Brokers' stood out with a 24 per cent increase over the last three years – essentially the absolute core areas of South Kensington, Knightsbridge and Mayfair (within the London Boroughs of Westminster and Kensington and Chelsea). Taken as a whole the total value of property sales in London between 2011 and 2013 was £124 808 124 549 (£124 billion), while nearly two-thirds of this was in the 'non-Alpha Territory' areas (£80 661 878 973) (£80 billion), around one-third was generated by the relatively small area (only 16 per cent) of the Alpha Territories (£44 146 245 576) (£44 billion). The ten most expensive land registry recorded sales of the past three years ranged from £20 500 000 (in the postcode SW1X) up to £54 959 000 (SW10). What makes these figures starker is that the volume of transactions was also much lower in the Alpha areas (only 51 360 over the three years compared with 229 624 in the rest of London). Such data reveal the intense value of this market to real estate agents and as a relative source of property sales tax (stamp duty) revenues that helps to reveal why the story of wealth pouring into the city is so valued by politicians. It is this political economy of property and wealth that we turn to later. London appears increasingly to be a city that has detached itself from the wider country (see Chris Paris, Chapter 12 in this volume) and as a site of some of the profoundest wealth and poverty (Dorling, 2014).

In order to get a more nuanced understanding of the housing market dynamics that underpin this broad mapping, it is perhaps helpful to focus on a more circumscribed territory that is unambiguously emblematic of 'super-prime' London. A focus on Westminster as a case study certainly fulfils this requirement, covering as it does, core 'super-prime' neighbourhoods such as Knightsbridge and Belgravia, the West End, Hyde Park, Regents Park, Maida Vale, Little Venice and St James's. We are fortunate in that a recent report for Westminster Council (Ramidus Consulting, 2014), to which we contributed, provides a detailed statistical and qualitative analysis of the situation as it pertained at the end of 2013.

In 2012 the population in Westminster was estimated to be 223 900 and in 2013 122 608 properties were recorded as being liable for council tax. This compared with 2011 census estimates that found 120 066 households residing in 118 318 dwellings. The population of Westminster is certainly affluent. In 2012 it was estimated that average incomes were some 61.7 per cent higher than those for Great Britain as a whole and almost one-third higher than those for London as a whole. The population also clearly has a different form of attachment to their accommodation compared to most of the rest of the population. It is estimated that in 2011, 11.9 per cent of addresses (14294 homes) possessed no usual resident (an increase of about 10 per cent compared to 2001), but this figure was highly variable with some 40 per cent of properties in the West End and over one-quarter in Knightsbridge and Belgravia appearing to be (at least partially) empty. Across Westminster as a whole some 16 per cent of households were recorded as possessing a second home, but in neighbourhoods such as Knightsbridge and Belgravia some 40 per cent owned other accommodation and 23 per cent had a second (or more) home overseas. In general then such figures do seem to confirm many of the popular stereotypes about how the wealthy utilize expensive central London properties.

But what is the scale and the scope of the 'super-prime' housing market in an area such as Westminster and how does the functioning of this market relate to the local housing market more generally? Of course, this notion could be operationalized in any number of different ways and one of the major methodological problems of work on this topic is the highly variable manner in which this occurs between different studies.

If we consider the 118318 dwellings in Westminster as a whole we know that about 30 per cent are still in the public rented sector and are thus (at the moment at least) not part of the local private housing market. So this means that the total private housing stock in 2011 was about 83000 dwellings. Between the beginning of 2003 and the end of 2012 about 43000 dwellings were bought and sold (some more than once) giving a rough annual turnover over the period of between about 4 and 5 per cent. The highest volume of transactions were in 2006 and 2007, with a marked decline in 2008 and 2009 following the financial crisis, but with a marked recovery since then. However, although the number of transactions has been volatile, at no time did the average price of property fall. Indeed, compared to 2003, average prices have tripled and, compared to 2008, average prices have increased from £875000 to an eye-watering £1575000.

Of course not all of the dwellings in Westminster should be considered as 'super-prime'. A sensitizing definition of this sector of the market might be those properties classified for council tax purposes as being in the top band, H. As at the end of 2013 this roughly equates to properties valued at £2 million or more – some 13 per cent (14679) of the housing stock in Westminster. A detailed analysis of transactions between 2003 and 2013 reveals that different segments of the market – especially those above and below this rough threshold – have been functioning in very different ways.

As property values in general have been increasing the number of sales below £1 million have been in secular decline. Sales between £1 million and £2 million fell around the period of the 2008 crisis but subsequently rose. However, sales about £2 million have escalated. The devil here is in the detail. In the decade up to the end of 2013 there were some 3335 transactions of properties valued at £2 million or more. The total value of these sales was some £15.4 billion. These transactions only accounted for some 8 per cent of all sales in Westminster but a huge 41 per cent of total value. There were 791 transactions involving properties valued at £5 million or more – just 2 per cent of all transactions but a full one-fifth of the total value of sales. Finally, there were just 90 transactions involving properties valued at £15 million or above – just 0.2 per cent of sales but a hugely disproportionate 6 per cent of the total value of all sales.

Sales below £1 million in Westminster show no correlation at all with average earnings. Sales between £1 million and £2 million show only a very weak correlation. However, the pattern of sales above £2 million shows a strong correlation, not with average earnings, but with the price of gold. It is important to note here that above £2 million it is the volume of sales that have increased but not prices. Below this rough threshold prices have increased significantly but the volume of transactions has been variable. Above this threshold it is the number of transactions that have increased but not the average price of property. Although the numbers are small, until the last four years there were, on average, about eight sales of properties valued at above £15 million per annum, but in the last four years this has risen to between 12 and 16 per annum, but the mean value of such properties has, in fact, gone down from £30 million to £21 million.

There is strong evidence to suggest that within the 'super-prime' segment of the Westminster housing market the majority of purchasers are from overseas. For example, according to the prime estate agent Savills (2012), 69 per cent of buyers in Knightsbridge in 2011 originated from overseas, while in both Belgravia and St John's Wood the figure was 60 per cent. In Mayfair, the proportion was lower, at just over one-half. On balance the data suggest that over half of all buyers in much of prime central London between 2011 and 2013 were of overseas origin but that the proportion was significantly higher in Knightsbridge, Belgravia, Mayfair and St John's Wood. In Knightsbridge and Mayfair the principal reason for purchase appeared to be as a second home, whilst in Belgravia and St John's Wood it was to use as a main residence. Across London as a whole there is also strong evidence to suggest that overseas purchasers are more likely to favour new-build properties than second-hand homes; again, according to Savills, 73 per cent of sales of new build prime property were to overseas buyers, whilst the figure for second-hand homes was just 38 per cent.

THE BIGGER PICTURE: WEALTH, STRESS AND POVERTY IN LONDON

The story of London's riches only makes full sense when contextualized in the broader story of loss, stress and housing decline that marks the lives of a significant fraction of the city's wider population. Slender pockets of public housing remain in many of these locations. Inside and out of the super-prime areas we can find stories of rising social distress as welfare changes, rising rents and house prices, low levels of house building and a growing population create intense competition for housing resources. Before we turn to indicators of these problems we might note in passing the data kept on the scale of homelessness in the capital with provisional figures for the fourth quarter of 2013 (the latest available) showing there were 42 430 households placed in temporary accommodation in London (pending enquiries over their homelessness claim or after being accepted as homeless), compared with 14 500 throughout the rest of England. A low was reached in the second quarter of 2011 of 35 620, after which figures have begun to rise again (as they have across the rest of England too) (Gov.uk, 2015a). In addition, overcrowding in London, analysed using the bedroom standard in the English Housing Survey, showed that for 2011/12 England as a whole had 2.8 per cent of households with insufficient bedrooms for their needs but in London the figure was 7.0 per cent.

Surviving the excesses of the housing market is strained by a lack of supply and prices that by far exceed prevailing incomes and has become a recurrent theme in the somewhat hollow posturing of political campaigning around the capital in the run up to the 2015 general election. The inadequacy of responses to these problems is highlighted by Land Registry House Price Index data for February 2014. As already noted, this showed an annual change of +5.3 per cent for house prices in England and Wales, with an average property price of £170 000. Yet in Greater London the annual increase was +13.8 per cent, with an average property price of £414 356. The story of stress in the private rental sector is even more marked; here we see that private rents in the mainstream lettings market in London are more than double that of most other regions. A one-bed flat in Greater London had a median monthly rent of £1000 in the 12 months to Q3 2012, compared with £495 for England as a whole. For two-bed flats the figure was £1250, compared with £570 for the whole of England (Gov.uk, 2015b).

The media and the city's public are simultaneously split between being appalled and subtly in awe at the eye-watering prices offered by plutocrats. These responses tend to mask the wider realities and housing stresses of the city, the second face of a city that has tended to present its 'best side' to investment by industry and the wealthy. In such narratives London is open for business, even if this means that poorer households are sidelined by the hegemony of these narratives, so often presented in terms of the realities and necessities of London's position at the apex of a global economy now focused on international finance and services – this, after all, is a dynamic economy whose activities are seen as the lifeblood of the nation's fortunes.

We suggest that to understand, locate and challenge these underlying and systemic material inequalities, and many of the housing problems in the city, we need to turn our attention to and locate the role and impact of the very wealthy, less because of their direct impact (even in terms of the housing market) and more because of the way in which London's financial institutions, property developers, estate agents and many politicians appear to have lost a wider ability to focus on the care and maintenance of the wider working city. If the rich keep London going then the homeless, workless, poorly housed, overcrowded and stressed may see little in the way of direct benefits. This point is further exemplified by the sense of a disproportionate take on housing resources by the more affluent in which (using the 2011 census) we can see that the number of homes lying vacant was around 1 in 20 (5.4 per cent) in the Alpha Territories compared with 2.9 per cent for the rest of London (dwellings that did not have a usual resident on the night of the census). In the most transitory and international Alpha areas this figure is much higher at 11.1 per cent of dwellings in areas dominated by the Global Power Brokers. At a time of such pronounced housing stress these contrasts are increasingly noted by a more politically active and fractious constituency looking for answers to the housing crisis.

CONCLUSION

In this chapter we have explored some of the reasons for London's increasingly pre-eminent position as a global hub attracting the world's most wealthy households in pursuit of lifestyles and investment opportunities. We have argued that a number of key factors, including global regional instability, the desirability of its homes as unrivalled positional goods and the desire to invest in both the life or bricks and mortar of the city are major reasons for the growing conspicuous presence of the wealthy in the city. Such factors have combined to help explain a resurgent opulence to the wealthy districts of central London that suggests a new *belle époque* for those who have benefited from new rounds of capital accumulation across the global economy. These changes signal a shift in the symbolic character now seen in key changes in the city's architecture, modes of dress, streetscapes, commercial aesthetics and other urban re-makings that are being worked through London's housing market, economy and the wider flow of its urban culture and life. These forms of physical rending and social change have also generated a range of wider impacts. Chief among these are new rounds of social conversation, political debate, anxiety, anger and analysis focused on whether the rich bring net gains to the wider city they, or their money, have come to colonize. Crucially these arguments are inflected by the positions of those involved, for estate agents that take a percentage of house sales rising prices are hard to complain about and generate calls for unfettered transactions and the embracing of the very wealthy. For the diverse communities touched by this gentrification of gentrification in the city's prime property markets, on the other hand, we find narratives of loss and displacement as major themes.

At the core of these contestations lie much deeper questions about what London is, both as a city with a declining sense of social citizenship and as an economy with its own cold logic connected only to motives of financial gain. Our analysis in this chapter has therefore focused upon some of the broad-brush transitions and general problems that London faces in terms of housing and social stress and the kinds of space being produced by switches of capital investment into its super-affluent areas from other regions and assets globally. Undergirding these changes and rapt commentaries around record-breaking nine-figure house prices there remains a working capital, a city of the working classes (in the widest sense) whose populace are increasingly opposed to the interests seeking open pathways to the colonization of the city by capital. This social politics is focused around a resentment of the ways in which wealth has created nesting spaces for money as much as for people and with the full spectrum of the wider political class offering only a callous politics generated in large part by a mandate legitimated by the financial crisis. This charged emotional landscape, focused on housing, its cost and exclusion from opportunity, has perhaps also begun to connect with a synoptic impression of the workings of capitalism and its flows to wherever pays best or whichever asset class generates the best return. In this sense London shows a kind of reverse polarity to earlier waves of empire and accumulation with London the epicentre of a capital expansion that shows itself returning home with the power to roost and supplant.

Programmes of demolition and heavily cut welfare policies have appeared at a time of increasingly visible wealth that appears to have further challenged the relative vitality of lower-income communities. What is interesting here is the way in which the changes wrought by massive injections of capital have touched the traditional heartlands of the upper classes and the way that this has generated feelings of displacement among them.

The relative emptiness of areas colonized by the very wealthy has been one of the most notable features of debates about the rich themselves and of their unnecessary 'take' on housing resources at a time when government appears very weak in the wake of housing need across the capital. London's property market has generated a system of opportunities and constraints that generates related winners and losers (Hamnett, 2005). The narrative of unchecked growth and gains for owners has, however, begun to falter in the context of the excesses at the top of the system while access to affordable housing, job prospects and security have been destabilized for the wider population. One serious implication of these shifts has been to generate anger as a result of the unwillingness of the state to offer appropriate tariffs to help fund and better coordinate the supply of affordable housing. In addition to these changes the conspicuous presence and consumption of the very affluent distracts political attention from the often-concealed presence of significant social distress, much of it in the wider sectors of the city beyond the boundaries of the 'Alpha Territories'. The impression that a core human need, the domestic dwelling, can produce money as though it grew on trees or is such a mere adornment of wealth that it is barely or never lived in presents challenges for social cohesion and the common project underlying a narrative of London as a diverse and tolerant society.

REFERENCES

Atkinson, R. (2000), 'Measuring gentrification and displacement in Greater London', *Urban Studies*, **37** (1), 149–65.

Atkinson, R. (2015), 'Limited exposure: social concealment, mobility and engagement with public space by the super-rich in London', *Environment and Planning A*, pp. 1–16.

Burrows, R. (2013), 'The new gilded ghettos: the geodemographics of the super-rich', *Discover Society*, 3 December, accessed 25 August 2015 at http://discoversociety.org/2013/12/03/the-new-gilded-ghettos-the-geodemographics-of-the-super-rich/.

Burrows, R. (2106), '"Studying up" in the era of Big Data', in L. McKie and L. Ryan (eds), *An End to the Crisis of Empirical Sociology?: Trends and Challenges in Social Research*, London: Routledge, pp. 65–80.

Burrows, R. and N. Gane (2006), 'Geodemographics, software and class', *Sociology*, **40** (5), 793–812.

Butler, T. (2003), 'Living in the bubble: gentrification and its "others" in North London', *Urban Studies*, **40** (12), 2469–86.

Capgemini and RBC Wealth Management (2015), *World Wealth Report 2015*, accessed 1 August 2015 at https://www.worldwealthreport.com/.

Dorling, D. (2013), *The 32 Stops: The Central Line*, London: Penguin.

Dorling, D. (2014), *All That is Solid: The Great Housing Disaster*, London: Allen Lane.

Fenton, A. (2011), *Housing Benefit Reform and the Spatial Segregation of Low-Income Households in London*, London: LSE ePrints.

Fenton, A., R. Lupton and R. Arrundale et al. (2013), 'Public housing, commodification, and rights to the city: the US and England compared', *Cities*, **35**, 373–8.
Freeland, C. (2012), *Plutocrats: The Rise of the New Global Super Rich*, London: Allen Lane.
Froud, J. and K. Williams (2007), 'Private equity and the culture of value extraction', *New Political Economy*, **12** (3), 405–20.
Gov.uk (2015a), 'Live tables on homelessness', *Gov.uk*, 9 July, accessed 17 August 2015 at https://www.gov.uk/government/statistical-data-sets/live-tables-on-homelessness.
Gov.uk (2015b), 'Private rental market statistics', *Gov.uk*, 28 May, accessed 17 August 2015 at https://www.gov.uk/government/statistics/private-rental-market-statistics-may-2015.
Graham, S. and S. Marvin (2001), *Splintering Urbanism: Networked Infrastructures, Technological Mobilities and the Urban Condition*, London: Routledge.
Hamnett, C. (2005), *Winners and Losers*, London: Routledge.
Harvey, D. (2005), *A Brief History of Neoliberalism*, Oxford, UK: Oxford University Press.
Harvey, D. (2014), *Seventeen Contradictions and the End of Capitalism*, Oxford, UK: Oxford University Press.
Hay, I. and S. Muller (2012), '"That tiny, stratospheric apex that owns most of the world" – exploring geographies of the super-rich', *Geographical Research*, **50**, (1), 75–88.
Haywood, A. (2012), *London for Sale? An Assessment of the Private Housing Market in London and Impact of Growing Overseas Investment*, London: The Smith Institute, accessed 17 August 2015 at http://socialwelfare.bl.uk/subject-areas/services-activity/housing-homelessness/smithinstitute/london12.aspx.
Mirowski, P. (2013), *Never Let a Serious Crisis Go to Waste: How Neoliberalism Survived the Financial Meltdown*, London: Verso.
Paris, C. (2013), 'The homes of the super-rich: multiple residences, hyper-mobility and decoupling of prime residential housing in global cities', in I. Hay (ed), *Geographies of the Super-Rich*, Cheltenham, UK and Northampton, MA, USA: Edward Elgar Publishing, pp. 94–109.
Parker, S., E. Uprichard and R. Burrows (2007), 'Class places and place classes: geodemographics and the spatialization of class', *Information, Communication and Society*, **10** (6), 901–20.
Piketty, T. (2014), *Capital in the Twenty-First Century* [French edition published in 2013 as *Le capital au XXI siècle*, Editions du Seuil], Cambridge, MA/London: Belknap Press of Harvard University Press.
Pinçon, M. and M. Pinçon-Charlot (1999), *Grand Fortunes: Dynasties of Wealth in France*, New York: Algora Publishing.
Ramidus Consulting (2014), *The Prime Residential Market in Westminster*, accessed 21 January 2015 at http://transact.westminster.gov.uk/docstores/publications_store/news/prime_residential_research_report_140722.pdf.
Sassen, S. (2010), 'A savage sorting of winners and losers: contemporary versions of primitive accumulation', *Globalizations*, **7** (1–2), 23–50.
Savills (2012), *Prime London Index*, London: Savills.
Sayer, A. (2014), 'Postscript: elite mobilities and critique', in T. Birtchnell and J. Calatrío (eds), *Elite Mobilities*, London: Routledge, pp. 251–62.
Shaxson, N. (2012), *Treasure Islands: Tax Havens and the Men who Stole the World*, London: Random House.
Slater, T. (2006), 'The eviction of critical perspectives from gentrification research', *International Journal of Urban and Regional Research*, **30** (4), 737–57.
Slater, T. (2009),'Missing Marcuse: on gentrification and displacement', *City*, **1** (2–3), 292–311.
Sunday Times Magazine (2014), *The Super-Rich List*, 11 May.
Wilkinson, R. and K. Pickett (2010), *The Spirit Level: Why Equality is Better for Everyone*, London: Penguin.

12. The residential spaces of the super-rich

Chris Paris

We are all in the gutter, but some of us are looking at the stars.
(Oscar Wilde, *Lady Windermere's Fan*, 1893)

INTRODUCTION

The study of the super-rich – especially the billionaire 'stars' – is somewhat different from other areas of social scientific enquiry because we are considering rare but often identifiable individuals whose anonymity cannot be guaranteed through generalizations about aggregated large-scale survey or census data. Unlike the vast bulk of humanity, we can name and show pictures of most of the world's richest people and their homes. The annual *Forbes* list of billionaires recorded a new peak of 1645 in 2014, up by over 25 per cent since 2011. The 2014 *Sunday Times* Rich List showed that the wealth of the super-rich in the UK had doubled over five years so that assets of £85 million were needed to be in the top 1000. This is good news for people whose incomes derive from servicing the needs and desires of the super-rich, including the global residential real estate industry, but not so good for the rest of us, as inequalities in income and wealth have increased at all spatial scales and the richest 1 per cent of people are said to own more than the poorest 50 per cent (Oxfam, 2014).

The study of the residential spaces of the super-rich provides a valuable corrective to much housing-related scholarship. The focus of housing research has been primarily on policy-related issues and debates, especially poverty and social housing provision, with research priorities largely driven by the preferences and desires of government funders or charitable foundations (Paris, 2013a). Housing researchers have paid little attention to transnational dimensions of housing markets, typically treating housing markets as interactions *within* nation-states and ignoring transnational investment in housing (Paris, 2013b). Their analyses of 'housing finance' almost always utilize data relating to officially recorded mortgages or loans, but as the purchase and sale of housing by the global super-rich is largely unrecorded within these statistics, the very idea of 'national' housing markets has become increasingly problematic. Even those housing scholars who have considered globalization have focused mainly on the impacts on nation-states, especially aspects of welfare states (Clapham, 2006; Wu, 2012).

My analysis of the residential spaces of the super-rich shows that the super-rich have disproportional impacts on national and international housing systems far greater than their mere numbers. This chapter considers sources of data on the super-rich and their residential spaces, with the latter comprising both the dwellings they own and the wider environments within which their dwellings are located. I review the sources of their wealth and their countries of origin as there is a high degree of covariance between these variables and the nature of their residential spaces. I consider the types of residential spaces owned and consumed by the super-rich and how their participation in transnational housing and leisure markets has remade places through the decoupling of prime residential real estate from national housing markets and in zones of exclusive hyperconsumption.

DATA SOURCES

The global super-rich, also described as high net worth individuals (HNWIs) or ultra HNWIs (UHNWIs), are annually listed by *Forbes* magazine and *The Sunday Times*. The format of *Forbes* magazine's online lists changed in 2014 from an annual one-off list to regularly updated information and commentary on the changing fortunes of the super-rich. The data is based on extensive research and analysis and the numbers are checked with a range of outside experts as well as the individuals in the lists (though not all of the latter cooperate). The *Forbes* team includes a specialist covering real estate 'from ultra-luxury homes to foreclosures to the people making the deals happen'. The *Forbes* focus is on individuals rather than families and is designed to reflect 'individual, entrepreneurial wealth that could be passed down to a younger generation or truly given away' (Kroll, 2014). The list tries to identify all sources of wealth as well as debts to establish a baseline on 12 February each year. The picture is complicated by diverse family relationships, especially after multiple inheritances, and the existence of many non-profit foundations. The list is changing constantly as fortunes wax and wane, individuals die and others inherit. *Forbes* has turned the study of the super-rich into a spectator sport, with daily updates of 'today's biggest winners and losers' and real-time changes in personal fortunes; thus on 22 May 2014 it was reported that the net worth of Mukesh Ambani was up by US$804 million but Ma Huateng was down by US$1.3 billion.

Material from the *Sunday Times* has been available electronically only by subscription since early 2014 (though Wikipedia repeats some of this material). The lists and other electronic sources, such as www.billionaire.com, Bloomberg, *The Wall Street Journal* and other online print media, provide extensive information on super-rich individuals and families and the sources of their wealth (Paris, 2013a). There is also extensive evidence about the residential real estate investments of the super-rich in a wide range of printed and electronic sources, with research and commentary from the international real estate and luxury industries, especially global real estate agents including Knight Frank, Savills and Christie's. All publish regular commentary and market analyses on cities, regions and the global economy both for their clients and, in effect, *about* their clients. In addition, there is widespread commentary on the homes of the super-rich and the wealthy more generally in published media and across the Internet.

SOURCES OF WEALTH

There were essentially three routes into great wealth in pre-industrial societies: birth and inheritance, imperial patronage and force of arms (Beresford and Rubenstein, 2007). Kings, emperors and aristocrats from the very earliest times can be conceptualized as positional super-rich, typically deriving their wealth from *who* they are rather than what they have done. The extent to which royal status confers rights to use and dispose of wealth, however, varies considerably, as most constitutional monarchs are de facto caretakers of non-marketable wealth and assets that are owned by the state. Thus the *Sunday Times* and *Forbes* lists exclude most royalty, despots and dictators.

The growth of colonialism, mercantilism and industrial capitalism widened the scope for individual and family wealth acquisition, as merchants, industrialists, financiers, entrepreneurs, miners and agri-businesses acquired vast fortunes. The surging growth of a globalized economy, the emergence of new structures of production and distribution and the ICT revolution have expanded the bases of extreme wealth acquisition, with new forms of finance, technological and information industries, and the rapid rise of high-earning 'superstars', most obviously in the media but also including the increasingly rich providers of services to the super-rich, especially banking and real estate (Freeland, 2012). The scale of markets is enormous and innovations have spread almost instantaneously, especially new informational and social media.

Many of the contemporary super-rich have acquired their fortunes rapidly through their own work in industry, especially ICT and related sectors, retail and finance (but see Sayer, Chapter 5 in this volume for a critical account of the sources of wealth of the very rich). Inheritance of great wealth still figures strongly in Europe and North America and the origins of some contemporary fortunes elsewhere were not entirely unlike plunder, with the privatization of former state assets, especially minerals and telecommunications, in post-communist and developing countries (Freeland, 2012). Overall, however, *Forbes* suggested that around two-thirds of the billionaires had built their own fortunes, some 13 per cent had inherited and 21 per cent had inherited but then built substantially on their inheritances (Kroll, 2014).

Many super-rich individuals and families have multiple sources of wealth, and all can transfer their wealth between categories and/or diversify among categories. Despite such caveats, it is still instructive to review the main sources of wealth of the 100 leading entries in the 2014 *Forbes* list of billionaires. The top 100 is dominated by citizens of the USA: 39 per cent overall with 18 of the top 25; the next highest sharers are eight each in Germany and Russia. The fortunes of six out of the top 25 were derived from technology, with all based in the USA, including the current number one, Bill Gates, as well as Larry Page and Sergey Brin, founders of Google, and Facebook's founder Mark Zuckerberg. Metals, mining and energy industries accounted for nearly 20 per cent of the billionaires, including Australia's Gina Rinehart, at number 44, who built substantially upon her inheritance from her father, Lang Hancock. Fashion and retail was another major contributory sector with 16 of the top 100 billionaires overall; eight were in the top 25 including three members of the Walton family, heirs to the Wal-Mart fortune.

The fastest recent growth in the numbers of billionaires has been in Russia and China (Freeland, 2012), and these in turn are playing major roles in transforming transnational housing markets. Moreno (2012) suggested that there are many similarities between super-rich Russians and Chinese in contrast to longer-established North American or European HNWIs. Neither country had any billionaires before the 1990s but their numerical growth has been extremely rapid, on average 15 years younger than American or German billionaires, and with higher levels of self-made fortunes (100 per cent in Russia and 83 per cent in China compared to 43 per cent in France and 33 per cent in Germany). Billionaires in both countries tended to be 'lone wolves', with low rates of family involvement in asset management, as well as relatively high numbers of self-made women among the ranks of billionaires, especially China. Moreno (2012) claimed there was little philanthropic giving in Russia or China, in contrast to the situation in the USA where many of the richest are energetic philanthropists, especially Bill Gates and Warren Buffett (for a critical discussion of super-rich philanthropy, see Ilan Kapoor, Chapter 6 in this volume). However, the 2014 *Forbes* list shows that numerous Russian billionaires, including Roman Abramovich, *have* been generous supporters of charitable foundations.

GLOBALIZATION AND RESIDENTIAL REAL ESTATE

> Continued global wealth creation, particularly in emerging markets, has been a key driver for prime property markets. (Knight Frank, 2014, p. 6)

Residential real estate is a vital element of the global economy. For example, Barkham (2012) estimated that the residential property market in the UK was ten times more valuable than the commercial property market in 2007, accounting for 65 per cent of the UK's residents' wealth. Prime housing has scarcity value and a luxury good: 'as societies get wealthier, people want more and better residential real estate' (Barkham, 2012, p. 213).

Huge geopolitical transformations over the last 30 years have had massive implications for residential real estate (Barkham, 2012), both within countries and across borders, especially the break-up of the Soviet empire and rapid rise of China, India, Brazil and Russia (the 'BRIC' countries). Former state-dominated housing systems have been deregulated and liberalized, in some case literally overnight. The abandonment of communism by China resulted rapidly in 'the incorporation into the global trading system of nearly 1.5bn additional workers' (2012, p. 18). Knight Frank (2013b, p. 3) estimated that the annual GDP growth in the BRICs countries was 'equivalent to the creation of a new economy the size of Italy each year'. These changed relations have had profound impacts on the development of globalized elements of housing markets by freeing up outward movement of capital from former socialist and rapidly developing economies and opening new markets for inward investment.

Much recent work on globalization and housing has been stimulated by aspects of the global financial crisis (GFC) especially the packaging of mortgage finance into bundles that were bought and sold internationally. Political economists from Marxist and other perspectives have argued that deregulation and securitization of mortgage finance increased risk, and that the switching of capital from 'primary' circuits (manufacturing) into secondary circuits (the built environment, including housing) inevitably results in periodic financial crashes (Harvey, 1978, 2001; Logan, 1993;

Bartelt, 1997; Mandelbrot and Hudson, 2005; Harrison, 2007; Krugman, 2008; Turner, 2008):

> The GFC, which was due to excessive real estate lending, was the direct linear descendant of the dot-com boom and slump and the Asian currency crisis in 1998, which preceded it. These highly unstable economic conditions are still in play today and will substantively impact the real estate research agenda for the next 10 years. (Barkham, 2012, p. 12)

Oblivious to such commentary, prime residential real estate remains a key element in super-rich investment portfolios, both for private use through luxury consumption *and* as investment items with anticipated long-term capital gain, often untaxed as properties are owned by companies rather than individuals. Most of the homes of the super-rich are purchased using cash, specialized financial instruments and/or through companies, and 'the higher the price of the property, the less likely buyers were to arrange traditional mortgage financing for the home acquisition. Whether buyers are foreign or domestic, cash transactions predominate at the higher end of the market' (Christie's, 2013, p. 14). Such transactions, therefore, never enter 'national' housing accounting systems and play no part in many accounts of aggregate 'national' house price trends. For example, the analysis of house price trends in the Joseph Rowntree Foundation UK Housing Review is based on data relating to transactions using mortgages or loans, and EU and OECD comparisons between countries are based on the same kinds of data (Paris, 2013b).

The 2014 edition of Knight Frank's annual *Wealth Report* reported a survey of 600 private bankers and/or wealth advisors across the world representing over 20000 UHNWIs. Residential property accounted for nearly 30 per cent of UHNWI wealth, with an average holding of 2.4 'second homes' in addition to their main residences. About 20 per cent were considering purchasing another property and 15 per cent were considering changing their country of residence. Quality of life was cited most frequently as the main factor affecting where they were considering buying additional properties, with the UK named as the most popular option. The prime residential 'hot spots' over the previous year had been Jakarta and Auckland with annual house price growth of 38 per cent and 28 per cent respectively. At the same time both Dublin and Dubai, both of which had experienced substantial falls previously, appeared to be at risk of building up into bubbles about to burst again! Overall, the survey noted rapidly growing wealth across Asia and interest in purchasing residential property in other countries.

An earlier Knight Frank survey (2010) of the global 'cross-border, non-domestic, luxury residential market' identified that the main factors motivating purchasers of luxury housing were lifestyle, security and investment (60 per cent). Other factors including business, tax and education were relatively less important, though purchases reflect *combinations* of factors. Most future demand for prime homes was expected to be *into* 'mature established regions' in countries with stable governments, whilst coming from all world regions. Demand was expected to be very strong in high-end European sun and snow resort and leisure areas (Alps, French Riviera), and for homes in low-tax jurisdictions (Monaco, Channel Islands, Switzerland, Dubai and some Caribbean nations). The high value placed on political stability recurs throughout expert commentary on transnational purchase of prime residential property, especially in global cities of countries seen as 'safe havens' for investment, and in exclusive locales of hyperconsumptive leisure, often involving 'horsification'[1] (e.g., Newmarket, UK; St. Bart's; the French Riviera):

> Following the economic downturn, Miami, London and New York came to epitomise the so-called safe haven market, with overseas buyers looking to escape currency, economic, political and security crises by putting equity into tangible assets that appeared safe from government sequestration so global capital flows continue to concentrate on a few key hubs. (Knight Frank, 2013a, p. 2)

The super-rich are globally oriented and rarely constrained within particular nations (see Short, Chapter 18 this volume for a discussion), though national policies and practices can affect *how* they operate transnationally. For example, many commentators suggest that Chinese government restrictions on domestic property purchase have stimulated surging growth of Chinese investment in residential real estate in other countries.

THE MANY HOMES OF THE GLOBAL SUPER-RICH

Unlike the richest households and dynasties of pre-industrial Europe, the twenty-first-century super-rich have almost unlimited mobility, typically in their own or chartered aeroplanes or First Class cabins of premier airlines (for a discussion, see Lucy Budd, Chapter 15 in this volume). They own many dwellings and country estates in numerous countries, topped by the 27-storey mansion 'Antilla' in Mumbai (Mawani, 2011), as well as luxury yachts and aircraft (Divergilio, 2011) (Figure 12.1).

They own residences in global cities, high-end resorts, remote fortified estates and high-amenity leisure regions, including islands. Political stability and transparency of commercial and legal systems are key factors affecting their residential purchase decisions (Freeland, 2012; Savills,

Note: *Clockwise from top left: John Travolta's home-cum-airport; 'One Hyde Park'; 'Antilla' (Mumbai); one of Roman Abramovich's yachts; Bill Gates's residence; one of Lakshmi Mittal's homes in London.*

Source: Collage prepared by Scott Johnson for use by Chris Paris in scholarly publications.

Figure 12.1 *Some of the many homes of the super-rich*

2012; Christie's, 2013). Many choose residential bases in tax havens such as Monaco, the Cayman Islands or Guernsey (Shaxson, 2011a). They can also effectively choose their nominal country of residence as they face few of the limitations on access that may be imposed on poorer prospective immigrants or asylum seekers. As John Rennie Short discusses in Chapter 18 in this volume, countries *compete* with each other to attract the attention and investments of the global super-rich:

> As wealth grows in emerging markets, so does the desire to protect it. Increasingly wealthy individuals are looking for places where their assets will be safe. Investor visas, which grant residency or citizenship in exchange for investments, account for a small but growing percentage of immigration worldwide. (Knight Frank, 2014, p. 37)

The super-rich amass large portfolios of residential properties for three overlapping reasons, over and above the simple consideration that everybody needs at least one 'home'. I propose a typology of the residential spaces of the super-rich, using three broad types of residential spaces: trophy homes, boltholes and strategic asset management. Some places contain a range of prime residential real estate with all three qualities: many homes have global trophy status; buyers can feel confident that their assets will at least maintain their value; and non-domiciles are relatively safe from potential risks in more volatile countries.

The term 'trophy homes' is used widely in the upmarket real estate literature to describe the most exclusive and expensive residential properties – typically within the top 2 per cent of prime housing sub-markets. They are sites of consumption, hyperconsumption and reproduction. They are occupied at least occasionally by their owners though the duration and frequency of occupancy varies depending on how many homes they own and their (often changing) preferences. Many are newly built in major global cities for buyers from a *global* market: 'At the very top end of the market. . .projects such as 432 Park Avenue in New York, Tour Odéon in Monaco, Faena in Miami and OPUS in Hong Kong, the mix of buyers is very global' (Knight Frank, 2013c, p. 3).

Others are longer-established mansions, albeit often updated and expanded, as in Kensington Palace Gardens, colloquially known as 'Billionaires Row', a gated, semi-fortified and heavily guarded street containing numerous embassies and with homes owned by a cosmopolitan mix of billionaires including two of Britain's four wealthiest men in the 2014 *Sunday Times* Rich List, Len Blavatnik (Russian) and Lakshmi Mittal (Indian). Listed as the second most expensive street in the world in *Billionaire* (online), beaten only by Pollock's Path on the Peak in Hong Kong (Wilkinson, 2012), Kensington Palace Gardens also contains one of many British royal homes, Kensington Palace, occupied by the Duke and Duchess of Cambridge (*Forbes*, 2012).

Trophy homes are repositories of assets rather than places in which to dwell. They are functionally identical to valuable up-market works of art:

scarce or unique objects with high value consumption characteristics plus real asset value growth over time, especially with growing numbers of the super-rich and restricted supply of space and art (even given the productivity of Damien Hurst). The parallel between trophy homes and artworks was articulated explicitly by Christie's International Real Estate:

> As the only real estate network owned by a fine-art auction house, Christie's International Real Estate has unparalleled access to the HNWIs around the globe who procure assets such as art, wine, jewellery, and, of course, real estate. Location, lifestyle, and provenance, particularly at the top end of the luxury residential real estate market, are the hallmarks of value and often equally as important as price when HNWIs consider a property purchase. (Christie's, 2013, p. 3)

Although trophy homes have parallels with works of art as stores of value, one aspect of their production differs fundamentally. Canvas, stone or other art materials typically have little inherent value, are substitutable and easily moved. Their key market attributes derive from the value of artworks based on the fame of the artist and their relative scarcity. A key feature of trophy homes, by way of contrast, is their location and its attributes: land is not easily moved and retains its value when homes are demolished.

Unlike the homes of most owner-occupiers whose purchases are recorded in national housing finance statistics, few trophy homes are the main residences of their owners. Savills (2013) argued that many overseas purchasers of ultra-prime homes in London do intend to use them as primary residences, but this would mean by definition that their *other* residences will be used less often.

American billionaires figure strongly among the lists of trophy home owners, often within the USA rather than overseas, especially *Forbes* 2014 number five, Larry Ellison, whose fortune reportedly grew by US$556 million on 19 May 2014 – such are the extravagant vagaries of the stock market! Reputedly 'the nation's most avid trophy-home buyer', he has numerous homes across mainland USA from Rhode Island to California and over the Pacific to Hawaii, as well as a villa in Japan (Stone, 2015). Hedge fund proprietor Steve Cohen, at 115 in the 2014 *Forbes* list, is reported to own numerous homes in New York and the Hamptons (Brennan, 2013). One of his Manhattan apartments was on sale at the time of writing for US$98 million, reportedly reduced from an initial asking price of US$115 million (Business Insider, 2015).

The term 'bolthole' is often used to refer to a place of relative secrecy where people can hide or find anonymity. The term is used here to refer to large exclusive residences purchased as an element of risk management strategies relating to two analytically different types of risk: primarily political risks (regime change, revolution, loss of favour or exposure of corruption) but also physical environmental risks. These residential spaces tend often to be unoccupied, especially those acquired to mitigate potential political risks, as exemplified in recent commentary on decaying mansions in London. The main purchasers of boltholes come from outside Western Europe and North America, though many Americans and Europeans obtain personal or corporate domicile for their assets in overseas tax havens, albeit often without the inconvenience of actually having to spend much time in such places.

Many purchasers of boltholes are from potentially unstable oil-rich countries; others are dictators and despots from around the globe. Some, no doubt, are heads of major organized crime groups, laundering their money in safe havens but I have no knowledge of such groups, nor do I seek any. Some of the most notorious examples of boltholes are in Hampstead, inner north London, where ten mansions in The Bishops Avenue 'have stood almost entirely vacant since they were bought a quarter of a century ago, it is believed on behalf of members of the Saudi Arabian royal family' (Booth, 2014). The *Guardian* newspaper reported that at least 16 mansions were empty 'with their windows shuttered with steel grilles and overgrown grounds patrolled by guard dogs' (Booth, 2014). These house wrecks are the most blatant examples of 'a property culture in which the world's richest people see British property as investments' and over half of the most expensive 55 properties in The Bishops Avenue are registered to offshore companies or foreign owners (Booth, 2014). The occupancy and maintenance of environmental boltholes tends to be higher than that of political risk management investments.

Strategic asset management may be considered more an approach to property investment than as a type of home of the super-rich, though in many instances it is the main factor driving acquisition of homes. This includes the acquisition of dwellings as part of mixed asset portfolios, with dwellings located in different countries to reduce market uncertainties, though this also can include an element of entrepreneurial risk-taking with strategic investment in potentially high-earning assets. Such dwellings also tend largely to be left empty, though they are more likely to be well maintained than boltholes and are often occupied during peak seasons; in the UK, for example, during Ascot or Wimbledon weeks (Freeland, 2012).

There is a strong element of risk avoidance and search for safe haven investment in the strategies of many super-rich (and merely-rich) Chinese. Whereas super-rich Russians were well-established purchasers of prime residential properties in London and New York by 2010, the impact of HNW Chinese purchasers of overseas prime property has been more recent. Knight Frank (2013a) remarked on 'the widening demand for luxury property', as 'the real story of 2012 was the rise of demand from China' both in the UK, essentially London, and the USA. Chinese super-rich demand is growing rapidly and having major impacts on those cities, as well as others in the Pacific region, including Hong Kong, Los Angeles, San Francisco, Vancouver, Melbourne, Sydney and Auckland (Ley, 2010; Rapoza, 2011; Roberts and Zhao, 2011; Shen, 2012).

Unlike many super-rich Russians who leave the country permanently, most Chinese super-rich maintain a presence within China, especially if their main business interests lie within the country (Roberts and Zhao, 2011). However, many Chinese super-rich also acquire citizenship and residential assets in overseas safe havens, especially where there are good educational opportunities for their children (see Short, Chapter 18 this volume). Many commentators emphasize fears about possible future unrest, as many of the global super-rich are buying boltholes, especially those who fear the discovery of some illegality relating to the development of their assets. They would be at potential risk both from the government and/or anti-government forces. One émigré in Boston was cited as saying that the Chinese government spent more on internal security than on defence, so 'if things got ugly, the rich would be targets not just for being rich but for their close connections with the government' (Roberts and Zhao, 2011).

Other studies have identified strong demand for exclusive leisure activities, especially conspicuous consumption of extensive leisure-related properties in the 'elite countryside' (Woods, 2013), often in areas associated with horse racing and polo (McManus, 2013; Roberts and Schein, 2013). The leisure theme was explored in Savills' *World Cities Review* (2012), which included a section on global leisure properties for the super-rich residents of its ten 'world cities'[2] from which 'limousines and Learjets make the weekend escape easy', with contrasts between the modes of travel to favoured leisure locations. Limousines enabled super-rich Muscovites, Londoners and New Yorkers to get to Razdory, Oxfordshire and the Hamptons, but their counterparts from Shanghai and Hong Kong were more likely to use jet planes, in both cases to get to Phuket.

Purchases of luxury residential property reported in the Knight Frank (2010) survey took place mainly *within* some regions, especially Russia, South America and the Middle East. By way of contrast, the origins of the super-rich buyers of European and Caribbean residential property were more diverse, with high proportions from the USA and Canada plus strong and growing involvement of Asian buyers. Asian HNWIs were also active in Africa, Europe, the Middle East and North America. Anticipated future demand was significantly different in terms of the origins of buyers, with lower growth in demand from established rich countries of Europe and North America but stronger growth likely from South America, Asia and Africa, albeit from low bases. The biggest anticipated growth of demand was from Asia, where it was *already* high.

Recent commentary on prime global residential property markets has identified diversity and differences between cities and regions. For example, Knight Frank (2013b) argued that 'mainstream' middle class housing markets had out-performed prime residential properties in many Asian countries, partly due to growing demand generated by the spectacular growth in absolute numbers of middle class home buyers. In addition, many Asian super-rich households were more inclined to invest in prime global property markets 'outside their domestic arena':

> With an increasingly mobile, educated and well-travelled class of property owners in the Asian region, the lifestyle choice of having a second home abroad, for personal or children's educational use is proving to be one of the key narratives for HNW Asian buyers. Purely from an investment point of view, as a diversifier away from the steamy and controlled Asian markets, it has been seen as a sensible strategy for their wealth portfolios. (Knight Frank, 2013b, p. 4)

Elliott and Urry (2010, p. 82) conducted in-depth research on the lives of global business leaders; they argued that mobile global elites pursue 'mobile life strategies' and are fascinated with space, especially 'private, exclusive luxurious space'. The ownership of many homes enables the global super-rich to be 'at home wherever they find themselves'. They also identified specific residential spaces where the super-rich can be found 'consuming to excess', citing Sheller's (2008) analysis of 'the neo-liberal respatialization of the Caribbean for the benefit of the super-rich, yacht-owning, automobile global elite'. Islands, they argue, provide 'natural' barriers to the outside world. Many Caribbean islands are locales for exclusive leisure and untaxed investments of the global super-rich, forming 'part of the unregulated, tax-free, offshore economy, the "offshore world". . .constituted through the stitching together of physical space with cyberspace' so that formerly 'differentiated neighbourhoods morph into mobile de-differentiated consumption zones of excess' (Elliott and Urry, 2010, p. 127).

This kind of 'elite casino capitalism' may only apply to a small proportion of the world's population, but Elliott and Urry argue that it is of wider economic, social and political importance because such places are exemplars from which mass-market versions are developed and zones of excess for the super-rich are typically places where many other people work in poor conditions and for low wages. Such workers themselves often are mobile, whether flown from poor countries to labour on construction projects in oil-rich deserts or cross-border workers entering 'offshore' Monaco on a daily basis. Elliott and Urry (2010, p. 127) also argue that as the super-rich become based for tax purposes in 'offshore' locations, this 'has the effect of reducing the tax-take of states and lowering the level of scale of public provision, both in the island resort and in the country from which visitors travel'. Elliott and Urry (2010, p. 127) describe these 'dream-worlds' as paradigm examples of splintering urbanism, whereby public space is reduced and access to beaches, rivers or other amenities is denied to locals or poorer visitors, reinforcing and extending the mobility field, producing even greater inequalities 'between the economic and network capital rich and the economic and network capital poor'. Finally, they argue that the spaces of luxury hyperconsumption constitute further extensions of 'the hyper-high-carbon societies of the twentieth century' as the economies of whole cities, regions or islands' economies 'are transformed into centres of "wasteful" production and consumption' (2010, p. 129).

THE GLOBAL SUPER-RICH AND THE DECOUPLING OF PRIME RESIDENTIAL REAL ESTATE

The residential investment decisions by the super-rich have *huge* impacts on prime real estate in cities and regions across the globe (Paris, 2013a, 2013b). Sassen (2006, p. 10) argued that the emergence of an international property market 'means that real estate prices at the center of New York City are more connected to prices in central London or Frankfurt than to the overall real estate market in New York's metropolitan areas'. Leading real estate agencies operating at the top end of the housing market are well aware of these developments and play key roles in servicing the requirements of super-rich investors and their purchase of many homes in global cities and leisure regions (Cousin and Chauvin, 2013; McManus, 2013; Paris, 2013b; Roberts and Schein, 2013; Woods, 2013).

The market analyses and marketing activities of international real estate agencies provide ample documentation of the dynamic nature of the international real estate market and the varying positions of different cities and regions. Critically, global luxury real estate is almost entirely insulated from developments in 'national' housing markets, as Savills and Christie's point out: 'Billionaires pay money for real estate in world-class markets quite independently from the national residential markets in which they sit' (Savills, 2012, p. 18). The luxury housing market remains insulated from money flows and political shifts, as these concerns are less likely to determine the purchase of a trophy home for UHNW population around the globe (Christie's, 2013, p. 18).

Prices of prime residential real estate in London have wildly outstripped the rest of the UK housing market since 2008 as foreign purchasers have transformed the property market. London has been 'for sale' to all-comers (Heywood, 2012), 'the first choice for second homes' of the global super-rich (Hipwell, 2011) as foreign investors 'snap up 70 per cent of all central London new build homes' (Boyce, 2013). UK and London house prices have been skewed, boosted enormously by surging growth in prime inner city residential real estate, accentuated by massive inflows of overseas investment. The average price of a super-prime property increased by 26 per cent between 2012 and 2013 as London became the 'billionaire capital of Europe' (Withnall, 2014). Savills (2014) reported that homes in the ten wealthiest boroughs in London 'are now worth a total of £609 billion, 9 per cent higher than the value of Scotland, Wales and Northern Ireland combined'. Two London boroughs particularly favoured by foreign investors, Westminster and Kensington and Chelsea, had 'a combined value of just over £200 billion, 15 per cent more than the total housing stock of Wales'. The March 2013 edition of *High Net Worth*, 'a unique monthly publication that provides a digest of data and insights for professionals who market and sell to the wealthy'[3] celebrated because the increase *alone* in prime London home values between 2007 and 2013 was worth more than all of the dwellings in the northeast of England.

Super-prime New York is also affected ('afflicted'?) by the same process of decoupling, with property values also soaring in the prestigious Hamptons. Other celebrated global sites of hyperconsumption of mansions include: Pollock's Path on the Peak in Hong Kong, Avenue Princess Grace in Monaco, Boulevard du General de Gaulle in Cap Ferrat ('the most sought-after piece of land along the Mediterranean Côte d'Azur' where Charlie Chaplin once owned a holiday home), Chemin de Ruth in Geneva (Switzerland), Romazzino Hill in Sardinia ('the quintessential billionaire's holiday playground'), and Ostozhenka in Moscow (home to Alisher Usmanov, second in the 2014 *Sunday Times* Rich List, who also owns a mansion in Hampstead, London and Tudor manor house in Surrey, UK) (Wilkinson, 2012).

One of the attractions of prime residential areas of London and New York, in addition to their role as safe havens, is the scarcity of land and severe supply constraints. They are not only 'safe' places to store assets, but also 'demand is ever rising while the stock of desirable locations remains virtually static, and so global capital flows continue to concentrate on a few key hubs' (Knight Frank, 2013b, p. 4). Constraints on supply are *good* things for super-rich buyers, because trophy homes benefit as positional goods (Freeland, 2012). Restricted supply sustains and enhances land and housing values, as in Sydney where 'underlying restrictions of land availability, due to zoning regulations and limited land release, ensure that land values, and in turn house prices, remain the highest in the country' (Savills, 2012, p. 12). Thus planning restrictions on land supply or new development *add* to the lustre of existing dwellings or new prestige developments.

Some developments in London have much in common with island tax havens, especially 'One Hyde Park' (OHP), 'the most expensive apartment block ever built anywhere on Earth' (Arlidge, 2011, p. 18). The name is symbolic, not an actual address as it is located in Knightsbridge, not in Hyde Park, and linked to the Mandarin Oriental Hotel. This development has been described as a form of 'offshore island' functioning largely outside the UK tax system, 'largely an empty shell for overseas investors to park their assets' (2011, p. 23) and 'a stunning symbol of the excess, raw cynicism, artificiality, market distortions and sheer opaque lifelessness of offshore finance' (Shaxson, 2011b). It combines supreme luxury, security and amenities:

> Its unique location, with the elegance and excitement of bustling Knightsbridge to the south and the glorious romance and serenity of Hyde Park to the north, offers an incomparable London living experience for those who want the ultimate in elegance, peace and style. The exceptional collaboration between leading architects, designers, artists and hoteliers ensures that One Hyde Park delivers ultimate perfection and a unique experience on every level. (Knight Frank, 2012)

OHP is the paradigmatic residential real estate development in London purchased by super-rich investors, often through overseas companies based in offshore tax havens, with most apartments left empty (Arlidge, 2011; Heywood, 2012). It was developed by a joint venture combining the CDC Group (run by the brothers Nick and Christian Candy) and Waterknights, the private company of Qatar's Sheikh Hamad Bin Jassim Bin Jaber Al Thani (ABC News, 2014). An OHP penthouse property was sold in May 2014 for £140 million to an unnamed buyer reputedly of Eastern European origin, establishing a new London property price record (ABC News, 2014). The previous UK record price for an apartment, also in OHP, was set by a Ukrainian billionaire who purchased a penthouse and apartment to knock together into one property.

CONCLUSIONS

The numbers and aggregate wealth of the super-rich have grown strongly over the last ten years, together with the significance of residential real estate as a key element of their wealth portfolios. Residential real estate is crucial because particular sites take on additional value due to their location in cultural cores of 'safe haven' cities and desirable exclusive leisure locales. In many instances, moreover, these investments in residential spaces are enhanced by restrictive and exclusionary planning and zoning regulations. Whilst globalization and deregulation have been accompanied by ever faster and easier movement of capital, and expensive labour in developed countries and regions can easily be replaced by cheaper labour elsewhere, prime global sites are fixed in space.

The purchase of many homes in different countries has been a vital element of super-rich asset management and household life course strategies, including personal physical relocation and the acquisition of citizenship or residency rights, as well as intergenerational transfers of assets through the education of their children and inheritance management. The distinction between trophy homes and boltholes helps explain the different patterns of residential investment by billionaires from different countries as well as where they have bought such properties. American billionaires have tended to buy trophy homes largely within the USA and Europeans mainly buy within their home countries or elsewhere in Europe. By way of contrast, the super-rich from oil-rich and many developing countries acquire boltholes in Europe in case of a possible future need to make a rapid departure from turbulent local circumstances. Many Russian and Asian billionaires buy in the USA, Europe or English-speaking Pacific Region countries for a combination of reasons, especially asset management and as possible boltholes.

Prime residential housing markets are unambiguously transnational, with global buying and selling of residential properties and leisure spaces. Specific sites of global luxury property ownership are decoupled from regional and national housing markets, so analyses of housing market performance based on national statistics using mortgage data bear little relationship to what is actually happening within regions containing such sites. Commercial assessments of UK market values show London exemplifying wider decoupling: massive foreign investment within a limited area has greatly inflated land and property values, protected by an exclusionary planning system, so that houses in a few dozen streets are worth more than all of the residential properties in Wales.

These developments have many potential political implications and many contradictions, not least because much of the mobility of the super-rich has been based on oil resources, which are steadily running out.

NOTES

1. Horsification is a term borrowed from rural landscape studies to refer to the transformation of formerly productive countryside into a zone of urbanized leisure use, typically involving horses (Antrop and van Eetvelde, 2008).
2. New York, London, Paris, Moscow, Mumbai, Singapore, Shanghai, Sydney, Hong Kong and Tokyo.
3. This was available by subscription from https://www.ledburyresearch.com.

REFERENCES

ABC News (2014), 'One Hyde Park's $255 million penthouse apartment sale breaks London property record', *ABC News*, 3 May, accessed 20 May 2014 at http://www.abc.net.au/news/2014-05-02/london-sets-record-with-255-million-dollar-apartment-sale/5427988.

Antrop, M. and V. van Eetvelde (2008), *Mechanisms in Recent Landscape Transformation, WIT Transactions on the Built Environment, Vol. 100*, pp. 183–92.

Arlidge, J. (2011), 'Anybody home?' *The Sunday Times Magazine*, 20 November, 16–23.

Barkham, R. (2012), *Real Estate and Globalisation*, London: Wiley.

Bartelt, D. (1997), 'Urban housing in an era of global capital', *Annals of the American Academy of Political and Social Science*, **551** (1), 121–36.

Beresford, P. and W. Rubenstein (2007), *The Richest of the Rich*, Petersfield, UK: Harriman House.

Booth, R. (2014), 'Inside 'Billionaire's Row': London's rotting, derelict mansions worth £350m', *The Guardian*, 1 February, accessed 10 March 2014 at http://www.theguardian.com/society/2014/jan/31/inside-london-billionaires-row-derelict-mansions-hampstead.

Boyce, L. (2013), 'Foreign investors snap up 70% of all central London new build homes fuelling a surge in prices', *This is Money*, 16 August, accessed 18 August 2015 at http://www.thisismoney.co.uk/money/mortgageshome/article-2394704/Foreign-investors-snap-70-new-build-homes-central-London.html.

Brennan, M. (2013), 'The many mansions of hedge fund billionaire Steve Cohen', *Forbes*, 4 September, accessed 2 February 2014 at http://www.forbes.com/sites/morganbrennan/2013/09/04/billionaire-property-portfolio-steve-cohens-castles/.

Business Insider (2015), 'No one wants to buy Steve Cohen's unbelievable $98 million Upper East Side Penthouse', *Business Insider*, 12 May, accessed 2 June 2014 at http://www.businessinsider.com/steve-cohen-cant-sell-his-penthouse-2014-5?IR=T.

Christie's International Real Estate (2013), *Luxury Defined Around the World*, accessed 25 April 2013 at http://www.christiesrealestate.com/eng/luxury-defined.

Clapham, D. (2006), 'Housing policy and the discourse of globalization', *European Journal of Housing Policy*, **6** (1), 55–76.

Cousin, B. and S. Chauvin (2013), 'Islanders, immigrants and millionaires: the dynamics of upper-class segregation in St Barts, French West Indies', in I. Hay (ed.), *Geographies of the Super-Rich*, Cheltenham, UK and Northampton, MA, USA: Edward Elgar Publishing, pp. 186–200.

Divergilio, A. (2011), 'Luxury yachts of the super-rich', *Bornrich.com*, 10 June, accessed 22 January 2015 at http://www.bornrich.com/luxury-yachts-of-the-super-rich.html.

Elliott, A. and J. Urry (2010), *Mobile Lives*, London: Routledge.

Forbes (2012), 'London's Billionaires' Row', 23 May, *Forbes.com*, 23 May, accessed 23 January 2015 at http://www.forbes.com/global/2012/0604/companies-people-trophy-homes-london-billionaire-row.html.

Freeland, C. (2012), *Plutocrats*, London: Allen Lane/Penguin Books.

Harrison, F. (2007), *Boom, Bust: House Prices, Banking and the Depression of 2010*, London: Shepheard-Walwyn.

Harvey, D. (1978), 'The urban process under capitalism', *International Journal of Urban and Regional Research*, **2** (1–4), 101–31.
Harvey, D. (2001), *Megacities Lecture 4*, accessed 5 March 2013 at http://www.kas.de/upload/dokumente/megacities/MegacitiesLectur4Worlds.pdf.
Heywood, A. (2012), 'London for sale: an assessment of the private housing market in London and the impact of growing overseas investment', *Social Welfare*, accessed 2 September 2012 at http://socialwelfare.bl.uk/subject-areas/services-activity/housing-homelessness/smithinstitute/london12.aspx.
Hipwell, D. (2011), 'London, the first choice for second homes', *The Times*, 6 April, 37.
Knight Frank (2010), 'Global property wealth survey 2010', *Knight Frank*, accessed 18 August 2015 at http://www.knightfrank.com/research/global-property-wealth-survey-2010-23.aspx.
Knight Frank (2012), 'One Hyde Park, Knightsbridge W1', *Knight* Frank, accessed 4 January 2012 at http://search.knightfrank.com/gb0203.
Knight Frank (2013a), 'Global prime residential performance', *Knight Frank blog*, 6 March, accessed 20 March 2013 at http://www.knightfrankblog.com/wealthreport/current-edition/safety-first/.
Knight Frank (2013b), 'Asia-Pacific residential review', *Knight* Frank, accessed 20 March 2013 at http://my.knightfrank.com/research-reports/asia-pacific-residential-review.aspx.
Knight Frank (2013c), 'International residential investment in London', *Knight* Frank, accessed 21 January 2015 at http://my.knightfrank.com/research-reports/international-residential-investment-in-london.aspx/.
Knight Frank (2014), *The Wealth Report 2014*, accessed 2 March 2014 at http://www.thewealthreport.net/.
Kroll, L. (2014), 'Inside the 2014 Forbes billionaires list: facts and figures', *Forbes.com*, 3 March, accessed 20 April 2014 at www.forbes.com/sites/luisakroll/2014/03/03/inside-the-2014-forbes-billionaires-list-facts-and-figures.
Krugman, P. (2008), *The Return of Depression Economics and the Crisis of 2008*, London: Penguin.
Ley, D. (2010), *Millionaire Migrants: Trans-Pacific Life Lines*, Malden, MA: Wiley-Blackwell.
Logan, J. (1993), 'Cycles and trends in globalization of real estate', in P. Knox (ed.), *The Restless Urban Landscape*, Englewood Cliffs, NJ: Prentice Hall.
Mandelbrot, B. and R. Hudson (2005), *The (Mis) Behaviour of Markets: A Fractal View of Risk, Ruin and Reward*, London: Profile Books.
Mawani, V. (2011), 'Larger than life: Antilla Mumbai – the world's MOST expensive home', *Industry Leaders Magazine*, 19 January, accessed 30 November 2011 at http://www.industryleadersmagazine.com/larger-than-life-antilla-mumbai-the-worlds-most-expensive-home.
McManus, P. (2013), 'The sport of kings, queens, sheikhs and the super-rich: thoroughbred breeding and racing as leisure for the super-rich', in I. Hay (ed.), *Geographies of the Super-Rich*, Cheltenham, UK and Northampton, MA, USA: Edward Elgar Publishing, pp. 155–70.
Moreno, K. (2012), 'Why China's rich and Russia's rich share many traits', *Forbes.com*, 30 March, accessed 29 April 2013 at http://www.forbes.com/sites/kasiamoreno/2012/03/30/why-chinas-rich-and-russias-rich-share-many-traits-2/.
Oxfam (2014), 'Working for the few', *Oxfam Briefing Paper No. 178*, accessed 21 January 2015 at http://www.oxfam.org/en/policy/working-for-the-few-economic-inequality.
Paris, C. (2013a), 'The homes of the super-rich: multiple residences, hyper-mobility and decoupling of prime residential housing in global cities', in I. Hay (ed.), *Geographies of the Super-Rich*, Cheltenham, UK and Northampton, MA, USA: Edward Elgar Publishing, pp. 94–109.
Paris, C. (2013b), 'The super-rich and the globalisation of prime housing markets', *Housing Finance International*, Summer, 18–27.
Rapoza, K. (2011), 'Where rich Chinese are buying real estate', *Forbes.com*, 20 June, accessed 22 April 2013 at http://www.forbes.com/sites/kenrapoza/2011/06/20/where-rich-chinese-are-buying-real-estate/.

Roberts, D. and J. Zhao (2011), 'China's super-rich buy a better life abroad', *Bloomberg Businessweek*, 22 November, accessed 10 March 2014 at http://www.businessweek.com/magazine/chinas-superrich-buy-a-better-life-abroad-11222011.html.
Roberts, S. and R. Schein (2013), 'The super-rich and the transformation of a rural landscape in Kentucky', in I. Hay (ed.), *Geographies of the Super-Rich*, Cheltenham, UK and Northampton, MA, USA: Edward Elgar Publishing, pp. 137–54.
Sassen, S. (2006), *Cities in a World Economy*, 3rd edition, London: Sage.
Savills (2012), *Savills Insights World Cities Review 2012*, accessed 2 December 2012 at http://magazines.savills.com/spring_2012.html.
Savills (2013), *Savills Insights World Cities Review 2013*, accessed 3 September 2013 at http://pdf.euro.savills.co.uk/residential---other/world-cities.pdf.
Savills (2014), 'UK housing stock value climbs to £5,205,000,000,000 but the gap between the haves and the have nots grows', *Savills*, 24 January, accessed 21 January 2015 at www.savills.co.uk/_news/article/72418/172125-0/1/2014/uk-housing-stock-value-climbs-to-%C2%A35-205-000-000-000-but-the-gap-between-the-haves-and-the-have-nots-grows.
Shaxson, N. (2011a), *Treasure Islands: Tax Havens and the Men who Stole the World*, London: Bodley Head.
Shaxson, N. (2011b), 'One Hyde Park: another offshore island in London', Treasureislands.org, 21 November, accessed 12 November 2013 at http://treasureislands.org/one-hyde-park-an-offshore-island-in-london.
Sheller, M. (2008), 'Infrastructures of the imagined island: software, mobilities and the new architecture of cyberspatial paradise', *Environment and Planning A*, **41** (6), 1385–403.
Shen, A. (2012), 'China's super-rich flock into the US property market', *Forbes.com*, 13 July, accessed 2 May 2014 at http://www.forbes.com/sites/china/2012/06/13/chinas-super-rich-flock-into-u-s-property-market.
Stone, S. (2015), 'The incredible real estate portfolio of Oracle billionaire Larry Ellison', *Business Insider*, 22 April, accessed 20 August 2015 at http://www.businessinsider.com.au/larry-ellison-real-estate-2015-4.
Turner, G. (2008), *The Credit Crunch*, London: Pluto Press.
Wilkinson, T. (2012), 'Good neighbours: the world's top 10 most exclusive residential streets', *Billionaire*, 3 July, accessed 2 May 2014 at http://www.billionaire.com/ambience/622/good-neighbours-the-worlds-top-10-most-exclusive-streets.
Withnall, A. (2014), 'London named billionaire capital of Europe', *The Independent*, 18 August, accessed 1 May 2014 at www.independent.co.uk/news/uk/home-news/london-named-billionaire-capital-of-europe-9225318.html.
Woods, M. (2013), 'The elite countryside: shifting rural geographies of the transnational super-rich', in I. Hay (ed.), *Geographies of the Super-Rich*, Cheltenham, UK and Northampton, MA, USA: Edward Elgar Publishing, pp. 123–36.
Wu, F (2012), 'Globalisation', in S. Smith (ed.), *International Encyclopedia of Housing and Home, Vol. 2*, Oxford, UK: Elsevier, pp. 292–7.

13. Reconfiguring places – wealth and the transformation of rural areas

Michael Woods

INTRODUCTION

The rural landscape of the Western world has been shaped by the wealth and power of elite groups from the landed aristocracy to contemporary affluent amenity migrants. For centuries, wealth has provided elites with the capacity to transform rural places, from the quasi-autocratic ability of rich landowners to redesign the appearance of the rural landscape, to the political influence of affluent groups on fiscal and land use planning policies, to the effect of differentiated property prices in segregating residential geographies, to the tailoring of services and amenities to the lifestyle and consumption preferences of wealthy residents. This chapter examines the structures and dynamics that lie behind such articulations of power, and in particular the close and historically embedded relationship between wealth and landownership in rural societies. In doing so, the chapter focuses on the examples of three distinct forms of wealth that have had an impact on rural places – the old wealth of the landed gentry in Europe; the new wealth of gentrifying middle classes; and the global wealth of the super-rich. It also considers the mobilization of opposition to the power of the wealthy in rural areas. The intersection of wealth, land and power in rural societies has deep historical roots.

The first enclosure of land as private property in early agrarian civilizations created a commodity that was to become both a source and an expression of wealth in rural society. Although the earliest forms of land acquisition were not through monetary exchange but through force, displacement, patronage or the expedient act of claiming 'unowned' land, the access afforded by private land ownership to mineral, forest and agricultural resources facilitated the primary means of wealth accumulation in pre-industrial economies. As the value of rural land as a basis for wealth thus became apparent, legal conventions to record and codify land title were established as well as markets for the sale of land titles, such that acquisition of land became increasingly dependent on wealth. The primacy of private ownership of rural land – and consequently the social stratification of rural society according to the means of an individual to purchase or rent land – was consolidated in Europe by acts of enclosure of common land during the liberal ascendency of the seventeenth, eighteenth and nineteenth centuries, and has achieved global reach (Richards, 2002; Hall, 2013).

The relationship between wealth and land in rural society is not, however, fixed, and shifts according to geographical, political-economic and historical context. Three key dimensions of the relationship can be identified. First, as noted above, land can be a source of wealth. This occurs most directly through the exploitation of minerals and other natural resources and production of agricultural commodities. However, land is unusual as a resource in that it is commonly leased and frequently worked or occupied by others. As such, wealth is also generated from land through the rents of tenants, and in feudal times, from tithes and taxes imposed on the production of subservient farmers. Furthermore, as trade in land inflated property values, so a further means of wealth generation from land emerged in the form of speculation.

Second, land ownership allows the display of wealth. The ability to purchase private property – land and/or housing – is an expression of relative wealth, and the larger and more prestigious the property the greater the implication of wealth. At the same time, rural land provides a platform for the ostentatious exhibition of wealth through extravagant spending, for example on elaborate and expensive building or landscaping projects, or in the pursuit of expensive hobbies or lifestyles (e.g., hunting, horse-riding, aviation, collecting classic cars, throwing lavish parties, and so on). The significance as a status symbol of owning rural land or large country properties is further reinforced by the cultural cachet that is attached to it in many countries, romanticizing the landed gentry as a pinnacle of exclusivity and privilege.

Third, wealth can be deployed as a means of controlling access to land, with land ownership in turn enforcing class divisions. In agricultural societies the denial of land ownership to those who could not afford to purchase property constrained the potential for tenant farmers and farm-workers to achieve economic independence and perpetuated their subservience to landowners. Today, the purchase of rural estates by the global super-rich, or of properties in gated communities by the wealthy middle classes in Western countries, buys not only an asset and status, but also privacy and security. More broadly, affluent middle classes engage in local politics to limit the supply of property in sought-after villages by resisting new developments, thus inflating house prices and ensuring exclusivity.

Wealth and land thus mutually reinforce each other in constructing power in rural societies, including the capacity of select actors to transform rural places. The exercise of this capacity by three distinct wealthy groups is discussed in detail in the following parts of this chapter, starting with the 'old wealth' of the European landed gentry.

OLD WEALTH: THE ARISTOCRACY IN RURAL EUROPE

The power of the aristocracy and landed gentry in pre-twentieth-century Europe was founded on their ownership of (predominantly rural) land and the wealth that they derived from it. Rural land allowed the generation of wealth directly through the sale of agricultural commodities and the exploitation of natural resources such as minerals and timber, and indirectly from rents and the taxing and tithing of tenants. The capacity for wealth generation had been reinforced by the enclosure of common land in the seventeenth and eighteenth centuries, which consolidated the commercialization of agriculture and food supply but also limited the potential for rural workers to support themselves independently of landowners. The resulting inequality of wealth and power in the countryside was targeted by revolutionary and reform movements, with serfdom abolished across Central and Eastern Europe during the late eighteenth and nineteenth centuries (it had already become obsolete in most of Western Europe), and limited land reform introduced in post-revolutionary France (Beaur, 2008).

Nonetheless, the majority of rural land in Europe remained in the hands of the aristocracy and gentry. James Bateman's pioneering analysis of the 1872 return of landownership in Britain and Ireland found that 16.5 per cent of land was owned by 400 peers and peeresses, 24.6 per cent by 1288 'great landowners' (or the landed gentry), and 12.5 per cent by 2529 'squires' (Cahill, 2001). In other words, 53.6 per cent of the land of Britain and Ireland was owned by 0.02 per cent of the population. Furthermore, Cahill calculates from Bateman's analysis that there were 27 aristocratic landowners whose holdings exceeded the equivalent of £100 million in value at 2001 prices (Table 13.1).

Rural land was also important as a platform on which wealth could be displayed, notably through the construction of large and elaborate mansion houses and landscaped parks. The installation of new technologies such as electric lighting or glasshouses was also an exposition of wealth, as was the display of expensive artworks and antiquities. At the same time, such works were justified by the discourse of the country gentleman as acts of 'improvement' and expressions of the 'duty' of the landowner towards the environment (Everett, 1994; Woods, 1997). Simply maintaining an estate was an expensive undertaking, with the Earl of Ancaster, for instance,

Table 13.1 Wealthiest landowners in Britain and Ireland, 1872, by value of land

Landowner	Home County	Total Area (Acres/ Hectares)	Estimated Land Value 1872 (£)	Equivalent Land Value 2001 Prices (£)
Duke of Westminster	Cheshire	19749 (7992 ha)	7250000	507500000
Duke of Buccleuch & Queensbury	Northamptonshire	460108 (186199 ha)	5800000	406000000
Duke of Bedford	Bedfordshire	86335 (34938 ha)	5625000	393750000
Sir John Ramsden	Lincolnshire	150048 (60722 ha)	4532350	317264500
Duke of Devonshire	Derbyshire	198572 (80539 ha)	4518750	316312500
Duke of Northumberland	Northumberland	186397 (75432 ha)	4401200	308084000
Earl of Derby	Lancashire	68942 (27899 ha)	4081825	285727750
Marquess of Bute	Bute	116668 (47213 ha)	3778375	264486250
Duke of Sutherland	Sutherland	1358545 (549783 ha)	3541675	247917250
Earl Fitzwilliam	Yorkshire	115743 (46839 ha)	3470025	242901750

Source: Based on Cahill (2001).

spending more than £1 million between 1872 and 1893 on the upkeep of his lands in Rutland, Derbyshire and Huntingdonshire (Beard, 1989).

The same duty was extended to the landowners' tenants, and wealth facilitated the altruistic establishment of almshouses, schools and hospitals. The duty did not, however, prevent the occasional displacement of tenants when in the way of new building projects, as at Milton Abbey in Dorset, England, where the model village of Milton Abbas was built by the Earl of Dorchester in 1780 to replace the older settlement of Middleton that was demolished to make way for an ornamental lake. Through these interventions, the old wealth of the aristocracy and gentry shaped the landscape and villagescape that is familiar across much of rural Europe today.

The value of rural estates as status symbols was recognized by the new class of industrialists and capitalists whose wealth had started to exceed that of the traditional aristocracy, and who from the late nineteenth century began to buy country houses and estates (Woods, 1997). Engineering magnate Weetman Pearson (later Lord Cowdray) spent more than £1 million buying land between 1909 and 1924, including Cowdray estate in Sussex and Dunecht Castle in Aberdeenshire (Beard, 1989); soap manufacturer Lord Leverhulme bought the island of North Harris in Scotland, whilst Guinness brewer Lord Iveagh purchased the Elveden estate in Suffolk, meat trader Lord Vestey bought Stowell Park in Gloucestershire, and tobacco manufacturers the Wills family became significant landowners in Somerset (Beard, 1989; Woods, 1997). Members of the American industrial and mercantile elite were also lured by the status of the European landed gent, with William Waldorf Astor, for example, purchasing the Cliveden estate near London for US$1.2 million in 1893. Others sought to replicate the European landed estate in North America, most notably William Randolph Hearst's construction of Hearst Castle in California.

The entry of industrialists into the landowning classes reflected the weakening significance of rural land as a source of wealth as agricultural prices were squeezed by competition from imports. As Rothery (2007) shows, many landed gentry families started to diversify their wealth base, investing in industries, railways and colonial bonds, whilst the high costs of maintaining country estates forced the sale of some land. This process was accelerated by World War I and its ensuing political upheavals, including the break up of the Austro-Hungarian, German and Russian empires. Even in Britain, military casualties from landed families had broken lines of inheritance for some estates and burdened others with substantial death duties that could often only be paid by the sale of land. Between the end of the war in November 1918 and March 1919, over half a million acres (202 000 ha) of land had been put on the market, and it is estimated that a quarter of the land surface of England changed hands between 1918 and 1922 (Beard, 1989; Howkins, 1991).

The majority of this land was bought by sitting tenant farmers, who became the new elite in many villages. As the most extensive landowners, employers of agricultural labour, and agricultural producers, large farmers were looked to for local leadership and exercised influence in their communities through structures of paternalism (Newby et al., 1978). Yet, their wealth was not necessarily substantially more than that of other rural residents, and certainly not many local professionals and business owners. As such, the redistribution of land in post-World War I Britain contributed to a narrowing of the wealth gap in rural areas in the mid-twentieth century.

However, whilst diminished, the European aristocracy and landed gentry have not disappeared as significant rural landowners and wealth holders. Cahill (2001) notes that there were 83 hereditary peers among the 1000 richest people in Britain in the 1998 *Sunday Times* Rich List, and that the 25 dukes collectively owned over 1 145 000 acres (463 000 ha) of land valued at nearly £15 210 million in total (including 133 602 acres [54 062 ha] held by the Duchy of Cornwall). Elsewhere in Europe, aristocratic landowners continue to own around 15 million acres (6.07 million ha) in France, 1 million acres (405 000 ha) in Belgium and 750 000 acres (303 500 ha) in the Netherlands, whilst the Prince of Thurn und Taxis alone owns 500 000 acres (202 000 ha) in Austria (Cahill and McMahon, 2010).

NEW WEALTH: GENTRIFICATION AND THE MIDDLE CLASS COUNTRYSIDE

The restructuring of rural regions during the early to mid-twentieth century, and especially the economic decline of agriculture and traditional manufacturing and mining industries, prompted a pattern of rural out-migration across the developed world. As rural populations contracted, properties and land lost value and fell into disuse, creating an opportunity for a new type of rural resident: ex-urban professionals, managers and business owners who were not rich enough to own rural estates, but who were sufficiently affluent to afford to buy and maintain former manor houses, vicarages and other substantial rural properties without depending on the precarious rural economy for their income. This trend emerged in the 1920s and 1930s in rural areas close to major cities, such as the Southeast of England (Burchardt, 2012), and spread after World War II with the growth of the middle class and improvements in transport. As competition for desirable rural properties intensified, house prices were inflated, initiating a process of rural gentrification and laying the foundations for a 'turnaround' in the rural–urban migration flow in countries such as Britain and the United States.

Patterns of rural gentrification, linked to the settlement of 'new wealth' middle class migrants, have been recorded across much of the developed world, including in Britain (Cloke et al., 1998; Phillips, 2002; Stockdale, 2010), the United States (Travis, 2007; Nelson et al., 2010; Marcouiller et al., 2011; Hines, 2012), Canada (Whitson, 2001; Guimond and Simard, 2010), Australia (Connell and McManus, 2011), New Zealand (Collins, 2013), Germany (Dirksmeier, 2008), Spain (Solana-Solana, 2010), Sweden (Turner, 2013) and the Alpine countries (Perlik, 2011). Whilst gentrification is represented as being fairly widespread in rural Britain, in other countries studies have pointed to concentration in selected ex-urban, coastal and mountain resort areas, the latter being associated with amenity migration. Quantitative analysis by Nelson et al. (2010) identifies 86 non-metropolitan counties in the United States with strong evidence of gentrification, disproportionately located in the west, but with notable clusters in Colorado, Texas and around the Great Lakes. There is also some evidence of rural gentrification in parts of the developing world, in some cases linked to tentative migration by emerging middle classes in expanding economies such as China (Qian et al., 2013) and South Africa (Spocter, 2013), but more commonly associated with international amenity migration from Europe and North America, especially in Latin America (Moss, 2006; Matarrita-Cascante and Stocks, 2013).

Drivers of Rural Gentrification

Research on rural gentrification has tended to emphasize cultural explanations, perceiving rural gentrifiers to be footloose post-industrial middle classes attracted to both the recreational possibilities of rural amenities and the nostalgic security of rural and small town communities (Hines, 2010a, 2010b, 2012; Guimond and Simard, 2010). As such, rural gentrification is commonly associated with amenity migration, especially to mountain and coastal locations, but it has also been linked with the discursive pull of the 'rural idyll', especially in studies in Britain (Cloke et al., 1998; Smith and Phillips, 2001; Phillips, 2002). Indeed, although Cotswold villages or Montana mountain towns may seem a long way from Camden or Greenwich Village, some commentators have argued that both urban and rural gentrification destinations reflect a similar search for authenticity (Bruegmann, 2005; cf. Zukin, 1982 and Rose, 1984 on urban gentrification).

The desired authentic rural may include the continuing appeal to the moneyed middle classes of the 'country gentleman' as a model of an elite lifestyle. Heley (2010) argues that the pursuit of an idealized gentrified lifestyle is a motivating factor for middle class migration to selected rural areas in England, noting that 'longstanding appetites for Barbour jackets, Hunter wellies, Harris Tweed jackets and other aspects of "country chic" have been joined by a seemingly deep hunger for the very bricks and mortar of the old country set, with landed estates becoming the new must-have property' (p. 321). As one member of this 'new squirearchy' interviewed by Heley observes, buying into the lifestyle is a 'marker' of achievement among individuals who have become wealthy through work in the financial sector of the City of London:

> All I am saying is that buying a country house and going shooting is like buying somewhere abroad, buying a yacht, or living in London and having a

> nice car. . . They are all kind of like middle class marker posts, and I think they relate to certain times of your life. You know, you have a Porsche or Ferrari in your late 20s or 30s, then you get shot of it and get something a bit less. . .well, vulgar. . .something more practical when the kids arrive. Then you go to the country house because it is better for the kids to grow up in. . .it's all very middle class. (Christopher, quoted by Heley, 2010, p. 328)

For many of the new squirearchy the lifestyle is closely entwined with business, and game shooting events and horse riding can be opportunities for entertaining clients. The balance of business and pleasure hence restricts the new squirearchy to certain rural settings: villages within commuting distance of London and international airports, with availability of large houses with sufficient land for horse paddocks, and access to gun clubs, shooting estates, golf clubs, a local fox hunt and a traditional village pub.

Members of Heley's new squirearchy are not actively engaged in farming, unlike a variant group of 'new country gentlemen' identified by Sutherland (2012) and defined as newcomers to farming who have purchased medium to large-scale farms of more than 300 hectares (740 acres), 'producing agricultural goods without the intention or ambition of making a living from it' (p. 571). Farming is approached as a hobby, as something to 'play at', with the farmland itself often valued more for its recreational opportunities, as one of Sutherland's interviewees described, echoing the lifestyle concerns of the new squirearchy in emphasizing how land-owning facilitated participation in certain country pursuits:

> If you want to go to play golf you have to pay. If you want to go fishing you have to pay. If you want to shoot you have got to be prepared to pay. . .to shoot a pheasant just now you will pay £25/£26 on an organised shoot. It doesn't cost us that so. . .plus I have a lot of friends that shoot, I get invited to their shoot, they come to my shoot. It's just a very social thing, for me it is anyway. (Andrew, quoted by Sutherland, 2012, p. 572)

A parallel can be found in 'gentleman' or 'hobby' ranchers in the United States (Friedberger, 1996; Travis, 2007). Roughly half of the ranches in the Western United States had become hobby operations, held for amenity potential or landscape value rather than livestock farming, by 2002 (Travis, 2007). The trend has been driven by poor markets for agricultural products, environmental conditions, family dynamics and financial problems, as well as the ability of non-traditional buyers to bid more than the agricultural value of the land. The attraction of the ranches to these buyers, however, is in the cultural model of the cowboy. Travis notes that many were baby boomers brought up on Western movies and TV shows, and that 'many of these baby boomers are now living out their dreams as the owners of ranchette-sized "horse properties" or even full-scale ranches' (2007, pp. 159–61).

At the same time, however, rural gentrification can also be viewed as an exercise in capital accumulation based on the valorization of undercapitalized property. The direct translation of political-economic models of urban gentrification (Smith, 1979, 1987, 1996) is problematic, as Darling (2005) discusses, but there are parallels. Rural gentrification proceeds more commonly through the inflation of freehold prices than closing of the 'rent gap', but 'marginal gentrification' where value is added to lower-cost property through the 'sweat equity' of purchasers is important in both settings (Cloke et al., 1998; Hines, 2012). Private householders and real estate agents are arguably more significant agents of rural gentrification than speculative developers, but new-build is important especially in resort developments, gated communities and substantial mansion and ranch-type properties.

Perhaps the most notable difference of rural gentrification is the role played by second or vacation home purchases, especially in high-amenity areas. As Darling observes in the Adirondacks in New York State:

> What gets produced in the process of urban gentrification is residential space. What gets produced in the process of wilderness gentrification is recreational nature. The production of gentrified housing in the Adirondacks has little to do with the construction of permanent residential abodes; indeed, the vast majority of gentrified housing lies empty for the majority of the year because it is occupied on a temporary and sporadic basis by tourists. (Darling, 2005, p. 1022)

Fundamentally, though, rural gentrification like urban gentrification involves disinvestment in land uses that bring relatively low capital returns – not just low-cost housing, but also increasingly agricultural production – and reinvestment in land uses with higher returns, such as high-cost property and tourism-linked activities. Thus, as Phillips (2000a, p. 6) argues, 'gentrification can be seen as one form of the revalorisation of resources and spaces that have become seen as unproductive or marginal to agrarian capital, and indeed a variety of other rural capitals'.

Impacts of Rural Gentrification

The transformative impact of middle class wealth channelled through gentrification on rural areas is two-fold. By inflating property prices and limiting the supply of low-cost housing in rural areas, gentrification leads to a reconfiguration of the social structure of rural communities. At the same time, by involving modifications to properties, fuelling new housing developments, and introducing new lifestyle and consumption choices that require particular supporting infrastructure, the new wealth of gentrifying middle classes also changes the appearance of the rural landscape.

The distortion of rural property markets is the first and arguably most important impact. Research in the United States has suggested that in-migrants to rural high-amenity counties in the 'New West' have per capita incomes that are 22 per cent higher than those of out-migrants from the same counties (Shumway and Otterstrom, 2001), with the disparity exaggerated further in the most in-demand areas, such as Sun Valley, Idaho, where Nelson (2005) notes that in-migrants have per capita levels of investment income that are 2.5 times those of out-migrants. The availability of investment capital to middle class in-migrants has an inflationary effect on property prices that rise to match the buyer's ability to pay. Average house prices in most of rural England increased by at least 70 per cent between 1998 and 2003 (Woods, 2005a), whilst in North America studies recorded a 93 per cent increase in the mean value of a single family home in Bayfield, Wisconsin between 1990 and 2000 (Barcus, 2011), a tripling of average house prices in Missoula, Montana over the same period (Ghose, 2004) and a doubling of property prices in Telluride, British Columbia over five years in the late 1990s (Whitson, 2001).

In such areas, the escalation in rural property prices has far outstripped increases in local incomes, thus further reducing the availability of the housing stock to lower-income local residents. In Wales, for example, average house prices in rural areas were 5.9 times the mean household income in 2005, up from 4.2 times in 2003 (Wales Rural Observatory, 2006). House price inflation affects not only higher-end properties, but all housing types and can be more exaggerated at lower levels. The ratio of average lower-quartile house prices to average lower quartile earnings in rural districts of England, for example, was 8.9:1 in 2007, increasing to 13:1 in the most expensive districts, such as East Dorset and Chichester (Taylor, 2008). This lower-end inflation is not only due to drag effects but also because of the impact of 'marginal gentrification' whereby in-migrants with less investment capital buy and renovate cheaper properties, thus increasing their value (Cloke et al., 1998). Furthermore second-home buyers often target smaller and cheaper properties, as well as areas with lower house prices. Gallent et al. (2005), for example, point to ten communities that are among the top third most deprived wards in Wales, but have more than 5 per cent of their housing stock occupied as second homes.

As such, it is not only the affordability of housing to lower-income residents that has been reduced, but also the supply of lower-cost housing, compounded by the privatization of social housing in countries such as Britain, the preference of developers to build new higher-end housing than affordable housing, and in resort areas, the tendency for rental properties to be oriented towards tourist lets rather than long-term tenancies, thus achieving higher capitalization (Darling, 2005). Rural working class residents have hence been progressively shut out from access to rural housing, contributing to increased levels of homelessness in rural areas, and out-migration by lower-income earners.

Hidden flows of out-migration by young people and lower-income residents from rural areas that otherwise have net flows of (predominantly middle class) in-migrants intensify the restructuring of rural populations towards more elderly and more middle class profiles. For instance, 'professionals and managers' comprised 16.1 per cent of the working population in remote rural districts of England in 1971, but 30.1 per cent in 1991 (Hoggart, 1997). However, both Hoggart (1997) and Phillips (2007) argue that the inference from these statistics that the countryside has become predominantly middle class is overly aggregative and obscures the spatial distribution of classes in rural areas; whilst Stockdale (2010) cautions that not all counterurbanization is gentrification, recognizing that the expansion of urban-to-rural migration from the 1970s onwards was founded on a diversification of the social base of migrants.

Indeed, whilst the first wave of migrants required significant financial resources to buy into rural life, a combination of factors including easier availability of mortgage finance, enhanced pension provisions, the growth of professional service sector employment and the mass supply of new 'family' houses in rural areas on new-build housing estates, opened up rural migration to individuals that whilst occupationally part of the middle class did not enjoy the income levels of the early counterurbanizers. Many were members of the so-called 'service class' of intermediate managers and professionals that service rather than own capital, and who were estimated to constitute around 40 per cent of urban-to-rural migrants in Britain during the 1970s and 1980s (Woods, 2005a). Although these migrants had higher purchasing power than working class rural residents and contributed to property price inflation, and whilst they introduced cultural ideas and expectations of rural life that sometimes created tensions in rural communities (Woods, 2005b), their actual wealth was not significantly different from that of the extant rural population.

At the same time, higher-income in-migrants sought to differentiate themselves socially and spatially by reinforcing the exclusivity of the most sought-after rural communities. The further escalation of property prices in these localities was permitted by the capital available to wealthy middle class residents, but also facilitated by political action to limit the supply of housing through controls on the volume and type of new development. Murdoch and Marsden (1994), for example, document the influence of middle class residents in shaping land use planning policies in Buckinghamshire, England, with the effect of producing:

> a delightful stretch of countryside in the midst of an urban region. With Milton Keynes to the north, green belt and London to the south, the struggle to hold onto the rural in Aylesbury Vale is by no means easy. But the social make-up of the place means that a formidable array of actors can usually be assembled to orchestrate opposition to unwelcome development. As its positional status grows, the area will become even more attractive to those would-be residents who are trapped on the 'outside'. Thus competition for resources, notably housing, will continue to increase, making it more and more difficult for those on low incomes either to stay in, or move to, such areas. The middle class complexion of the locality is thus assured – at least in the short term – especially as political action and planning policy are likely to reinforce and reflect the prevailing social composition. (Murdoch and Marsden, 1994, p. 229)

Separation is also achieved through the proliferation of gated communities in rural areas, specifically developed as locales for middle class investment. Gated communities were initially developed in urban areas as means of providing security and privacy to middle class residents in neighbourhoods that were otherwise perceived as unsafe, with high crime rates. They provide a similar function for both domestic and international migrants to some rural areas in the developing world, notably in South Africa (Spocter, 2013) and parts of Latin America (Roitman and Giglio, 2010). Yet, gated communities have also started to appear in rural areas of the United States and Europe where the security imperative is less evident, including in Montana, North Carolina and Texas (Phillips, 2000b; Ghose, 2004), and the English Cotswolds (McGhie, 2005). Here, the appeal of the gated community appears to be more the promise of exclusivity, with reassurance about the social status of neighbours, private access to recreational amenities, and a manufactured sense of community (Phillips, 2000b).

Counterurbanization and rural gentrification has therefore produced a trizonal geography of wealth in the contemporary Western countryside. In the inner zone are the exclusive enclaves of high-income middle class in-migrants, forming a metaphorical archipelago of pristine country villages, high-class coastal and mountain resorts, and gated communities. They are protected by the shield of exceptionally high property prices and include places such as East Horsley, Cookham and Alderley Edge in England, all with multiple house sales at more than £1 million (Table 13.2) (McGhie, 2011), and small towns such as Sagaponack, Jupiter Island and Kings Point in the United States, with median house values of over US$2 million (Table 13.3) (Wong, 2011). These enclaves are also the sites of the greatest concentrations of wealth in the countryside, including

Table 13.2 Most expensive villages in England, by house sales over £1 million to 2011

Village	County	Number of Houses Sold for Over £1 Million (to 2011)
East Horsley	Surrey	46
Cookham	Berkshire	41
Alderley Edge	Cheshire	36
Chalfont St Peter	Buckinghamshire	35
Prestbury	Cheshire	32
Penn	Buckinghamshire	24
West Clandon	Surrey	24
Stoke Poges	Buckinghamshire	23
Bramley	Surrey	22
Wonersh	Surrey	21

Source: Based on McGhie (2011).

Table 13.3 Most expensive small towns in the United States, 2011, by median house value

Town	State	Population	Median House Value 2011 (US$)
Sagaponack	New York	582	3406640
Jupiter Island	Florida	875	2810434
Kings Point	New York	5132	2379905
Los Altos Hills	California	7981	2161255
Water Mill	New York	2137	2111688
Belvedere	California	2101	2100453
Rolling Hills	California	1921	2063917
Hidden Hills	California	2003	1871182
Sands Point	New York	2782	1823677
Woodside	California	5316	1792837

Source: Based on Wong (2011).

English small towns and villages such as Bray, Burford, Chipping Camden, Dartmouth, Ilkley, Lyndhurst and Windermere – each home to over 200 sterling millionaires (Mannan, 2013), as well as American exurban[1] and 'equestrian' districts that Higley (1995) identifies as home to multiple individuals listed in the exclusive *Social Register*, including Locust Valley and Millbrook in New York State, Ligonier and Sewickley

Heights in Pennsylvania, and Middleburg and Albemarle County in Virginia.

Beyond the inner zone are those rural areas that have been recipients of the large majority of in-migration, where demand has had an inflationary effect on property prices, but the purchasing power of incomers has not been sufficient to obtain the level of exclusivity of the inner zone. Finally, outside these are those parts of the rural world that are less attractive to investors and continue to experience depopulation, notably more remote regions and industrial and agricultural-dependent districts.

The differentiated geography is also evident in the landscape impacts of gentrification. In the inner zone, wealth buys space and perceived authenticity. Restored and refurbished original property is often favoured over new-build, and new constructions tend to be spacious with expensive materials and frequently architect-designed. Green space is preserved in large gardens and paddocks, in part for aesthetic and privacy reasons, but also to facilitate recreational activities such as horse-riding. In the intermediate zone, by contrast, demand has driven development with extensive new housing estates in regions such as Southern England, and the subdivision of ranches into smaller, more affordable amenity 'ranchettes' in the American West. Harner and Benz (2013), for example, plot an increase in the number of ranchettes in La Plata County, Colorado, from 1561 in 1988 to 2610 in 2008.

Land use change is also driven by consumption demands, with undercapitalized farmland converted for amenity purposes including golf courses, shooting ranges, stables and riding schools; whilst Travis (2007, p. 169) observes that, 'the big difference between amenity and traditional ranches shows up at the residence (or residences), which is almost always large and elaborate on amenity ranches. Guest cabins, horse arenas, trout ponds, and maybe even airstrips and hangars also tend to sprout on recreational spreads'. Related environmental impacts can include reduced predator control, greater hospitality to wildlife, but also increased demands for water in often water-stressed areas (Glorioso and Moss, 2006; Gosnell et al., 2007; Travis, 2007).

The consumption preferences of wealthy incomers also change the service structures of small towns and villages, attracting gourmet coffee shops and restaurants, gastropubs, boutique shops, antique dealers and agents for expensive car brands (Riebsame, 1997; Decker, 1998), sometimes displacing established local shops, bars and businesses. The transition is particularly pronounced in localities with high concentrations of second homes, where the depletion of the year-round population can undermine the viability of local grocery stores, post offices and schools, and weaken community cohesion (Gallent at al., 2005).

The social and economic impact of rural gentrification is hence mixed. Nelson (2005) calculates that the net gain from affluent in-migration to Sun Valley, Idaho, to be an injection of US$45 million to the area, and many rural localities in North America have actively courted wealthy (and especially retired) migrants.

GLOBAL WEALTH: RURAL GEOGRAPHIES OF THE TRANSNATIONAL SUPER-RICH

The millennial rise of the global super-rich, individuals with investable assets of more than US$1 million, has been extensively documented (Hay and Muller, 2012; Volscho and Kelly, 2012; Hay, 2013). A characteristic of this group is ownership of multiple properties in different localities around the world, with portfolios commonly including at least one rural property. Rural properties are valued by the super-rich as investments, but also for the amenity opportunities, privacy and status they afford. Like wealthy industrialists a century ago, the super-rich perceive rural landownership as bestowing cultural legitimacy on new money. These priorities mean that super-rich purchases tend to be concentrated on certain types of rural property in certain localities: landed estates in England and Scotland; chateaux and vineyards in France and Italy; ranches in the American West and high country stations in New Zealand; chalets in exclusive ski resorts in the Rockies and Alps; villas on the Mediterranean and New England coasts; and private islands in the Caribbean and Aegean.

The attraction of becoming a 'laird', for example, has attracted transnational super-rich investment in the Scottish Highlands, where large estates continue to occupy a significant proportion of the land. Wightman (2010) calculated that there were around 80 estates in foreign ownership, covering around 905000 acres (366000 ha) in total. The largest of these landowners included capitalists from the Netherlands, Malaysia, Switzerland, Sweden, the United States, and Arab royalty (Table 13.4), whilst Cramb (1996) observed that 'much of Wester Ross is Dutch, Argyll is part-owned by Danes and Belgians, and there are Arabs in Easter Ross and Perthshire' (p. 21).

English country estates, especially those in close proximity to London, have similarly become sought after by super-rich investors, notably Russians. Oligarchs Roman Abramovich and Boris Berezovsky reportedly paid £12 million and £10 million for estates in West Sussex and Surrey, and Russian entrepreneur Leon Max purchased the Easton Neston estate in Northamptonshire for £15 million in 2008 (Yeomans, 2011). Investors taking advantage of new opportunities to purchase islands in the Aegean

Table 13.4 Major foreign landowners in Scotland, 2010

Landowner	Nationality	Business Interests	Estate and Area (Hectares)
Van Vlissingen family (Clyde Properties BV)	Dutch	Food and energy	Letterewe 35 262
'Mr Salleh' (Andras Ltd)	Malaysian	Unknown	Glenavon 28 910
Mohammed bin Rashid al Maktoum (Smech Properties Ltd)	UAE	Crown Prince of Dubai	Glomach 25 094
Ulrich Kohli (Argo Invest Overseas Ltd)	Swiss	Banking	Loch Ericht 22 887
Sigrid Rausing	Swedish	Tetra Pak	Coignafearn 16 022
Mohamed Al-Fayed (Bocardo SA)	Egyptian	Media, trade	Balnagowan 13 077
Roesner family (H H Roesner Ltd)	USA	Unknown	Sallachy 9945
Van Beuningen family	Dutch	Shipping, private investments	Foich and Inverlael 9857
De Spoelberch family	Belgian	Brewing	Altnafeadh 9268
Count Adam Knuth	Danish	Land, farming	Ben Loyal 8100 hectares
Mahdi Al Tajir (Park Tower Holdings)	UAE	Retail, property, water	Blackford 5900

Source: Woods (2013).

following the Greek economic crisis in 2008 have included the Emir of Qatar, who bought six islands for €8.5 million, and Russian heiress Ekaterina Rybolovleva, who paid €117 million for the island of Skopios (Smith, 2013a, 2013b). In the Western United States, super-rich purchasers of ranches have followed earlier wealthy owners such as Theodore Roosevelt and William Randolph Hearst, and include entertainment figures and business leaders. Travis (2007) quotes a journalist listing ranch owners in Carbon County, Montana:

> Michael Keaton has a place here. The novelist Tom McGuane has one on the West Fork of the Boulder River. So do Dave Grusin, the musician, and Robert

> D. Haas, chairman of Levi Strauss & Company. Just north of Yellowstone National Park, Dennis Quaid and Meg Ryan have a house. Jeff Bridges is a neighbour. So is Peter Fonda. (*New York Times*, 21 March 1990, quoted by Travis, 2007, p. 163)

Ownership of a country estate or a ranch provides the super-rich with a stage to live out aristocratic fantasies and indulge in the pursuits of the leisured landed classes. Entrepreneur Leon Max reportedly commented that 'I like the idea of being a country gentleman. . . I am looking forward to shuffling to my atelier in my monogrammed slippers' (Yeomans, 2011, p. 34), whilst entertainer Madonna – who bought a country estate in Wiltshire, England, with then husband Guy Ritchie – was praised by upmarket magazine *Country Life* for participating in shooting and riding and bringing 'rural living into the metropolitan and international consciousness' (quoted by Woods, 2013, p. 129).

For some, this performance extends to playing the squire and sponsoring local activities, but more frequently, the super-rich and their entourages fly in and out of properties, to the frustration of local communities who feel that they receive little benefit from elite neighbours but are confronted by construction work, heightened security and occasional media intrusions (Yeomans, 2011). Entertainers Madonna and Shania Twain have both been embroiled in conflicts with local residents over attempts to divert footpaths and restrict access to their land in England and New Zealand respectively, whilst Nicholas van Hoogstraten, the owner of England's largest new-build mansion, Hamilton Palace in Sussex, gained notoriety for his aggressive attempts to block public rights of way on his estate (Kennedy and Weaver, 2004).

The super-rich also have a transformative impact on rural places through the footprint that supports their lifestyles. Chinese investment in French vineyards, for instance, has normally not been for residence but for status and to secure a supply of coveted French wine. Both McManus (2013) and Roberts and Schein (2013), meanwhile, document the impact of thoroughbred racehorse breeding and training in reshaping rural economies and landscapes in selected regions, including notably Kentucky. Furthermore, for some high-wealth individuals, especially Russian oligarchs, the rural is still also a site of wealth generation through investment in farmland, forests, resort developments and rural-based energy businesses (Visser et al., 2012). However, in contrast to the estates of the historic aristocracy, the rural sites of wealth creation for the global super-rich are spatially separated from their rural spaces of display and recreation, contributing further to the differentiation of the global countryside.

CONTESTING WEALTH: RESISTANCE TO THE RURAL RICH

Inequalities of land ownership and hence wealth in the countryside has historically been a source of conflict, prompting protest movements such as the Diggers in seventeenth-century England and the nineteenth-century Irish Land League, and acts of disobedience such as poaching. Pressures for land reform developed across Europe, including unsuccessful campaigns for land taxation and land nationalization in early twentieth-century Britain. With the contraction of the landed gentry and the narrowing of the wealth gap in rural society in many parts of the developed world after World War I, however, the new rural elite often succeed in persuading rural working classes that their interests were the same as wealthier agrarian farmers (Woods, 2005b). Not only was radical class politics discouraged, but perceptions of inequality were also diluted. Research by Cloke et al. (1995) in England recorded the tendency of rural residents to deny the presence of poverty in their area, even if they were by objective measures living in poverty themselves.

Political consciousness of rural inequality has been restoked by the growing transformative impact of wealthy minorities in rural areas. Anti-gentrification protests have emerged in high-pressure localities such as Hebden Bridge, Yorkshire, whilst in rural Wales nationalist campaign groups have targeted affluent incomers for their cultural as well as their social and economic impact. At their most extreme, the Welsh protests included the fire-bombing of second homes during the 1980s (Gallent et al., 2005). Yet, it can be difficult to disentangle progressive protests against gentrification from the mobilization of affluent middle class residents against new housing developments to protect the exclusivity of their communities (Murdoch and Marsden, 1994; Woods, 2005b).

The complexity of these dynamics is illustrated by the mountain resort of Queenstown, New Zealand (Woods, 2011). Protests in the early 2000s against extensive new housing development – including the subdivision of farmland into 'lifestyle blocks' – were led by affluent in-migrants, who the pro-development mayor portrayed as a wealthy elite intent on keeping lower-income residents out. As he told a local newspaper: 'We don't want to become the Aspen of the South Pacific. We. . .shouldn't become a community of millionaires and multi-millionaires' (quoted by Woods, 2011, p. 378).

Conflicts have also arisen around exclusive recreational developments, such as Donald Trump's controversial golf resort in Northeast Scotland and his attempts to block an adjacent off-shore windfarm project because it would spoil the view (Carrell, 2014). Even in such cases, however, local opinion can be divided over the economic benefits. Proposals to redevelop the depressed resort town of Misasa in Japan with vacation homes for wealthy Chinese, for example, have been supported strongly by local residents, but attracted opposition nationally (Tabuchi, 2010).

Such cases have reopened debates about rural landownership, and especially foreign ownership, most notably in Scotland. Many of the larger Scottish estates include whole communities as tenants, and tensions mounted during the late twentieth century over the sale of such estates without consultation with residents, including to 'absentee' foreign buyers. The sale of North Lochinver estate by English businessman Edmund Vestey to a Swedish company in 1989 was such a case, and when the company went into liquidation three years later, residents seized the opportunity to buy the estate for the community (Wightman, 2010). In 1995, the sale of the Isle of Eigg by controversial owner Keith Schellenberg to a German artist, Marlin Eckhard Maruma, for £1.6 million stoked existing antagonisms, such that 'soon the islanders were in open revolt' (2010, p. 148), eventually leading to their own community buy-out. A right of communities to buy the land on which they lived was subsequently introduced by the Land Reform (Scotland) Act 2003, and whilst Wightman (2010) observes that take up of the power has been quite limited, the legislation had created an opportunity for rural communities in Scotland not only to divest themselves of wealthy landlords, but crucially to also take control of their own economic development, exploiting resources such as wind energy (Mackenzie, 2006a, 2006b, 2013).

CONCLUSION

Wealth, land and power have always been entwined in rural societies, and the ideal of the 'country gentleman' created by the landed classes of the eighteenth and nineteenth centuries has remained potent as a model wealthy lifestyle. It is the status embodied in the model that leads the new global super-rich to add rural estates to their property portfolios, and affluent middle classes to buy substantial rural houses, farms or ranchettes as they pursue the dream of the 'new squirearchy', 'new gentleman farmers' or 'gentleman ranchers'. Equally the recreational amenities that attract the wealthy to exclusive coastal and mountain resorts are pursuits that were pioneered by the leisured landed classes – horse-riding, shooting, skiing, sailing – and which are still expensive hobbies.

The requirements of these aspirational lifestyles favour certain types of rural property in certain localities, thus creating a differentiated geography that is reinforced by the efforts of wealthy residents to protect the exclusivity of their settlements by restricting access and limiting housing supply. Islands of affluence have thus emerged where stratospheric property prices buy space, authenticity and privacy; whilst outside, rapid development in rural communities has been driven by the demand from middle-income middle classes for a cut-price rural idyll, but with still sufficient purchasing power to inflate property prices and displace endogenous working class residents. Wealth therefore is a driver of change in rural places, but in complex and geographically divergent ways.

NOTE

1. The ring of prosperous communities beyond the suburbs, which are commuter towns for an urban area.

REFERENCES

Barcus, H. (2011), 'Heterogeneity of rural housing markets', in D. Marcouiller, M. Lapping and O. Furuseth (eds), *Rural Housing, Exurbanization, and Amenity-driven Development*, Farnham, UK/Burlington, VT: Ashgate, pp. 51–74.
Beard, M. (1989), *English Landed Society in the Twentieth Century*, London: Routledge.
Beaur, G. (2008), 'Revolution and the redistribution of wealth in the countryside: myth or reality?', *Annales Historiques de la Révolution Française*, **352**, 209.
Bruegmann, R. (2005), *Sprawl: A Compact History*, Los Angeles, CA: University of California Press.
Burchardt, J. (2012), 'Historicizing counterurbanization: in-migration and the reconstruction of rural space in Berkshire (UK), 1901–51', *Journal of Historical Geography*, **38** (2), 155–66.
Cahill, K. (2001), *Who Owns Britain: The Hidden Facts Behind Landownership in the UK and Ireland*, Edinburgh: Canongate.
Cahill, K. and R. McMahon (2010), *Who Owns the World*, New York: Grand Central Publishing.
Carrell, S. (2014), 'Donald Trump buys Irish golf resort after losing Scottish court battle', *The Guardian*, 11 February, accessed 18 August 2015 at http://www.theguardian.com/uk-news/2014/feb/11/donald-trump-irish-golf-scotland-windfarm.
Cloke, P., M. Phillips and N. Thrift (1998), 'Class, colonization and lifestyle strategies in Gower', in P. Boyle and K. Halfacree (eds), *Migration into Rural Areas*, Chichester, UK: Wiley, pp. 166–85.
Cloke, P., M. Goodwin, P. Milbourne and C. Thomas (1995), 'Deprivation, poverty and marginalisation in rural lifestyles in England and Wales', *Journal of Rural Studies*, **11** (4), 351–66.
Collins, D. (2013), 'Gentrification or "multiplication of the suburbs"? Residential development in New Zealand's coastal countryside', *Environment and Planning A*, **45** (1), 109–25.
Connell, J. and P. McManus (2011), *Rural Revival? Place Marketing, Tree Change and Regional Migration in Australia*, Farnham, UK/Burlington, VT: Ashgate.
Cramb, A. (1996), *Who Owns Scotland Now?* Edinburgh/London: Mainstream Publishing.
Darling, E. (2005), 'The city in the country: wilderness gentrification and the rent gap', *Environment and Planning A*, **37** (6), 1015–32.
Decker, P.R. (1998), *Old Fences, New Neighbors*, Tucson, AZ: University of Arizona Press.

Dirksmeier, P. (2008), 'Strife in the rural idyll? The relationship between autochthons and in-migrants in scenic regions of South Bavaria', *Erkunde*, **62** (2), 159–71.

Everett, N. (1994), *The Tory View of Landscape*, New Haven, CT/London: Yale University Press.

Friedberger, M. (1996), 'Rural gentrification and livestock raising: Texas as a test case, 1940–1995', *Rural History*, **7** (1), 53–68.

Gallent, N., A. Mace and M. Tewdwr-Jones (2005), *Second Homes: European Perspectives and UK Policies*, Aldershot, UK: Ashgate.

Ghose, R. (2004), 'Big sky or big sprawl? Rural gentrification and the changing cultural landscape of Missoula, Montana', *Urban Geography*, **25** (6), 528–49.

Glorioso, R.S. and L. Moss (2006), 'Santa Fe, a fading dream: 1986 profile and 2005 post-script', in L. Moss (ed.), *The Amenity Migrants*, Wallingford, UK/Washington, DC: CABI, pp. 73–93.

Gosnell, H., J.H. Haggerty and P. Byorth (2007), 'Ranch ownership change and new approaches to water resources management in Southwestern Montana: implications for fisheries', *Journal of the American Water Resources Association*, **43** (4), 990–1003.

Guimond, L. and M. Simard (2010), 'Gentrification and neo-rural populations in the Québec countryside: representations of various actors', *Journal of Rural Studies*, **26** (4), 449–64.

Hall, D. (2013), *Land*, Cambridge, UK/Malden, MA: Polity.

Harner, J. and B. Benz (2013), 'The growth of ranchettes in La Plata County, Colorado, 1988–2008', *Professional Geographer*, **65** (2), 329–44.

Hay, I. (ed.) (2013), *Geographies of the Super-Rich*, Cheltenham, UK and Northampton, MA, USA: Edward Elgar Publishing.

Hay, I. and S. Muller (2012), '"That tiny, stratospheric apex that owns most of the world" – exploring geographies of the super-rich', *Geographical Research*, **50** (1), 75–88.

Heley, J. (2010), 'The new squirearchy and emergent cultures of the new middle classes in rural areas', *Journal of Rural Studies*, **26** (4), 321–31.

Higley, S.R. (1995), *Privilege, Power and Place: The Geography of the American Upper Class*, Lanham, MD: Rowman and Littlefield.

Hines, J.D. (2010a), 'In pursuit of experience: the postindustrial gentrification of the rural American West', *Ethnography*, **11** (2), 285–308.

Hines, J.D. (2010b), 'Rural gentrification of permanent tourism: the creation of the "New" West Archipelago as postindustrial cultural space', *Environment and Planning D: Society and Space*, **28** (3), 509–25.

Hines, J.D. (2012), 'The post-industrial regime of production/consumption and the rural gentrification of the New West Archipelago', *Antipode*, **44** (1), 74–97.

Hoggart, K. (1997), 'The middle classes in rural England 1971–1991', *Journal of Rural Studies*, **13** (3), 253–74.

Howkins, A. (1991), *Reshaping Rural England: A Social History, 1850–1925*, London: Routledge.

Kennedy, M. and M. Weaver (2004), 'Touch of fantasy: the right style of modern mansion', *The Guardian*, 4 August, 3.

Mackenzie, A.F.D. (2006a), 'S Leinn Fhèin am Fearann (The Land is Ours): re-claiming land, re-creating community, North Harris, Outer Hebrides, Scotland', *Environment and Planning D: Society and Space*, **24** (4), 577–98.

Mackenzie, A.F.D. (2006b), 'A working land: crofting communities, place and the politics of the possible in post-Land Reform Scotland', *Transactions of the Institute of British Geographers*, **31** (3), 383–98.

Mackenzie, A.F.D. (2013), *Places of Possibility: Property, Nature and Community Land Ownership*, Chichester, UK: Wiley-Blackwell.

Mannan, T. (2013), 'Revealed: the wealthiest towns and villages in the UK', *Yourmoney.com*, 13 May, accessed 6 May 2014 at www.yourmoney.com/your-money/news/2267741/revealed-the-wealthiest-towns-villages-in-the-uk.

Marcouiller, D., M. Lapping and O. Furuseth (eds) (2011), *Rural Housing, Exurbanization, and Amenity-driven Development*, Farnham, UK/Burlington, MA: Ashgate.

Matarrita-Cascante, D. and G. Stocks (2013), 'Amenity migration to the Global South: implications for community development', *Geoforum*, **49** (2), 91–102.
McGhie, C. (2005), 'Cape Cotswolds', *The Sunday Telegraph*, 20 February, *House and Home Supplement*, 1.
McGhie, C. (2011), 'Britain's richest villages', *The Daily Telegraph*, 28 March 2011, accessed 6 May 2014 at www.telegraph.co.uk/property/luxuryhomes/8410974/Britains-richest-villages.html.
McManus, P. (2013), 'The sport of kings, queens, sheikhs and the super-rich: thoroughbred breeding and racing as leisure for the super-rich', in I. Hay (ed.), *Geographies of the Super-Rich*, Cheltenham, UK and Northampton, MA, USA: Edward Elgar Publishing, pp. 155–70.
Moss, L. (ed.) (2006), *The Amenity Migrants*, Wallingford, UK/Cambridge, MA: CABI.
Murdoch, J. and T. Marsden (1994), *Reconstituting Rurality*, London: UCL Press.
Nelson, P. (2005), 'Migration and the spatial redistribution of nonearnings income in the United States: metropolitan and nonmetropolitan perspectives from 1975–2000', *Environment and Planning A*, **37** (9), 1613–16.
Nelson, P., A. Oberg and L. Nelson (2010), 'Rural gentrification and linked migration in the United States', *Journal of Rural Studies*, **26** (4), 343–52.
Newby, H., C. Bell and D. Rose et al. (1978), *Property, Paternalism and Power*, London: Hutchinson.
Perlik, M. (2011), 'Alpine gentrification: the mountain village as a metropolitan neighbourhood', *Revue de Géographie Alpine*, **99** (1), 118–36.
Phillips, M. (2000a), 'Making space for rural gentrification', paper presented to the Anglo-Spanish Symposium on Rural Geography, Valladolid, Spain, July.
Phillips, M. (2000b), 'Landscapes of defence, exclusivity and leisure: rural private communities in North Carolina', in J.R. Gold and G. Revill (eds), *Landscapes of Defence*, Harlow: Pearson Education, pp. 130–45.
Phillips, M. (2002), 'The production, symbolization and socialization of gentrification: impressions from two Berkshire villages', *Transactions of the Institute of British Geographers*, **27** (3), 282–308.
Phillips, M. (2007), 'Changing class complexions on and in the British countryside', *Journal of Rural Studies*, **23** (3), 283–304.
Qian, J.X., S.J. He and L. Liu (2013), 'Aestheticisation, rent-seeking, and rural gentrification amidst China's rapid urbanisation: the case of Xiaozhou village, Guangzhou', *Journal of Rural Studies*, **32**, 331–45.
Richards, J.F. (2002), 'Towards a global system of property rights in land', in *Land, Property, and the Environment*, Oakland, CA: ICS Press, pp. 13–37.
Riebsame, W.E. (ed.) (1997), *Atlas of the New West*, New York/London: W.W. Norton & Company.
Roberts, S.M. and R.H. Schein (2013), 'The super-rich, horses and the transformation of a rural landscape in Kentucky', in I. Hay (ed.), *Geographies of the Super-Rich*, Cheltenham, UK and Northampton, MA, USA: Edward Elgar Publishing, pp. 137–54.
Roitman, S. and M.A. Giglio (2010), 'Latin American gated communities: the latest symbol of historic social segregation', in S. Bagaeen and O. Uduku (eds), *Gated Communities: Social Sustainability in Contemporary and Historical Gated Developments*, London/Washington, DC: Earthscan, pp. 63–78.
Rose, D. (1984), 'Rethinking gentrification: beyond the uneven development of Marxist urban theory', *Environment and Planning D: Society and Space*, **2** (1), 47–74.
Rothery, M. (2007), 'The wealth of the English landed gentry, 1870–1935', *Agricultural History Review*, **55** (2), 251–68.
Shumway, J.M. and S.M. Otterstrom (2001), 'Spatial patterns of migration and income change in the Mountain West: the dominance of service-based, amenity-rich counties', *Professional Geographer*, **53** (4), 492–502.
Smith, D. and D. Phillips (2001), 'Socio-cultural representations of greentrified Pennine rurality', *Journal of Rural Studies*, **17** (4), 457–69.

Smith, H. (2013a), 'Going, going gone. . . Qatari ruler snaps up six islands in Greek bargain basement sale', *The Guardian*, 5 March, 5.

Smith, H. (2013b), 'Sold for £100m: the idyllic Greek island where billionaire Onassis wed Kennedy', *The Guardian*, 17 April, 3.

Smith, N. (1979), 'Gentrification and capital: theory, practice and ideology in Society Hill', *Antipode*, **11** (3), 24–35.

Smith, N. (1987), 'Gentrification and the rent gap', *Annals of the Association of American Geographers*, **77** (3), 462–5.

Smith, N. (1996), *The New Urban Frontier: Gentrification and the Revanchist City*, London/New York: Routledge.

Solana-Solana, M. (2010), 'Rural gentrification in Catalonia, Spain: a case study of migration, social change and conflicts in the Empordanet area', *Geoforum*, **41** (3), 508–17.

Spocter, M. (2013), 'Rural gated developments as a contributor to post-productivism in the Western Cape', *South African Geographical Journal*, **95** (2), 165–86.

Stockdale, A. (2010), 'The diverse geographies of rural gentrification in Scotland', *Journal of Rural Studies*, **26** (1), 31–40.

Sutherland, L.-A. (2012), 'Return of the gentleman farmer? Conceptualising gentrification in UK agriculture', *Journal of Rural Studies*, **28** (4), 568–76.

Tabuchi, H. (2010), 'In Japan, deep fear over China's deep pockets', *The New York Times* in *The Observer*, 17 October, 5.

Taylor, M. (2008), *Living Working Countryside: The Taylor Review of Rural Economy and Affordable Housing*, London: Department for Communities and Local Government.

Travis, W.R. (2007), *New Geographies of the American West*, Washington, DC: Island Press.

Turner, L. (2013), 'Hunting for hotspots in the countryside of northern Sweden', *Journal of Housing and the Built Environment*, **28** (2), 237–55.

Visser, O., N. Mamonova and M. Spoor (2012), 'Oligarchs, megafarms and land reserves: understanding land grabbing in Russia', *Journal of Peasant Studies*, **39** (3–4), 899–931.

Volscho, T.W. and N.J. Kelly (2012), 'The rise of the super-rich: power resources, taxes, financial markets and the dynamics of the top 1 percent, 1949 to 2008', *American Sociological Review*, **77** (5), 679–99.

Wales Rural Observatory (2006), *Housing Need in Rural Wales: Towards a Sustainable Solution*, Cardiff, UK: Wales Rural Observatory.

Whitson, D. (2001), 'Nature as playground: recreation and gentrification in the Mountain West', in R. Epp and D. Whitson (eds), *Writing Off the Rural West*, Edmonton, AB: University of Alberta Press, pp. 145–64.

Wightman, A. (2010), *The Poor Had No Lawyers: Who Owns Scotland (and How They Got It)*, Edinburgh: Birlinn.

Wong, V. (2011), 'America's 50 most expensive small towns 2011', *Bloomberg Businessweek*, 21 January, accessed 6 May 2014 at images.businessweek.com/slideshows/20110120/america-s-50-most-expensive-small-towns-2011#slide49.

Woods, M. (1997), 'Discourses of power and rurality: local politics in Somerset in the 20th century', *Political Geography*, **16** (6), 453–78.

Woods, M. (2005a), *Rural Geography*, London/Thousand Oaks, CA: Sage.

Woods, M. (2005b), *Contesting Rurality: Politics in the British Countryside*, Aldershot, UK: Ashgate.

Woods, M. (2011), 'The local politics of the global countryside: boosterism, aspirational ruralism and the contested reconstitution of Queenstown, New Zealand', *Geojournal*, **76** (4), 365–81.

Woods, M. (2013), 'The elite countryside: shifting rural geographies of the transnational super-rich', in I. Hay (ed.), *Geographies of the Super-Rich*, Cheltenham, UK and Northampton, MA, USA: Edward Elgar Publishing, pp. 123–37.

Yeomans, P. (2011), 'Who owns our green and pleasant land?', *The Observer Magazine*, 7 August, 32–7.

Zukin, S. (1982), *Loft Living: Culture and Capital in Urban Change*, London: Radius.

14. Performing wealth and status: observing super-yachts and the super-rich in Monaco

Emma Spence

INTRODUCTION

Contemporary and emerging studies of the super-rich within geography, and across the social sciences, continue to develop significant research agendas in terms of wealth creation, economic and social polarization, and hypermobility (see Elliot and Urry, 2010, p.65; Birtchnell and Caletrío, 2013; Hay, 2013; as well as Koh, Wissink and Forrest, Chapter 2 in this volume). However, the burgeoning work in the field of super-rich geographies generally overlooks leisure and lifestyle as noteworthy areas of critique (McManus, 2013, p.155). The ability to analyse and measure wealth status of the super-rich has advanced in recent years with readily available and comprehensive quantitative analyses of the super-rich such as *The Times* Rich List, *Forbes* Rich List and the Bloomberg Billionaire Index (Bloomberg, 2014). In addition, the inequality and social injustice associated with the super-rich is well documented by the media and a growing number of scholarly and popular publications, such as Danny Dorling's (2014) *Inequality and the 1%*, Thomas Piketty's (2014) *Capital in the Twenty-First Century*, Shaxson's (2011) *Treasure Islands* and Armstrong's (2010) *The Super-Rich Shall Inherit the Earth*. Nonetheless, significant methodological challenges for scholars remain. Barriers such as access, time and financial resources have contributed to limited empirical engagements with the super-rich and their various leisure practices (Rojek, 2000, p.8; Featherstone, 2013, p.5; see also Schulz and Hay, Chapter 8 this volume). The result is that we assume that the super-rich exist between a global archipelago of luxury yachts, private islands, multiple mansions, luxury hotels and exclusive country clubs (Caletrío, 2012), yet we know relatively little of the social and spatial significance of such lifestyles (Atkinson and Blandy, 2009, p.95). This chapter goes some way to addressing this shortcoming.

In the following pages I use the notion of performance to illustrate how wealth and status are expressed socially and spatially in the Principality of Monaco. Following Goffman (1956) and Butler (1993) the notion of performance is utilized across the social sciences to illustrate how meanings and identities are produced and enacted in particular settings. Performance is increasingly used in cultural geography as a means through which to examine how spaces are practised, experienced and embodied (Rogers, 2012, p. 60). Drawing from personal experiences working within a Monaco-based yacht brokerage I introduce the super-yacht as a symbol of wealth and cultural status and explore the associated performances and experiences of the super-rich. From my key intermediary role as a facilitator of super-rich lifestyles I am able to provide an insight into the social and spatial practices of super-rich lifestyles as practised in Monaco and to illustrate ways in which scholars can utilize intermediaries engaged with the super-rich as gatekeepers to an otherwise inaccessible group.

GAINING ACCESS TO THE SUPER-RICH

In this chapter I focus upon the super-yacht as a key tool for exploring how performances of wealth are made visible in Monaco. A super-yacht is a privately owned and professionally crewed luxury vessel over 30 metres in length. An average super-yacht, at approximately 47 metres in length, costs around €30 million to buy new, operates with a permanent crew of ten, and costs around €1.8 million per year to run. Larger super-yachts such as Motor Yacht (M/Y) *Madame Gu* (99 metres in length), or the current largest super-yacht in the world M/Y *Azzam* (180 metres in length) cost substantially more to build and to run. The price to charter (rent) a super-yacht also varies considerably with size, age and reputation of the shipyard in which it was built. For example, a typical 47-metre yacht can range between €100 000 to €600 000 per week to charter, plus costs.[1] At the most exclusive end of the super-yacht charter industry costs are much higher. M/Y *Solange*, for example, is an 85-metre newly built yacht (2013) from reputable German shipyard Lürssen, which operates with 29 full-time crew, and is priced at €1 million plus costs to charter per week.[2] The super-yacht industry is worth an estimated €24 billion globally (Rutherford, 2014, p. 51).

As an exemplar of Thorstein Veblen's ([1889] 1994) conspicuous consumption, super-yachts are exceptional not only in terms of the vast cost of ownership, but also in the way they are used as tools for performing wealth status between super-rich peers. Use and ownership of a yacht can be employed as a financial and social indicator of super-rich lifestyles (see also McManus, 2013, pp. 157–9), and demonstrates a person's membership of the often exclusive and elusive super-rich and as such is one of the world's 'few pinnacle cultural products of identity expression and image projection' (Jennings et al., 2015, p. 116). The super-yacht is a highly visible asset that not only incurs substantial running costs, but that also depreciates rapidly in value with age. The luxury vessels are emblematic of super-rich lifestyles, and the sheer cost of owning and operating a super-yacht and the motivations for doing so warrants further scholarly investigation in their own right.

Despite the rich avenues of academic enquiry presented by luxury and exclusive spaces such as the super-yacht, academic critique remains scarce. Scholars' limited engagement with the super-rich can be attributed, in large part, to the difficulties in obtaining access. To illustrate how barriers to access can be partially overcome I draw upon an ethnographic study conducted within a yacht brokerage in the Principality of Monaco. With over six years experience working on board super-yachts as crew (see Spence, 2014), I was able to make use of existing industry contacts and overcome typical barriers to access in ways that allowed me to conduct this ethnographic study over the period May–September 2014.

The role of the yacht broker is to negotiate the sales process during yacht acquisitions on behalf of buyers, to identify vessels on the market suitable for their client, and to manage logistics and arrangements for those wishing to charter a yacht on a short-term basis (typically one to two weeks). From my sea-based experience I had acquired appropriate knowledge of super-yachts, including terminology, practices, rules and regulations that allowed me to make a useful contribution within the brokerage. For example, my experience of, and frequent interaction with, super-rich individuals on board enabled me to meet, greet and engage with the company's super-rich clients appropriately. In exchange for database research and office administration tasks I was able to shadow yacht brokers at otherwise inaccessible spaces and activities such as business meetings, shipyards, yacht shows, a private aviation terminal, aboard a private jet, a commercial helicopter and numerous super-yachts. Frequent engagements with clients and associates of the yacht brokerage facilitated insights to super-rich lifestyles.

Aside from yacht brokers, other key intermediaries to super-rich lifestyles include those providing concierge services, luxury goods representatives, air and yacht crew, legal staff and real estate brokers, to name but a few. They hold in common close access to and relationships with their super-rich clients, and the knowledge and experience to anticipate their needs, motivations and willingness to pay for such services. It is self-evident that these gatekeepers could prove invaluable to future scholarly research of the super-rich in terms of securing access and providing empirical depth to our knowledge of the corporate ecology surrounding the super-rich.

My trajectory in securing access to the brokerage was admittedly largely determined by my previous knowledge of and experience in the industry. However, this is not to say that other scholars could not secure similar access in other domains of the super-yacht industry, or within other luxury leisure sectors. For example, Sherman (2007) was able to conduct in-depth ethnographies in two luxury hotels in exchange for administrative duties and sharing her analysis of employee–guest relations within the hotels to senior management. Pow (2011) utilized gatekeepers to gain access to exclusive communities in Singapore and to interview their wealthy residents in his analysis of super-rich enclaves and elite urbanism. And Rogers (2015) attended public luxury housing conventions in his analysis of super-rich real estate investment. There are evidently many ways to gain access to the networks of companies, agencies and brokerages that enable luxury lifestyle practices for the super-rich. Thus, not only can we learn more about how super-rich leisure practices are shaped by various intermediaries as a significant area of study in its own right, but also new routes of access and exposure to the super-rich themselves become apparent and available.

As the discussion above suggests, my role in the brokerage enabled unique access opportunities to otherwise inaccessible places and people. From my privileged observational position to the yachting industry, I was able to gain insight into performances of wealth and status as displayed in public and private spaces throughout the Principality.

PERFORMANCE AND WEALTH STATUS

Places are not simply encountered, but are performed through embodied actions (Sheller and Urry, 2004, p. 4). Successful performances by actors require appropriate 'props'. In the case of super-rich leisure practices, these tools come in forms such as luxury wearable goods and luxury modes of transportation. The prestige car, super-yacht and private jet are emblematic of the symbolic and functional performance of super-rich lifestyles (Beaverstock and Faulconbridge, 2013, p. 50).

Take, for example, the super-yacht. During an interview with Frank,[3] who spoke as a previous owner and regular charterer of yachts, he described the super-yacht as a 'lifestyle enhancing asset', driven by the desire to have the super-rich lifestyle experience rather than as a way to engage with the elements. He deemed the yacht a prop to enhance lifestyle experience, as a leisure pursuit, rather than offering any functional enhancement to the super-rich as might be argued for the time saved by travelling on a private jet rather than by commercial aviation. Ownership and use of a yacht allows the super-rich to perform their wealth status, rather than functioning as an investable asset or timesaving mode of transportation.

Ownership of a yacht, however, is not enough in itself to enhance the status or profile of a yacht owner within the super-rich super-yachting community. The super-rich 'perform' wealth and project status where there are others who can 'read' these displays. Thus, *where* the yacht is used is equally important in the display of wealth and status. As I have argued elsewhere (Spence, 2014), typical super-yacht use follows similar patterns in terms of itineraries, anchorages and ports of call at certain times of the year, whether concentrated along the Côte d'Azur region of the South of France/Mediterranean in the Northern Hemisphere's summer, or between the Caribbean Islands in the winter. In the summer, popular French anchorages such as Beaulieu-sur-Mer and Villefranche-sur-Mer, and port towns St. Tropez, Cannes and Monaco attract high concentrations of super-rich, enabling highly visible performances of wealth and social status amongst super-rich peers through the image of the super-yacht. The super-yacht is considered to be a highly visual articulation of wealth, yet the true worth of a yacht is understood only to those actively engaged in the industry. Typically, those who can read the wealth status of a super-yacht owner by the size, brand and age of his or her vessel are already engaged with the industry. These actors include other super-rich yacht owners, shipyard executives and yacht brokers. Knowledge of the vessel's value, operating costs, the duties of the crew on board, the nationality of the owner, the industry in which they have accumulated vast wealth, the amount of time spent onboard, and the cruising itineraries of the yacht adds deeper empirical knowledge of the yacht and helps to map the leisure practices of the super-rich. And such knowledge allows us to better understand the extent of super-rich wealth, the distribution of wealth and the social and spatial implications of such extreme wealth. In the pages that follow I endeavour to sketch out some of the key practices of super-yacht use in Monaco, and to illustrate the performances of wealth and status by the super-rich in the Principality.

MONACO: 'WORLD CAPITAL OF YACHTING'

Built between the steep French hills of La Turbie, Roquebrune le Fort and the expansive Mediterranean Sea, the Principality of Monaco is an iconic super-rich haven. Monaco offers 0 per cent income tax and favourable corporation tax rates to its residents. Monaco's geographical centrality within Europe, excellent infrastructure (including accessibility to Nice airport) and Mediterranean climate conjoin attractively with the city-state's status as a competitive tax haven (see Urry, 2014a, p. 46). Situated a few kilometres from the Italian border and surrounded on three sides by France, and on an increasingly English-speaking coastline, it is no surprise that the languages and cultures of Monaco represent an eclectic mix of nationalities. According to government statistics, out of the population of 36 000 just 8000 are of Monegasque nationality, with the remainder made up of expatriates, primarily from other European countries.[4]

Monaco's population is noteworthy given the high concentration of super-rich residents. According to the Wealth-X and UBS *World Ultra Wealth Report* (2014, p. 45), Monaco has 210 ultra-wealthy individuals who have a combined fortune of €23 billion. This is a very high population of super-rich residents, particularly given the Principality's small geographic area of 2 square kilometres. As a comparison, Switzerland at over 40 000 square kilometres hosts 635 ultra high net worth residents, with a combined wealth of near €750 billion.

In an interview with previous yacht owner and super-rich businessman, Harry, I asked why Monaco was so attractive for the super-rich, given the multitude of destinations available to them. He replied: 'Monaco is unique in all the world. . .you have a wonderful location, you have generally wonderful weather. You have very liberal tax law that brought people here initially. You have the aura and the charm. . .unlike other enclaves, Monaco delivers'. In a separate interview, Frank – a previous yacht owner and retired London banker – suggested that Monaco's centrality in Europe and ease of accessibility allows European super-rich to maintain residence in Monaco for the officially required number of days, yet spending most of their time in their native homes such as the UK or Russia. Given the high concentration of super-rich residents and visitors, and given its accessibility, Monaco is a convenient and culturally rich destination for the super-rich.

The high concentration of wealth and the growing infrastructure and accommodation of super-yachts to Monaco has obvious advantages to those targeting a super-rich client base – such as yacht brokerages, luxury goods outlets and private wealth management companies. As a result, prime office space surrounding the central Port Hercules in La Condamine area of Monaco hosts a wealth of such companies, in prominent and visible locations in a bid to attract the attention of the potential clients who enter the port by yacht. These companies and intermediaries take full advantage of Monaco's appeal and the summer influx of super-rich visitors in an attempt to attract more business. The summer yachting season in Monaco launches immediately after the annual Cannes Film Festival in May, as yachts and their owners make their way from Cannes to the annual Formula 1 Grand Prix in Monaco.

From my experience working on board various yachts as a crew member, and on the basis of observations made during my time ashore with the yacht brokerage, it is evident that Monaco is perhaps the most prestigious port of call in the summer itineraries of yacht charters in the Mediterranean, if not the world. The perceived prestige is manifest through the inclusion of Monaco in most yacht charter itineraries (as established during my time working within the yacht brokerage), the subsequent over-demand for berths and high docking fees, and also in the performances of crew when docking in Monaco's central Port Hercules.

As Edensor (2006, p. 484) explains: 'performances might be scrutinized by fellow performers to minimize any diversions from conventions'. In the case of yachting practices, the off-duty performances of the crew, how the yacht is presented and maintained, and the prominence of the berth in the port all come under intense scrutiny from other yacht owners who may choose to report back to fellow super-yacht owners, or to make assumptions about the owner's control over and involvement in practices on their yacht. For example, prior to docking in Monaco, many captains will instruct their crew to change from their day uniforms of polo shirt, shorts and deck shoes into their smarter uniforms of shirts, trousers and epaulettes – attire usually reserved for the arrival and departure of guests and the start and end of their charter. Changing into a smarter uniform helps to signify and reproduce the prestige and reputation of Monaco for the super-rich yacht guests. Furthermore, when yachts are berthed in Monaco without guests it is common for the crew to be forbidden to smoke or drink ashore in their uniform, or to loiter on the dock close to the yacht. These restricted practices are otherwise commonplace in less prominent neighbouring ports such as Nice, Antibes and Golfe-Juan, where the chances of being observed by other yacht owners are minimized. The prestige associated with Monaco enhances the display of wealth and status for yacht users, as there is present a higher concentration of super-rich and those able to read the value of the yacht and thus interpret the wealth status of the owner or users. Thus, the ways in which super-yachts and their crew are presented become integral to the performance of wealth and status.

Although yacht users perform their wealth and leisure practices to a knowing audience of super-rich, there are, in fact, multiple actors and audiences entwined in the social and spatial composition of Monaco. Additional actors such as non-wealthy tourists serve to further dramatize performances of wealth throughout the city-state.

TOURISM AND THE SUPER-RICH IN MONACO

To avoid the significant traffic congestion and hordes of tourists, many super-rich residents opt to leave Monaco during the busiest summer months, often taking up residence in holiday villas in France or elsewhere. However, the exodus of super-rich residents during the busy summer months does not render the city-state empty or undesirable to the super-rich. Rather, many more travel to Monaco eager to perform their wealth and status to other visiting super-rich. Given the difficulty in gaining access to the Principality by road during this time, many opt to arrive in Monaco via helicopter (private and commercial) or yacht. As viewed from the brokerage office and as confirmed using vessel tracking technologies[5] it became clear that super-rich visitors arriving by yacht typically spend one to two nights in port before heading off to their next destination. Despite the absence of many of Monaco's super-rich residents, the high numbers of visiting super-rich enhance the opportunities to consume luxury goods and prestige modes of transportation conspicuously. For example, the berths in Port Hercules are mostly full throughout the summer with chartered and private super-yachts, private helicopters are displayed at the heliport in Fontvieille and supercars fill the congested streets.

The super-rich perform their wealth status not only to their peers through the conspicuous consumption of luxury goods and modes of transportation but also to onlooking tourists. In this section I turn my attention to the role of the tourist in shaping and responding to the performances of super-rich wealth and status in the Principality, illustrating how the ability to express wealth both within and beyond the super-rich community is an integral component for super-rich lifestyle performances.

According to national statistics, tourism accounts for around 25 per cent of annual revenue in Monaco. Accommodating and attracting less wealthy tourists has therefore become as much a priority for the government as attracting the super-rich. However, rather than create spaces or entertainment specifically for the non-wealthy tourist group, Monaco sells to this group its greatest asset: experience of, and exposure to, super-rich lifestyles. Non-wealthy tourists to Monaco are motivated to visit by the close proximity to, and experience of, wealth and the super-rich. The highly visible expressions of wealth performed by the super-rich are concentrated in a relatively small space within Monaco, and allow non-wealthy tourists to experience and to get close to the super-rich and their luxury props such as supercars and yachts, often invisible to them in their daily lives. As Urry explains: 'The tourist gaze is directed to features of the landscape and townscape which separate them off from everyday experience. Such aspects are viewed because they are taken to be in some sense out of the ordinary' (Urry, 1990, p. 3; see also Baerenholdt et al., 2004). As Harry explained further: 'Tourists [in Monaco] live vicariously through the images of the rich. They'll never experience the kind of luxury and glamour that the celebrities and the wealthy have. Being in Monaco, if even for a day, makes a good story back home. . .the associations with glamour, richness, celebrity is passed on through the tourist'.

Tourists are able to feel a part of or close to super-rich lifestyles through guided walking tours, open-top tour buses, and road trains that operate specifically to cater for large groups of day-trippers. Visitors walk along the dock and pose for photographs at the aft of super-yachts. They cluster around Casino Square and photograph guests leaving Hôtel de Paris. And they take advantage of the pedestrian crossings to photograph the stopped supercars and their drivers. Visitors to Monaco thus play the part of the audience to the displays of wealth amongst the super-rich. Despite their assumed inability to participate in super-rich activities, or to gain access to private spaces, tourists are not passive bystanders in Monaco. Rather Monaco's tourists play an integral and entwined role as an interactive audience to the super-rich who can perform to them (see MacCannell, 2001 for tourist agency). Tourists remind the super-rich of their wealth and their social status, thereby reinforcing their performances or projections of wealth. Without a crowd to film and cheer the supercars driven through tourist hotspots in Monaco, there is simply less motivation for the super-rich drivers to travel there.

Other super-rich enclaves such as St. Barts, as discussed by Cousin and Chauvin (2013), resist mass tourism and instead concentrate on preserving access and experiences exclusively for the super-rich. By restricting cruise ships' passengers, St. Barts allows access primarily to those on board luxury yachts and the existing wealthy residents. This allows St. Barts to create and sell an island of insularity, exclusivity and luxury, where the super-rich can relax out of sight of the public. Monaco on the other hand embraces the demand from tourism, welcoming in port up to two large cruise liners per day in the summer season. The ability to express their wealth both within and beyond the super-rich community is an integral component to leading a super-rich lifestyle for some. The insularity and privacy offered by St. Barts does not appeal, at least long term, to many who own a super-yacht.

Accounting for a quarter of Monaco's annual revenue, tourism is a powerhouse that will continue to sustain Monaco's economic future. In addition to the revenues they generate for the Principality, tourists play a vital role in the meanings and identities enacted for and with the super-rich. Some of the glamour, luxury and sense of space created for (and by) the super-rich is priced out of reach for the non-wealthy visitor, yet the appeal of the Principality still holds. Arguably, if the government were to cater fully to the mass tourist group they would inadvertently devalue the experience for those tourists, who come to Monaco to experience the super-rich lifestyle. Rather, visitors to Monaco observe and participate temporarily in super-rich lifestyle performances. The visibility of and physical accessibility to individual super-rich people and their lifestyles by non- or aspiring wealthy draw many thousands of visitors to Monaco every year. Almost paradoxically, this ability to attract mass tourists in vast numbers is complementary to the government's strategy to attract the super-rich (see Short, Chapter 18 in this volume for a discussion of other strategies intended to attract the super-rich).

Despite performances of wealth and status being directed beyond the super-rich community, there remains a desire to retain a degree of spatial exclusivity for the super-rich in Monaco. Such spatial exclusivity is achieved in a number of ways. First, high prices for rooms, drinks and food in key luxury destinations such as Café de Paris, Hôtel de Paris, the Fairmont and the Hôtel Hermitage serve to restrict entry and engagement by non-wealthy visitors to the Principality. Second, such places, together with the famed Monte Carlo Casino, create an illusion of spatial exclusion, demonstrated, for example, by the presence of smartly dressed and imposing doormen. The doormen at the Casino bar the entrance doors and signify to crowds of onlookers that the site is not accessible. As I observed during my time in Monaco, many potential Casino patrons stop short of the doormen on the stairs and end up taking a photograph before descending again to join the throngs of passing tourists. The Casino is in fact publicly accessible, with an entrance fee (€10) charged beyond the main entrance hall. The impression given by unofficial bouncers at the entrance to such sites is that tourists are not welcome. However, assuming that they were willing to pay the high prices, if these tourists were to make it as far as the door they would be greeted and welcomed in. Private members' clubs also seek to retain a degree of spatial exclusivity for the super-rich in the Principality. The most prominent of these is the Yacht Club de Monaco.

Yacht Club de Monaco

In 2014, the municipality launched a directive to (re)emphasize Monaco's position as 'world capital' of yachting in an attempt to attract the owners of the world's largest super-yachts. In his speech during the official opening ceremony, Prince Albert described the intended role of the new building in achieving long-term economic objectives via the super-rich: 'With this architectural masterpiece we are affirming our Monaco yachting identity, our ambition being to continue to orient our country's future prosperity to the sea' (Sovereign Prince Albert II in Yacht Club de Monaco, 2014). Monaco has long been an elite yachting hub, fuelled by the Royal family's tradition for sailing and their passion for the ocean. Yet the need to reaffirm Monaco's yachting identity, as pledged by the Prince, is to reposition the Principality in the minds of the super-rich who own, or are thinking about owning, a super-yacht.

In recent years increased competition from Mediterranean ports and marinas has threatened Monaco's share of super-yacht (and thus super-rich) revenue. The subsequent rebranding campaign is driven by the desire for Monaco to reclaim its reputation as the key global super-yachting destination on the world stage of super-rich lifestyles. Following the lead of competitor Port Adriano in Palma, and the success of its new marina facilities designed by Philippe Starck, Monaco enlisted the expertise of Lord Norman Foster (and Partners) to design the new Yacht Club de Monaco (hereafter YCM) (see McNeill, 2005 for star architect brands). The cornerstone of the government's rebranding campaign is estimated to have cost €99 million to complete (Rutherford, 2014, p. 50). The YCM is an elite and restrictive members-only club with a high membership fee and annual costs. Through contacts made via his client base, Paul, the senior sales broker at the yacht brokerage I was associated with, was able to secure membership to the Yacht Club prior to the completion of the new premises. His active membership of the YCM enabled chance meetings and introductions with potential clients, helping to enhance his professional reputation and that of the brokerage in the Principality. Paul accepted the high annual membership fees and substantial monthly hospitality bills because his association with the club and his regular attendance enabled him to increase his status and appeal as a successful and desirable sales broker. As a result of Paul's membership I was able to join him at YCM for frequent meetings, networking opportunities and social visits that we made weekly following the Club's official opening in July 2014. Through my regular presence at the Yacht Club I was able to observe and participate in the experiences, performances and rituals on display inside and in close proximity to the building.

The spatial and social exclusivity of the Yacht Club – ensured by its closed and selective membership practices – serves to reinforce affiliation with yachting practices in Monaco and to facilitate interactions and engagements with ostensibly likeminded patrons. Full membership to the club is considered highly prestigious, and the new premises offer members enhanced opportunities for connecting with fellow wealthy business people. The subsequent connections and interactions between members enable a form of social capital. As Field (2003, p. 1) explains: 'People connect through a series of networks and they tend to share common values with other members of these networks; to the extent that these networks constitute a resource, they can be seen as forming a kind of capital', whether that is a chance meeting at the lunch buffet or conversation in the lift, or whilst waiting on the valet to bring up one's car, the excuse to exchange business cards and become a familiar face at the Yacht Club combine to enhance business and social relations.

For the super-rich this can enhance their profile amongst peers, within and across industries. And for intermediaries such as yacht broker Paul, this enables positive exposure to potential clients. For Paul, YCM membership means that he is an active player in the exclusive spatiality: by performing shared leisure practices through this membership he is taken more seriously by the club's super-rich clientele and is thus more likely to encounter new business. Much like those New York boutique hotels discussed by McNeill and McNamara (2010) Monaco's new Yacht Club is now a destination and an experience in itself for those who can gain access, thus strengthening the appeal of patrons to spend more time, and money, in the Principality. Those waiting to secure membership rely on friends and associates to grant them access to what is one of the most exclusive and luxury yachting spaces in the world.

To conclude, my privileged access as an observer to super-yachts and associated practices in Monaco has allowed me to make sense of some of the performances of wealth and status in the Principality's public and private spaces. Through the example of the super-yacht, I have argued that to advance and deepen our knowledge of the super-rich, scholars should consider carefully the lifestyles and leisure practices of the super-rich. There is another axis to my discussion and that concerns the role of intermediaries in making sense of the super-rich. Whilst he acknowledges the need to know more about super-rich lifestyles and spending, Andrew Sayer (2014, p. 251 and Chapter 5 this volume) implores scholars to nurture critical analysis of the super-rich and to focus on how the super-rich obtain their vast wealth in the first place. However, because those wealth accumulation practices are often disguised through a network of tax havens, international corporations, family trusts and complex off-shore financial arrangements (see Chapter 20 by Beaverstock and Hall and Chapter 21 by Palan and Mangraviti in this volume), uncovering the sources of individual wealth amongst the super-rich is a daunting if not impossible task for academic researchers (Urry, 2014b, p. 227). I have suggested here that in order to help overcome these significant barriers to access, scholars can engage usefully and productively with intermediaries who work with/for the super-rich and who actively shape their lifestyle practices. And while it does not offer a comprehensive account, this chapter does go some way to illustrating the value of overcoming such barriers to shed additional light on the spatial and social implications and engagements of the super-rich.

Attracting the super-rich to Monaco is not without its tensions. For example, there has long been a challenge between managing the extreme pressure on Monaco's land space without compromising the main asset, the sea. Efforts to provide space for the super-wealthy generate friction between the economic and the environmental goals of the Principality. Monaco is facing trade-offs between environmental preservation and sustainability agendas in its continuing provision for super-yachts and cruise ships. Driven by the topography and Monaco's small size, the municipality has welcomed high-density developments and significant land reclamation. Therein lies a pressure to balance attractiveness for the global super-rich and the ability to accommodate a growing, richer population of expatriates. Maintaining super-rich interest in the Principality through enhanced super-yacht infrastructure (deeper and longer berths to accommodate the largest super-yachts, for example) does somewhat contradict the Palace's campaign for marine conservation, and the government's sustainability pledges. The municipality will inevitably have to address the long-term sustainability of the super-yachting hub it has created both economically and environmentally.

NOTES

1. See charter listings displayed on www.ypigroup.com.
2. See www.camperandnicholsons.com.
3. Pseudonyms have been used to protect the identity of respondents.
4. See www.monacostatistics.mc.
5. Such as www.marinetraffic.com.

REFERENCES

Armstrong, S. (2010), *The Super-Rich Shall Inherit the Earth*, London: Constable.

Atkinson, R. and S. Blandy (2009), 'A picture of the floating world: grounding the secessionary affluence of the residential cruise liner', *Antipode*, **41** (1), 92–110.

Baerenholdt, J., M. Haldrup and J. Larsen et al. (2004), *Performing Tourist Places*, Aldershot, UK: Ashgate.

Beaverstock, J.V. and J. Faulconbridge (2013), 'Wealth segmentation and the mobilities of the super-rich. A conceptual framework', in T. Birtchnell and J. Caletrío (eds), *Elite Mobilities*, London: Routledge, pp. 40–61.

Birtchnell, T. and J. Caletrío (eds) (2013), *Elite Mobilities*, London: Routledge.

Bloomberg Billionaires (2014), 'Bloomberg billionaires: today's ranking of the world's richest people', accessed 1 September 2014 at www.bloomberg.com/billionaires.

Butler, J. (1993), *Bodies That Matter: The Discursive Limits of Sex*, London: Routledge.
Caletrío, J. (2012), 'Global elites, privilege and mobilities in post-organized capitalism', *Theory, Culture Society*, **29** (2), 135–49.
Cousin, B. and S. Chauvin (2013), 'Islanders, immigrants and millionaires: the dynamics of upper-class segregation in St. Barts, French West Indies', in I. Hay (ed.), *Geographies of the Super-Rich*, Cheltenham, UK and Northampton, MA, USA: Edward Elgar Publishing, pp. 186–200.
Dorling, D. (2014), *Inequality and the 1%*, London: Verso.
Edensor, T. (2006), 'Performing rurality', in P. Cloke, T. Marsden and P. Mooney (eds), *Handbook of Rural Studies*, London: Sage, pp. 484–95.
Elliot, A. and J. Urry (2010), 'The globals and their mobility', in *Mobile Lives*, Abingdon, UK: Routledge, pp. 65–84.
Featherstone, M. (2013), 'The rich and the super-rich: mobility, consumption and luxury lifestyle', in N. Mather (ed.), *Consumer Culture, Modernity and Identity*, New Delhi: Sage, pp. 3–44.
Field, J. (2003), *Social Capital*, London: Routledge.
Goffman, E. (1956), *The Presentation of Self in Everyday Life*, London: Penguin.
Hay, I. (2013), *Geographies of the Super-Rich*, Cheltenham, UK and Northampton, MA, USA: Edward Elgar Publishing.
Jennings, J., T. Edwards and P. Devereaux Jennings et al. (2015), 'Emotional arousal and entrepreneurial outcomes: combining qualitative methods to elaborate theory', *Journal of Business Venturing*, **30** (1), 113–30.
MacCannell, D. (2001), 'Tourist agency', *Tourist Studies*, **1** (1), 23–37.
McManus, P. (2013), 'The sport of kings, queens, sheiks and the super-rich: thoroughbred breeding and racing as leisure for the super-rich', in I. Hay (ed.), *Geographies of the Super-Rich*, Cheltenham, UK and Northampton, MA, USA: Edward Elgar Publishing, pp. 155–70.
McNeill, D. (2005), 'In search of the global architect: the case of Norman Foster (and partners)', *International Journal of Urban and Regional Research*, **29**, 501–15. doi: 10.1111/j.1468-2427.2005.00602.x
McNeill, D. and K. McNamara (2010), 'The cultural economy of the boutique hotel: the case of the Schrager and W hotels in New York', in M. Goodman, D. Goodman and M. Redclift (eds), *Consuming Space: Placing Consumption in Perspective*, Farnham, UK: Ashgate, pp. 147–62.
Piketty, T. (2014), *Capital in the Twenty-First Century* [French edition published 2013 as *Le capital au XXI siècle*, Editions du Seuil], Cambridge, MA: Belknap Press of Harvard University Press.
Pow, C. (2011), 'Living it up: super-rich enclave and transnational elite urbanism in Singapore', *Geoforum*, **42** (3), 382–93.
Rogers, A. (2012), 'Geographies of the performing arts: landscapes, places and cities', *Geography Compass*, **6** (2), 60–75.
Rogers, D. (2015), 'Becoming a super-rich foreign real estate investor. Globalising real estate data, publications and events', paper presentation at 'The 1% City – Reconsidering the Super-Rich' workshop at The Department of Public Policy, City University of Hong Kong, 15–16 January 2015.
Rojek, C. (2000), 'Leisure and the rich today: Veblen's thesis after a century', *Leisure Studies*, **19** (1), 1–15.
Rutherford, T. (2014), 'The super centre', *PrivatAir Magazine*, Autumn, 47–54.
Sayer, A. (2014), 'Postscript: elite mobilities and critique', in T. Birtchnell and J. Caletrío (eds), *Elite Mobilities*, Abingdon, UK: Routledge, pp. 251–62.
Shaxson, N. (2011), *Treasure Islands*, London: Random House.
Sheller, M. and J. Urry (2004), 'Places to play, places in play', in M. Sheller and J. Urry (eds), *Tourism Mobilities. Places to Play, Places in Play*, London: Routledge, pp. 1–10.
Sherman, R. (2007), *Class Acts. Service and Inequality in Luxury Hotels*, Berkeley and Los Angeles, CA: University of California Press.

Spence, E. (2014), 'Unraveling the politics of super-rich mobility: a study of crew and guests on board luxury yachts', *Mobilities*, **9** (3), 401–13.
Urry, J. (1990), *The Tourist Gaze: Leisure and Travel in Contemporary Societies*, London: Sage.
Urry, J. (2014a), *Offshoring*, Cambridge, UK: Polity.
Urry, J. (2014b), 'The super-rich and offshore worlds', in T. Birtchnell and J. Caletrío (eds), *Elite Mobilities*, London: Routledge, pp. 226–40.
Veblen, T. ([1899] 1994), 'The theory of the leisure class', in *The Collected Works of Thorstein Veblen*, London: Routledge, pp. 1–404.
Wealth-X and UBS (2014), *World Ultra Wealth Report 2014*, accessed 19 August 2015 at https://www.private-banking-magazin.de/uploads/fm/1416410395.Wealth-X__UBS_World_Ultra_Wealth_Report_2014_Final.pdf.
Yacht Club de Monaco (2014), 'HSH the Sovereign Prince Albert II officially opens the Yacht Club de Monaco's new premises', accessed 1 August 2014 atwww.yacht-club-monaco.mc/en/clubhouse-du-ycm-2014-in63.html.

15. Flights of indulgence (or how the very wealthy fly): the aeromobile patterns and practices of the super-rich

Lucy Budd

INTRODUCTION

Following the first heavier-than-air powered human flight in December 1903, aircraft have been important symbols of socioeconomic progress and technological modernity. In a little over a century, civil aviation has evolved from being an elite and elitist mode of mobility, utilized only by a very small minority of society's most affluent members, into a multibillion-dollar commercial enterprise that facilitates the routine global mobility of over 3 billion passengers and 50 million tonnes of airfreight every year (ATAG, 2012).

During the latter half of the twentieth century, progressive regulatory reform of the global air transport industry, combined with the introduction of more fuel-efficient aircraft and the formation of new airline business models, has reduced the financial cost of flying and enabled more people to fly to more places more frequently than ever before. However, the emergence and subsequent rapid expansion of low-cost carriers in the deregulated and liberalized markets of North America, Europe and parts of Asia, the Middle East, Australasia, Africa and Latin America, and the swift response of extant full-service operators to this new competitive threat, has led to growing social segregation and stratification in the skies as airlines seek ever more elaborate ways to differentiate their in-flight products and service offerings in what has become a highly contestable and price-sensitive market. The emphasis on fares, service and personal utility that has resulted has arguably polarized the passenger experience of commercial flight between those who travel on private jets or in the comfort of First Class cabins on award-winning full-service airlines and passengers who utilize the cheap(er) 'no frills' services offered by low-cost and charter operators for whom price, rather than service, is often the primary consideration.

Crucially, although considerable scholarly attention has been paid to the patterns and practices of global passenger (and, to a lesser extent, cargo) air transport (see Goetz and Budd, 2014), other forms of civil aviation, including (but not limited to) business, corporate and general aviation activities, have received far less attention despite their importance to the super-rich (see Freeland, 2012 and Hay, 2013). This is a significant omission. First/Business Class and private business aviation, for example, while only accounting for a small proportion of total passenger traffic, exert a considerable social-cultural and economic influence because they are very high-yielding segments and cater to the mobility needs of society's most affluent and influential members (Beaverstock and Faulconbridge, 2013 offer a valuable conceptual framework for examining the mobilities of the super-rich).

In order to interrogate this diverse and high-yielding subcategory of civil aviation practice, the chapter is divided into five sections. The next section details the historical evolution and principal characteristics of First Class aviation while the second explores private business aviation. Section three then examines why these forms of aerial mobility offer such attractive propositions for users. The fourth section presents data on the spatial distribution of business aviation activities worldwide, focusing not only on the location of the aircraft themselves but also on the provision of supporting infrastructures and services. The chapter concludes by discussing what these elite 'flights of indulgence' say about practices of consumption, the globalization of wealth and expressions of affluent mobility in the modern world.

AERIAL ELITES

Civil aviation has historically been an elite and elitist mode of transportation. Aircraft are expensive to purchase, maintain and operate and, in the early years of passenger aviation only the most affluent, including 'exceptional businessmen [sic], clever tourists, romantic honeymoon couples, fast-moving directors [and] modern lawyers' (cited in Hudson and Pettifer, 1979, p. 44) could afford to fly and airlines unashamedly protected and promoted the exclusivity of flight.

By the early 1930s, Britain's Imperial Airways was offering two service classes on its London–Paris route 'so that the famous and influential might have an opportunity to avoid the company of their social inferiors' (1979, p. 46). The development of long-haul intercontinental air services later that decade saw airlines invest in in-flight comfort and service to entice wealthy travellers into the air. Despite the bodily discomforts encountered on the early flights, air travel was portrayed and promoted as an exciting and adventurous form of mobility that reflected the personal and professional success of its users (Budd, 2011).

This unashamed elitism continued until the outbreak of World War II when most commercial flying was suspended. After 1945, progressive regulatory reforms enabled some US airlines to offer a number of 'coach' (or economy) class seats on selected flights. Initially, these tickets were accompanied by complex fare restrictions but, over time, airlines were able to offer a wider range of cabin products. In order to physically and psychologically separate the high-yielding full-fare passengers from those in economy, cabin dividers and curtains were employed and the aerial hierarchy further reinforced by the provision of superior seating and in-flight service for First Class passengers who were seated at the front of the cabin (away from the worst of the engine noise and vibration).

By the 1970s, airlines began offering three distinct cabin classes: First, Business and Economy. While not enjoying the full range of luxuries afforded to First Class passengers, Business Class travellers could nevertheless enjoy levels of comfort and service that were not afforded to Economy Class customers. The mid-1970s also saw the inauguration of supersonic transatlantic passenger services by Air France and British Airways. Flying at twice the speed of sound and at an altitude from which its 100 First Class passengers could view the curvature of the Earth, Concorde epitomized all that was modern and glamorous about elite aerial travel from its inauguration into commercial service 1976 until its eventual withdrawal in 2003.

The cessation of Concorde services meant that passengers were once again forced to endure subsonic journey times (which extended the flight duration from London to New York from three to over seven hours). In an effort to improve passenger comfort and productivity, airlines developed new lie-flat beds, purchased designer cosmetics for aircraft washrooms, and employed Michelin-starred chefs to design new menus.

By the mid-2000s, a number of new all Business Class airlines were competing for custom on the lucrative North Atlantic routes. However, despite introducing a range of service innovations, oil price volatility and falling consumer demand during the global economic downturn of 2008–09 forced them out of business. Yet, as British Airways' all Business Class services between London City Airport and New York demonstrate, demand for elite all Business Class services remains. Airlines based in the Middle East, in particular, are currently (re)inventing the First Class experience. Qatar Airways' A380s feature chandeliers while Abu Dhabi-based Etihad is transforming its First Class product with the introduction of 'The Residence' and 'First Apartments' on their A380. 'With a living room, separate bedroom and en suite bathroom, The Residence by Etihad is the only three-room suite in the sky' and features a double seat leather sofa, two dining tables, a personal butler, chilled drinks cabinet and a 32 in (81 cm) flat screen television (Etihad, 2014).

Yet while First and Business Class cabins vie for the custom of wealthy travellers, the fact remains that passengers have to travel at times that suit the airline, not the individual, and the flights only serve a relatively limited range of destinations. Owing to these restrictions, another expression of elite air travel – business aviation – has evolved to fulfil a market need for bespoke 'go now' mobilities, luxury aerial travel and complex itineraries.

BUSINESS AVIATION AND AERIAL DIFFERENTIATION

According to the US-based International Business Aviation Council (IBAC), business aviation activities concern the operation or use of private civil aircraft by companies or individuals to transport passengers or goods for business purposes (IBAC, 2013). Unlike commercial aircraft, business aircraft are generally *not* available for public hire. Business aviation includes a wide range of aeronautical activity and aircraft types, including aerial surveying and crop spraying in single-seat piston-powered aircraft; scheduled twin-turbine helicopter flights to offshore oil and gas installations; on-demand corporate shuttles in long-range 12+ seat business jets; and 'go now' services in high-capacity 'bizliners' (a conflation of 'business' and 'airliner'), such as Airbus's A319CJ (Corporate Jet) and the B737-derived Boeing Business Jet (NEXA, 2010).

The diversity of aircraft types and operating practices notwithstanding, business aviation can be classified into one of three principal types of activity according to who owns and operates the aircraft. The first, commercial business aviation, describes a situation in which professional flight crew operate a private aircraft on behalf of a third-party client. These flights may involve one-off on-demand charters where an aircraft is used once for a specific mission or is operated as part of a fractional ownership scheme in which a business or an individual buys a block of flight time for a set price on an aircraft owned by a fractional operator. Unsurprisingly, the cost of hiring an aircraft varies by type, mission length, service level required and the nature of the lease agreement and can range from around €3800 an hour on the smallest jets to over €20 100 an hour on a bizliner (Table 15.1).

Although fractional schemes and commercial leasing arrangements offer one way for companies and individuals to gain access to private aerial travel, they are not the only business model available. The second principal type of business aviation operation, corporate business aviation,

Table 15.1 Approximate price per hour of operating different types of business aircraft

Price (Per Hour Flight Time in €)	Aircraft Type
20 183	Bizliner
13 287	Ultralong-range business jet
10 650	Heavy business jet
7192	Mid-size business jet
4186	Turboprop
3852	Very light jet

Source: EBAA (2014).

describes a situation in which an aircraft is owned by a company and flown by that company's employees as an aid to the conduct of their business. This type of operation incurs higher capital start-up and ongoing support and maintenance costs but it also ensures that the company or individual has sole and exclusive use of the aircraft. This mode of operation is relatively common among international finance institutions, investment and asset management firms, international business consultancy organizations and mining/mineral companies that require flexible aerial mobility solutions to gain rapid access to the diverse, dynamic and (in the case of mining companies) sometimes inaccessible international markets in which they operate.

The third and final category, owner-operated business aviation, describes a situation in which the aircraft's owner is also its pilot. This type of operation usually involves smaller fixed or rotary wing aircraft that can be flown by a single pilot (as larger airframes usually require a minimum of two qualified flight crew).

Owing to the different patterns of ownership and operation, each of the three categories of business aviation activity has evolved its own distinct set of operating practices and spatial and temporal characteristics. Yet, they all share one important element: the enduring cultural allure of business aviation, which is built on powerful notions of prestige, exclusivity, wealth, desirability and social and professional success. Although business aviation is often considered a sign of achievement, in times of diminishing shareholder returns and rising concern about aviation's environmental effects, it has also been considered to represent an extravagant and unsustainable form of capitalist consumption (Shomko, 2012). In order to better understand the social, spatial and operational nuances of these different forms of elite business aviation mobility, it is necessary to explore the motivations that underpin the use of these aircraft.

ELITE FLYERS: BUSINESS AVIATION USERS EXAMINED

One of the key challenges associated with conducting research into business aviation is the lack of independently verifiable data on the scale and scope of the sector and the number and profile of its users. Unlike surveys of commercial passengers, which are routinely conducted at major airports, and the wealth of data on airline schedules, seat capacity and origin/destination traffic, information on the business aviation sector is sparse, often anecdotal, and (where it does exist) often published by commercial enterprises pursuing a particular corporate agenda. There are, of course, good reasons for this. Many thousands of individual private companies worldwide are involved in the delivery of business aviation services and the airports used by business aviation are often located far away from the principal commercial facilities. They may also be privately owned and operated, making standard national data collection problematic, expensive and time consuming. In addition, the privacy business aviation affords its users is undoubtedly one of its principal attractions and discourses of commercial security and client confidentiality dictate that obtaining information on the origin/destination and purpose of business aviation flights as well as details about the number and professional/social status of aircraft occupants is inherently challenging.

For high net worth individuals (HNWIs) and those in positions of relative political or corporate power and responsibility there may be personal security imperatives to segregate themselves socially and spatially from other (potentially dangerous) travellers when flying. Celebrities too may wish to travel away from the glare of the flying public and their seemingly ubiquitous camera phones and to avoid the long lenses of the paparazzi. Corporations involved in politically sensitive and/or commercially secret missions may need to eliminate the potential for industrial espionage by moving their human and material assets by private aircraft. At the other end of the scale, self-employed doctors, farmers and small business owners may pilot their own aircraft for business purposes between small, often private, airstrips, when no viable ground transport alternative exists and from where no monthly passenger figures or air traffic movements are recorded. These factors, combined with the diversity of potential business aviation users, present a problem for elite aeromobilities research but also an opportunity to uncover some of business aviation's hidden characteristics.

According to a 2014 National Business Aviation Association of America (NBAA) report, *corporate* and *commercial* business aviation aircraft are primarily used to transport company employees and individuals but they may also on occasion carry customers, clients, cargo and humanitarian supplies (NBAA, 2014). In the United States an estimated 11 000 companies use business aircraft every year but only 3 per cent of US registered business aircraft are flown by Fortune 500 companies (NBAA, 2014). By far the majority (85 per cent) are operated by small and medium-sized enterprises (SMEs), 70 per cent of which employ fewer than 1000 personnel and 59 per cent of whom employ fewer than 500 staff, while the remaining 12 per cent are used by individuals, charities and education establishments (NBAA, 2014). Furthermore, despite the prevalence of claims that business aircraft are only used by senior executives, surveys by NEXA (2009) and NBAA (2014) suggest that business aviation supports the mobility needs of employees throughout the professional hierarchy and as many as 72 per cent of users are reported to be non-executive employees and sales/service staff.

Another feature that differentiates business aviation from commercial aviation is the number of aircraft in each fleet. Whereas major full service legacy carriers like American Airlines, British Airways and Lufthansa operate several hundred aircraft, the fleet sizes of commercial and corporate business aviation users are generally more modest. Among US-based corporate business aviation users, 75 per cent own just one aircraft, 12 per cent have two and only 13 per cent have three or more (NBAA, 2014). In the USA, the biggest users of corporate and commercial business aviation include companies in the consumer discretionary, industrial, ICT, consumer goods, materials and utilities sectors (NEXA, 2009) and it is reported that business aircraft users are not only more resilient to economic downturns than non-users but that they also generate higher shareholder returns than equivalent size companies that do not (NEXA, 2010, 2012).

While these data provide a useful insight into the sorts of companies that utilize corporate and commercial business aviation, it does not reveal anything about the situation beyond the USA or reveal anything about the individual people who fly or the individual owner-operators. Indeed, while the majority of business aircraft are owned and/or operated on behalf of SMEs and major multinational companies, a small but significant number are registered to individuals. A very small number are owned by celebrities such as actor-turned-pilot John Travolta, while the rest are owned by members of the global super rich. One of the few places where it is possible to obtain publicly available data on the registered owners of business aircraft is the Isle of Man Aircraft Registry. This aircraft registry was established in May 2007 as a place where aircraft owners worldwide could register 'high-quality' twin-turbine helicopters and fixed-wing business aircraft that have a maximum take-off weight exceeding 5700 kg (smaller aircraft that are owned by residents or businesses based on the Isle of Man are also accepted).

Of the 670 aircraft that have been registered since its inception, fewer than 5 per cent (31 aircraft) are registered to named individuals or groups, with the rest being owned by commercial companies and institutions based either in the British Isles or overseas. Of these 31 aircraft, 27 were registered to single men, one was registered to a female owner and the rest were registered to groups of named individuals. In addition to offering a stable political and regulatory environment, the Isle of Man Registry also allows owners to adopt personalized aircraft registration markings (in just the same way as you can acquire personalized number plates for cars) in the series M-xxxx that reflect their business, lifestyle or personal initials. As a result, registrations that have been taken out include obvious ones such as M-IDAS, M-YJET and M-LEAR (on a Learjet). While the majority of registrations are for twin-engine business jets in the 5–12 seat range, much larger aircraft, including a number of wide-bodied bizliners, are also registered.

Although this data source provides some information about the companies and the business sectors that own and utilize business aircraft we still know very little about how these aircraft are used, where, or by whom and so it is necessary to triangulate data from other sources. Although confirmatory data do not exist, it is arguably not unreasonable to assume that the majority of users are male and economically active. This assumption would appear to be confirmed by a crude content analysis of the 39 photographs that appear on the front covers of online issues of *Business Jet Traveller* magazine as all but two of them are of men (*Business Jet Traveller*, 2014). Leaving challenging and perhaps irresolvable questions of gender aside, in order to better understand who is likely to use business aircraft it is necessary to understand the myriad lifestyle and business factors that stimulate demand for business aviation.

PRIVATE AEROMOBILITIES – THE BENEFITS OF BUSINESS AVIATION

For those with adequate personal and financial resources, the decision to utilize a business aircraft for a particular mission depends on a variety of factors. These will include considerations of cost and perceived value for money, convenience, trip itinerary (including intended destinations and the sequence in which they must be visited), the number of travellers, whether proprietary business matters need to be discussed en route and whether commercially sensitive or specialist equipment must be transported and issues of personal/employee safety and comfort.

Cost

Although the capital outlay associated with purchasing, operating or hiring a business aircraft for a mission may initially appear prohibitive, once employee downtime and airport delays are factored in, travelling by private aircraft can be more cost efficient than utilizing commercial services, particularly when several employees need to travel multiple sectors during the course of a day at short notice. Business aviation operators claim that their services can actually save companies money as aircraft can operate at a time and between locations of the client's choosing, thus minimizing travel time to/from the airport, reducing employee dead time once at the airport, and eliminating overnight accommodation bills.

In an effort to highlight the cost savings that can be made by chartering a business aircraft, many operators provide free-to-use cost calculators on their websites that enable companies to compare the cost of flying on a private aircraft with the 'total' cost of sending their employees on a commercial service once all the additional time penalties associated with the latter are factored in. Some also include the ability to input employees' salaries into the algorithm in an apparent effort to demonstrate that employee time really does translate into money.

Convenience

A second factor that may influence an individual's or a company's decision to utilize business aircraft concerns convenience. Owing to their smaller size and unique operating characteristics, business aircraft can operate at airports that cannot be served by scheduled passenger flights and fly routes that are not commercially viable. In the United States alone, it is estimated that business aviation serves ten times as many airports as the commercial airlines do. This is partly a result of processes of airline deregulation that prompted airlines to reconfigure their services around hub-and-spoke networks to concentrate flights on key routes and consolidate their operations at a few major hubs to protect their market share. As a consequence, of the 5171 public use airports in the USA, only 499 have commercial services and 70 per cent of all passenger enplanements in the country occur at just 29 large hubs (FAA, 2012). Similarly in Europe, the European Business Aviation Association (EBAA) estimates that business aircraft serve 103 000 city pairs in the continents, three times as many as the scheduled airlines serve, and business aircraft access 966 airports that are not regularly served by scheduled passenger flights (EBAA, 2014). The spatial configuration and concentration of scheduled airline services at a limited number of airports is such that many destinations and airport pairs are either underserved or not served by commercial airlines. It is this physical gap in the market that business aviation seeks to exploit. According to NEXA (2010), the principal motivations for using these smaller airports are to get close to customer offices, exploit new business opportunities and to access company-owned offices or facilities.

As well as sating otherwise unfulfilled demand for flights between these smaller airports, business aviation's use of these facilities offers additional advantages. Smaller airports are less prone to delays; they are less congested, and do not suffer from the capacity constraints and slot restrictions that are often encountered at large airports. This means that business aircraft can fly at a time of the client's choosing and can offer a more personalized bespoke and private service. According to one user, 'A business jet is not a luxury, it is a necessity' as it enables him to bypass congested and delay-prone commercial airports (cited in NEXA, 2013, p. 7).

Business aviation also offers enhanced flexibility and gives users control over their flight schedules. Unlike commercial flights that operate to a strict timetable, business aircraft fly on demand and can be airborne in a matter of hours. This enables users to rapidly respond to emerging business opportunities and close deals before their competitors arrive on site. Furthermore, by flying directly between origin and destination on demand, business aircraft can reach multiple destinations quickly and efficiently enable users to accomplish missions in a day that might be impossible to accomplish using scheduled commercial services even if they were available. By minimizing business hours that are spent away from home by negating the need for hotel stopovers, business aircraft can help to support a healthy work/life balance and it is claimed that this may aid employee retention (NBAA, 2014). Accordingly, business aircraft are described as 'time-savers and productivity multipliers that enable users to do more, faster' (2014, p. 17).

These attributes collectively result in a higher degree of flight reliability and schedule adherence. Users can be personally escorted through security, immigration and customs and be directly driven out to, and collected from, the door of their aircraft in chauffeur-driven limousines. This not only minimizes ground time and reduces the time spent on official formalities but also enables users to negotiate the spaces of the airport well away from the eyes of commercial passengers and intrusive airport paparazzi.

Once in the air, business aviation users may be able to access satellite communications systems and the Internet and participate in teleconferences during a flight or engage in more restful sleep. This improves productivity and enables users to maximize their travel time and minimize their ground time. Seamless communications extend to the ground where personal surface transport provision by road minimizes down time and ensures users are quickly on their way to a meeting. In addition to transporting key personnel, business aircraft may also be used to transport time-critical and commercially sensitive or valuable cargo and equipment quickly and securely. Business aircraft are also used for humanitarian relief flights as they can quickly reach remote regions that larger commercial aircraft cannot.

Comfort

The third and final factor motivating the use of business aviation concerns personal comfort. As Bissell (2010) and Budd (2011) have shown, travelling by air can subject the human body to a range of potentially unpleasant physical sensations that may variously result from combinations of turbulence, seat pitch, cabin configuration and the conduct and proximity of fellow passengers and cabin crew. Despite the best efforts of commercial airlines to improve the in-flight experience through the provision of lie-flat beds, on-demand catering services and privacy screens, commercial air travel obliges passengers to fly in the company of strangers and in aircraft that are usually uniformly configured. The commercial aircraft cabin, therefore, may be a space that is not conducive to productive work or rest.

In contrast, numerous industry reports claim that business aircraft improve employee productivity. According to the US-based NBAA, productive in-flight collaboration between work colleagues reportedly occurs eight times more frequently in business aircraft than it would between the same combination of employees flying on commercial services and productive collaboration with customers is seven times more likely to happen aboard a business aircraft (NBAA, 2014). The rationale for this is that the environment of the cabin is quieter and therefore more conducive to work and that if employees feel they are valued enough to travel on a private aircraft then they will engage in more productive work for longer. Indeed, according to the same NBAA survey, business aircraft users spend 36 per cent of their time on board a business aircraft in work-related meetings as opposed to 3 per cent of their time on commercial flights (NBBA, 2014).

In addition to providing a more productive work environment, flying only with company colleagues and business associates reduces the potential for industrial espionage and virtually eliminates the risk of physical threat posed by other passengers. Reassuringly for users, accident rates for business jets are reportedly comparable with commercial airlines (NBAA, 2010).

Owning or using a private aircraft, particularly a business jet, has become an important indicator not only of wealth but also of social prestige and differentiation and increasingly *how* one flies has become very important and there has been growing fragmentation *within* the business aviation sector to differentiate the 'entry-level' aircraft from the 'higher-end' products. As a result, numerous companies have been established to serve the needs of the most demanding and discerning clients. Lifestyle magazines, including *Business Jet Traveller*, report on the latest innovations in interior aircraft design and exterior finishing and create aspirational desires to customize aircraft using the latest designer fabrics, paints and wood veneers. Aircraft interiors can be configured in many different ways, from boardrooms to bedrooms, according to the needs of users. Some of the world's major luxury consumer brands, including BMW and Versace, have designed cabin interiors. Although the fabrics, fixtures and fittings have to conform to strict aeronautical regulations concerning fire retardance, users can select materials that reflect their personal tastes and organizational culture. As a result, some aircraft interiors are reportedly bedecked in gold while others are more restrained.

While some facilities can be installed in aircraft if money is no object, others are prohibited or restricted for reasons of safety or weight. Water is heavy, so cabin showers and Jacuzzis, while technically possible, are expensive and can only be installed on the most powerful aircraft and cannot be used during inflight turbulence. Bedrooms are generally fine, as are boardrooms, but galleys are limited by aeronautical regulations concerning open flames and sources of heat generation.

Given the bewildering array of different options, numerous print and online publications provide reviews of different business aircraft and onboard equipment to support the decision-making process and also offer profiles of famous business jet users from the world of sport, commerce and entertainment. As well as detailing interior and exterior aircraft styling options, these publications also provide a tantalizing glimpse into the range of business and personal activities that can be provided during a flight, either to enhance user well-being or to maximize the productivity of time spent in the air. Business aviation users can thus stipulate the level of cabin service and the in-flight catering they require (NBAA, 2010) as well as utilize the services of personal masseurs, who provide bespoke inflight spa treatments, and tailors who measure people up for suits and advise them on fit and fabrics during a flight (Burger, 2006).

According to the results of a 2009 Harris Interactive Survey (reported by the NBAA, 2014), 64 per cent of US-based business aviation users used business aircraft, as their trip itineraries could not be fulfilled by scheduled airline services. Nineteen per cent of respondents reported that they used business aircraft to reach places that scheduled airlines did not serve, while 9 per cent employed them for industrial or personal security reasons. Similar results were reported by an EBAA survey, which discovered that 77 per cent of users employ business aircraft to save time and 69 per cent to reach destinations that scheduled airlines do not serve. Interestingly, 44 per cent also confirmed that they used business aircraft to ensure a more comfortable flight; 37 per cent cited enhanced privacy as a motivating factor; 34 per cent spoke of the importance of being able to work effectively en route; and 29 per cent said they flew on business aircraft as they afforded a greater degree of personal and commercial security (NEXA, 2013).

Given that accessibility and connectivity were cited as the most important reasons for using business aircraft, it is salient to explore the scale and scope of the global business aviation sector in order to better understand the geographies of this elite form of mobility.

AEROMOBILE ASSETS – THE GLOBAL DISTRIBUTION OF BUSINESS AIRCRAFT

One of the main reasons companies and individuals repeatedly give for using business aviation is the spatial and temporal freedom and flexibility private aircraft afford. Given that business aircraft are, by definition, used to take people from where they are to where they want to be by wealthy corporations and individuals, the geographic distribution and utilization of business aircraft closely reflects patterns of international trade, economic prosperity and globalization. On certain metrics, the scale and scope of the global business aviation sector dwarfs the commercial aviation industry. For even though the former transports fewer people it has far more aircraft and airports at its disposal. Worldwide, there are currently in excess of 32 700 registered fixed-wing business aircraft compared with 23 800 commercial aircraft (ATAG, 2012; AIN, 2014). Of these 32 700 business aircraft, almost 60 per cent (over 19 300 units) are business jets and approximately 40 per cent (about 1400) are turboprops (AIN, 2014) yet these aircraft are not uniformly distributed around the world.

Figure 15.1 shows the current distribution of business aircraft by world

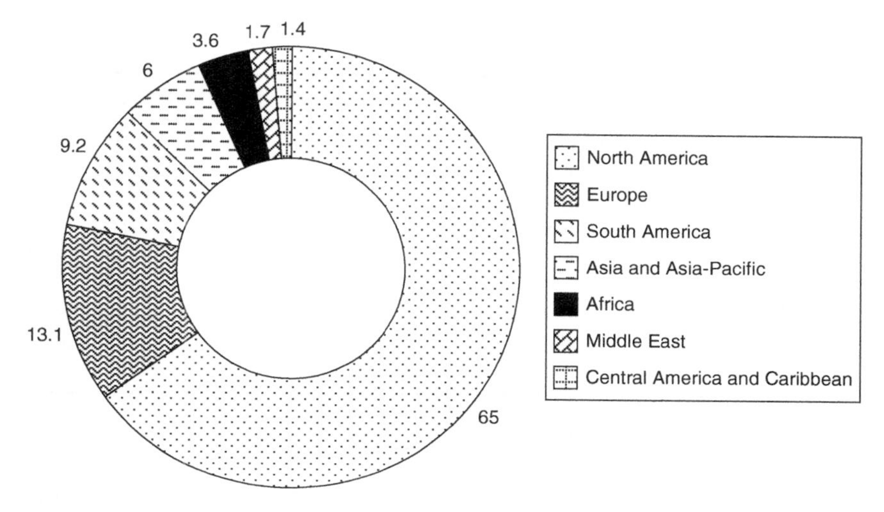

Note: n = 32 700.

Source: Data derived from NBAA (2013).

Figure 15.1 Proportion of fixed-wing business aircraft by world region, 2014

region. North America currently represents the world's largest market for business aircraft, with 65 per cent of the total, followed by Europe (13 per cent) and South America (9.2 per cent). Significantly, the proportion of the world total North America accounts for has declined from 72 per cent in 2008 (see Budd and Graham, 2009) owing to the effects of the 2008–09 recession and growing demand for business aviation in emerging markets in Africa, Asia-Pacific and China. Most tellingly perhaps, the figure for 2011 was 45 per cent, indicating the extent to which the 2008–09 recession affected the North American (and, in particular, the US) market, with many companies cutting back on their corporate travel or 'downgrading' from business aviation to commercial airlines (Budd, 2014).

Table 15.2 shows the distribution of business jets and turboprops by world region in 2014. In North America, Europe, Asia, the Middle East and the Caribbean there are more business jets than there are turboprops, a situation that may reflect the relative prosperity of these regions (business jets are considerably more expensive to purchase and operate than turboprops), the geographical characteristics of these regions, and the nature of business aviation demand.

At an individual country level, the USA still has by far the highest

Table 15.2 Distribution of business aircraft by world region, 2014

World Region	Jets	Turboprops	Total Fixed Wing
North America	12953	8297	21250
Europe	2991	1305	4296
South America	1311	1709	3020
Asia	768	546	1314
Africa	442	730	1172
Asia-Pacific	211	437	648
Middle East	429	122	551
Central America	72	163	235
Caribbean	142	80	222

Source: Data derived from AIN (2014).

number of business aircraft registrations in the world and the highest number of annual registrations. This is partly due to the early development of business aviation in the country as well as its relative economic prosperity, physical size and the dispersed nature of business trip origins. However, the increasing need for speed and flexible mobility has resulted in a steady rise in the use of business aircraft worldwide. In 2012, the majority of business aircraft were purchased in the United States followed by Brazil, Canada and Australia (Table 15.3). The UK was the biggest European market with 176 new aircraft deliveries (NEXA, 2013). However, the fastest growth rates since 2004 have been recorded in countries in Africa, the Middle East and Asia (NEXA, 2013).

Nevertheless, while the country of registration might provide an indication of where in the world the owner/operator is based and where the aircraft might be most intensively used, the relationship is not straightforward. Indeed, aircraft registered in one country for reasons of tax efficiency, political neutrality or business expedience may actually be based in a second country and operated primarily between a third and a fourth one. For example, over 120 aircraft that appear on the Isle of Man Aircraft Registry were/are registered to owners with addresses in the British Virgin Islands and aircraft on the registry are registered to owners originating in over 40 different countries across Europe, Africa, Asia, Australasia and the Caribbean.

Despite the 2008–09 recession temporarily depressing consumer demand for certain types of business aviation activity in particular world markets, in the 12 months between 30 September 2011 and 2012 the world's business aircraft fleet increased by 2.5 per cent to over 29000 units. The biggest increase was in business jets (up 565 in the year to 17974 aircraft) but the

Table 15.3 Business aviation aircraft purchases by country, 2012

Rank	Country	Number of Business Aircraft Purchases
1	USA	3845
2	Brazil	445
3	Canada	369
4	Australia	323
5	Mexico	177
6	South Africa (equal with Mexico)	177
7	UK	176
8	Germany	131
9	New Zealand	120
10	China	117

Source: NEXA (2013).

number of turboprops also increased by 180 to 11 700 airframes. However, globally this growth was very uneven with a largely stagnant North American market being offset by growth in Asia-Pacific. As a consequence, the average age of business aircraft in Asia-Pacific is the youngest in the world at 10 years for jets and 14 for turboprops compared with an average age of 16 and 22 years respectively for North American registered business aircraft (Sarsfield, 2012).

Unlike commercial aircraft, which can be in the air on revenue-generating flights for eight hours or more in any 24-hour period, business aircraft are generally used far less intensively. Indeed, it is estimated that the average business aircraft is only in the air for 10 per cent of the time a commercial aircraft is (NBAA, 2014). This means that while aircraft are available for 'go now' departures, expensive (and continually depreciating) assets are often parked up at airfields rather than undertaking missions. In terms of business aircraft utilization, the US business aircraft fleet performed over 3 million flight hours a year in the period 2004–07 before the 2008–09 recession softened demand (Figure 15.2).

Relative levels of global economic prosperity clearly have a demonstrable effect on levels of business aviation demand. As with commercial aircraft, deliveries of business aircraft are cyclical with periods of strong growth (coinciding with periods of global economic prosperity) being punctuated by sharp declines in times of recession. The number of business jets shipments worldwide between 1996 and 2013 is shown in Figure 15.3. Over this time period, 12 974 new business jets were delivered, an average of 649 a year. Deliveries grew steadily between 1996 and 2001 before falling to 518 units in 2003 and then steadily growing to a high of 1306

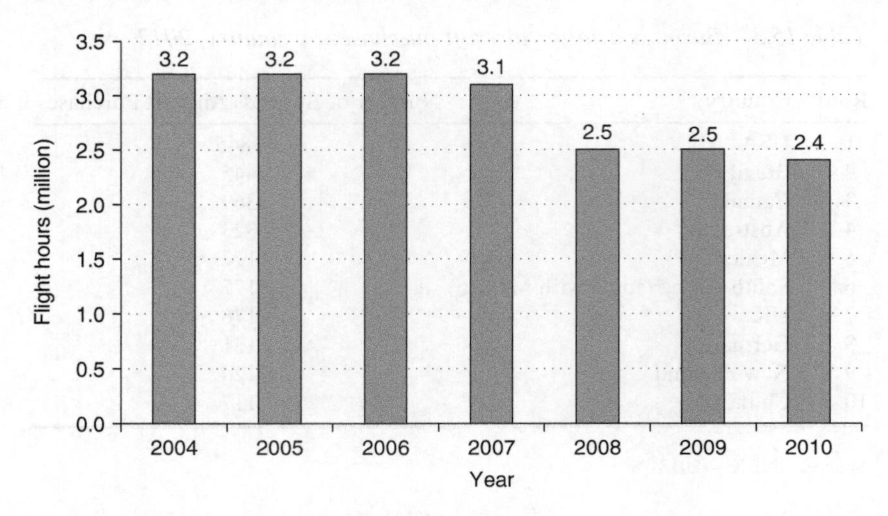

Source: Derived from NBAA (2010, p. 8 and 2014, p. 11).

Figure 15.2 Business aviation hours flown in the USA, 2004–10

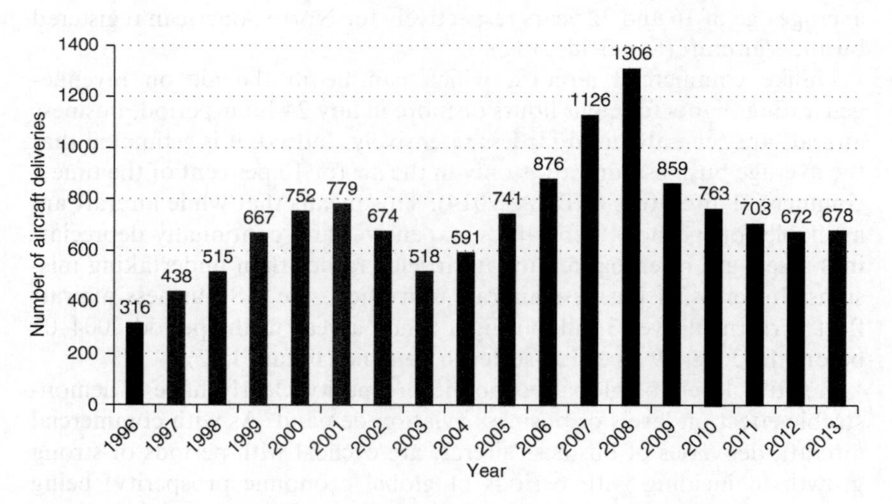

Source: Data derived from GAMA (2014).

Figure 15.3 Number of business jet shipments worldwide, 1996–2013

in 2008. There was then a large drop in 2009 and a fall in annual deliveries until a 0.9 per cent rise in 2013 when 678 new airframes were shipped. Interestingly, the financial crisis only appeared to affect certain sectors of the business aviation market. Wild and Raggi (2013) report that while sales of aircraft in the US$4–25 million price range declined by 56 per cent between 2008 and 2012, deliveries of jets costing in excess of US$26 million actually rose over the same period and there was evidence of existing customers upgrading to larger aircraft (Wright, 2013). Certainly it would appear that confidence is continuing to grow as the value of all business aircraft sales (including jets, pistons and turboprops) rose from US$18.9 billion in 2012 to US$23.4 billion in 2013 (GAMA, 2014), with as many as 625 'super mid-size' bigger business jets expected to be delivered in 2014 (Wild and Raggi, 2013).

CONCLUSION

This chapter has offered an insight into the elite aeromobilities that are undertaken by the world's super-rich. The expressions of, motivations for, and users of these services are very diverse and the sector has variously attracted criticism for being a profligate manifestation of the inequities of capitalist consumption while being described as a necessary market response to the (perceived or real) inadequacies of mass air travel. Irrespective of one's personal view, there is little doubt that future investigations into the geographies of the super-rich will need to consider the implications of these elite forms of aeromobility for individuals, economies and societies. As with commercial air traffic, complaints have been registered about the noise and atmospheric pollution that is caused by business aircraft operations and there have been protracted debates about the extent to which (and how) business aviation activities should be taxed. Certainly, with highly aeromobile assets at their disposal, aircraft owners and operators can (and do) move their aircraft around the world to take advantage of new markets, lifestyle opportunities and favourable political and regulatory environments.

Given the scale and scope of elite aeromobilities, as well as their relative 'invisibility' from academic and public debates, it is imperative that they are not excluded from debates concerning practices of consumption, sustainable mobility and socioeconomic inequalities. For while practices of elite aviation undoubtedly make a valuable contribution to certain sections of the economy and society, they remain an exclusive and exclusionary expression of contemporary mobility whose benefits can only be enjoyed by the most rich and powerful in society.

REFERENCES

AIN (2014), 'Business aircraft by region of the world interactive guide', accessed 30 March 2014 at www.ainonline.com.

Air Transport Action Group (ATAG) (2012), *Aviation Benefits Beyond Borders*, Geneva: ATAG.

Beaverstock, J.V. and J.R. Faulconbridge (2013), 'Wealth segmentation and the mobilities of the super rich. A conceptual framework', in T. Birchnell and J. Caletrío (eds), *Elite Mobilities,* London: Routledge, pp. 40–61.

Bissell, D. (2010), 'Vibrating materialities: mobility–body–technology relations', *Area,* **42** (4), 479–86.

Budd, L. (2011), 'On being aeromobile: airline passengers and the affective experiences of flight', *Journal of Transport Geography*, **19** (5), 1010–16.

Budd, L. (2014), 'Aeromobile elites: private business aviation and the global economy', in T. Birtchnell and J. Caletrio (eds), *Elite Mobilities*, London: Routledge, pp. 78–98.

Budd, L. and B. Graham (2009), 'Unintended trajectories: liberalization and the geographies of private business flight', *Journal of Transport Geography*, **17** (4), 285–92.

Burger, J. (2006), 'The personal clothier in the sky', *Business Jet Traveller*, December.

Business Jet Traveller (2014), *Business Jet Traveller* issue archive, accessed 3 April 2014 at bjtonline.com/issue-archive.

Etihad (2014), 'The Etihad experience. The residence on board our A380', *Etihad Airways*, accessed 15 October 2014 at www.etihad.com/en-gb/experience-etihad/flying-reimagined/the-residence/.

European Business Aviation Association (EBAA) (2014), *EBAA State of the Industry 2012*, accessed 3 April 2014 http://www.ebaa.org/documents/document/20130513153145-state_of_industry_new_%282%29.pdf.

Federal Aviation Administration (FAA) (2012), *National Plan of Integrated Airport Systems 2013–2017*, Washington, DC: Federal Aviation Administration.

Freeland, C. (2012), *Plutocrats: The Rise of the Global Super Rich*, Harmondsworth, UK: Penguin.

General Aviation Manufacturers Association (GAMA) (2014), 'GAMA releases 2013 year-end aircraft shipment and billing numbers at annual state of the industry press conference', *GAMA*, 19 February 2014, accessed 30 March 2014 at http://www.gama.aero/media-center/press-releases/content/gama-releases-2013-year-end-aircraft-shipment-and-billing-number/.

Goetz, A. and L. Budd (eds) (2014), *The Geographies of Air Transport*, Farnham, UK: Ashgate.

Hay, I. (ed.) (2013), *Geographies of the Super-Rich*, Cheltenham, UK and Northampton, MA, USA: Edward Elgar Publishing.

Hudson, K. and J. Pettifer (1979), *Diamonds in the Sky: A Social History of Air Travel*, London: Bodley Head.

International Business Aviation Council (IBAC) (2013), 'IBAC definitions of business aviation, accessed 2 November 2015 at http://www.ibac.org/about-ibac/ibac-definitions-of-business-aviation.

International Business Aviation Council (IBAC) (2014), 'About IBAC', accessed 30 March 2014 at http://www.ibac.org/about-ibac.

Isle of Man Aircraft Registry (2014), website accessed 20 August 2015 at https://www.gov.im/ded/Aircraft/.

National Business Aviation Association (NBAA) (2010), *2010 Business Aviation Fact Book*, Washington DC: NBAA.

National Business Aviation Association (NBAA) (2013), *Business Aviation and the World's Top Performing Companies, NEXA and NBAA*, accessed 25 August 2015 at https://www.nbaa.org/business-aviation/NEXA-Report-Part-5-2013.pdf.

National Business Aviation Association (NBAA) (2014), *2014 Business Aviation Fact Book*, Washington DC: NBAA.

NEXA (2009), *Business Aviation: An Enterprise Value Perspective. The S&P 500 from 2003–2009*, Washington, DC: NEXA Advisors.
NEXA (2010), *Business Aviation: An Enterprise Value Perspective. S&P Smallcap 600 Companies from 2005–2010 – Small and Medium Enterprises*, Washington, DC: NEXA Advisors.
NEXA (2012), *Business Aviation Maintaining Shareholder Value Through Turbulent Times. The S&P 500 During the Great Recession 2007–2012*, Washington, DC, NEXA Advisors.
NEXA (2013), *Business Aviation and the World's Top Performing Companies*, Washington, DC, NEXA Advisors.
Sarsfield, K. (2012), 'Business aircraft census 2012', *Flightglobal.com*, 22 October, accessed 30 March 2014 at www.flightglobal.com/news/articles/focus-business-aircraft-census-2012-377607/ 22/10/12.
Shomko, D. (2012), *Ethics in Business and Corporate Aviation*, London: Aurum Press.
Wild, J. and R. Raggi (2013), 'Expansion is on the flight path but will be patchy', in *The Financial Times Corporate Aviation Report*, 22 October, p. 1.
Wright, R. (2013), 'Competition rises in tough marketplace', in *The Financial Times Corporate Aviation Report*, 22 October, p. 2.

16. Looking at luxury: consuming luxury fashion in global cities

Louise Crewe and Amber Martin

INTRODUCTION

This chapter explores the luxury fashion industry,[1] an empirically important but theoretically neglected area of scholarship and one with a pronounced geography that requires scrutiny. In conceptual terms, our research lies in recent debates about global economic austerity (Pollin, 2005; Foster and Magdoff, 2009; McNally, 2009), the second Gilded Age (Short, 2013; Piketty, 2014), and the future of consumption under conditions of precarity and polarization. It is also one of the first studies to explore luxury fashion within broader geographical scholarship on retailing, consumption and space. Empirically, the chapter has three key foci. First, the chapter focuses on the remarkable resilience and growth of the luxury market in the wake of global recession and the slowdown in consumer spending, looking specifically at the dramatic geographical expansion of luxury retailers into emerging markets. With an estimated worth of US$263 billion in 2007, the luxury brand market increased by 31 per cent from 2004 to 2009 and is predicted to grow by 71 per cent in the next five years (Tynan et al., 2010). Second, the chapter explores the ways in which luxury fashion houses maintain aura and grow their markets whilst retaining brand value and signature under increasingly complex global conditions. We argue that the luxury fashion market is a clear illustration of the powers of aesthetic capitalism in the contemporary era (Gasparina, 2009) in which luxury is increasingly traded in symbolic terms rather than being a sector defined by high-skilled and artisanal craft production and by a fixed geographical manufacturing identity. Third, the chapter reveals how global luxury brands build and fix their value spatially, and addresses the role that location, labels, architecture and design play in the making of luxury markets. We argue that a key element of luxury retailers' competitive strength lies in their selectively located flagship stores that act as the spatial manifestation of aura, power and exclusivity. Flagship stores are global spaces of desire, places for the display and consumption of hyper-visible opulence and an important vehicle for the enhancement of brands' symbolic capital. The global flagship store is a spatial tactic employed to affirm the geographical presence of a brand and to endorse the auratic qualities of luxury in an expanding and highly competitive global market.

THE GROWTH OF LUXURY

In one of the longest and deepest periods of global recession, which has seen unemployment rise, GDP figures in the USA and Western Europe plummet, and retail stores and sales stagnate, luxury retailing and consumption has not been exempt from market vulnerability. In 2009, at the heart of the global financial crisis, the world's three largest luxury consumer markets (Europe, the United States and Japan) showed negative growth rates (Zhang and Kim, 2013) with a generalized reduction in demand for luxury goods from Western consumers (Roper et al., 2013). However, as Western economies began to emerge from the financial crisis and show continued signs of growth from 2011 onwards, the demand for luxury goods was fuelled once again (Kapferer, 2012). The rate of expansion has been driven by a variety of factors, including increasing disquiet over the economic, environmental and social impacts of throwaway fashion, a desire for more responsible investment purchasing, and renewed interest in the creative capacities of experiential retail spaces. In addition, the development of luxury fashion retailing can be viewed as symptomatic of the second Gilded Age (or second *belle époque*) in which the increased polarization of wealth, and the subsequent rise of the '1 per cent' (Piketty, 2014), has resulted in an increase in the number of high net worth individuals (HNWIs) with the emotional desire and economic capital for luxury brand consumption (Bourdieu, 1984; Capgemini, 2013). The luxury fashion market is thus significant not only in terms of its value but also in terms of its rate of growth, which 'has significantly outpaced that of other consumer goods categories over recent decades' (Fionda and Moore, 2009, p. 347) and is predicted to continue to rise. The UK luxury fashion sector, for example, is forecast to almost double in size from £6.6 billion to £12.2 billion between 2012 and 2017 (Ledbury Research and Walpole, 2013). At a finer level of granularity, a number of luxury fashion houses are continuing to grow in spite of the difficult economic climate – Burberry, for example, has seen global revenues triple (2006–13) and opened its largest Asian flagship store in April 2014, using store openings as a way of raising brand awareness among Chinese customers (Sharman and Robinson, 2014) and the French fashion house Yves Saint Laurent reported a 59 per cent increase in annual sales during 2013 (Butler, 2013).

One of the key reasons for the dramatic and sustained growth of the world's largest luxury fashion conglomerates (Hermès, Kering, Richemont and LVMH have, for example, all grown ten-fold in the past two decades) is geographic expansion in a range of emerging markets, but particularly in China. In 2012, China was the world's second largest market for luxury fashion (Zhan and He, 2012) and it is predicted that, in 2015, China will account for approximately 20 per cent (US$27 billion) of global sales in luxury goods (Atsmon et al., 2011). The appetite for luxury consumption is also apparent in the emerging economies of Brazil, Russia, India and China (the so-called BRIC economies) and more generally in Asia and the Middle East (Tynan et al., 2010; Shukla, 2012; McKinsey & Company, 2013) alongside increased demand from 'smaller emerging markets such as Malaysia, Egypt, Thailand and Turkey' (Shukla, 2012, p. 577). Increased personal wealth in emerging markets is fuelling an exponential rise in the demand for luxury items (Roper et al., 2013, p. 377). The resurgence in the acquisition activity of luxury brands by investment firms in Asia and the Middle East, and by luxury conglomerates such as LVMH and Richemont since 2011, is further testament to the buoyancy of the sector and to the apparent immunity of high-end branded products to the more deleterious effects of global crisis. This speaks to the continued polarization of wealth in the second Gilded Age with the increase of HNWI, and a growing middle class in emerging markets such as China, able to buy into the luxury fashion market (Zhan and He, 2012). It is this demand and growth in emerging markets that has maintained and developed the demand for luxury goods and enabled the continued expansion of the industry (Kiessling et al., 2009).

As the global mosaic of HNWI and luxury consumers shifts, luxury retailers are responding to growing demand for their products in emerging markets through dramatic international store expansion strategies. It is predicted that 85 per cent of all new luxury retail stores will be opening in emerging markets over the coming decade (Shukla, 2012, p. 576). Strong growth figures are particularly reported in China, which offers enormous opportunities for global luxury brands and is predicted to become the world's largest luxury market by the end of 2013, worth £16 billion (McKinsey & Company, 2013). The Chinese luxury sector continued to grow in spite of the global recession and is now the second largest luxury consumer market after Japan (Zhan and He, 2012, p. 1452). Women's luxury apparel, for example, experienced a 10 per cent increase in revenue between 2012 and 2013, valued at RMB6 billion in 2013 (Bain & Company, 2013). Rapid urbanization and growing wealth outside China's largest cities is driving the emergence of new geographic markets for luxury in China and global brands such as Louis Vuitton, Gucci, Coach and Burberry are all expanding into third-tier Chinese cities in order to take advantage of continued rising demand (Kapferer, 2012). It is predicted that smaller urban areas, such as Wuxi and Qingdao, will soon 'become large enough to justify the presence of stores catering to them' and 'by 2015, consumption in such cities will approach today's levels in Hangzhou and Nanjing – now two of China's most developed luxury-goods markets' (Atsmon et al., 2011, p. 4). Chinese customers are expected to buy in excess of 44 per cent of the world's luxury goods by 2020 (CLSA Asia-Pacific, in *China Daily*, 2011) and are key drivers in global luxury markets.

It has been argued that high GDP in emerging middle class markets has increased the demand for luxury goods (Kapferer, 2012). Particularly in China, middle class consumers are being targeted as key consumers for whom luxury products are seen as aspirational commodities, a key means through which to increase social status via their associations with affluent, cosmopolitan Western lifestyles given the relaxing of social mores that previously sanctioned excessive displays of wealth (Zhan and He, 2012, p. 1453; Zhang and Kim, 2013). The targeting of luxury beyond the spaces of the super-rich has been readily apparent in America, Europe and Asia as well as in emerging economies in the BRIC countries and the Middle East (Atwal and Williams, 2009; Kiessling et al., 2009) and has resulted in a trend of middle market consumers trading up and reaching out for products that meet their aspirational needs (Atwal and Williams, 2009, p. 339). Luxury brands are recognized globally as being icons signalling a certain level of achievement and success, and luxury fashion firms are key players in orchestrating this tiering of the luxury fashion market across space. In a sluggish world economy, both high-end and more affordable 'premium luxury' markets are growing apace. We are thus witnessing what has been termed the 'luxurification of society' (Silverstein and Fiske, 2003) or the 'democratization of luxury' (Atwal and Williams, 2009) as once-exclusive luxury goods have become increasingly accessible (Roper et al., 2013). Quite how luxury firms can continue to be exceptional in spite of the regularity of the brand and the rapidity of their global expansion and penetration is a key conceptual question to which we now turn.

THE GROWING COMPLEXITY OF THE GEOGRAPHIES OF LUXURY RETAIL

Within this complex tapestry of global retail investment and variegated consumption practice, other less discernible transformations in luxury markets are at work. In particular, there has been a notable global shift in the geographies of luxury production, with a number of fashion houses outsourcing production to offshore locations including China in the case of Burberry and Prada. The example of Gucci is instructive here. The Italian heritage of the brand was formerly secured and promoted by the 'Made in Italy' appellation. Since the 1990s production has been met through increasingly complex supply chains including the offshoring and outsourcing of production. This resulted in Gucci shifting its marketing message away from the primacy of the geographical origin of production towards the 'context of consumption' (Tokatli, 2013, p. 253) by appropriating 'the place image of Los Angeles (without actually being located there)' (p. 239). This was an adept move that enabled Gucci to maintain its luxury credentials and aura by blurring the lines between country of origin (or manufacture) and brand origin – which 'can be thought of as the country a brand is associated with by its target consumers regardless of where it is manufactured' (Shukla, 2011, p. 243). Such developments begin to obfuscate the boundary lines between luxury and mass markets and problematize the definition of luxury products that traditionally created and maintained exclusivity and value through transparent 'Made in' labelling, craft production, quality and scarcity.

There are also tensions between the continued expansion and growth of luxury retailers and the premise of luxury brands being exclusive and rare (Kapferer, 2012). In order to construct themselves as exceptional and exclusive, whilst simultaneously catering for the demands of more inclusive and larger markets, luxury firms are conjoining the creative and commercial elements of their business and are emphasizing the symbolic and immaterial qualities of their brand. Brands become repositories of meaning, a means of conveying distinction and value (Simmel, 1957; (Bourdieu, 1984). Drawing on devices and techniques that suggest metonymy, luxury labels have created an economy of qualities whereby a finish, logotype or print evokes the essence of the brand. By employing the cult of the creative designer, liaising with celebrity architects and artists and constructing themselves as art, luxury brands have successfully maintained their creative and distinctive aura whilst expanding geographically by de-emphasizing country of origin, a skilful obfuscation of the places and means of production in what amounts to a strategy of super-commodity fetishism. Here we see how the luxury strategy is increasingly one of social and symbolic constructionism. As Kapferer (2012, p. 452) argues, the 'rarity of ingredients or craft has been replaced by qualitative rarity. . . Today, brands in the luxury sector are actually selling symbolic and magic power to the masses'. Central to the evolution of the luxury market are the ways in which retailers actively put geography to work in their creation of value. Key to this business strategy is the role of the flagship store that stands as a highly prominent spatial manifestation of the brand. One way of achieving a strong geographical presence, in the wake of increased global markets and complex geographical expansion, is through the symbolic and authoritative presence of the flagship store. It is to this that we now turn.

FLAGSHIP STORES: SCALING FASHION'S LUXURY SPACES

The retailscape of the twenty-first century is suffused with a number of innovative, coalescent elements that compete for consumption. The flagship store acts as the material expression of the brand and offers a place of seduction and desire. For the retailer, a flagship store serves to showcase the entire brand story to the consumer under one roof and to make use of all of the tools available to highlight the brand statement and philosophy. The luxury flagship store's origins can be traced back to Paris's Couture and Ready to Wear Ateliers, which were multilevel buildings (consisting of offices, workshops and a shop on the ground floor). These 'Maisons de Mode' acted as the creative hearts and brains of the brands (Barreneche, 2008) in which collections would be both produced and consumed (Tungate, 2008; Nobbs et al., 2012). As the number of luxury fashion producers increased, this format spread from Paris to the cities of London, New York and Milan (Nobbs et al., 2012).

In the last decade, as this store format has trickled down to the middle and mass market, luxury brands have evolved to cater for today's ever-discerning customer (Tungate, 2008) by creating differentiated branded experience (Nobbs et al., 2012). This has ensued the development of the 'uber' or 'mega-flagship', in which key dimensions of stores are enhanced to become larger, improved and more memorable (Nobbs et al., 2012). Examples of these top-level flagship stores include Louis Vuitton's 'Global Maisons' and Prada's 'Epicentre' stores (Passariello and Dodes, 2007). 'These stores are characterized by their large scale, cutting edge architecture, offering of cultural events' (Nobbs et al., 2012, p. 923) and, increasingly, a technologically mediated spectacular consumption experience.

Situated in prestigious retail sites within global cities,[2] flagship stores represent a key means through which luxury fashion houses internationalize (Fernie et al., 1997). Their location in luxury enclaves serves to increase exclusivity and prestige (Doherty and Moore, 2007); these luxurious streets of style 'form communities of affluence which appear to support and feed-off each other in terms of their sense of exclusivity and style' (Nobbs et al., 2012, p. 931). These clusters of luxury stores are quite literally 'economies of icons' (Sternberg, 1999) where brands are enshrined in exclusive spaces, thus allowing 'the brand to re-enforce its image communication through establishing a physical presence in a prestige shopping location' (Jackson, 2004, p. 177).

The flagship store is a key (and much neglected) site for the constitution and representation of contemporary consumption. A magical space of possibility where culture and commerce merge and meld, the flagship store is the physical manifestation of the soul and signature of the brand. Flagship stores act as powerful spatial landscapes that set the stage in the contemporary city for the performance of everyday life. They serve to develop the global reputation and presence of the brand, to leverage brand status and awareness and act a means of communicating and enhancing the image and personality of a luxury retailer's brand identity (Nobbs et al., 2012, p. 921). Once placed in its flagship setting, a brand draws meaning from architectural form. These forms are indexical symbols of the world of art. The flagship brand store becomes an 'autonomous presence' (Habraken and Teicher, 1998, p. 233) contributing to the brand's persona 'at least as much through connotation as it does explicitly' (Kozinets et al., 2002, p. 28).

The flagship store is often referred to as 'the brand cathedral' as it acts as an emblem of visible and tangible power of the label, paralleling the ways in which the cathedral is seen as the geographical emblem of the power of the church (Cervellon and Coudriet, 2013, p. 874). Flagships clearly demonstrate power and prestige though their status as visitor attractions that elevate and enhance the eminence of the city in which they are located (Nobbs et al., 2012) and become tourist destinations that are commonly frequented by 'non-traditional customers of the brand' (2012, p. 923). Flagship stores are founded on significant financial investment and are considered crucial to a brand's marketing communication process and reputation. The intention of the stores is maintain and embody the image and symbolic capital of the brand, rather than generate profit directly (Kozinets, 2002; Doherty and Moore, 2007), with stores commonly not being required to show a typical return on investment. Flagship stores are used as a key entry-to-market strategy in emerging markets in the early stages of luxury retailers' business development and as a method of gaining direct entry into, and demonstrating commitment to, foreign markets (Doherty and Moore, 2007).

Flagships are intended to impose themselves powerfully onto the exclusive streetscapes in which they are located. Brand names are embossed on store fronts, buildings and canopies to tangibly and visibly superimpose the brand on the urban landscape. Similarly, flags displaying company logos hang authoritatively from the store fronts, a stark mechanism through which to quite literally brand the city. The brand flags allude to the notion of expedition and colonization, suggesting that the brand has physically and metaphorically marked its territorial claim on the exclusive retailscape. The territorial claim of the flagship is reproduced further through the imposing nature of the size of flagship stores that are intentionally larger than their non-flagship counterparts to serve 'as a physical manifestation of their premier status' (Nobbs et al., 2012, p. 932). The inclusion of 'unproductive space' (2012, p. 932) in prime retail locations emphasizes the excessive wealth and prestige of the brand and therefore their luxurious and exclusive characteristics.

The art of designing a luxury store is an exercise in communication, in making concrete the imaginative energy and creative power of artists and designers. In part this is achieved through architecture and visual design and display that both embodies and builds the brand. Exclusivity, exceptionality and prestige are reflected through the positioning of flagship stores 'in historic structures or landmark buildings' (2012, p. 926) and through the blurring of art, architecture and retail in elaborate store spaces 'which themselves are conceived of as works of art' (Dion and Arnould, 2011, p. 511), increasingly designed by star artists and architects commissioned by luxury retailers. Notable examples include Renzo Piano who designed for Hermès, Frank Gehry for Louis Vuitton and John Pawson for Calvin Klein (Doherty and Moore, 2007, p. 280; Crewe, 2010).

Casting one's eye over the dramatic fashionscapes of world cities reveals the profound ways through which fashion retail and architectural design are combining to aestheticize, project and (re)present the city, drawing particularly on materiality, colour and sensory geographies in their development of a global luxury strategy. The architect Rem Koolhaas for Prada has argued, somewhat controversially, that global expansion via spectacular flagship stores can be employed as a means of stretching, bending and perhaps permanently redefining the brand. When the flagship is recast as an Epicentre store it can become a device that renews rather than dilutes the brand by counteracting and destabilizing any received notion of what Prada is, does, or will become. The Epicentre store acts as a conceptual window, a space of anticipation, spectacle and desire. Next-generation flagship stores have the potential to be simultaneously perceptual and physical, symbolic and material. Both have the capacity to communicate in non-dialogic ways. Progressive architecture and iconic fashion houses combine to shape the metropolis socially and spatially. As cities become adorned with fashion signs, symbols and logos, retail architecture is rebranding urban space (Quinn, 2002, p. 29); the spectral nature of fashion is exposed through the exterior built form and interior retail spaces. Architecture and fashion have converged to aestheticize urban space via dazzling displays, staged performances, fantastic spectacles and dramatized city skylines. Flagship fashion stores iconicize the city not merely through the surface features of glamour and glitz but also via their shared understanding of the affective power of space, form, materiality and colour (Antonelli, 2007).

Chanel is an interesting example of how the elements of colour, material and light fuse to capture the essence of the brand and quite literally project into the cityscape. Coco Chanel long recognized the affective and symbolic affordances enshrined in colour. From the 'little black dress' that has become a fashion classic, to her use of the black sans serif logotype throughout her store and product designs, Chanel understood well the timeless aesthetic appeal of achromaticity. The Chanel store in New York, designed by Peter Marino, reveals the sensual and captivating power of colour and light. Whilst the exterior resembles a white cube, the interior surfaces are uniformly finished in high black gloss and have hundreds of tiny backlit perforations randomly cut into the surface. The effect is magical and enchanting. The visual collision of white light and black gloss is a tantalizing example of the achromatic chic and monochromatic materiality that has characterized both fashion and architecture for many decades (Ojeda and McCown, 2004). In Marino's Chanel store in Tokyo, Japan, the interplay between white and black, light and dark is again revealed to dramatic effect through the use of technology, colour and textures. Through a fusion of ceramics, glass and iron, the store reveals an exterior surface that is illuminated by 700 000 LED backlights. Built as part of the building's skin, dynamic videoscreens enable luxury brand building in its most literal form (Crewe, 2010). Chanel can project an infinite number of corporate images and texts onto the streets of the city. This dramatic use of mediatecture through cladding buildings with visual, branded screens changes not only the aesthetic of the city but also the way in which buildings occupy space. The building itself, through new technological architectures and sensory stimuli, becomes a representational feature of both architect and brand. It is the materialization of luxury.

Shop Windows

Alongside architectural prowess, another key instrument through which retailers communicate strategy is through their store fronts and store window displays (Sen et al., 2002, p. 277). Store windows communicate between the interior and the exterior of the store and bring dreams and fantasy into tangible view. The signs, symbols and products displayed in flagship store windows 'play a key role in defining global fashion culture and in charting its discursive space' (Shinkle, 2008, p. 1). As Bingham (2005) argues, 'glass is a solid liquid, a magical paradox, which links the real world outside the world of luxury inside. The displays behind it are magical too – far more than its individual parts'. The symbolic power of the window design can be spectacular and is designed 'to reflect the essence of what the store represents, its product range and symbolism' (Pereira et al., 2010, p. 2). Flagship stores have highly design-focused and luxurious shop fronts that intentionally exude style, sophistication, exclusivity and luxury. This is exemplified by the flagship luxury stores on London's Bond Street where the colours white, black and gold are a key motif and recurrent theme. Black and white have long been key signifiers of fashion, taste and style and gold is iconic of luxury, expense and prestige. Store fronts and windows exude elegance and exclusivity through the communication of the store interior and brand essence to the public space of the street outside. The Louis Vuitton store in Roppongi, Tokyo, for example, is constructed of more than 20 000 glass tubes that are expertly arranged to form a vast pixelated screen that transmits changing images of the Louis Vuitton logo and motifs. Similarly, the Louis Vuitton store front in Seoul is blanketed by a translucent mesh over its façade with the intention of mimicking the fabric that is used to cover the iconic Louis Vuitton travel chests (Curtis and Watson, 2007). It is through such mechanisms that shop fronts become melded as part of the architectural and artistic structure of the flagship store. Seducing the street outside through imagery and advertising, the glass shop window brings dreams and fantasy into reality by linking the 'real' world outside to the world of luxury within (Bingham, 2005). The emotional and affective charge of the window design can be engaging and powerful as 'the pane of glass separating the object of desire from the shopper forms an imaginary screen not unlike the mirror, a surface for receiving and reflecting' (Oswald, 1992, p. 52), tantalizing the consumer with the possibilities of purchase. Window displays produce a sense that there is something more, some more intense experience or a wider horizon to be found. Store windows act as a powerful spatial landscape that sets the stage in the contemporary city for the performance of everyday life, acting as theatres of signs and symbols in which representation is not the opposite of materiality but rather its alter ego, a space that both constitutes and reflects commerce and culture, transaction and imagination.

Gateways to Luxury

The luxury flagship window expresses worlds of excitement, luxury and indulgence that are intended to be highly visual and inviting. However, in contrast to the high visibility of the shop windows, and the products displayed there, is the relative (in)visibility of price tags. Hiding the prices of products in shop windows is a common tactic employed by luxury retailers to invest both the products and the associated brand with an air of exclusivity and expense. This offers entry into a secret world of privilege and exclusivity for those for whom the price of goods for sale is not a concern. The invisibility of product prices in shop windows is a way of simultaneously flaunting the products for sale in store whilst providing an invisible barrier to entry. It elevates the products further into a beyond-market status, valuable, priceless.

Similarly, the selective and exclusive entry into luxury retail spaces is physically, as well as symbolically, significant. Although flagship stores are located on the public space of the street, they are highly securitized, patrolled and surveilled. This is no more readily apparent than by the doorways of luxury retailers, which are always closed, symbolizing the gateway 'between the mundane (the street) and the sacred (the brand universe)' (Cervellon and Coudriet, 2013, p. 875). These gateways are guarded by uniformed door staff who play a shifting role between doorman and bouncer: welcoming 'desirable' customers into the store, whilst simultaneously intimidating others by their mere presence. And yet this is a kind of servile labour that one might not associate with luxury: their job is long and boring and consists largely of standing, waiting, doing nothing, looking the part. Like the prominent goods with no price tags, here are highly visible workers who remain largely invisible to consumers – door staff, butlers, the servants to the retail brand.

If granted access into the store through its physical and symbolic gateways, the notions of exclusivity, rarity and expense are emphasized further throughout the retailing tactics employed to encase the products for sale with auratic-like qualities. This is achieved through the display of products as 'treasures' displayed behind glass, in sleek cabinets, placed delicately on pedestals and lit directly – echoing the artistic tradition of exhibition space. Just as in a museum, distance is maintained between the viewer and the object (Dion and Arnould, 2011) – touching is strictly forbidden. This creation of spaces of display is an additional means by which luxury brands legitimize their power (Cervellon and Coudriet, 2013, p. 880) and emphasizes the sacredness of the products and ultimately the brand. The rarefied positioning of goods is further endorsed by sales assistants that act as brand ambassadors (2013, p. 869) and touch products with precaution and care as part of a carefully choreographed 'selling ceremony' (2013, p. 869), which is designed to confer the superiority of the sales person over the customer, the product over its price. It is through such rituals of careful commodity handling that a form of symbolic dominance is intended to be enacted over the consumer, making efforts to invoke awe in both the product and shopping experience.

Experiential Retailing

The presentation of products through mechanisms typically derived from museums and art galleries, and the sacralization rituals performed by sales assistants, not only assists in maintaining exclusivity and sacredness, but also creates a consumption experience that is ultimately just as important as the creation and display of rare and expertly crafted products. Consumers in flagship stores can expect to be met with a consumption experience intended to epitomize the company ethos and immerse the consumer in a complete branded experience (Kozinets et al., 2002), thus imprinting the consumer experience, not just the product (Fionda and Moore, 2009). It is in this way that luxury retailers expand the notion of luxury as being above and beyond ownership to encompass the sensory experience of luxury (Roper et al., 2013). This is no more apparent than in the flagship store where the highest levels of investment correlate with the highest levels of experience, interaction, performance and spectacle (Fionda and Moore, 2009).

At one level, this translates into a form of experiential service where the consumer is both indulged and engaged with the consumption experience. For example, in Cartier's Paris flagship store, customers can design their own made-to-order perfumes (Nobbs et al., 2012) and in the Louis Vuitton flagship in Tokyo customers can join a private club that allows them to browse products whilst sipping champagne (Bingham, 2005). At another level, this manifests itself as a form of immersive experiential retailing that extends to create an all-encompassing brand lifestyle experience far beyond the mere purchase of fashion items. For example, the flagship Chanel store in Tokyo has its own restaurant that is intended to evoke the elegance of Coco Chanel, and the spirit of the Chanel brand, through the delicate white, cream and black colour palette and the use of Chanel's iconic tweed for its interior furnishings (Atwal and Williams, 2009). At the Dolce & Gabbana flagship store in London there is a traditional Sicilian barber where male customers can have a shave and a haircut, whilst Kenzo in Paris offers a massage service (Nobbs et al., 2012). These offerings are designed to create luxury, hedonistic, pleasurable, multisensory consumption experiences that are 'affect-rich' (Dion and Arnould, 2011, p. 503) and enable consumers to 'interact with and touch the brand' (Nobbs et al., 2012, p. 926).

That luxury shopping is spectacle, theatre, performance and performativity is nowhere more apparent than in the Burberry flagship store at 121 Regent Street, London. Opened in 2012, this is a space that is intentionally experiential: it brings the runway to the store, the catwalk to the shop. The store is tactile, visual, aural, immersive. The consumer is drawn in and whirled around along grand staircases where the wooden interior design co-exists and coalesces with the most innovative virtual worlds, products reflected and reflecting, interiors melting into exteriors. Mannequins bear gifts, products are packages, an opulent covering for the unknown commodities inside that lie hidden, enticing, waiting to be discovered. Boxes and bags, draped in ribbon and rained upon by cascading gold leaves, are redolent, enticing. The sheer scale of the store and its architecture reveal its place within this city of signs. Crucially, Burberry has transformed and enlivened conventional in-store structures and strategies through its use of digital technologies. In addition to hosting live concerts by the Kaiser Chiefs and Jake Bugg, for example, at its flagship Regent Street store, Burberry developed a 'Runway to Reality' in-store event that included the live-streaming of its Spring Womenswear Collection across a number of digital platforms. Burberry Regent Street has used disruptive digital technologies to create an immersive multisensory environment that bridges the online–offline experience: RFID (radio-frequency identification) chips inside garments bring product information to consumers in innovative ways as the chips trigger RFID-enabled mirrors to become digital screens, which in turn display information about production, craft and catwalk footage. At set times a choreographed 'digital rain storm' is synchronized across the store's 50 speakers and 100 digital screens, starting as a gentle shower and building into a downpour with a climactic thunder crack shown on every screen and echoing in every space, simulating the iconic London downpour and exposing customers to an immersive sensory brand experience.

CONCLUSIONS

Collectively, these examples of experiential tactics employed by flagship stores demonstrate the importance of the physical and symbolic space of the store to luxury retailing. In flagship stores, the expertly designed and architecturally impressive shop spaces, sacralized product displays, carefully manufactured service encounters, and consumption experiences, collectively create a powerful stage for the creation and reproduction of the brand. Flagship store designs are as much about creation and art as about sales and persuasion; their effect on the brand and the city is one of communication as much as it is one of commerce.

Flagships, we argue, are a symbolically efficacious form of branding. These physical exemplars of brand power expressed and embedded in the flagship store are difficult to replicate on a virtual platform online (Jackson, 2009; Okonkwo, 2009). Therefore, despite the rise of extensive online retailing, the flagship store remains as an important medium for communicating brand messages, asserting retailer dominance and maintaining 'auratic power' (Cervellon and Coudriet, 2013, p. 869) in an ever-increasing virtual retailscape. In contrast to many new economy prophets that imagined physical stores to become a dead weight, the global flagship stores discussed here reveal the enduring power and potency of the material retail form, created as seamless steps into a new retail era that has the capacity to recast luxury in line with shifting consumption practices. The maintenance of their auratic power is increasingly important as luxury retailers continue to compete in global markets, particularly in emerging economies and in new luxury markets where the symbolic resonance of the brand may not yet be well established. It is in these conditions of global expansion and competition that luxury retailers must maintain their brand dominance and enchantment.

In this chapter we have argued that brand aura, value and signature are fixed and manifested spatially by flagship stores – which act as physical beacons, embassies and embodiments of brands, in which luxury is increasingly defined and appropriated in symbolic terms rather than merely being a sector defined by high-skilled and artisanal craft production. We have argued that the flagship store is a tactic employed to give geographical presence and a physical sense of exclusion, power and rarity in an increasingly expanding and inclusive luxury market. At the micro-scale the flagship asserts brand dominance and symbolic sacralized power through exclusive shop spaces, selling rituals and experiential retailing. At the macro-scale, flagships act as global markers of brand strength and commitment to new markets through asserting their dominance in an increasingly globalized luxury marketplace. The flagship store can be seen therefore as a tactic employed by luxury retailers to give a strong geographical presence and a physical sense of exclusion, power and rarity in an increasingly expanding and inclusive luxury market. The flagship store acts as the material expression of the brand that offers a place of seduction and desire. Store spaces act as powerful signifiers, windows on fantastic worlds. Flagship stores function as microeconomies of icons where brands are enshrined in exclusive spaces.

More broadly we suggest that a close interrogation of luxury flagship stores opens up a space to explore critically how new representational worlds are brought into being and offer new ways to understand how creative activity can be rooted in (and reflective of) broader social, economic and cultural concerns. Luxury fashion and its complex geographies act as an early warning system for major cultural transformations. This requires that we need to better understand the means and mechanisms for sustaining luxury brands. This in turn depends on exploring more critical approaches that engage and question the desires and demands for commodities with a high level of intangible value so that we can more fully understand their affective capacities and emotional reach.

NOTES

1. We define luxury fashion as commodities that are intended to be worn and displayed on the body: that is, clothing and accessories (including shoes, bags, watches and jewellery), affiliated with famous designers and fashion houses and that place an emphasis on uniqueness, exclusivity and the high desirability of their products. Key examples include Prada, Gucci, Hermès, Chanel, Burberry, Fendi and Cartier.
2. Bond Street/Sloane Street (London); 5th Avenue/Madison Avenue (New York); Rue du Faubourg Saint-Honoré/Avenue Montaigne (Paris's Triangle d'Or); Via Manzoni/Via Montenapoleone (Milan); Harumi Dori (Ginza)/Aoyama Dori (Tokyo); Queen Street/Canton Road (Hong Kong).

REFERENCES

Antonelli, P. (2007), 'Bias-cut architecture', in M. Gabellini (ed), *Architecture of the Interior*, New York: Rizzoli, pp. 60–70.

Atsmon, Y., V. Dixit and C. Wu (2011), 'Tapping China's luxury-goods market', *McKinsey & Company* accessed 21 August 2015 at http://www.mckinsey.com/insights/marketing_sales/tapping_chinas_luxury-goods_market.

Atwal, G. and A. Williams (2009), 'Luxury brand marketing – the experience is everything!', *Journal of Brand Management*, **16**, 338–46. doi:10.1057/bm.2008.48.

Bain & Company (2013), *Luxury Goods Worldwide Market Study 2013*, Milan: Bain & Co.

Barreneche, R.A. (2008), *New Retail*, London: Phaidon Press.

Bingham, N. (2005), *The New Boutique*, London: Merrell.

Bourdieu, P. (1984), *Distinction*, London: Routledge.

Butler, S. (2013), 'Chinese demand for luxury goods boosts Kering', *The Guardian*, 25 July.

Capgemini (2013), *World Wealth Report*, London: Capgemini.

Cervellon, M.C. and R. Coudriet (2013), 'Brand social power in luxury retail: manifestations of brand dominance over clients in the store', *International Journal of Retail & Distribution Management*, **41** (11/12), 869–84.

China Daily (2011), 'Chinese to buy 44% of luxury goods by 2012: CLSA', accessed 21 August 2015 at http://www.chinadaily.com.cn/hkedition/2011-02/11/content_11980217.htm.

Crewe, L. (2010), 'Wear: where. The convergent geographies of fashion and architecture', *Environment and Planning A*, **42** (9), 2093–108.

Curtis, E. and H. Watson (2007), *Fashion Retail*, Chichester, UK: Wiley.

Dion, D. and E. Arnould (2011), 'Retail luxury strategy: assembling charisma through art and magic', *Journal of Retailing*, **87** (4), 502–20.

Doherty, C. and A. Moore (2007), *The International Flagship Stores of Luxury Fashion Retailers*, Oxford: Butterworth Heinemann.

Fernie, J., C. Moore and A. Lawrie et al. (1997), 'The internationalization of the high fashion brand: the case of central London', *Journal of Product & Brand Management*, **6** (3), 151–62.

Fionda, A.M. and C.M. Moore (2009), 'The anatomy of the luxury fashion brand', *Journal of Brand Management*, **16** (5), 347–63.

Foster, J.B. and F. Magdoff (2009), *The Great Financial Crisis: Causes and Consequences*, New York: NYU Press.
Gasparina, J. (2009), '33 colours', in J. Gasparina et al., *Louis Vuitton: Art, Fashion & Architecture*, New York: Rizzoli, pp. 40–63.
Habraken, N.J. and J. Teicher (eds) (1998), *The Structure of the Ordinary: Form and Control in the Built Environment*, Cambridge, MA: MIT Press.
Jackson, T. (2004), 'A contemporary analysis of global luxury brands', in M. Bruce, C. Moore and G. Birtwistle (eds), *International Retail Marketing: A Case Study Approach*, Oxford, UK: Elsevier Butterworth Heinemann, pp. 155–69.
Jackson, T. (2009), 'Virtual flagships', in T. Kent and R. Brown (eds), *Flagship Marketing*, London: Routledge.
Kapferer, J. (2012), 'Abundant rarity: the key to luxury growth', *Business Horizons*, **55** (5), 453–62.
Kiessling, G., C. Balekjian and A. Oehmichen (2009), 'What credit crunch. More luxury for new money: European rising stars & established markets', *Journal of Retail & Leisure Property*, **8** (1), 3–23.
Kozinets, R.V., J.F. Sherry and B. DeBerry-Spence et al. (2002), 'Themed flagship brand stores in the new millennium', *Journal of Retailing*, **78** (1), 17–29.
Ledbury Research and Walpole (2013), *The UK Luxury Benchmark Report*, London: Ledbury Research.
McKinsey & Company (2013), *Understanding China's Growing Love for Luxury*, London: McKinsey & Company.
McNally, D. (2009), 'From financial crisis to world-slump: accumulation, financialisation, and the global slowdown', *Historical Materialism*, **17** (2), 35–83.
Nobbs, K., C.M. Moore and M. Sheridan (2012), 'The flagship format within the luxury fashion market', *International Journal of Retail and Distribution Management*, **40** (12), 920–34.
Ojeda, O. and J. McCown (2004), *Colors: Architecture in Detail*, Beverly, MA: Rockport Publishers.
Okonkwo, U. (2009), 'Sustaining the luxury brand on the internet', *Journal of Brand Management*, **6** (5), 302–11.
Oswald, L. (1992), 'The place and space of consumption in a material world', *Design Issues*, **12** (1), 48–62.
Passariello, C. and R. Dodes (2007), 'Art in fashion: luxury boutiques dress up as galleries', *Wall Street Journal*, 16 February, 8.
Pereira, M., S. Azevedo and V. Bernardo et al. (2010), 'The effect of visual merchandising on fashion stores in shopping centres', paper at the Fifth International Textile, Clothing and Design Conference – Magic World of Textiles.
Piketty, T. (2014), *Capital in the Twenty-First Century* [French edition published 2013 as *Le capital au XXI siècle*, Editions du Seuil], Cambridge, MA: Belknap Press of Harvard University Press.
Pollin, R. (2005), *Contours of Descent: US Economic Fractures and the Landscape of Global Austerity*, London/New York: Verso.
Quinn, B. (2002), *The Fashion of Architecture*, Oxford, UK: Berg.
Roper, S., R. Caruana and D. Medway et al. (2013), 'Constructing luxury brands: exploring the role of consumer discourse', *European Journal of Marketing*, **47** (3/4), 375–400.
Sen, S., L. Block and S. Chandran (2002), 'Window displays and consumer shopping decisions', *Journal of Retailing and Consumer Services*, **9** (5), 277–90.
Sharman, A. and D. Robinson (2014), 'Burberry rises on Chinese sales', *Financial Times*, 16 April.
Shinkle, E. (2008), *Fashion as Photograph*, London: I.B. Tauris.
Short, J.R. (2013), 'Economic wealth and political power in the second Gilded Age', in I. Hay (ed.), *Geographies of the Super-Rich*, Cheltenham, UK and Northampton, MA, USA: Edward Elgar Publishing, pp. 26–42.
Shukla, P. (2011), 'Impact of interpersonal influences, brand origin and brand image on

luxury purchase intentions: measuring interfunctional interactions and a cross-national comparison', *Journal of World Business*, **46** (2), 242–52.
Shukla, P. (2012), 'The influence of value perceptions on luxury purchase intentions in developed and emerging markets', *International Marketing Review*, **29** (6), 574–96.
Silverstein, M. and N. Fiske (2003), *Trading Up: The New American Luxury*, New York: Penguin.
Simmel, G. (1957), 'Fashion', *American Journal of Sociology*, **62** (6), 541–58.
Sternberg, E. (1999), *The Economy of Icons: How Business Manufactures Meaning*, Westport, CT: Praeger.
Tokatli, N. (2013), 'Doing a Gucci: the transformation of an Italian fashion firm into a global powerhouse in a "Los Angeles-izing" world', *Journal of Economic Geography*, **13** (2), 239–55.
Tungate, M. (2008), *Fashion Brands: Branding Style from Armani to Zara*, London: Kogan Page.
Tynan, C., S. McKechnie and C. Chhuon (2010), 'Co-creating value for luxury brands', *Journal of Business Research*, **63** (11), 1156–63.
Zhan, L. and Y. He (2012), 'Understanding luxury consumption in China: consumer perceptions of best-known brands', *Journal of Business Research*, **65** (10), 1452–60.
Zhang, B. and J.H. Kim (2013), 'Luxury fashion consumption in China: factors affecting attitude and purchase intent', *Journal of Retailing and Consumer Services*, **20** (1), 68–79.

17. The luxury of nature: the environmental consequences of super-rich lives

Aidan Davison

> 'The thing is,' Walter said, 'the land is disappearing so fast it's hopeless to wait for governments to do conservation. The problem with governments is that they're elected by majorities that don't give a shit about biodiversity. Whereas billionaires do tend to care. They've got a stake in keeping the planet not entirely fucked, because they and their heirs are going to be the ones with enough money to enjoy the planet'.
>
> (Franzen, 2010, p. 225)

SUSTAINABILITY AND THE QUESTION OF INDIVIDUAL WEALTH

The argument that a combination of population growth, resource consumption and environmental degradation is making current forms of development unsustainable is now influential. While central to environmental discourse, management and activism, conclusions drawn from this argument vary widely.

The boom after World War II saw global population grow by almost half between 1950 and 1970 and resource consumption accelerate. Most population growth occurred in poor regions emerging from long histories of European colonial oppression; regions tellingly classed as 'underdeveloped' by United States President Harry Truman in 1949 (Sachs, 1992). In contrast, consumption of a dazzling array of mass-produced goods – automobiles, refrigerators, televisions and the rest – was concentrated in an expanding middle class, predominantly in Europe and European settler societies.

In this context, environmentalists in high-consumption societies[1] set the early terms of debate about sustainability, warning of the perils of population growth and economic growth on a finite planet (Ehrlich, 1968; Meadows et al., 1972). In response, they advocated frugality, simplicity and self-sufficiency (Schumacher, 1973). By the late 1980s, however, talk of steady-state societies was countered by a case for the ecological credentials of modern development, an argument powerfully put in the 1987 report of the Brundtland Commission, *Our Common Future* (WCED,

1987). The concept of sustainable development broadcast by this report has been used to argue that, far from being the problem, economic growth is a precondition for environmental protection (Pearce et al., 1989; Lomborg, 2001).[2]

Today, the environmental movement draws on diverse political interests, bodies of knowledge and practical contexts. Arguments about economic wealth and sustainability stretch along a continuum from claims they are antithetical to claims they are coincident. The United Nations Environment Programme (UNEP) concludes that

> [t]he relationship between income and environmental change is ambiguous. On the one hand, high income tends to coincide with high consumption levels, leading to further environmental degradation. On the other, an increase in income can also coincide with lower population levels, an increasing appreciation of a clean environment and rapid technological change. These trends may lead to a decrease in environmental pressure as incomes rise. (UNEP, 2012, pp. 427–8)

This ambiguity is at the centre of debate about the relationship between modernization – as expressed in economic growth, technological innovation, scientific management and liberal democracy – and environmental sustainability (Redclift, 1987; Hajer, 1997; Mol and Sonnenfeld, 2000). There is consensus that poverty is incompatible with sustainability, but not about the causes of poverty or the means of its overcoming. And disagreement is profound about whether 'developed' nations have cut a path to social progress that 'developing' nations can follow, or whether colonial legacies and capitalist dynamics in the wealth of rich nations are inseparable from poverty creation elsewhere.

The question of what is physically necessary for a human life is notoriously easier than the question of what might be sufficient for a richly sustaining one. Often framed in terms of needs versus wants, disagreement about the sustainability of modern development effectively turns on ideas of sustainable affluence, although it is rarely presented this way. It might be presumed that assertions about the environmental merits or otherwise of economic growth would translate into assertions about personal wealth, but this is not so. Despite an escalation in financial inequality in many countries since the mid-1970s – enabling the amassing of private fortunes to rival any ancient dynasty (Atkinson et al., 2011; Alvaredo et al., 2013; Piketty, 2014) – little attention has been given by environmental scholars to individual wealth. This neglect is surprising given interest in the relative contribution of population, affluence and technology in environmental impact[3] and emphasis on principles of equity in discourses of sustainability. It is partly explained by crude contrasts between developed and developing nations that overlook both disparity within high-consumption nations and the globalized agency of wealthy elites. While the neoliberal political economy[4] underpinning recent growth of extreme private wealth is regular fare in environmental debate, the implications of this wealth are often masked beneath abstractions about development, sustainability and equity.

As indicated by the present book, study of the social, political and cultural implications of super-rich lives[5] has gained momentum (Freeland, 2012; Hay, 2013a; Birtchnell and Caletrío, 2014). Several have speculated about the environmental implications of these lives, mainly in relation to conspicuous consumption and what the economist Robert Frank (1999) dubs 'luxury fever', a contagion spreading from the richest consumers leading to increased consumption by all. For Mike Davis and Daniel Monk (2007, p. xv), wealthy lives enflame desire for 'infinite consumption' in ways 'clearly incompatible with the ecological and moral survival of humanity'. In more measured tones, Hay and Muller (2012, p. 83) argue that 'it is almost self-evident that the impact of "luxury fever" – with its attendant excessive consumption – has significant ecological implications'. Yet there has been little systematic treatment of questions of sustainability in light of the resurgence of extreme wealth. In this chapter, I make a preliminary start on this task, drawing together interest in the super-rich with interest in the relationship between environment and development. In so doing, I take seriously the reality, reflected in the epigraph to this chapter from Jonathan Franzen's novel, *Freedom*, that the super-rich are amongst the most powerful actors shaping environmental futures.

For reasons of space I focus on high-consumption societies, and particularly on the USA, although issues related to low-consumption societies are vital (Myers and Kent, 2003). I first sketch out the rise of the super-rich and assess their role in conspicuous consumption and resource use. Convinced of the need to consider the amassing as well as the spending of super-rich fortunes, I then explore the role of the super-rich in aligning environmental concern with neoliberal political economy. I argue this alignment has privileged private action in the market, deflected environmental critique of capitalism and undermined political responses to a worsening and collective environmental predicament.

THE RESURGENCE OF CONSPICUOUS WEALTH

In *Luxury Fever*, Frank (1999, p. 3) argues that

> the spending of the superrich, though sharply higher than in decades past, still constitutes just a small fraction of total spending. Yet their purchases are far

> more significant than might appear, for they have been the leading edge of pervasive changes in the spending patterns of middle- and even low-income families. The runaway spending at the top has been a virus, one that's spawned a luxury fever that, to one degree or another, has all of us in its grip.

Frank takes his cue from Thorsten Veblen ([1899] 1925), whose treatise, *The Theory of the Leisure Class*, introduced the concepts of conspicuous leisure and conspicuous consumption. Like that delivered a few decades earlier by Karl Marx, Veblen's critique was motivated by unequal economic growth in Europe and North America associated with industrialization. Coming after generations of blood-soaked democratic struggle to weaken aristocratic privilege, this Gilded Age saw the re-emergence of elites whose opulence is hard to overstate (Short, 2013). Veblen's critique was that the power of individual wealth lies not just in the accumulation of money, and the political influence this buys, but also in the display of conspicuous lives emulated by the masses. Arguing that what made rich lives conspicuous was 'the element of waste', Veblen ([1899] 1925, p. 85) charted a historical shift from conspicuous leisure – 'a waste of time and effort' – to conspicuous consumption – 'a waste of goods'.

The Gilded Age of industrial capitalism came to an abrupt end with World War I and the subsequent turmoil of the Great Depression and a second global war. The wiping away of vast fortunes in the market was followed by the redistributive ambition of Keynesian states, who oversaw relatively low and stable high-end incomes (Atkinson et al., 2011; Alvaredo et al., 2013). The decisive return of inequality in the late 1970s in some but not all high-consumption nations, notably the USA, Australia, the United Kingdom and Canada, has instituted a Gilded Age of neoliberal capitalism (Short, 2013).[6] Industrialization remains important in this new Gilded Age, especially in emerging economies. However, extreme personal wealth is increasingly linked to financialization – 'a pattern of accumulation in which profits accrue primarily through financial channels rather than through trade and commodity production' (Krippner, 2005, p. 174) – and to the financial elites that propel and benefit from these processes (Beaverstock et al., 2013). Whereas the share of income received by the top 10 per cent of US income earners grew from 33 to 50 per cent between 1976 and 2007, the share of the top 10 per cent of this group increased from 8.9 to 23.5 per cent; that is, the top 1 per cent not only now claim around a quarter of the entire population's income, they also earn almost half of the income of the top 10 per cent. The top 0.1 per cent in turn earn over half of the income of this top percentile (Atkinson et al., 2011, pp. 6–7). As Thomas Piketty (2014) argues in his landmark *Capital in the Twenty-First Century*, understanding this hugely skewed distribution requires recognition of the differential contribution of accumulated wealth (what he calls capital) and labour in dynamics of income and growth. Consider the contrast between the USA and France, a country experiencing a small rise in inequality. Between 1976 and 2006, incomes increased 32.2 per cent in the USA but only 27.1 per cent in France. However, excluding the top percentile, US incomes grew at 17.9 per cent and those in France grew at 26.4 per cent (Atkinson et al., 2011, p. 6). Consider also that the recovery in US incomes in 2009–10 of 2.3 per cent is even more unequal than the lopsided growth that contributed to the 2007–08 global financial crisis, with the incomes of the top percentile lifting 11.3 per cent and those of the rest only 0.2 per cent (Freeland, 2012, p. 5).

This picture of inequality is even starker in global view. A study of net household wealth, comprising income and assets minus debt, found that over two-thirds of the global adult population hold just 3 per cent of total private wealth (Shorrocks et al., 2013, p. 94). At the razor-sharp apex of this distribution are 32 million adults (0.7 per cent), each with investable wealth in excess of US$1 million and an astonishing 41 per cent of global household wealth (2013, p. 94). Disaggregating households from the nation-state in this way is useful in understanding individual wealth in a supranational economic order. However, correlation between private and national geographies of wealth is far from incidental, or innocent, as shown by the gap between household wealth held by US residents (29.9 per cent) and that in the next largest household economies: Japan (9.4 per cent), China (9.2 per cent) and Germany (5.4 per cent) (2013, pp. 22–5).

THE GLOW OF GILT LIVES

Although super-rich spending is much less robustly documented than income – with top incomes often deliberately left out of consumer surveys to avoid skewing median data – lavishly wasteful super-rich lives are indeed conspicuous in popular culture (Frank, 2007; Irvin, 2008; Hay, 2013a; Birtchnell and Caletrío, 2014). A confluence of celebrity, fashion, advertising, entertainment, media and luxury markets ensures super-rich lives are vivid in the imagination of many consumers. Ironically, the 'secessionary affluence' (Atkinson and Blandy, 2009, p. 93) of many super-rich lives – their retreat from the social mass – heightens this visibility, if only via voyeuristic media. For while the super-rich are dispersed – 64 countries now sport billionaires (Shorrocks et al., 2013, p. 94) – and hyper-mobile within a global space of 'networked extra-territoriality' (Atkinson and Blandy, 2009, p. 92), many move within a tight circuit of iconic locales (Pow, 2011; Hay and Muller, 2012). From 'The City' and 'Wall Street' to havens, enclaves, paradises and playgrounds, the super-rich cluster in geographies of opulence that are the everyday fare of movies and magazines (Davis and Monk, 2007; Beaverstock et al., 2013; Hay, 2013a).

Without doubt, the super-rich constitute an inordinate amount of consumer spending. An oft-cited study found correspondence between income and spending in the USA, with the top quintile accounting for 57.8 and 57.5 per cent of income and spending, respectively (Kapur et al., 2006, p. 11). Comprehensive, cross-national data is required to consolidate (or contest) this finding. It is nonetheless clear that super-rich spending exerts strong influence in consumer markets, comprising a disproportionate slice of retail profit. The essence of Frank's (1999) analysis is that this influence is cultural as well as financial. Extending Fred Hirsch's (1977) explanation of futile leap-frogging inherent in positional consumption, Frank argues that luxury markets create a stratospheric point of reference that inflates prices and desires across the board. In capitalist growth dynamics, richer consumers ensure that those on the next rung down on the ladder of comparative consumption do not feel rich, driving capital accumulation. Furthermore, the aspiration of super-rich consumers to differentiate themselves from peers – as in the case of the leap-frogging of the size, speed and resplendence of super-yachts and private jets (Frank, 2007; Beaverstock and Faulconbridge, 2014; Budd, 2014) – combined with skewed income growth, ensures that the referent of luxury keeps soaring. This extends the ladder of comparative consumption and intensifies dissatisfaction with prosaic goods and services.

In *Consuming the Planet to Excess*, sociologist John Urry (2010) places escalating consumer desire at the heart of the unfolding predicament of human-induced climate change. Linking luxury fever stemming from the hypermobility of the super-rich to the 'binge mobility' inherent in high-carbon societies, Urry (2010, p. 200) suggests that even though awareness of the environmental dangers of overconsumption and carbon-intensive practices is growing, consumers remain caught in an accelerating dynamic of consumption. Hay (2013b, p. 13) concludes from this analysis that

> luxury fever and the associated 'arms race' of possessions in which many of us find ourselves now occurs against a crumbling and putrefying environmental backdrop. So troubling is our predicament that it has yielded an uprising of catastrophist thinking in the social and environmental sciences and suggestions. . .that the super-rich are core parties to the excessive global consumption that is the prospective 'gravedigger' for twenty-first century capitalism.

Many environmentalists see only suicidal greed and vacuous mimicry in this escalation. Yet an implication of Frank's argument is that the increasing intensity of consumption practices is as much as anything the result of efforts to ward off a sense of declining quality of life. Narratives of impending planetary doom, then, may conceivably exacerbate consumer desire to retreat from a dangerous world and to hang on to the good life here and now.

In light of the rise of a new Gilded Age, the role of conspicuous consumption in environmental problems deserves serious attention in scholarship about sustainability. So too does the potential role of conspicuous consumption in facilitating sustainable practices. In particular, there is a need to investigate whether the rise of 'green celebrity' – which includes figures of popular culture, such as Leonardo DiCaprio, Cate Blanchett and Gisele Bündchen (CNBC, n.d.), and super-rich business leaders such as (arguably) Bill Gates, Richard Branson and Ted Turner (Callahan, 2010) – is influencing consumer choice. More generally, while super-rich lives may contribute to environmental destruction at large, one of the prime commodities wealth buys is the ability to live in safe and attractive environments and to avoid hazards such as dangerous water, air and food (conversely, one of many risks poverty imposes is proximity to environmental hazards: Walker, 2012). Evident in popular culture, the emerging market in luxury 'green' or 'eco-chic' consumption has not yet been documented, although, as I later discuss, the more general rise of sustainable consumption has (Jackson, 2006a; Lane and Gorman-Murray, 2011). Take, as one among many possible examples, the conspicuous 'ecomansions' of Bündchen – the world's highest paid super-model (Bergin, 2014) – and Laura Turner Seydel, daughter of billionaire environmentalist Ted Turner (EcoManor, n.d.). In Australia, the intersection of environmentalism, housing and luxury can be seen in the popular magazine of the Alternative Technology Association, *Sanctuary – Sustainable Living with Style* (Davison, 2011). I first became aware of this intersection in the late 1990s through my involvement in a non-profit permaculture garden centre.[7] The centre was engaged to build a living swimming pool (in which plants, animals and microorganisms maintain water quality) in the garden of an expensive riverside home. This pool caught the attention of others and the centre was soon overwhelmed with requests from wealthy householders for similar, if bespoke, pools. One declared their prime motive to be lifting the resale value of their home. This led to heated argument within this barely solvent organization about the merits of focusing on these lucrative consumers when its mission was encourage organic food production within the wider population and among disadvantaged groups.

BEYOND FEVERED CONSUMPTION

Despite being worthy of further inquiry, the environmental impacts of luxury fever – and 'green fever' – should not detract attention from other drivers of consumption in high-consumption societies since the mid-1970s. Such drivers include the role of private debt accumulation in offsetting wage repression associated with growing income inequality; entry of women into the labour force and consequent new domains of consumption; price suppression related to global economic integration, including geographical expansion of the consumer base, specialization of production, exploitation of disenfranchised labour and weak regulatory regimes, and weakening of the labour movement; creation of consumer markets through technological, financial and cultural innovation; and integration of media and advertising industries within corporate production chains and related influence of private media interests in political processes. This list is not exhaustive and beyond my scope to elaborate here.[8] The upshot is that excessive consumption is systemic within neoliberal political economy.

Undue emphasis on luxury fever may obscure complex cultural dynamics in consumption practices, including the role of social norms, or informal and implicit codes of conduct. As Pierre Bourdieu (1977) long argued, our dispositions, tastes and aspirations are, in part, the product of embodied learning within specific contexts (fields) of social practice. As part of this, acts of consumption constitute and are constitutive of social position (class) and identity. As Elizabeth Shove (2010, p. 1279) explains, the 'meanings of normal and the patterns of consumption associated with them require constant reproduction. The. . .conventions that are often taken to constitute the context of behaviour have no separate existence: rather, they are themselves sustained and changed through the ongoing reproduction of social practice'. This claim is also beyond my present scope (see Davison, 2001). The key message is that positional consumption does not take place in a vacuum; it is mediated through norms specific to diverse social fields. These norms are constantly renegotiated within trajectories of social change, of which environmental concerns are now an important part. Any response to luxury lives will be strongly influenced by the norms that bind social peers. Among my friends and colleagues, for example, the purchase of a house with four bathrooms or a prestige sports utility vehicle would probably attract opprobrium, not emulation, while my eco-efficient housing renovation and penchant for the latest bushwalking equipment are well received.

The 1950s and 1960s saw a steep rise in resource use on the back of cheap oil and mass production (Krausmann et al., 2009), despite being lean years for the super-rich. It is worth speculating, then, about the environmental consequences of redistributing super-rich wealth to others in high-consumption societies, while otherwise leaving political economic arrangements intact. Resource use and waste production would almost certainly rise alarmingly. For although many super-rich lives depend on profligate if not obscene resource use, much of their spending buys quality, status and exclusivity more than physical resources. Thus, while the environmental impact of gold mining implicated in a US$100 000 luxury watch is greater than that involved in the production of a US$100 steel watch, it is not likely to be 1000 times greater. Private aviation is a conspicuously unsustainable aspect of luxury lives (Callahan, 2010; Budd, 2014), and a person travelling First Class on a civil flight arguably uses more fuel than someone in Economy Class as a result of occupying more space. However, the vast expansion in middle-class aeromobility in recent decades (Adey et al., 2007) suggests that redistribution of wealth from the super-rich may increase the environmental impacts of aviation, or at least sustain these impacts in the face of efforts to price carbon. And what of redeploying to a McDonald's franchise the US$1000 previously spent on a (organic, local, artisanal) degustation in a Michelin-starred restaurant?

My point here is threefold: first, disproportionate super-rich spending need not equate to similarly disproportionate use of physical resources; second, luxury fever may exert a stronger pull on spending than on resource use; and third, assuming income inequality is associated with reduced (or negative) growth in low and middle incomes, super-rich spending may be linked to reduced spending power of the majority.[9] This is not to say the only environmental consequence of super-rich lives is in direct resource use – far from it. Capital flowing to the super-rich is fully implicated in the political economy of environmental practices, and the circulation of capital resulting from super-rich spending reinforces these practices. It is to say that the relevance of the super-rich for questions of sustainability may not lie primarily with their resource consumption practices, but with their role in aligning environmental concern with neoliberal political economy. I return to this shortly, but look first at the environmental consequences of resource use.

CONSUMING THE PLANET

The ecological footprint indicator (Rees, 1992) compares the use of renewable resources to the 'biocapacity' of planetary systems to renew these resources using a spatial equivalence measure – a global hectare of 'bioproductive' land or water. This measure shows that, in 1961, humanity consumed 75 per cent of the annually renewed resource base. By 2008, consumption had doubled to 150 per cent, or one and a half planets (WWF, 2012). Ecological footprints surpassed biocapacity in 1970, creating a state of 'overshoot' akin to 'overdrawing from a bank account' (2012, p. 40). This overdraft has increased steadily since, a dire trajectory compounded by two dynamics. First, despite a substantial slowdown in the rate of population growth since 1970, it is expected that 3 billion people will be added this century, pushing the global population past 10 billion (UNPD, 2013). Second, greenhouse gas emissions create grave risk of systemic and unmanageable changes to planetary systems (Urry, 2010). If emissions were to cease today, considerable momentum for global heating would remain. The present trend of increasing emissions, however, raises the real prospect of a dramatic, nonlinear decrease in biocapacity (IPCC, 2013).

Shifting from this global view to the case of the USA, ecological footprint data indicate that this population was entrenched in overshoot by 1961, with per capita footprint more than double biocapacity (Global Footprint Network, n.d.). From there the USA's ecological footprint increased sharply, peaked in 1973, declined a little before remaining steady until 2005, then declined to the level of the early 1960s by 2010. Although per capita biocapacity declined 40 per cent during this period, per capita resource use did not increase in step with the resurgence of extreme wealth. This finding is mirrored in US per capita energy consumption, which grew strongly between 1949 and 1973 and then declined a little before relative stasis during the 1980s and 1990s (Diamond and Moezzi, 2004).[10]

None of this obscures the fact that at least four Earths are required to enable everyone to match US lifestyles. Although the multiple varies, this disparity is true of all high-consumption nations, for there is close correspondence between ecological footprint and modern development. Take the UN Development Programme's (UNDP) Human Development Index (HDI), a moderately broad measure incorporating indices for life expectancy, educational attainment and per capita gross domestic product. The positive correlation of HDI and footprint is such that no nation presently passing the threshold of high human development has a sustainable footprint (one below per capita biocapacity) (WWF, 2012, p. 424).

Nonetheless, given continued economic growth in the USA, lack of growth in this nation's per capita ecological footprint over the past 35 years is evidence of a decoupling of resource use and economic productivity. Global footprint data indicate this decoupling is not limited to the USA, with overshoot growing primarily as the result of per capita decrease in biocapacity rather than per capita increase in footprint (2012, p. 40). Supporting this, a comprehensive study of renewable and non-renewable resource extraction (Krausmann et al., 2009) found that while global annual per capita resource use doubled during the twentieth century, per capita income increased seven-fold (Krausmann et al., 2009).

Hope that such decoupling of resource use and economic growth will make wealthy lives ecologically benign, and available to all, is central to business advocacy for sustainable development. For all but the end of the twentieth century, this decoupling was a largely unrecognized result of technological innovation.[11] Advocates of sustainable development argue that an active pursuit of 'eco-efficiency' will drive a process of ecological modernization, reconciling economic wealth and environmental sustainability (Davison, 2001). As expressed in Hawken et al.'s (1999) *Natural Capitalism*, this is hope for a 'factor ten' increase in eco-efficiency based upon profitable dematerialization, decarbonization and detoxification of production and a shift towards a service-based (and away from a goods-based) economy. As evident in faith placed in 'green' economic growth at the 2012 'Rio+20' Earth Summit, efforts to rehabilitate capitalism as an engine of environmental sustainability have progressed apace in recent years (UN, 2012), helping and helped along by, so I now argue, the super-rich.

SUSTAINABLE DEVELOPMENT AND THE SUPER-RICH

The first Rio Earth Summit, in 1992, the largest gathering of heads of state to that point, aimed to galvanize national governments around the Brundtland vision of sustainable development. Intractable dispute between low-consumption (high population growth) and high-consumption (low population growth) nations in two years of preparatory negotiations, and subsequent failure of high-consumption nations to act on their commitments (UNEP, 2002), make it easy to overlook a passing of the baton in the early 1990s from governments to producers, consumers and philanthropists in pursuit of sustainable development.[12] I focus on three ways the super-rich are implicated in the aligning of sustainable development with neoliberal reforms that transfer agency from states to markets, and from the public to the private sphere.

First was the role of the super-rich in the Earth Summit itself, and particularly that of Summit Secretary-General Maurice Strong, a Canadian entrepreneur who first made his fortune in Alberta oil. Strong, also Secretary-General of the 1972 UN Conference on the Human Environment, did much to engage business in the Summit, principally by partnering with the Business Council for Sustainable Development (BCSD). Formed in 1991 by Swiss billionaire Stephan Schmidheiny

(1992)[13] to represent industry at the Summit, the BCSD (now the WBSCD) comprised 48 leaders of transnational corporations including Royal Dutch Shell, DuPont, Dow Chemical Company, ALCOA, Mitsubishi, Ciba-Geigy and 3M. Schmidheiny, chair of the BCSD, was appointed an advisor to Strong and ensured the Summit received substantial corporate sponsorship (Chatterjee and Finger, 1994, p. 114).

It was the BCSD that coined the neologism 'eco-efficiency' to encapsulate the goal of 'creating more goods and services with ever less use of resources, waste and pollution' (WBCSD, 2000). Related confidence that an invisible green hand will guide free markets toward sustainability was encapsulated by then Chairman of Dow, Frank Popoff: 'If we view sustainable development as an opportunity for growth and not as prohibitive, industry can shape a new social and ethical framework for assessing our relationship with the environment and each other' (cited in Schmidheiny, 1992, p. 72). Acting on this confidence, the BCSD and the International Chamber of Commerce lobbied in pre-Summit negotiations to block a regulatory code for corporations proposed by the UN Centre on Transnational Companies (UNCTC). This lobbying was successful, with the centre closed down (Chatterjee and Finger, 1994), arguably launching an era of voluntary codes of conduct.

A second way in which neoliberal capitalism was rehabilitated in the face of environmental problems – thereby contributing to super-rich prosperity – was through the rise of neoclassical economic arguments about environmental protection. Of particular note was the World Bank's introduction in 1992 of the environmental Kuznets curve (EKC) (Stern, 2004). Based upon economist Simon Kuznets' post-war dictum that economic growth is accompanied first by a rise and then by a decline in inequality, the EKC is the proposition that: '[in] the early stages of economic growth degradation and pollution increase, but beyond some level of income per capita. . .the trend reverses, so that at high-income levels economic growth leads to environmental improvement' (2004, p. 1419). In effect, the EKC deflects attention away from environmental damage associated with wealth to that associated with poverty.[14] It justifies claims that free markets are sustainable in the long run and that ecological modernization is inevitable (Torras and Boyce, 1998; Cole, 2007). The EKC is commonly explained by some variant of the post-materialist thesis associated with political scientist Ronald Inglehart (1997) and colleagues. This thesis asserts that populations with high material security and economic wealth are characterized by non-material values related to morality, aesthetics and qualitative satisfaction that heighten environmental concern. This argument was popularized by *The Skeptical Environmentalist*, Bjorn Lomborg's (2001) account of a causal relationship between increasing affluence, decreasing population growth, increasing environmental concern and improving environmental quality.

Aspects of Inglehart's analysis are contested, with Riley Dunlap and Richard York (2008) countering that some forms of environmental concern are more correlated with wealth than others and that the poor do express environmental concerns. Related to this, empirical evidence for the EKC is limited principally to local or regional issues of air and water pollution in high-consumption nations (Torras and Boyce, 1998; Stern, 2004; Cole, 2007). The strong correlation of national affluence and ecological footprint noted earlier clearly contradicts the EKC thesis (Caviglia-Harris et al., 2009), as does the case of greenhouse gas emissions. Challenging claims that population growth is the major impediment to reduced emissions, Satterthwaite (2009, p. 556) shows that, between 1980 and 2005, low-consumption nations accounted for 52.1 per cent of population growth but only 12.8 per cent of emissions growth. High-consumption nations, in contrast, contributed 29.1 per cent of emissions growth, yet only 7.2 per cent of population growth. In the absence of data allowing direct correlation of individual wealth and climate change, Satterthwaite (2009, p. 563) speculates that 'it is likely that the very rich would have GHG emissions per person that were thousands of times those of large sections of the poorest groups'. In line with this, a recent study of UK household emissions found income to be the 'main driver' and top incomes to contribute very disproportionately (Gough et al., 2012, p. 11).

Despite limited empirical support, the EKC has enabled neoliberal advocates of sustainable development to interpret evidence of worsening environmental conditions as a sign of improvement to come.[15] It reinforces an assumption embedded in the operation of the World Bank, International Monetary Fund and World Trade Organization – the troika of global trade neoliberalization (Harvey, 2005) – that free trade and foreign investment are the best way to overcome environmental destruction wrought by the poor. These efforts have obscured issues of affluence in discourses of sustainable development while, of course, swelling super-rich coffers.

The third alignment of sustainable development and neoliberal reforms I address is the shift from public and collective to private and individual forms of environmental action. Linked to the success of super-rich producers in displacing questions of equity with those of efficiency at the Earth Summit, this shift first returns my focus to super-rich consumers. A key part of this shift is emphasis on 'sustainable consumption', which, by the mid-1990s, had become an "overriding issue" and a "cross-cutting theme" in the sustainable development debate' (Jackson, 2006b, p. 3). Whereas 'sustainable production' internalized the environment within markets, most notably through price signals (Pearce et al., 1989),[16] sustainable consumption has relied on the capacity of post-materialist consumers to lead market change through demand signals (Maniates, 2002; Jackson, 2006a). However, the shift from public to private forms of environmental action extends beyond issues of production and consumption to the rationale of political agency more generally.

A tension between public and private interests has been evident in environmental movements at least since Garrett Hardin's 1968 essay, 'The Tragedy of the Commons', which proposed that collective ownership of environmental goods leads to exploitation and that private self-interest is the best guarantee of environmental prudence (Robbins, 2012). As exemplified in the life of Henry David Thoreau, ideals related to solitude and self-sufficiency have been influential in Anglophone environmental movements whose core constituency has been the land-owning suburban middle class (Peterson del Mar, 2006; Davison, 2011). From the late 1970s, this cultural privatism coexisted with neoliberal privatization of public assets. By the time of the Earth Summit, the scene was set for the focus of privatization to turn to the environment. In low consumption societies this has led to corporate enclosure of resources such as freshwater (Roberts, 2008), opening up new domains of capital accumulation for the neoliberal super-rich. In high-consumption societies, a prominent example of environmental privatization has been the rise of what Holmes (2012) calls neoliberal conservation.

The juxtaposition of extreme personal wealth and environmental concern is not new.[17] In the USA, David Callahan (2010, p. 64) argues that the 'eco-rich' have been central to environmentalism at least since Theodore Roosevelt became the 'first environmentalist president' and 'huge land acquisitions' by the Rockefellers enabled the creation of national parks during the first Gilded Age. What is distinctive about the new eco-rich is their pioneering of nature conservation outside public policy and public institutions. Consider the billionaires Ted Turner and Brad Kelley, whose fortunes were made in media and cigarettes respectively, and whose conservation interests have led them to become the second and fourth largest private landholders in the US (Fay Ranches, 2013). While Turner has focused on native biodiversity – he has the distinction of owning more bison than anyone else (Weiner, 2007) – Kelley focuses on breeding endangered exotic animals such as the black rhino (Callahan, 2010; Keates, 2012). Turner's land management practices, in particular, are controversial, with his wilderness used not just for conservation but for capital accumulation – he has 1000 gas wells in New Mexico alone – and exclusive forms of eco-retreat for the super-rich (Weiner, 2007).

The rise of eco-rich landholders raises many questions. For instance, can Turner and Kelley be understood as part of a wider trend towards 'drawbridge sustainability'? That is, are some super-rich (and perhaps some just-plain-rich) shielding themselves from the environmental consequences of their wealth by establishing eco-enclaves in which they enjoy contact with nature denied to others and can seek private redemption through conservation? Are eco-rich landholders giving rise to a green fever that enflames consumer desire for private ways in which to find redemption and protection in the face of 'nature's revenge'? And are the eco-rich cultivating wider social acceptance of neoliberal claims about the compatibility of sustainability and private wealth?

To this point I have considered the agency of the eco-rich as consumers. However, their agency is increasingly expressed in private acts of super-rich philanthropy. As indicated by the Rockefellers' gift of national parks in the early twentieth century, philanthropy has long been central to environmentalism, although the super-rich have not always been central to this philanthropy. Vital to the success of organizations such as Greenpeace and Friends of the Earth has been their middle-class donors. Yet the past decade has seen philanthropy of unprecedented scale amongst the world's richest individuals (Hay and Muller, 2014; see also Kapoor, Chapter 6 this volume), starting with the pledges of Bill Gates and Warren Buffett in 2006 – then, the world's two richest individuals – to dispense their fortunes for the greater good (Bishop and Green, 2008). The subsequent Giving Pledge (n.d.) project led by Gates and Buffett has seen over 80 US billionaires make similar pledges, with the project going global in 2013 (Hay and Muller, 2014, p. 636).

In the context of sustainable development, philanthropy is closely linked to the role of the super-rich as producers by focusing on profitable eco-efficient innovation. Having made much of his fortune through mass aviation, for example, Branson's Virgin Earth Challenge (n.d.) offers a US$25 million prize for a cost-effective way to remove greenhouse gases from the atmosphere.[18] Consider also the aim of the Gates Foundation to promote a food biotechnology 'revolution' in Africa, inspired by the 'Green Revolution' funded by the Rockefeller Foundation in the 1960s in Asia that proponents argue have 'saved over one billion lives' (Bishop and Green, 2008, p. 4).

In their upbeat assessment of a dawning age of 'philanthrocapitalism', Matthew Bishop and Michael Green (2008) argue that super-rich philanthropic investment is more than adequate compensation for the failing ambition and strength of the state in neoliberal times. In their view, the super-rich are 'hyperagents who have the capacity to do some essential things far better than anyone else' (Bishop and Green, 2008,

p. 12). They suggest that the 'retreat' of the state – which they happily assume is irreversible – means that it is no longer viable 'to see the state and philanthropy as alternatives' (2008, p. 10). As seen in the case of philanthropy driving eco-efficient production, it seems that the free market and philanthropy are not alternatives either (Hay and Muller, 2014). The agile, business-savvy hyperagency of super-rich philanthropists substitutes for cumbersome states in guiding and supplementing sustainable forms of neoliberal development. In the process, philanthropic investment keeps the state in retreat and attempts to rectify market failures, thereby maintaining dynamics of extreme wealth accumulation and translating extreme private wealth into a mode of governance.

CONCLUSION

I have argued that the super-rich are directing the pursuit for sustainable development in their roles as producers, consumers and philanthropists. While some attention has been paid to the way super-rich consumers fuel dynamics of conspicuous consumption, their wider influence in entrenching neoliberal political economic relations in the face of environmental problems deserves greater scrutiny. As noted at the outset, my progress in considering the environmental implications of super-rich lives in this chapter has been preliminary. I have, however, sought to demonstrate the benefit of holding together questions about how private fortunes are spent with questions about how they are made in the dynamics of neoliberal capitalism. I conclude from this that the super-rich are cause and consequence in ongoing transformation of a collective environmental predicament into matters of private concern.

The example of Al Gore exposes some of the issues that require attention. Having read his 1992 agenda for redressing climate change, *Earth in the Balance*, many citizens had cause to expect Gore would advance climate change policy during his vice-presidency of the USA between 1993 and 2001. Yet his political achievements on this front were modest, if not dismal. In contrast, his subsequent efforts as a private individual – who counts his wealth in the hundreds of millions of dollars – to raise awareness about climate change have arguably been more effective than those of any other. In my own suburb, privately owned rooftop solar technologies have proliferated in recent years in no small part as a result of Gore's 2006 documentary, *An Inconvenient Truth* (produced by the billionaire Jeffrey Skoll).[19] Yet, despite Gore's individual achievements, the global predicament created by rising greenhouse gas emissions has worsened considerably since 2006, especially with the collapse of intergovernmental negotiations in 2009.[20] I take from this example the fact that, under the leadership of the super-rich, efforts to resolve a fully evident and unfolding drama of environmental destruction by coordinating private interests in the market remain remarkably resilient. My hope is that I have made clear that these efforts also remain dangerously untenable.

NOTES

1. No nomenclature for nation-states is fully adequate. I adopt the convention of high-consumption and low-consumption societies, where consumption refers to per capita not total resource use.
2. Compare this with the arguments interrogated by Kapoor (Chapter 6 in this volume) about the relationships between capitalist growth, super-wealth and philanthropy.
3. Usually discussed in terms of the $I = P \times A \times T$ (environmental impact = population size × level of affluence × state of technology) formula introduced in the early 1970s (Chertow, 2000).
4. For an authoritative account of neoliberalism see Harvey (2005).
5. I follow the practice of defining super-rich (high net worth) individuals as those with investable assets of more than US$1 million (excluding owner-occupied housing and consumables). This category covers a vast diversity of wealth and power and is usually broken into subcategories (Hay and Muller, 2012; Hay, 2013b). Strong growth in the number of billionaires, which surpassed 1000 in 2010 (Shorrocks et al., 2013, p. 113), makes this group a prominent and distinct super-rich class.
6. This return to inequality is characterized by growth at the very tip of income distribution and by self-made fortunes made in the marketplace (rather than inherited) on the basis of production and speculation (rather than rentier entitlement). However, the recent turning of the generational wheel is seeing inheritance and dynastic power become increasingly important.
7. Permaculture, a contraction of permanent agriculture, is an ethos and method of organic-ecological food production founded by Australians Bill Mollison and David Holmgren.
8. Harvey (2005, 2011) discusses some of these structural factors.
9. There is some overlap here with Blake Alcott's (2008, p. 770) argument that any effort by rich consumers to reduce their resource consumption will be overwhelmed by a rebound effect whereby reduced demand would suppress prices for those on the next (more populous) rung down the ladder of comparative consumption.
10. This stasis is despite a 35 per cent increase in single-family housing floor area in the USA since the 1970s (Diamond and Moezzi, 2004, p. 3).
11. It was observed by the prescient Richard Buckminster Fuller ([1938] 1973) as early as the 1930s.
12. The exception here is local government, which took important steps to implement the Earth Summit's *Agenda 21*.
13. Schmidheiny was imprisoned for 16 years in 2012 for gross negligence relating to asbestos exposure of employees and customers. This sentence was cancelled due to the statute of limitations in 2014.
14. In effect, the EKC challenges assumptions underlying the $I = P \times A \times T$ formula (see note 3), by arguing that increased affluence decreases impact via improved technological efficiency and reduced population growth (see note 15).
15. The Kuznets curve and the demographic transition model (DTM) – which proposes that industrialization brings first rapid population growth then stabilization – arguably served a similar function in deflecting post-war criticism of development. Derived from

the experience of 'developed' nations, the DTM inadequately explains the experience of many 'developing' nations experiencing sustained population growth because the economic benefits of industrialization are accruing elsewhere.

16. In the context of environmental price signals, there is widespread evidence that resource savings achieved by eco-efficient production are often exceeded by a rebound effect of increased consumption driven by reduced price (UNEP, 2012, p. 428).
17. Think, for example, of the care with which feudal aristocracy in Europe tended their forests and hunting grounds, not least by excluding the hoi polloi.
18. Launched in 2007, 11 finalists for this prize have now been chosen and will be adjudicated by an esteemed panel including scientists Tim Flannery and James Hansen, and Al Gore. All finalists are from high-consumption societies.
19. This observation is based upon my involvement on the board of Sustainable Living Tasmania, the local organization overseeing Al Gore's Climate Ambassadors programme in Tasmania.
20. At the 15th Conference of the Parties of the UN Framework Convention for Climate Change in Copenhagen.

REFERENCES

Adey, P., L. Budd and P. Hubbard (2007), 'Flying lessons: exploring the social and cultural geographies of global air travel', *Progress in Human Geography*, **31** (6), 773–91.

Alcott, B. (2008), 'The sufficiency strategy: would rich-world frugality lower environmental impact?', *Ecological Economics*, **64** (4), 770–76.

Alvaredo, F., A.B. Atkinson and T. Piketty et al. (2013), 'The top 1 percent in international and historical perspective', *Journal of Economic Perspectives*, **27** (3), 3–20.

Atkinson, A.B., T. Piketty and E. Saez (2011), 'Top incomes in the long run of history', *Journal of Economic Literature*, **49** (1), 3–71.

Atkinson, R. and S. Blandy (2009), 'A picture of the floating world: grounding the secessionary affluence of the residential cruise liner', *Antipode*, **41** (1), 92–110.

Beaverstock, J.V. and J.R. Faulconbridge (2014), 'Wealth segmentation and the mobilities of the super-rich', in T. Birtchnell and J. Caletrío (eds), *Elite Mobilities*, London/New York: Routledge, pp. 40–61.

Beaverstock, J.V., S. Hall and T. Wainwright (2013), 'Servicing the super-rich: new financial elites and the rise of the private wealth management retail ecology', *Regional Studies*, **47** (6), 834–49.

Bergin, O. (2014), 'Inside Gisele Bündchen's $20m eco mansion', *The Telegraph*, 9 September, accessed 24 April 2014 at http://fashion.telegraph.co.uk/news-features/TMG10296271/Inside-Gisele-Bundchens-20m-eco-mansion.html.

Birtchnell, T. and J. Caletrío (eds) (2014), *Elite Mobilities*, London/New York: Routledge.

Bishop, M. and M. Green (2008), *Philanthropocapitalism: How the Rich Can Save the World*, New York: Bloomsbury Press.

Bourdieu, P. (1977), *Outline of a Theory of Practice*, trans. R. Nice, Cambridge, UK: Cambridge University Press.

Budd, L. (2014), 'Aeromobile elites: private business aviation and the global economy', in T. Birtchnell and J. Caletrío (eds), *Elite Mobilities*, London/New York: Routledge, pp. 78–98.

Callahan, D. (2010), *Fortunes of Change: The Rise of the Liberal Rich and the Remaking of America*, Hoboken, NJ: John Wiley & Sons.

Caviglia-Harris, J.L., D. Chambers and J.R. Kahn (2009), 'Taking the "U" out of Kuznets: a comprehensive analysis of the EKC and environmental degradation', *Ecological Economics*, **68** (4), 1149–59.

Chatterjee, P. and M. Finger (1994), *The Earth Brokers: Power, Politics and World Development*, London/New York: Routledge.

Chertow, M.R. (2000), 'The IPAT equation and its variants: changing views of technology and environmental impact', *Journal of Industrial Ecology*, **4** (4), 13–29.

CNBC (n.d.), 'Top 15 green celebrities', *CNBC*, accessed 20 April 2014 at www.cnbc.com/id/46997210.

Cole, M.A. (2007), 'Economic growth and the environment', in G. Atkinson, S. Dietz and E. Neumayer (eds), *Handbook of Sustainable Development*, Cheltenham, UK and Northampton, MA, USA: Edward Elgar Publishing, pp. 240–53.

Davis, M. and D.B. Monk (eds) (2007), *Evil Paradises: Dreamworlds of Neoliberalism*, New York: The New Press.

Davison, A. (2001), *Technology and the Contested Meanings of Sustainability*, Albany, NY: State University of New York Press.

Davison, A. (2011), 'A domestic twist on the eco-efficiency turn: technology, environmentalism, home', in R. Lane and A. Gorman-Murray (eds), *Material Geographies of Household Sustainability*, London: Ashgate, pp. 35–49.

Diamond, R. and M. Moezzi (2004), 'Changing trends: a brief history of the US consumption of energy, water, beverages and tobacco', *Proceedings of the 2004 Summer Study on Energy Efficiency in Buildings, American Council for an Energy Efficient Economy*, Washington, DC.

Dunlap, R. and R. York (2008), 'The globalization of environmental concern and the limits of the postmaterialist values explanation: evidence from four multinational surveys', *The Sociological Quarterly*, **49** (3), 529–63.

EcoManor (n.d.), website, accessed 10 April 2014 at www.eco.manor.com.

Ehrlich, P.R. (1968), *The Population Bomb*, New York: Ballantine.

Fay Ranches (2013), *The Land Report: The Magazine of the American Landowner*, accessed 12 April 2014 at www.landreport.com.

Frank, R.H. (1999), *Luxury Fever: Why Money Fails to Satisfy in an Era of Excess*, New York: The Free Press.

Frank, R. (2007), *Richi$tan: A Journey Through the American Wealth Boom and the Lives of the New Rich*, New York: Random House.

Franzen, J. (2010), *Freedom*, New York: Farrar, Strauss and Giroux.

Freeland, C. (2012), *Plutocrats: The Rise of the New Global Super-Rich and the Fall of Everyone Else*, New York: Penguin.

Fuller, R.B. ([1938] 1973), *Nine Chains to the Moon*, London: Jonathan Cape.

Giving Pledge (n.d.), 'The Giving Pledge is a commitment by the world's wealthiest individuals and families to dedicate the majority of their wealth to philanthropy', *Giving Pledge*, accessed 20 April 2014 at givingpledge.org.

Global Footprint Network (n.d.), 'Footprint for nations', *Global Footprint Network*, accessed 10 April 2014 at www.footprintnetwork.org/en/index.php/GFN/page/footprint_for_nations/.

Gore, A. (1992), *Earth in the Balance: Ecology and the Human Spirit*, New York: Plume.

Gough, I., S. Abdallah and V. Johnson et al. (2012), *The Distribution of Total Greenhouse Gas Emissions by Households in the UK, and Some Implications for Social Policy*, London: Centre for Analysis of Social Exclusion.

Hajer, M. (1997), *The Politics of Environmental Discourse: Ecological Modernization and the Policy Process*, New York: Oxford University Press.

Hardin, G. (1968), 'The tragedy of the commons', *Science*, **162**(3859), 1243–8.

Harvey, D. (2005), *A Brief History of Neoliberalism*, Oxford, UK: Oxford University Press.

Harvey, D. (2011), 'Roepke lecture in economic geography – crises, geographic disruptions and the uneven development of political responses', *Economic Geography*, **87** (1), 1–22.

Hawken, P., A.B. Lovins and L.H. Lovins (1999), *Natural Capitalism: The Next Industrial Revolution*, London: Earthscan.

Hay, I. (ed.) (2013a), *Geographies of the Super-Rich*, Cheltenham, UK and Northampton, MA, USA: Edward Elgar Publishing.

Hay, I. (2013b), 'Establishing geographies of the super-rich: axes for analysis of abundance',

in *Geographies of the Super-Rich*, Cheltenham, UK and Northampton, MA, USA: Edward Elgar Publishing, pp. 1–25.
Hay, I. and S. Muller (2012), '"That tiny, stratospheric apex that owns half of the world" – exploring geographies of the super-rich', *Geographical Research*, **50** (1), 75–88.
Hay, I. and S. Muller (2014), 'Questioning generosity in the golden age of philanthropy: towards critical geographies of super-philanthropy', *Progress in Human Geography*, **38** (5), 635–53.
Hirsch, F. (1977), *Social Limits to Growth*, Cambridge, MA: Harvard University Press.
Holmes, G. (2012), 'Biodiversity for billionaires: capitalism, conservation and the role of philanthropy in saving/selling nature', *Development and Change*, **43** (1), 185–203.
Inglehart, R. (1997), *Modernization and Postmodernization: Cultural, Economic, and Political Change in 43 Societies*, Princeton, NJ: Princeton University Press.
IPCC (2013), 'Summary for policymakers', in T.F. Stocker, D. Qin and G.-K. Plattner et al. (eds), *Climate Change 2013: The Physical Science Basis, Contribution of Working Group I to the Fifth Assessment Report of the Intergovernmental Panel on Climate Change*, Cambridge, UK: Cambridge University Press.
Irvin, G. (2008), *Super Rich: The Rise of Inequality in Britain and the United States*, Cambridge, UK: Polity.
Jackson, T. (2006a), *The Earthscan Reader in Sustainable Consumption*, London: Earthscan.
Jackson, T. (2006b), 'Readings in sustainable consumption', in *The Earthscan Reader in Sustainable Consumption*, London: Earthscan, pp. 1–23.
Kapur, A., N. MacLoed and N. Singh et al. (2006), 'The plutonomy symposium – rising tides lifting yachts', *The Citigroup Global Investigator*, 29 September, accessed 10 April 2014 at www.correntewire.com/sites/default/files/Citibank_Plutonomy_2.pdf.
Keates, N. (2012), 'The man with a million acres', *The Wall Street Journal*, 25 October, accessed 25 April 2014 at http://online.wsj.com/news/articles/.
Krausmann, F., S. Gingrich and N. Eisenmenger et al. (2009), 'Growth in global materials use, GDP, and population during the 20th century', *Ecological Economics*, **68** (10), 2696–705.
Krippner, G.R. (2005), 'The financialization of the American economy', *Socioeconomic Review*, **3** (2), 173–208.
Lane, R. and A. Gorman-Murray (eds) (2011), *Material Geographies of Household Sustainability*, Farnham, UK: Ashgate.
Lomborg, B. (2001), *The Skeptical Environmentalist: Measuring the Real State of the World*, Cambridge, UK/New York: Cambridge University Press.
Maniates, M. (2002), 'Individualisation: plant a tree, buy a bike, save the world?', in T. Princen, M. Maniates and K. Conca (eds), *Confronting Consumption*, Boston, MA: MIT Press, pp. 43–66.
Meadows, D.H., D.L. Meadows and J. Randers et al. (1972), *Limits to Growth*, New York: Universe Books.
Mol, A.P.J. and D.A. Sonnenfeld (eds) (2000), *Ecological Modernisation Around the World: Perspectives and Critical Debates*, London: Frank Cass.
Myers, N. and J. Kent (2003), 'New consumers: the influence of affluence on the environment', *Proceedings of the National Academy of Sciences of the United States of America*, **100** (8), 4963–8.
Pearce, D., A. Markandya and E.B. Barbier (1989), *Blueprint for a Green Economy*, London: Earthscan.
Peterson del Mar, D. (2006), *Environmentalism*, Harlow, UK: Pearson.
Piketty, T. (2014), *Capital in the Twenty-First Century* [French edition published 2013 as *Le capital au XXI siècle*, Editions du Seuil], Cambridge, MA: Belknap Press of Harvard University Press.
Pow, C.P. (2011), 'Living it up: super-rich enclave and transnational elite urbanism in Singapore', *Geoforum*, **42** (3), 382–93.
Redclift, M. (1987), *Sustainable Development: Exploring the Contradictions*, London/New York: Routledge.

Rees, W.E. (1992), 'Ecological footprints and appropriated carrying capacity: what urban economics leaves out', *Environment and Urbanization*, **4** (2), 121–30.
Robbins, P. (2012), *Political Ecology: A Critical Introduction*, Malden, MA: Wiley-Blackwell.
Roberts, A. (2008), 'Privatizing social reproduction: the primitive accumulation of water in an era of neoliberalism', *Antipode*, **40** (4), 535–60.
Sachs, W. (1992), *The Development Dictionary: A Guide to Knowledge as Power*, London: Zed Books.
Satterthwaite, D. (2009), 'The implications of population growth and urbanization for climate change', *Environment and Urbanization*, **21** (2), 545–67.
Schmidheiny, S. (1992), *Changing Course: A Business Perspective on Environment and Development*, Cambridge, MA: MIT Press.
Schumacher, E.F. (1973), *Small is Beautiful: A Study of Economics as if People Mattered*, London: Abacus.
Shorrocks, A., J.B. Davies and R. Lluberas (2013), *Credit Suisse Global Wealth Databook 2013*, 4th edition, Zurich: Credit Suisse Research Institute.
Short, J.R. (2013), 'Economic wealth and political power in the second Gilded Age', in I. Hay (ed.), *Geographies of the Super-Rich*, Cheltenham, UK and Northampton, MA, USA: Edward Elgar Publishing, pp. 26–42.
Shove, E. (2010), 'Beyond the ABC: climate change policy and theories of social change', *Environment and Planning A*, **42** (6), 1273–85.
Stern, D.I. (2004), 'The rise and fall of the environmental Kuznets curve', *World Development*, **32** (8), 1419–39.
Torras, M. and J.K. Boyce (1998), 'Income, inequality, and pollution: a reassessment of the environmental Kuznets curve', *Ecological Economics*, **25** (2), 147–60.
UN (2012), *Report of the United Nations Conference on Sustainable Development, Rio de Janiero, Brazil, 20–22 June 2012*, New York: United Nations.
UNEP (2002), *Global Environmental Outlook 3: Past, Present and Future Perspectives*, London: Earthscan/United Nations Environment Programme.
UNEP (2012), *Global Environmental Outlook 5: Environment for the Future We Want*, United Nations Environment Programme/GRID Arendal.
UNPD (2013), 'World population prospects: the 2012 revision', Population Division of the Department of Economic and Social Affairs of the United Nations Secretariat, online database, accessed 4 April 2014 at esa.un.org/unpd/wpp/index.htm.
Urry, J. (2010), 'Consuming the planet to excess', *Theory, Culture and Society*, **27** (2–3), 191–212.
Veblen, T. ([1899] 1925), *The Theory of the Leisure Class: An Economic Study of Institutions*, London: George Allen & Unwin.
Virgin Earth Challenge (n.d.), 'Earth Challenge: removing greenhouse gases from the atmosphere', *Earth Challenge*, accessed 24 April 2014 at www.virginearth.com.
Walker, G. (2012), *Environmental Justice: Concepts, Evidence and Politics*, London/NewYork: Routledge.
WBCSD (2000), *Eco-efficiency: Creating More Value with Less Impact*, Geneva: World Business Council for Sustainable Development, accessed 3 March 2009 at www.wbcsd.org/web/publications/eco_efficiency_creating_more_value.pdf.
WCED (1987), *Our Common Future: The Report of the World Commission on Environment and Development*, Oxford, UK: Oxford University Press.
Weiner, J. (2007), '"Hell is other people": Ted Turner's two million acres', in M. Davis and D.B. Monk (eds), *Evil Paradises: Dreamworlds of Neoliberalism*, New York: The New Press, pp. 199–206.
WWF (2012), *Living Planet Report 2012: Biodiversity, Biocapacity and Better Choices*, Gland, Switzerland: World-Wide Fund for Nature.

PART III

WEALTH AND POWER

18. Attracting wealth: crafting immigration policy to attract the rich

John Rennie Short

INTRODUCTION

The Rich Are Always With Us was the title used in a 1932 Hollywood romantic comedy. Not a classic by any means, but a useful starting point for this chapter. The movie dramatizes the international travels of a wealthy socialite from New York to Paris and alludes to the travels of her lover from Romania to Paris and then on to China and India. It employs a common theme: the intimate connections between the rich, travel and periods of overseas residence. But at the heart of the title and the story line is the notion that the rich are invariably immoveable from their main base. If this was ever true it is becoming less true as there are now more wealthy looking and able to migrate and more states trying to attract them.

This chapter examines the change in the regime of national immigration policies to attract the wealth of, and those with, capital. This shift towards a dual-track system, one for the wealthy and another for the rest, is more obvious in some countries than others but many states around the world are fundamentally changing as well as tweaking their immigration policies to attract the wealthy. The chronology of this shift is highlighted, the form of the policy shift is noted, and the effects are assessed.

MILLIONAIRE MIGRANTS

David Ley's (2010) categorization of 'millionaire migrants' reflects the fact that the rich are on the move. Not only in terms of their regular circuits around the globe from exclusive hangouts such as St. Bart's to enclaves in Paris, London and New York, but also in terms of permanent relocation. The trend is evident from an examination of the official statistics that show a doubling from 2010 to 2011 of the number of Chinese citizens applying to enter the USA through the investor programme or, in the reverse direction, the doubling in the same period of US citizens seeking to renounce US citizenship and its consequent tax burdens. A number of reasons lie behind this mobility: the creation of wealth in societies that either have high tax burdens or do not guarantee long-term economic and political stability, which in turn generates a supply of the mobile wealthy looking for a safe haven; the ease of capital mobility that makes asset transfer from one country to another a relatively simple proposition; and cash-strapped states that are on the look out to attract this growing pool of wealthy mobiles ostensibly and arguably to fill the national coffers, generate investment and create more jobs.

While the directions may vary the reasons remain the same. In what I have described elsewhere as the second Gilded Age (Short, 2013), there are now more rich people who want to protect their wealth through relocating to safer places. The protection comes in two forms: the safeguarding of asset protection that generates migration to countries with more stable political regimes; and the safeguarding of asset worth that makes some of the rich live in tax havens such as Monaco and Andorra that have a very low personal taxation burden. Eduardo Saverin is the poster boy for the mobile wealthy fleeing to escape tax burdens. His rich family emigrated from Brazil to the USA to escape the threat of kidnapping. While at Harvard he was involved in the start-up of Facebook. He moved to Singapore in 2009 and renounced his US citizenship in 2011. US citizens have to pay US tax no matter where they live in the world. His move occurred one year before the initial public offering of Facebook that assured him a windfall gain of over US$3 billion. Singapore residents pay no capital gains and enjoy very low personal tax rates capped at 20 per cent.

In some countries, such as Russia and China, while there are opportunities for making vast fortunes the possibility of holding on to them may be more difficult. Political instability and pervasive crime make the wealthy nervous and worry about their ability to hang on to their wealth (for a related discussion, see Hall, Chapter 10 this volume). Russia is the prime example of a country where millionaires and billionaires were produced after the privatization of state companies in 1991 but where a strong government and compliant legal system can crush even the richest (Freeland, 2012). In the course of a decade more than one in every six businessmen in Russia faced some form of prosecution and a 2012 poll revealed that more than 50 per cent thought that their property was not safe from a government takeover (Feifer, 2014). In the 1990s Mikhail Khodorkovsky was one of the wealthiest men in Russia, his fortune based on the privatization of the oil industry. After bankrolling liberal opposition parties he was arrested in 2003, found guilty of tax evasion and sentenced to eight years in prison. In 2010 he was tried again, this time for theft and money laundering and sentenced to another six years. He was released in 2013 and now lives in Switzerland with an estimated fortune of between US$110 and US$500 million. His trials and imprisonment are a strong message to the oligarchy that even the richest cannot escape political retribution and no matter how wealthy you are the politico-legal system may be turned against you. At least ten Russian billionaires now live in London, a city considered much safer than Moscow or St. Petersburg (see Atkinson, Burrows and Rhodes, Chapter 11 this volume).

The rich are attracted by four main things: limited or no taxation on their wealth, the rule of law so that their wealth is not easily appropriated, low levels of fiscal transparency that assure them some measure of financial anonymity and, for entrepreneurs, a business environment with low rates of corporate tax and flexible labour markets. These ideal requirements do not fit easily together. Countries with very low taxation levels and fiscal anonymity may lack a strong enough rule of law to guarantee asset protection. Guatemala has very low personal income tax of 7 per cent of income but is racked by endemic violence. And countries that have strong rule of law may have high taxation levels, such as the USA where combined federal, state and local tax may account for 55 per cent of personal income. Different countries represent compromises between the different requirements.

The wealthy have a new found mobility as the globalization of finance allows their financial assets to be more easily moved around the world rather than fixed in national space. They thus pose an attractive proposition to countries and regions eager to attract capital. Immigration regulatory regimes began to shift accordingly from around 1990 to attract and retain this increasingly large pool of mobile wealthy. Canada led the way in using immigration policies to attract investors. Other countries followed suit and we now have many countries competing to attract the wealthy. Behind this shift is the belief that there is a lot of mobile and flight capital looking for a secure home and a competitive sense that no country wants to miss out on this opportunity. As more countries try to attract the wealthy and their capital there is a downward pressure on the barriers to immigration as countries compete to provide the most generous terms and ease of entry. For countries faced with a fiscal crisis, attracting the mobile wealthy looks like an easy and quick way to generate capital. Changing the immigration regime to make it easier for the wealthy to relocate is now part of fiscal balancing and economic development policy measures. Immigration policies are initiated and justified by governments as a benefit especially through the possibility of job creation. Though often promoted as a job creation scheme there has also been in recent years a populist backlash with mounting criticism of the 'selling' of citizenship at too cheap a price with too few benefits to the host society (Young, 2014). The criticisms are most pronounced when and where there are racial and ethnic differences between the host country and the migrant streams. Paradoxically, this populist strain grows louder during economic slumps as does the putative rationale for attracting the investor class.

We should be careful, however, in seeing a direct relationship between wealth and propensity to move to another country. In a detailed study of 1625 billionaires over two decades Sanandaji (2012) found that few billionaires migrated, perhaps because they do not need to move to protect their wealth. The very richest, with assets of more than US$100 million, generally do not need to migrate. Their material wealth and political and cultural capital is often tied to specific countries and places. They have the money to be both mobile and fixed in place. They have the ability to move their assets so that their personal residence is disconnected from their asset distribution and the need for wealth protection. The rich, those with assets from around US$10 million to US$30 million, are more likely to move than the billionaire class especially from countries with more volatile political systems, hazier rules of law and more uncertain economic futures. The super-rich with assets of between US$30 million and US$100 million are somewhere between these two categories in terms of propensity to move. Sanandaji's study found that only 13 per cent of billionaires migrated, most of them from lower billionaire per capita countries to higher per capita countries and from higher tax countries to lower tax countries. Since most immigration policy programmes are centred on the investor class with more than US$1 million in liquid assets, I will use the term rich or wealthy throughout the remainder of this chapter as a description of those with around US$1 million to US$30 million in assets. The term super-rich will be restricted to those with more than US$30 million, while the term billionaire is self-explanatory: they have a lot of money.

BEGINNINGS

Canada introduced one of the earliest and most successful programmes to attract the wealthy. As a long-term immigrant society Canada had established a strong regulatory system well regarded for its transparency for potential immigrants and its efficiency in processing applications. It was thus well positioned to successfully tweak its smooth-running system to attract the wealthy. Under its Immigrant Investor Program, first introduced in 1989, foreign nationals could gain residency by loaning interest-free CAN$800 000 to any of the provinces for five years. Now more than 20 countries have similar types of programmes (Box 18.1). Canada's entry 'fee' was relatively cheap, cheaper than the US requirement of US$1 million, with the added advantage of a more generous social welfare and

BOX 18.1 COUNTRIES WITH IMMIGRATION POLICIES TO ATTRACT THE RICH

Antigua and Barbados
Australia
Canada
Cyprus
Greece
Hungary
Ireland
Latvia
Malaysia
Malta
New Zealand
Portugal
Puerto Rico
Singapore
St Kitts and Nevis
Spain
Switzerland
UK
USA

Source: Author's own data.

lower health and education costs. The programme was very attractive to Mainland wealthy Chinese as it was a relatively cheap method to gain residence in a secure, safe country with generous social benefits.

Since 1990 more than 130 000 individuals have entered Canada through the Immigrant Investor Program. The relatively low financial hurdle encouraged the younger cohorts of the rich. A government-sponsored report concluded that the programme was a net benefit to Canada with an annual contribution of around CAN$2 billion (Ware et al., 2010). The bulk of this contribution was in the purchase of assets, mainly property, rather than in direct job creation.

The programme was so successful in attracting Mainland Chinese that it resulted in a backlash as cities such as Vancouver, a very popular destination, saw steep rises in house prices – blamed rightly or wrongly on Chinese investments – and widely reported studies revealed that investor immigrants paid less tax than other immigrant categories (Ley, 2010; Grubel and Grady, 2011).

While the programme was a boon to wealthy Chinese it was seen increasingly in Canada as selling citizenship too cheaply with negative effects on property markets. There was also a racist undertone to criticism as the programme was seen as an exclusively Chinese or Asian phenomenon and its termination dovetailed with a particular form of rising populist politics. When the programme was cancelled in February 2014 it had 59 000 pending applicants, 45 000 from Mainland China. At its peak in 2005 the programme was responsible for almost 11 per cent of the roughly 250 000 immigrants allowed into the country each year.

The Canadian case highlights the tailoring of immigration policies to attract the investor class. This new class of newly rich emerged from the liberalization of formerly centrally planned economies, such as China and the former Soviet Union. China's liberalization began in the late 1970s but privatization accelerated after 1992. The fall of the Soviet Union in 1989 inaugurated a frenetic privatization of public companies and resources. The quick and large fortunes that were made in this privatization frenzy, sometimes in the dubious circumstances of crony capitalism, created a newly enriched class eager to seek safe refuge for their newfound and precarious wealth. In the case of Canada, by the 1990s and first decade of the twenty-first century the new investor immigration policies dovetailed with the creation of more wealthy Chinese seeking security and political stability.

NEW IMMIGRATION POLICIES TO ATTRACT THE WEALTHY

Since the early 1990s, and following Canada's lead, many more countries have tailored their immigration policies to attract the wealthy. Let us look at the new immigration regimes of selected countries: UK, USA, Australia, Singapore and Malaysia.

UK

Since 1994 the UK has a Tier 1 (Investor) visa for those from outside the European Economic Area (EEA) and Switzerland willing to invest £1 million.[1] It is a fast-track system with a visa decision within three weeks. It allows applicants to come to the UK for a maximum of three years and four months. If applicants invest £10 million they can apply to settle after two years, and after three years if they invest £5 million. Applicants cannot invest in property but can invest in UK government bonds, UK registered companies and in the applicant's own business. The funds can be held either in the UK or overseas at time of application if deposited in a regulated financial institution and available to spend in the UK. If applicants borrow the £1 million they also need to have £2 million in personal assets. The Tier 1 (Investor) visa programme is a classic example of encouraging flight capital to move to a secure country with limited expectation or measurement of the job creation or overall benefit to the UK economy.

During the 2011–12 financial year there were 594 investor applications; only about 8 per cent were refused (Vine, 2013). A government study of the Tier 1 (Investor) programme, based on 20 interviews and email responses from a further 26 found that the migrants were attracted to access to Europe, quality of life and the rule of law (Nathan et al., 2013). One recurring problem cited by the interviewees was the residency requirement of nine months when the programme was first launched. Applicants had to surrender their passports during this period. It is now reduced to six months. Many of these investors have global connections to lubricate and any restrictions on their mobility are experienced as a major constraint on their business affairs and financial dealings.

Tier 1 is a cheap way for the rich, especially for those from Russia, China and the Middle East to buy British citizenship. For a mere £1 million applicants get to stay in the UK indefinitely en route to full citizenship, bring immediate family members and spend up to six months outside the UK. And while applicants are in the UK they will not be taxed on non-UK income and if they bring this non-UK income into the country, it is relatively easy to reduce or eliminate UK tax. Funds remitted for investment in a UK commercial business are tax-exempt. Overall then, a sweet deal. Applications run, on average, about 600 a year with a 92 per cent success rate (Nathan et al., 2013; Vine, 2013). A recent review concluded that the entry fee was too low, most of the money was invested in government gilts, for which there is no lack of demand, and the greatest impact was on the London property market, raising prices (for a discussion, see Atkinson, Burrows and Rhodes, Chapter 11 this volume). This upward surge gave existing homeowners a higher-priced asset but made it more difficult for average income households to gain access to an overheated property market (Migration Advisory Committee, 2014). New proposals are expected to be put forward in response to concerns that the existing investor visa route is failing to benefit the UK and is simply a cheap way for some wealthy Russian, Chinese and Middle Eastern families to settle permanently in Britain. There are also plans to increase the minimum requirement of £1 million.

The UK has particular appeal to the wealthy as a place of residence because it is perceived as having a good quality of life, the rule of law and, while ostensibly a higher than average tax regime, it has the advantages of fiscal anonymity, which can at times easily elide into tax avoidance. Wealth in the UK can be transferred easily into tax havens. The UK banking and investment system, based in London, is hardwired into a global offshore network of tax havens – the financial fragments of empire in such places as the inner ring of Jersey, Guernsey and the Isle of Man and an outer ring of Bermuda, Cayman Islands and British Virgin Islands (BVI). The investor programme is of limited appeal to the super-rich because they can park their capital and hide their wealth in the UK banking and investment system without having to be UK citizens. Wealth protection and anonymity is assured in the UK without the need for residence. If applicants live in the UK for less than six months – and many do to avoid the dreary winters – they can have the benefit of capital protection, and tax minimization without paying UK property taxes, as properties are often owned by offshore companies rather than individuals. Take the case of Rinat Akhmetov, one of the richest men in the world whose wealth is valued at US$15 billion. He is a Ukrainian oligarch who made money from the privatization of state enterprise and crony capitalism. In 2013 his companies secured one-third of all state tenders. Akhmetov owns one of the most expensive properties in London, a penthouse in One Hyde Park composed of two apartments, which was bought for US$215 million. The property was purchased through his holding company System Capital Management (SCM) and a BVI company, Water Property Holdings. After significant improvements and asset value increase the property was then sold by SCM to Rinat Akhmetov as a way to avoid the initial purchase costs and also to privatize company assets. He is one of the super-rich who can invest and live in London without the need to secure UK citizenship or residence requirement. The Tier 1 (Investor) programme is more suitable for the wealthy who feel the need for greater protection of their assets.

USA

The USA has the rule of law and a functioning civil society so it has many attractions for the wealthy. But there are disadvantages. Its taxation system is less lax than that of the UK, since it is based on fixed citizenship rather than mobile residency and capital flows are now more transparent, a result of post-9/11 attempts to monitor the international flow of terrorist funding. So the USA is attractive but with the distinct possibility of a higher tax burden and less banking secrecy than say the UK.

Since 1990 the USA has an employment-based programme tailored for the wealthy entitled EB-5.[2] Under this programme, 10 000 visas each year are reserved for investors to receive permanent resident status if they invest at least US$1 million (only US$500 000 in high employment and rural areas, a key element as we will see) in a commercial enterprise that employs at least ten full-time US workers. The residence status is for two years to ensure that the investment is not withdrawn and the jobs were in fact created.

This immigration policy dovetails more directly with employment and regional development goals than the UK's Tier 1 programme, in part a response to the more sensitive climate of public opinion about the 'selling of citizenship'. The USA is an immigrant society so the notion of buying citizenship when there are so many people waiting to get in is more publicly sensitive than in many other countries.

Studies by the trade group Association to Invest in the USA estimates that the programme contributed US$3.39 billion to US GDP and resulted in 42 000 jobs in fiscal year 2012 (Kay, 2012) while a more critical review of the EB-5 programme comes to the conclusion that the visas are too cheap, the programme is badly run, and it compares unfavourably with the programmes of other countries (North, 2012). Another critic argues that the programme discriminates against Latino investors and entrepreneurs for whom the threshold is too high (Starr, 2012). So, the threshold, according to these two more critical studies, is too low for the Asian rich and too high for the Latino entrepreneur.

The vast majority of the investments under the EB-5 programme were in real estate, property development and construction. The bulk of applicants came from China and South Korea (Vidal, 2013). The system is now institutionalized so that property developers in selected regions plan for a certain amount of EB-5 funding in their investment decisions. A 2011 investigation by the *New York Times* showed that many projects in New York were gerrymandered into high unemployment areas. Two journalists, McGeehan and Semple, documented the case of a 34-storey glass tower in the middle of affluent Manhattan that was classified, through selective and creative use of census statistics, as an area of high unemployment. The total cost of the building was US$750 million with one-fifth coming through the EB-5 programme from foreign investors seeking green cards that permit residency before citizenship. More than US$1 billion in property investment in New York alone from 2008 to 2011 was located in 'phoney' high-unemployment areas in order to benefit from the lower investment threshold. Most of the overseas investment for these deals was from Mainland China (McGeehan and Semple, 2011).

The programme constitutes only a tiny proportion of foreign investments in the USA, which has no shortage of overseas investors. The programme grew from around 3000 applicants in 2011 to around 8500 in 2013. The increase was due to the narrowing of the Canadian immigration route and increasing amount of wealth in China. More than four out of every five applicants now comes from Mainland China (Vidal, 2013). There is growing competition between countries for immigrant investors as the USA joins Australia, Canada and New Zealand in offering attractive programmes.

As a footnote we can also note that some want to maintain the benefits of US citizenship but avoid the tax load of being a wealthy citizen. In other words, they want to maintain citizenship and residency but 'offshore' and hide their taxable assets. The Swiss bank UBS actively recruited US citizens to invest offshore to avoid taxes. In 2008 the Federal Bureau of Investigation made a formal request to the bank for the names of its US clients, estimated at around 52000 with assets estimated at between US$18 billion and US$20 billion. In 2009 the bank paid a fine of US$780 million to the US government for encouraging tax evasion by US citizens. The French and German governments also initiated similar investigations of tax evasion by their citizens using Swiss bank accounts. With greater financial transparency tax evasion is more easily uncovered. We may be seeing the wealthy having to move themselves rather than just their assets to protect their wealth.

Australia

Australia is an immigrant society with a long history, like Canada, of fine-tuning its immigration policies to meet perceived and articulated national needs. In 2012 the Significant Investor Visa was introduced, popularly known as the 'golden ticket'.[3] It was implemented to attract wealthy Chinese investors and provide a source of capital for start-up economic ventures. Under this programme investors must invest AUS$5 million over four years and spend a minimum of 160 days in Australia in any one calendar year. The money must be invested in managed funds, government bonds, infrastructure projects, start-up companies or agribusiness. After four years the applicant can then apply for a permanent visa. Within a year of the programme's launch, 170 applicants worth a potential AUS$850 million were lodged. The first successful applicant was a 36-year-old Chinese toymaker (News.com.au, 2013). The programme is at such an early stage that not enough time has elapsed to undertake detailed cost–benefit or impact analyses.

Singapore

Singapore makes a conscious attempt to lure the wealthy and the super-rich. It is turning itself into the Switzerland of Southeast Asia, attracting billionaire Australians and Chinese magnates (Pow, 2013). The attractions are obvious: it has banking secrecy laws, social and political stability, easy access to Asia, Australasia and the Middle East, low personal taxation – with a maximum of 20 per cent compared with 45 per cent for both Australia and the UK and 55 per cent for the USA – no inheritance tax and no capital gains. There are resorts in the city-state developed specifically for wealthy foreigners. Sentosa Cove is one, a luxury condominium development on a private island enclave just off the coast of Singapore. More than 2500 units are located on a 290-acre (117 hectares) waterfront gated community. In June 2012 a Chinese billionaire purchased a bungalow for S$36 million. Two Australian billionaires, Gina Rinehart and Nathan Tinkler, have residences. Wealthy foreigners purchasing 99-year leases are given an initial three-residence pass that is renewable every five years.

Choon-Piew Pow (2011) surveyed a sample of residents in the exclusive resort and identified people from Australia, Britain, China, France, India, Indonesia, New Zealand, South Korea and Taiwan with occupations including an owner of a shipping company, a commodity trader, a former investment banker and owners of various businesses including garment factories and a global logistics firm. Singapore actively courts the rich and super-rich through 'specific place-based strategies including customized urban policies such as creating a favourable tax (-free) environment and "fast-track" permanent residency status coupled with the deployment of dazzling real estate products and super-rich enclaves such as Sentosa Cove' (Pow, 2011, p. 392).

Singapore also encourages wealthy immigrants through its Global Investor Programme (GIP).[4] Under this programme immigration entry is assured to foreigners who invest at least S$2.5 million in a business or the same amount in a GIP fund that in turn invests in Singapore-based companies. Singapore's success in attracting the wealthy creates a virtuous cycle as the presence of the wealthy increases and deepens the range of services such as fine dining, spas, and financial management services, which in turn attract even wealthier households. Singapore is now an attractive and increasingly well-known haven for the wealthy, especially after the publicity given to Eduardo Saverin. There are a growing number of millionaires now resident in Singapore, an increase to 17 per cent of total resident households in 2012 from 14 per cent in 2010 (Pow, 2013).

Singapore is an attractive destination in Asia-Pacific for the very rich. It has banking secrecy, low taxation and a fast-track residency. There are also special facilities being created specifically for the global wealthy. In 2010 Changi, Singapore's international airport, opened a Freeport, a storage facility for the fine art and other treasures of the wealthy. Freeports are technically 'in transit' and are thus secure, secret sites that provide a wonderful opportunity to liquefy capital and evade taxes. Developed in Switzerland they are now found in Luxembourg as well as Singapore, with one planned for Beijing. Singapore also wants to attract gold holdings, recently reducing a 7 per cent sale tax on precious metals (*The Economist*, 2013).

Malaysia

There is segmentation in the market for attracting the wealthy. Malaysia has an active programme of attracting the retired wealthy. In 1996 it introduced what was termed the Silver Hair Programme, but this was changed to Malaysia My Second Home in 2002. The requirements for the Malaysia My Second Home programme include a minimum monthly income of over RM10 000, liquid assets of RM500 000 and a fixed deposit in a bank account in Malaysia of RM300 000. For those prospective residents aged over 50 the respective amounts are RM10 000, RM350 000 and RM150 000. Income from overseas is not taxed. The programme guarantees a renewable ten-year residency. To attract the rich, Malaysia touts its high standard of living and relatively low living costs. In 2002 818 households were given entry under this programme, 596 from Asia. By 2012 the number had increased to 3227, with 2728 from Asia (Abdul-Aziz et al., 2014). The fact that Chinese and English are commonly used languages makes Malaysia an attractive international destination while the significant Muslim population makes it a comfortable place for the Islamic wealthy from Bangladesh, Iran and the Middle East.

CRACKS AND CREVICES

Just as cracks and crevices in the biological world can provide opportunities for life forms so the fractures in the global polity can give opportunities for elite mobility channels of immigration and residency. Here I will consider the case of Puerto Rico and countries in the Schengen Area and the European Union (EU).

Puerto Rico

Puerto Rico is under the sovereignty of the USA, with easy and unrestricted access to the USA but with enough relative autonomy to initiate a distinct tax system. The island is eligible for the lower threshold of US$500 000 in the EB-5 programme. Since 2012 Puerto Rico allows a 100 per cent exemption on capital gains, dividends and interest. This is attractive for wealthy US individuals who do not have to renounce their citizenship to gain access to this opportunity, one that almost 100 people have taken up. The programme was given greater visibility with the publicity surrounding the wealthy hedge fund manager John Paulsen, net worth of around US$11.4 billion, who seriously considered the idea. He remained in the USA but invested heavily in the island to benefit from the tax advantages (Marino, 2013).

Schengen Area and EU

Much of Europe now counts as a single territorial unit in terms of freedom of movement and capital mobility. In 1985 five members of the European Economic Community – Belgium, France, West Germany, Luxembourg and the Netherlands – signed an agreement in the Belgian village of Schengen that guaranteed passport-free movement across their common borders. By 2014 the agreement allowed passport-free movement across the common borders of 26 European countries including 22 EU member states. The agreement also allows third-country visitors to travel freely once they are admitted into the Schengen Area. Access to the area varies by the point of entry. Some countries use their privileged position to attract the wealthy. In Latvia, for example, anyone who buys property worth at least LVL50000 (US$96000) in provincial cities, and LVL100000 (US$192000) in Riga, receives a five-year residency permit that allows them access to other countries in the Schengen Area. The programme, introduced to prop up the property market, allows easy access for wealthy people especially from China, Russia and Kazakhstan (mainly Russian speakers).

Greece, Spain and Hungary also have programmes that provide visas in exchange for money. Since 2012 Portugal has a 'golden visa' guaranteeing two-year residency in return for a €500000 investment in real estate investment or a €1000000 investment that creates 30 jobs. By March 2014, 542 visas were issued, with 433 going to Chinese applicants. Many of the applicants remain in China but the programme allows them to gain access to educational opportunities and the possibility of European residency for their children – a very important insurance against economic and political upheaval in China. In 2013 Spain and Greece adopted similar programmes for real estate investments of €500000 and €250000 euros respectively. The same year Hungary gave a residence permit in return for an investment of €250000 and a payment to 'partners' of the government for at least €40000. Cash-strapped nations in the Schengen Area can use their entry opportunity as a way to attract mobile capital in return for residency and fast-track citizenship that provides wider European mobility and the possibility of EU citizenship.

Malta proposed an Individual Investor Programme that offered citizenship for a straight fee of €650000. There were neither any investment nor residency requirements. One company, Henley and Partners, was given the monopoly of processing the applications and assured a fee of €70000 from each applicant as well as a 4 per cent commission. After heavy criticism, both domestically and from European partners that the programme was effectively selling European citizenship, the programme was placed on hold and then in November 2013 a revised programme offered citizenship in return for €1 150 000. The wider consequences of the decision are explored in Shachar and Bauböck (2014) who argue that the EU has little leverage against the selling of EU citizenship and that policies that link citizenship with ability to pay can easily lead to a corruption of democracy.

The Maltese case highlights the problem of a European-wide ease of mobility and access to citizenship with differing entry rules used by constituent national members. One way to plug a fiscal gap, generate revenue and fill the national coffers – especially for smaller, poorer countries and especially those undergoing property collapses, fiscal problems and economic uncertainties – is to effectively sell access to Europe and EU citizenship to the wealthy. And the competition especially for the wealthy Chinese investor class is driving down the entry barriers. In 2012 Portugal offered €500 000 for residency; the next year Greece asked for only €250 000.

Problems have arisen. Cyprus, a 'hot money' destination for Russian money and flight capital, as well as a member of the EU, also allowed passport and citizenship to the wealthy. The requirements were either €5 million in real estate or €3 million in bank deposits. For wealthy Russians in particular, this provided what seemed like a safe haven to move/hide/protect their investment while also holding out the possibility of residency and citizenship access to a wider Europe. However, Cypriot banks invested very heavily in Greek government bonds. The Greek financial crisis meant that these bonds were worthless and so the Cypriot banks were very vulnerable and in March 2013 the two main banks were closed. The European Central Bank bailed them out but deposits of over €100 000 were taxed at almost 10 per cent. Russian investors lost over US$3 billion from their total of US$31 billion in the two main Cypriot banks. Subsequently, most rich investors in Cyprus are taking the real estate rather than the bank deposit option.

Wealthy Chinese are particularly attracted to investment property; the Chinese rich have three to four times more of their capital invested in property than the non-Chinese rich. Despite the risks of volatile property markets, the attractive schemes of Greece, Spain and Portugal tap into the property bias of wealthy Chinese who can use the properties as rental units, vacation homes, or to lie empty for asset increase while they endow the ability to travel freely within the Schengen area for 90 days every six months.

It is not just the Southern European countries that encourage easy access. Belgium, for example, has no restriction on dual citizenship, no limitations on minimum stay and provides visa-free travel to Schengen Area countries. It is the only country that permits application for citizenship after only three years of legal continuous residence.

There is a backlash. The initial Maltese proposals were roundly condemned at home and overseas. As citizenship for sale becomes a more political issue it raises wider questions and ultimately becomes a matter of concern for the European Parliament, perhaps pitting 'looser entry' states against those with more stringent citizenship requirements.

CONCLUSIONS

Many of the rich are on the move and more states want to attract them. There is now a significant number of wealthy people looking for asset protection, safety and security. And there is a corresponding restructuring of national immigration regimes to favour the wealthy.

There is also the creation of an entire subsector of financial service that we may term the immigration-residency-financial protection sector that consists of immigration lawyers, accountants, tax attorneys, realtors, fixers and connectors who ease the path for citizenship, residency and property purchase for the rich (Beaverstock et al., 2013). This is an increasing part of the advanced producer services sector that attends to the needs and demands of the wealthy and the super-rich.

Since 1990 immigration policies in many countries have been tweaked to favour the wealthy and attract the rich. We now have global mobility corridors for the rich and super-rich, an undermining of equal access to citizenship, and in effect a dual path to citizenship, an easy one for the wealthy and a much more difficult one for the non-rich. We have migrant millionaires on the one hand and the permanently denied on the other.

One of the claims for those who promote and support cash for citizenship is that it provides benefit to the country. This is more of an assertion than a demonstrated fact. In truth, there are too few studies by independent researchers to make a qualified claim. We have consultancy reports galore from government bodies that show, without much real economic analysis, that the programmes make money and create jobs (see, for just one example, Kay, 2012). In fact the wider benefits to national economic growth have yet to be fully documented. Even most official reviews have now come to the conclusion that the entry prices are too cheap. The competition for the investor class has created a downward pressure on the price of entry. The growing competition between states, especially among those offering a cut-rate price to EU citizenship, is reducing the effective benefits. Malta's initial plan for €650 000 was ludicrously low yet even the revised limit of €1 150 000 is still too low a price for the benefits involved. As more countries compete for the mobile wealthy the cost of entry is lowered.

Attracting the investor class, while it fits into neoliberal ideas of the world, has in fact, provided less than the trumpeted benefits. The wider benefits are relatively small and few jobs are created. The biggest impact is on property markets. While this may boost sagging markets the longer-term effect may be to increase the price of accommodation for locals and nationals.

Residency and citizenship are now, in effect, for sale to the wealthy. However, there is a mounting realization that the programmes, hastily conceived in the rush to attract the newly wealthy citizens of Russia, China and other countries, are too cheap with few wider national benefits. Rich foreigners driving up property prices with full access to generous social welfare and public education opportunities is not an easy selling point for politicians seeking some form of democratic legitimacy. There is growing criticism of the programmes, a function of the mix of xenophobia, distaste at the selling of residency and citizenship and a realization that the benefits are less than touted. If the political costs exceed the putative economic benefits there may be a shift, prompted by national political forces, away from the creation of lubricated migration channels for the wealthy and the underpriced selling of citizenship.

NOTES

1. Official website: https://www.gov.uk/tier-1-investor/overview.
2. Official website: http://www.uscis.gov/working-united-states/permanent-workers/employment-based-immigration-fifth-preference-eb-5/eb-5-immigrant-investor.
3. Official website: http://www.immi.gov.au/FAQs/Pages/What-is-the-significant-investor-visa.aspx.
4. Official website: https://www.contactsingapore.sg/investors_business_owners/invest_in_singapore/global_investor_programme/.

REFERENCES

Abdul-Aziz, A., C. Loh and M. Jaafar (2014), 'Malaysia's My Second Home (MM2H) Programme: an examination of Malaysia as a destination for international retirees', *Tourism Management*, **40**, 203–12.

Beaverstock, J.V., S. Hall and T. Wainwright (2013), 'Servicing the super-rich: new financial elites and the rise of the private wealth management retail ecology', *Regional Studies*, **47** (6), 834–49.

Feifer, G. (2014), *Russians: The People Behind the Power*, New York: Hachette.

Freeland, C. (2012), *Plutocrats: The New Golden Age*, New York: Random House.
Grubel, H. and P. Grady (2011), *Immigration and the Canadian Welfare State 2011, Fraser Institute*, accessed 2 September 2014 at http://www.fraserinstitute.org/uploadedFiles/fraser-ca/Content/research-news/research/publications/immigration-and-the-canadian-welfare-state-2011.pdf.
Kay, D. (2012), *Economic Impacts of the EB-5 Immigration Program*, Chicago, IL: Association to Invest in the USA.
Ley, D. (2010), *Millionaire Migrants: Trans-Pacific Lifelines*, Chichester, UK: Wiley.
Marino, J. (2013), 'Law 22 attracting millionaire investors to Puerto Rico', *Caribbean Business*, **41** (11), accessed 2 September 2014 at http://www.caribbeanbusinesspr.com/prnt_ed/law-22-attracting-millionaire-investors-to-puerto-rico-8300.html.
McGeehan, P. and K. Semple (2011), 'Rules stretched as green cards go to investors', *The New York Times*, 18 December, accessed 28 April 2014 at http://www.nytimes.com/2011/12/19/nyregion/new-york-developers-take-advantage-of-financing-for-visas-program.html?pagewanted=all&_r=0.
Migration Advisory Committee (2014), *Tier 1 (Investor) Route: Investment Thresholds and Economic Benefits*, London: UK Government, accessed 28 April 2014 at https://www.gov.uk/government/uploads/system/uploads/attachment_data/file/285220/Tier1investmentRoute.pdf.
Nathan, M., H. Rolfe and C. Vargas-Silva (2013), *The Economic and Labour Market Impacts of Tier 1 Entrepreneur and Investor Migrants. Report to the Migration Advisory Committee*, London: UK Government, accessed 22 August 2015 at https://www.gov.uk/government/uploads/system/uploads/attachment_data/file/257258/economic-research.pdf.
News.com.au (2013), 'Chinese toy maker "buys" $5 million visa to bring family to Victoria', accessed 4 August 2014 at http://www.adelaidenow.com.au/news/national/chinese-toy-maker-buys-5-million-visa-to-bring-family-to-victoria/story-fncz7kyc-1226634922284.
North, D. (2012), *The Immigrant Investor (EB-5) Visa; A Program That Is, and Deserves to Be, Failing*, Washington, DC: Center for Immigration Studies.
Pow, C.-P, (2011), 'Living it up: super-rich enclave and transnational elite urbanism in Singapore', *Geoforum*, **42** (3), 383–93.
Pow, C.-P. (2013), '"The world needs a second Switzerland": onshoring Singapore as a liveable city for the super-rich', in I. Hay (ed.), *Geographies of the Super-Rich*, Cheltenham, UK and Northampton, MA, USA: Edward Elgar Publishing, pp. 61–76.
Sanandaji, T. (2012), 'The international mobility of the super-rich', *IFN Working Paper No. 904*, accessed 28 April 2014 at http://www.econstor.eu/handle/10419/81436.
Shachar, A. and R. Bauböck (2014), 'Should citizenship be for sale?', Florence: European University Institute, accessed 28 April 2014 at http://hdl.handle.net/1814/29318.
Short, J.R. (2013), 'Economic wealth and political power in the second Gilded Age', in I. Hay (ed.), *Geographies of the Super-Rich*, Cheltenham, UK and Northampton, USA: Edward Elgar Publishing, pp. 26–42.
Starr, A (2012), *Latino Immigrant Entrepreneurs*, New York: Council on Foreign Relations.
The Economist (2013), 'Uber-warehouses for the ultra-rich', *The Economics*, 21 November, accessed 4 August 2014 at http://www.economist.com/news/briefing/21590353-ever-more-wealth-being-parked-fancy-storage-facilities-some-customers-they-are.
Vidal, D. (2013), 'Startup immigration: stimulating startup communities with immigrant entrepreneurs', *McGeorge Law Review*, **45** (2), 319–46.
Vine, J. (2013), *An Inspection of Applications to Enter and Remain in The UK Under The Tier 1 Investor and Entrepreneur Categories of Points Based System 2012–2013*, London: Independent Chief Inspector of Border and Inspection Reports, accessed 22 August 2015 at http://icinspector.independent.gov.uk/wp-content/uploads/2013/09/An-Inspection-of-Tier-1-PBS-Investor-and-Entrepreneur-Applications.pdf.
Ware, R., P. Fortin and P.E. Paradis et al. (2010), *The Economic Impact of the Immigrant Investor Program in Canada*, Analysis Group Consultants, accessed 22 August 2015

at http://www.analysisgroup.com/uploadedfiles/content/insights/publishing/canada_iip_report_english.pdf.
Young, I. (2014), 'Anger over rich Chinese blamed for Canada migration scheme's axing', *South China Morning Post*, 20 February, accessed 2 September 2014 at http://www.scmp.com/news/world/article/1431991/anti-chinese-backlash-inspired-axing-canadas-investor-scheme-says-critic.

19. Sovereign wealth and the nation-state

Adam D. Dixon

INTRODUCTION

No comprehensive treatment of wealth and the super-rich would be complete without a presentation and analysis of sovereign wealth. Indeed, the great fortunes of high-tech entrepreneurs, hedge fund managers, and industrialists and their descendants, often pale in comparison to the assets controlled by some countries whose sovereignty is embodied by a single and often autocratic or quasi-autocratic ruling dynasty.[1] For the most part, these great fortunes are found in the Gulf States of the Middle East and Saudi Arabia – the product of vast natural resource wealth that has ballooned almost exponentially in the last decade from booming world commodity prices.

Although such fortunes have existed for more than a generation now – being recycled through international banks in the major financial centres of New York and London (Momani, 2008) – their organizational form and function have become more defined as established institutional investors, with a presence and self-awareness not seen previously. Even if much of the management of many of these great fortunes is still delegated to portfolio managers and private equity shops in the capitals of finance, these fortunes are not simply an accounting entry in the national treasury or a section of the central bank's balance sheet. Keeping pace with the globalization of financial services, and all that brings in terms of common practices and organizational forms, these great fortunes are increasingly represented by state-sponsored institutional investment funds, or what are now commonly referred to as sovereign wealth funds (SWFs).

The term, coined by fund manager and sovereign financial advisor Andrew Rozanov (2005) in the middle of the last decade, reflects the greater self-awareness these great fortunes have acquired, but more importantly the growing influence these funds have in global financial markets. A term was needed, moreover, because the growth in the number of SWFs, which include other state-sponsored investment funds from sovereign nations across the political spectrum, has been almost exponential in the last decade (Monk, 2011; Balding, 2012). Whereas at the end of the twentieth century there were only a handful of these funds, more and more countries have, or are in the process of establishing, their own SWFs.

While estimates vary depending on definition, by the author's account there are approximately 60 SWFs in existence around the world with around US$5–6 trillion in assets under management.[2] These numbers include SWFs from countries of different sovereign form, from social-democratic Norway and its Government Pension Fund – Global, to quasi-democratic city-state Singapore and its funds Temasek and GIC, and absolute monarchies like Qatar and its Qatar Investment Authority. While many SWFs are funded with the revenues from commodity production, mainly oil and gas, some SWFs, such as the China Investment Corporation are financed by balance of payments surpluses. Others, such as France's *Fonds stratégique d'investissement*, are financed by the proceeds of state asset privatizations or fiscal surpluses.

In an era of neoliberal and financialized global capitalism, where state involvement in the economy is discounted, at least in rhetoric, some in the West do not consider SWFs to be legitimate actors in financial markets (see, e.g., Beck and Fidora, 2008; Bremmer, 2010; Truman, 2010). This is not to say that SWF capital is unwelcome. Rather, they see SWFs as geopolitical power tools serving the interests of economic nationalism, instead of pure financial return-driven institutional investors. The actions of hedge funds and private equity shops, which have been compared to vultures or war machines (Engelen et al., 2011), are certainly not immune to criticism and questions of legitimacy, but their status as private actors affords them a freedom of movement in the world's largest economies and financial markets that is not always afforded to SWFs.

But such ideological purity as to who can participate and through which means comes up against a reality where the state, in its various sovereign forms, interacts with market capitalism in untold ways. In the United States, for example, the largest individual asset owners are public pension funds sponsored by state and local governments. Although their management is often through private means and following common theoretically informed beliefs on how assets should be managed, namely following the tenets of modern portfolio theory (Markowitz, 1952), which renders direct influence by the state and its representatives inert, the degree of separation between state and private capitalist market activity is relatively thin. While efforts to immunize the inherent political nature of SWFs renders them similarly inert, geopolitics adds a further layer of tension that subnational actors in the West, such as public employee pension funds, do not have to face.

This chapter chronicles the geography of sovereign wealth in contemporary globalization, drawing on examples from individual nations and the investment funds that manage such sovereign wealth. Although some SWFs are an extension of dynastic wealth, the wider field of SWFs is surveyed. Hence, this chapter takes some liberty with the definition of super-rich (for a critical discussion of this term, see Chapter 2 by Koh, Wissink and Forrest in this volume) to include large pools of capital directly controlled by the sovereign state, where sovereign form ranges from monarchies to democratic republics. The next three sections explain the rise of SWFs, considering different types of SWF and their different mandates. The penultimate section considers whether SWFs are a 'threat' to the economic and geopolitical interests of receiving countries, or are instead a form of salvation in the context of fiscal deficits and austerity, particularly in the advanced economies of Europe and North America. The final section concludes.

EXPLAINING THE RISE OF THE SOVEREIGN WEALTH FUND

Figure 19.1 shows the largest SWFs, excluding the pension reserve fund subvariety,[3] with assets under management greater than US$5 billion at year-end 2013. As is evident, the largest SWFs in the world are concentrated in the Middle East and East Asia, save for the Norwegian Government Pension Fund–Global (GPF-G). In the Western hemisphere there are few SWFs in comparison and the size of assets under management pales in comparison. This reflects, arguably, the limited scope national governments have in accumulating capital in the form of an SWF. In the United States, in particular, the federal government would have great difficulty in legitimizing a government-sponsored institutional investor, notwithstanding the legitimate role of public-sector pension funds and the Social Security Trust Fund – the latter of which is simply an accounting entry that the government owes itself. If the federal government were to accumulate capital surpluses in any form, it would likely be forced to redistribute such surpluses, probably through lower taxation, or it would likely be forced to retire national debts. The politics of sponsoring an institutional investor would be highly contentious where state intervention is already problematic and where financial markets and the financial services sector are already well developed. This feature of the US political economy is in many respects shared across other advanced economies. But the limited incidence of SWFs in the advanced economies is not simply a function of domestic political economy, but rather the relative economic and political power of the advanced economies in the global political economy.

Even though SWFs have gained increased notoriety in the last decade, not least because there are so many new ones, the SWF is not actually a

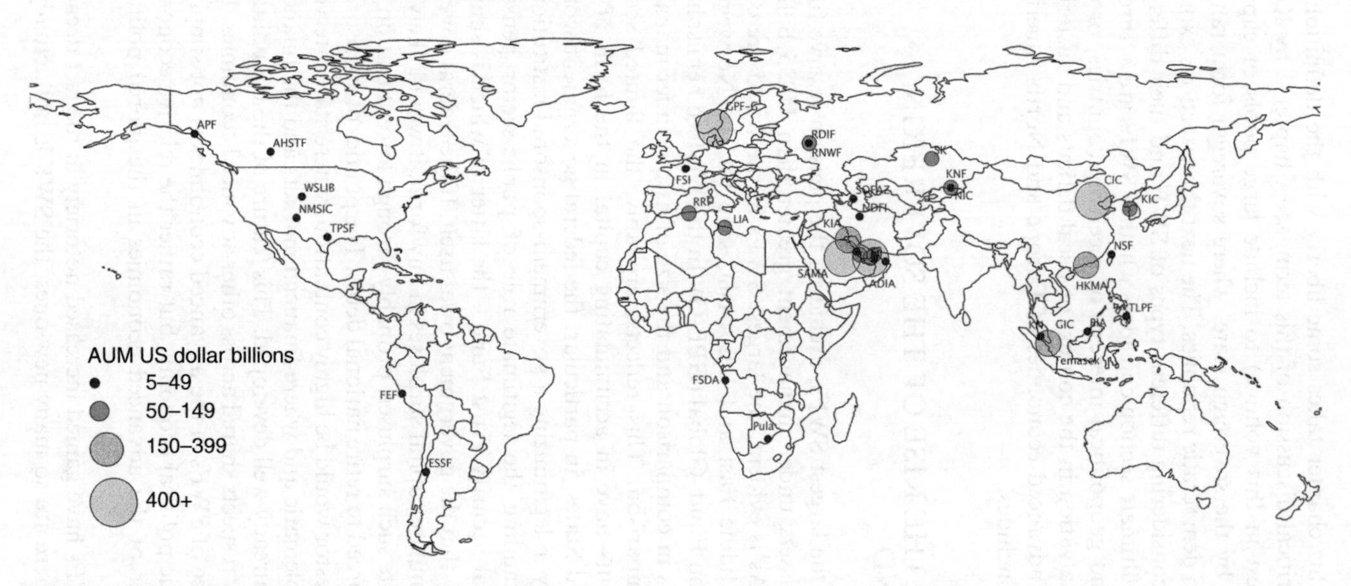

Note: AUM = assets under management.

Source: Author's compilation.

Figure 19.1 The geography of sovereign wealth funds

recent phenomenon. The organizational form of the SWF has developed in line with global financial integration and the dissemination of common asset management techniques, but the sovereign sponsorship of wealth funds has a long history in contemporary capitalism. For example, at the state level in the United States, government-sponsored wealth funds have existed since the nineteenth century. Some consider the first SWF to be the Texas Permanent School Fund, which was founded in 1854 and whose investment income is used to fund primary and secondary education in the state (Rose, 2011). But subnational SWFs are less significant in terms of geopolitics and the international political economy than SWFs sponsored by nation-states. The latter's history is more recent. One of the first still existing SWFs is the Kuwait Investment Authority, which was founded in London in 1954 as the Kuwait Investment Board while the country was still a protectorate of Great Britain. In that respect, the SWF was a creation of Western financial capitalism and not a *sui generis* creation derivative of some form of statist non-liberal capitalism (Clark et al., 2013).

Although Western financial capitalism appears biased against more statist and interventionist forms of capitalism, the state is still a crucial actor in the construction and maintenance of the global capitalist market economy (Wallerstein, 1983). As such, states in various ways engage with global capitalism, such as through supporting industrial development, financing R&D, or protecting intellectual property. The SWF can be classed as another tool for engaging in global capitalism (Dixon and Monk, 2012; Hatton and Pistor, 2012). Notwithstanding the underlying intent of the engagement via the SWF, the engagement ensures, in abstract terms, a certain degree of proximity for the state. Put slightly differently, states have various ways of engaging and following the leading edge of capitalist development, largely through regulation, supervision and macroeconomic management granted by fiscal and monetary authority. The SWF provides another avenue through which the state can participate (which does not necessarily entail active influence) in what is ostensibly private market activity.

Recognizing that global capitalism is a powerful force, with its expansionary logic driven by the search for new markets, technological change, the profit motive and the political ambitions of individuals and entire populations, engagement for the nation-state is by default a necessity. Some may argue that there is a choice (Gibson-Graham, 1996), but the collapse of alternative economic systems, namely the fall of communism, appears to limit the choice not to engage. Despite the necessity to engage, engagement does not mean that states are insensitive to the predations and risks associated with global market capitalism and the rapid changes and uncertainty it may bring. States employ a variety of tools to resist or tame the potential negative consequences of the market. Welfare state institutions and social insurance are the most obvious form of resistance. Others include anything from non-tariff barriers to the projection of military power and non-market industrial policy underwritten by a military-industrial complex, as the United States exhibits. Obviously, many states around the world are very limited in their ability to resist the incursions of global markets, in comparison to the high-income advanced economies of North America, Europe and Japan.

The SWF offers a tool of resistance to the free market. There are obvious limits to resisting the pressures of the free market, at least in the medium to long term. The force of the free market will force change eventually. But in the short term an SWF can provide a country the means to withstand an external shock, such as a balance of payments crisis. In that respect, the SWF can be utilized as a lender of last resort, underwriting a fledgling economy through the worst moments of a crisis. The SWF is therefore an insurance policy at the disposal of a country's economic managers. More importantly, the SWF is an insurance policy over which they have control and direction. Calling on the economic resources of the country's SWF does not come with conditions, as a country would face, for example, if it were to seek help from the International Monetary Fund – conditions that may be politically undesirable and injurious to some in society.

If resisting the economic and social consequences of short-term economic crises is important, particularly for sustaining higher potential economic growth and a more stable development trajectory, it is perhaps even more important to resist capitalist crises of long duration. These are the crises that result from the accumulation of change in social structures, such as demographic ageing, and the evolutionary technological changes that reshape, and in some cases ultimately eliminate, entire industries. They are also the crises that are produced by capitalist geographic expansion and consolidation of markets for products and services (Storper and Walker, 1989), when certain forms of resistance and protection a country has at its disposal become unviable or hopelessly inadequate. Without overstating the capacity, competence and power of any institutional investor and its financial resources to shape possibilities for resisting crises of a long duration, the SWF may still be an important component of a polity's answer to future economic and demographic contingencies. Perhaps the most obvious, an SWF, in the form of a pension reserve fund, functions to mitigate the effects of demographic ageing by buffering a pay-as-you-go social security system.

If the language of engagement and resistance provides a useful, yet partial, means to conceptualize the SWF as a policy device at the disposal of the sovereign state, it does not provide an adequate explanation for the conditions that make the creation of a SWF possible in the first instance. Nor does it provide sufficient clarity as to the different types of SWF. To understand the foundations of any SWF, it is necessary to consider its source capital. Although the source capital does not necessarily determine the mandate of an SWF in all cases, source capital is frequently correlated with its stated policy function. Source capital, as indicated previously, generally originates from commodity revenues, balances of payments surpluses and fiscal surpluses. In each case, there is wider connection to developments in the global economy.

THE COMMODITIES SUPER-CYCLE

Over much of the twentieth century the price of basic commodities, particularly petroleum, was relatively stable. Aside from the two oil shocks of the 1970s, the cost of petroleum for consuming countries and regions was relatively low, and therefore so were the revenues received by producers. As Figure 19.2 shows, the 2012 adjusted price of crude oil in 1900 was US$32 per barrel. In 1950 the 2012 adjusted price was US$16 per barrel. For much of the post-World War II economic boom years in the advanced countries, prices remained subdued. The first oil shock in 1973 saw the

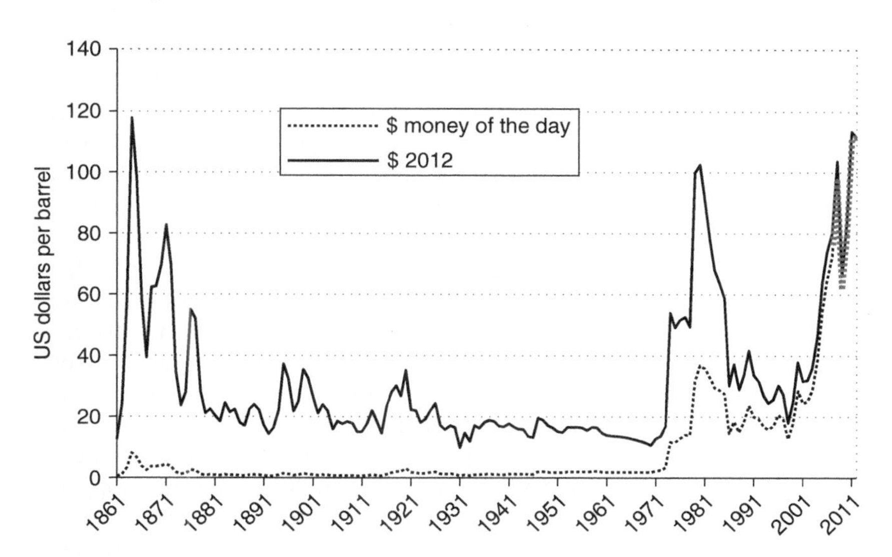

Source: *BP Statistical Review of World Energy 2013.*

Figure 19.2 Crude oil prices, 1861–2012

real (2012 dollars) price increase from US$17 per barrel to US$54 per barrel in 1974. The second oil shock in 1979 saw prices increase again to over US$100 (2012 dollars) in 1980. During that period oil-producing countries, particularly in the Middle East, received huge windfalls from the crisis, enriching government coffers, their political leaders and a new local super-rich class in the process (Khalaf, 1992). It was this period that set the groundwork for some of the world's largest current SWFs (see Figure 19.1).

After the 1970s oil shock petroleum prices decreased back toward their historical average. For petroleum-producing countries this still represented positive cash flow, but not to the degree experienced during the oil shocks. Yet, this stability in oil prices would be short-lived. At the turn of the century, prices would resume an upward trajectory, fuelled by increasingly strong demand from fast-growing emerging market economies like China and India coupled with an insatiable thirst for oil in the rich countries, as well as the decreasing availability of easily extractable petroleum resources. Consequently, higher prices and increasing resource exhaustion in certain geographies has pushed production out to new countries and regions, bringing possibilities for wealth generation to countries not generally associated with the oil-rich producing states (namely members of OPEC) that dominated the markets and the headlines in the twentieth century. Not surprisingly, this has resulted in the establishment or planned establishment of new SWFs. This is particularly the case in Africa, which has seen an expansion of oil and gas production as well as other primary commodities. Although estimates vary, the number of SWFs in Africa ranges from the low to mid-teens. Add to that the number of countries considering or constructing new SWFs, and Africa could soon be home to upwards of 20 or more SWFs (Barbary et al., 2011). As the size of many of these funds remains small, they are not shown in Figure 19.1.

Considering global population growth and an expanding middle class in emerging markets it would seem cogent to assume a trend of high long-term commodity prices, at least at levels higher than the long-term average of the twentieth century. This provides a strong incentive for governments controlling access to natural resources to exploit them and utilize the resource to finance government spending. If commodity prices are high and stable, this is a feasible fiscal policy. If commodity prices are more volatile or collapse in the short term, as they were following the subprime financial crisis of 2007–08, then this fiscal policy is unsustainable.

The other problem for commodity-producing economies is the problem of Dutch disease (Corden and Neary, 1982; Corden, 1984). This occurs when growth in the extractive sector leads to upward pressure on the value of the national currency on foreign exchange markets. Consequently, the traded manufacturing sector, inasmuch as one exists, becomes uncompetitive on global markets. The non-extractive sectors are furthermore under pressure as a larger proportion of economic resources flow to the extractive sector. Economic growth and development becomes highly asymmetric, which may be detrimental to the economy and society over the medium and long runs, as natural resources are extinguished or as prices collapse (Auty and Mikesell, 1998; Auty, 2001).

For sponsors of commodity-based SWFs, the ideal short-term economic function of the SWF is to provide stability to government resource revenues from one year to the next, while also serving as a buffer in the event of a collapse in prices (Davis et al., 2003; Das et al., 2009; Collier et al., 2010). In turn, by holding some of the proceeds of commodity production in foreign assets instead of injecting them in the domestic economy, the SWF limits appreciation of the national currency and thus the effects of Dutch disease. Once these functions are covered, the ideal long-term economic function of the SWF, or more appropriately a separately sponsored SWF than that of the stabilization fund, is to save for future generations when commodity exploitation and resource revenues will have decreased or have come to an end.

For developing countries, however, saving for the future is not necessarily the most beneficial route (Dixon and Monk, 2011; Gylfason, 2011). A diverse and modern economy that provides economic opportunities and well-being to current generations, underwritten by accessible and comprehensive education and healthcare systems, and basic provision of infrastructure, for example, is arguably a more effective way of ensuring the prosperity of future generations. To give an example, the strategy taken by Norway in the accumulation and saving of its petroleum wealth as an economic resource for future generations is not necessarily the best strategy for a capital-starved low-income country in Africa blighted by poverty, poor infrastructure and inadequate public services. For the former, there is a clear economic case for saving for future generations. For the latter, there is a clear case for putting the capital to work now by way of a sovereign development fund – a type of SWF devoted to purely developmental goals (Santiso, 2009).

There are few active sovereign development funds in the developing world, even though interest in them is growing. Moreover, such funds are not a replacement for parliamentary decision-making as to the allocation of state resources. Put simply, sovereign development funds, as institutional investors akin to a private equity fund, have limited influence. They can engage in pump-priming activities to support industrial and infrastructural development, but such pump priming is likely only to be sustainable in the long term if operationalized in an explicitly commercial manner. In other words, they must generate financial returns in the first instance to justify their existence. This does not negate possible double or triple bottom lines, and an explicit emphasis to find them. Rather, if financial returns are not a core operational and mandated concern, a different type of organizational form (e.g., a charity) is more appropriate.

Although the economic rationale behind the commodity-based SWF is fairly obvious for resource-rich economies, the underlying political rationale may be driven by a variety of reasons that have little to do with improving the prospects for the country's citizenry at large. The structure of political authority of sponsoring governments in the Middle East, as noted in the introduction, is generally that of an absolute monarchy (e.g., Saudi Arabia, Qatar, Oman, United Arab Emirates) or a constitutional monarchy with some notionally democratic institutions (e.g., Kuwait). In these countries vast natural resource wealth has produced a rentier social contract (Karl, 1997; Ross, 2012). As governments do not rely on popular taxation to finance the institutions of government, the government is essentially free from popular accountability. In modern democracies, in contrast, the power to tax and the legitimacy surrounding that power rests on government accountability and the existence of popular representation in the functions of the state. In a rentier state, governments have limited incentive to be accountable and to share decision-making across the citizenry. Moreover, resource revenues can be used to patronize different interests groups inasmuch as they exist.

As this state of affairs implies, the resilience of resource revenues is of critical importance to maintaining the existing structure of political power. Consequently, there is an incentive for the ruling elite to effectively manage resource wealth over the long term. Understanding this incentive provides a more nearly complete explanation for some commodity-based SWFs, particularly those associated with autocratic or oligarchic regimes (Dixon and Monk, 2012). In establishing a commodity-based SWF, the sovereign can diversify resource revenues into public and private securities and real assets around the world. Instead of relying on current and future commodity production alone, the sovereign authority is relying on the performance of the global economy.

With many resource-rich developing countries rushing to establish new commodity-based SWFs it is questionable whether this development is universally beneficial for the country. Although an SWF may provide a tool for managing resource revenues that is more in line with the realities of investing in the twenty-first-century global economy, an SWF should not necessarily be seen as an institutional innovation that necessarily counteracts the resource curse and the concentration of power in the hands of a few. SWFs are still creations of the sovereign authorities that sponsor them. While their economic effects as a policy tool managing resource revenues may prove beneficial for the wider economy, they may simply reinforce the existing balance of political power.

EXPORT-LED GROWTH AND DEVELOPMENT

To explain the source capital of a second variant of SWF, the reserve fund or those funds derivative of reserve funds, we need to understand the structure of the global political economy and how it developed over the latter half of the twentieth century, and the importance of export-led growth. At the end of World War II, the economies of Europe and Japan were in a shambles on the whole – save for a few isolated pockets that had avoided conflict (e.g., Sweden, Switzerland). Productive capital (i.e., factories and machinery) and major infrastructure had been destroyed or was in need of remodification to civilian uses. Europe and Japan were reliant on imports of goods and capital from the United States to cover consumption and reconstruction efforts, quickly exhausting their foreign exchange reserves and leading to a balance of payments crisis in need of correction (Scammell, 1980; Eichengreen, 2008).

The international monetary system established at Bretton Woods, New Hampshire in 1944, set to stabilize the international economy, establishing a system of fixed exchange rates centred on the US dollar pegged at US$35 to an ounce of gold. The US economy, acting as an open mass consumer market and a major supplier of capital, would become the core of the first-world global economy supporting the reconstruction and development of Japan and Western Europe (Dooley et al., 2004). The dollar peg, in effect, allowed Western Europe and Japan to redevelop through an export-led growth model, underpinned by undervalued currencies. Whereas at the end of World War II the United States was a major creditor with a large trade surplus, the three decades after the war completely reversed this, with Western Europe (primarily West Germany) and Japan having accumulated large trade surpluses and large foreign currency reserves. By the middle of the 1960s the United States began to have trouble maintaining the peg that was set at Bretton Woods, as its balance of payments deficit continued to grow. US gold reserves declined and surplus dollars held abroad grew. The system was in significant disequilibrium, and during the presidency of Richard Nixon the United States left the gold standard, in part to force a revaluation of the yen and the Deutschmark.

This major change in US policy represented the recognition that the destroyed economies of Europe and Japan had been rebuilt. Put simply, Europe and Japan graduated from the periphery to the core of the first-world global economy. It became increasingly difficult to support undervalued currencies that sustained the export-led reconstruction and development model. This does not mean that West Germany, in particular, and Japan ceased to be export-driven economies, amassing huge surpluses with regard to the United States and other economies. Despite the collapse of the Bretton Woods fixed exchange rate system in the early 1970s, the core periphery structure that characterized it is still present albeit with a different membership. The periphery now includes predominantly emerging economies in East Asia, which are following a growth and development model based on undervalued currencies and exports. This has supported rapid economic growth and prosperity for many.

The core, as during the pre-1970s Bretton Woods period, continues to provide unfettered goods and capital markets that support growth and development in periphery. As Figure 19.3 shows, this created significant global current account imbalances in the last decade, which only subsided as the global financial crisis took hold. Consequently, emerging Asian economies, particularly China, have amassed huge foreign exchange reserves, primarily held in US Treasuries. For example, China's foreign exchange reserves, managed by the State Administration of Foreign Exchange, topped US$3.31 trillion at the end of 2012.

Notwithstanding the key role of the export-led growth model in driving reserve accumulation, some have suggested that emerging economies in Asia have accumulated reserves in response to the 1997 Asian financial crisis. That crisis sapped growth and created generalized social hardship, particularly among the poor. Some countries, such as South Korea and Thailand, required assistance from the IMF to rectify their balance of payments problems. Yet, as mentioned in a previous section, help from the IMF comes with conditions that are not always politically and socially popular. Following that experience, reserve accumulation has become arguably a form of insurance that governments can utilize in the event of a crisis, thus eliminating (or seriously reducing) their need for external support (Griffith-Jones and Ocampo, 2011).

There is a problem, however, with reserve accumulation. On the one hand, the rate of return on foreign exchange reserves is usually below the rate of inflation of the domestic economy. In effect, there is a cost to holding reserves (Rodrik, 2006). On the other hand, as accumulating reserves is part of a strategy to hold back currency appreciation, the rate of return will be further depressed if currency appreciation is allowed to take hold (which would happen if the currency were allowed to float freely on the market). Considering that the reserves are held in foreign currency (primarily US dollars), the value of the reserves in the domestic currency would, in other words, be lower as they mature. Realizing this

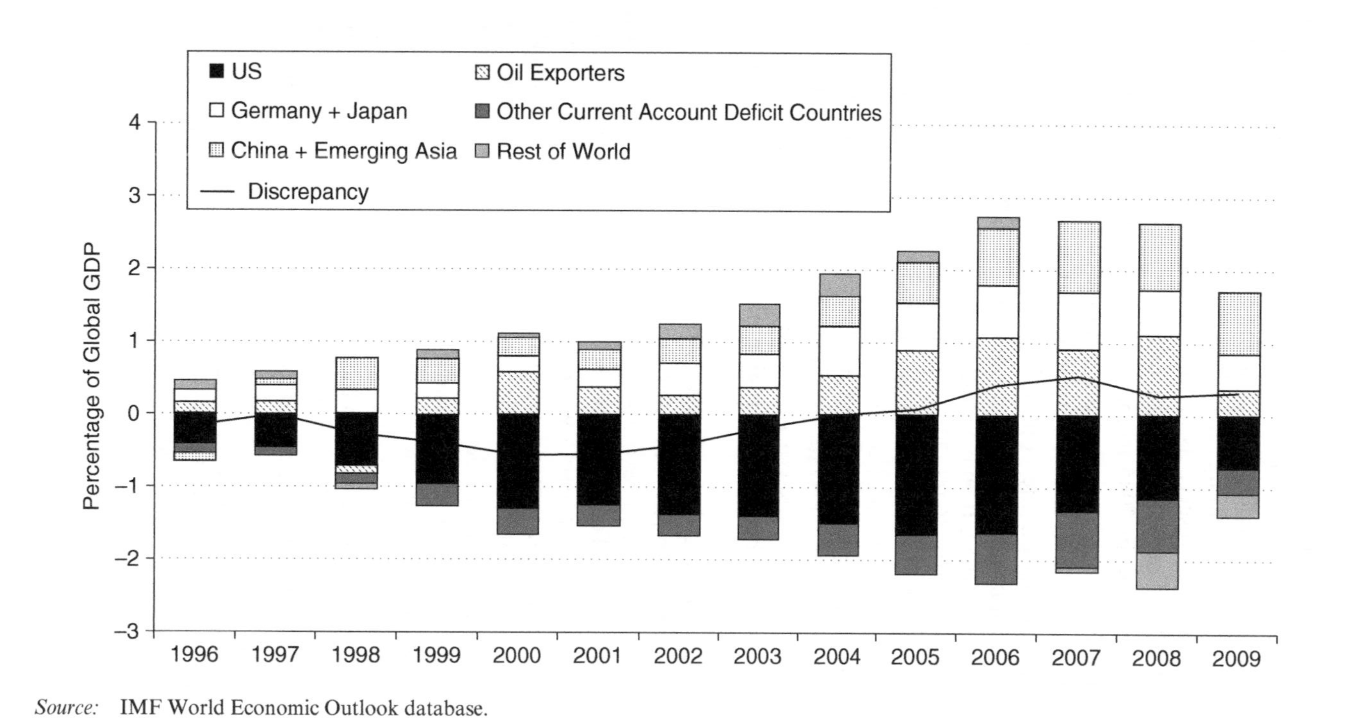

Source: IMF World Economic Outlook database.

Figure 19.3 Global imbalances, 1996–2009

implicit cost of reserve accumulation – notwithstanding the benefits it brings to export-led growth and the implied lender-of-last-resort insurance function – governments have typically sought to diversify the asset base of their reserves into higher-yielding securities and real assets.

This diversification of reserve assets has coincided with the establishment of new SWFs. South Korea, for example, established the Korea Investment Corporation in 2005 to manage funds entrusted to it by the government, the Bank of Korea and other public funds. China, likewise, established the China Investment Corporation (CIC) with the intent of maximizing the returns on the country's staggeringly and increasingly large reserves. Initially allocated approximately US$200 billion, assets under management at the CIC have continued to grow, reaching US$575 billion by the end of 2012 (CIC, 2012). But recent SWF development is not new to the region. In 1981 Singapore launched the Government of Singapore Investment Corporation to manage the city-state's reserves. Now called GIC Private Limited, the fund has offices around the world and is active in public and private markets. Whereas the largest SWFs were the almost exclusive reserve of commodity producers, export-led growth in East Asia has brought the region much higher in the ranks of SWFs (see Figure 19.1).

While not necessarily unique to SWFs from the East Asian region, the conduct and comportment of SWFs in the region appears to follow the developmental logic of the export-led growth model. There are those SWFs like Singapore's Temasek and Malaysia's Khazanah Nasional Berhad, which began as the asset owner of state assets and state-owned enterprises, that have explicit mandates as strategic investment funds that make investments in national and foreign companies in support of nation-building and national economic development (Yeung, 2011). These SWFs are not simply in pursuit of risk-adjusted returns. In effect, they are in search of a double bottom line, one that makes a return on investment and another that produces beneficial externalities that spill over into wider national economic development. Then there are those SWFs, like the CIC, whose mandate is not as explicit, but whose actions suggest otherwise. For example, the CIC has made strategic investments in the extractive sector, which appears geared toward supporting Chinese state-owned enterprises and ensuring stable energy and mineral supplies for the country's resource-intensive growth (Haberly, 2011). Not surprisingly, as will be discussed in the penultimate section, this has been a cause for concern among some Western countries.

THREAT OF SALVATION?

Prior to the global financial crisis that began in 2007 and spawned a generalized economic malaise in most advanced economies, there was some political consternation in the West, especially in the United States, surrounding the sudden rise of SWFs, fuelled in part by xenophobic posturing and the global war on terror rhetoric. Such consternation was certainly not new in Western politics. In the United States in the 1980s it was the massive incursion of Japanese capital buying up American assets. In the 2000s it was the Gulf States and, the elephant in the room, China. The global economy was booming and sovereign investors were scouring the world looking for assets to buy. For some, this would be a cause for concern with implications for national security. SWFs had to be viewed with caution, as so many of them remained opaque in terms of their investment strategies and operational governance (Truman, 2010). They were not seen as transparent Western financial institutions, subject to strict disclosure rules and robust principles of fiduciary duty. Even if the intent of an SWF was completely commercial and oriented toward obtaining financial returns, a lack of transparency would still leave many open questions.

Concerns came to a fever pitch when Dubai Ports World, a state-owned infrastructure operator from the United Arab Emirates, purchased UK-based ports operator P&O. P&O had contracts to operate a number of major US ports, but it did not own the ports themselves. Even though ownership of the ports remained with the local city or municipality, two members of the US Congress, Representative Peter T. King and Senator Charles Schumer, both from New York and a Republican and Democrat respectively, raised concerns over the potential national security implications of a state-owned operator from the Middle East. It did not matter that the Committee on Foreign Investment in the United States, which is an interagency committee that vets acquisitions by foreign parties of US-based assets for their national security implications, approved the deal (Rose, 2009). The negative publicity surrounding the deal was enough for DP World to divest P&O's US ports operations. The controversy seemed to signal that the United States, a normally faithful supporter (at least rhetorically) of free markets, was not open for business. Even though the DP World controversy did not pertain directly to an SWF transaction, it raised an important question as to whether SWFs would be welcome and whether they should even try to invest in the United States and other Western countries.

Yet in the autumn of 2007, the US Treasury Secretary at the time, Henry Paulson, moved quickly to get in front of the controversy. With rumblings in the capital markets from the emerging collapse of the subprime mortgage market, it was not the time to close the world's largest capital markets to SWFs flush with investable capital. Excluding SWFs from Western markets would be contrary to the liberal internationalism that has characterized much of US foreign policy since the end of World War II. In conjunction with the International Monetary Fund, the OECD and the World Bank, Paulson convened the International Working Group of SWFs, made up of 26 SWF sponsoring countries including the United States, to outline a set of generally accepted principles and practices (GAPP) (Norton, 2010a, 2010b; Bassan, 2011). Unveiled in the spring of 2008, the 24 voluntary GAPP, commonly referred to as the 'Santiago Principles', provided the groundwork for legitimizing SWF investment, by providing governance and disclosure guidelines for SWFs to follow that would drive behaviour toward exclusively commercial objectives. Put simply, the intent of the Santiago Principles is to recast SWFs as equivalent to other beneficiary financial institutions (e.g., supplementary occupational pension funds) and not as instruments of some mercantilist *realpolitik*.

In the years since the rollout of the Santiago Principles the working environment for SWFs has been hospitable, despite the rather half-hearted adoption of the Santiago Principles on the part of some SWFs (see Bagnall and Truman, 2011). No SWF investment has made any serious political waves; rather, SWF investment on the whole continues to be rather benign and conventional. On China, for example, the CIC does not draw anywhere near the concern that the internationalization of Chinese state-owned enterprises do (Gilligan and Bowman, 2014). Rhetoric on the supposed 'threat' that SWFs pose to national security seems to have been more hyperbole than reality (Kirshner, 2009). In reality, the Santiago Principles were less consequential for opening the door to SWFs than were the material outcomes of the global financial crisis and the ensuing recessions in the advanced economies (Langland, 2009). Instead of being shunned, SWFs were embraced and courted by companies and governments around the world (see, e.g., Haberly, 2014). Instead of being potential threats to the global capitalist order, they quickly became potential saviours.

CONCLUSIONS

The growing incidence of SWFs on the world stage is an intriguing development for the global political economy. Although SWFs do not fit nicely in the categorization of the super-rich given their public nature, even in cases where they are implicitly linked to the dynastic wealth of some countries' ruling families, in many ways they have emerged from the same sources that have contributed to the growth of the global super-rich in recent decades, while also reflecting the significance of financial markets and financial institutions in twenty-first-century global capitalism. On the one hand, SWFs reflect the changing global economy itself, one characterized by rapid export-led development in East Asia and the continued dependence the world over for natural resources. On the other hand, SWFs reflect the efforts of sponsoring governments to engage with the global financial economy, to reap the benefits of wealth concentration wherever it is produced, and the efforts of sponsoring governments to resist the potential predations and ever-pending crises and market failures that come with such engagement. For some countries an SWF may exist to help build the nation and underwrite collective well-being and financial security. For others, an SWF may simply be another tool at the hands of those that seek to maintain the status quo. Considering the different mandates that can be applied to an SWF and various sources of capital, SWFs cannot be classed into a singular group. Such heterogeneity and the existing local political-economic conditions of the sponsoring political economy should not be discounted.

For some, SWFs represent a threat to free-market capitalism. SWFs certainly form part of national economic interests. But the recent record does not indicate that they represent a force seeking to undermine the free market. Even if they are parastatal institutions that does not mean that they are necessarily reinforcing newly emergent forms of interventionist state capitalism. If anything, they are helping reinforce the growth, integration and development of global financialized capitalism. They simply reflect a global political economy that favours the holders of financial wealth and the power that financial wealth imparts. But just as fast as great fortunes can be made, they can also be lost. While some SWFs, particularly the largest, will likely continue to exist for some time, the recent interest in SWFs should be taken with caution.

The conditions in the global economy that allow for the creation of SWFs, namely high commodity prices and a tolerance of export-led growth and currency manipulation, could change. Likewise, accumulated capital could be poorly managed and invested without due care and diligence. And, domestic political instability and challenges to the existing political authority could repurpose a SWF or eliminate it completely. It is worth remembering that most countries of the world are actually quite young. Geopolitical history is not one of stasis, but rather dynamism and volatility. But until the global political economy ceases to be defined by finance and a preponderance of concentrated wealth, SWFs can be expected to be part of the picture.

NOTES

1. Consider, for example, that the world's richest individual in 2014 according to *Forbes* magazine is Mexican telecoms billionaire Carlos Slim Helú, with a fortune of US$83 billion. Contrast this with the Abu Dhabi Investment Authority (ADIA), which has roughly US$589 billion in assets under management. ADIA's assets belong to the state, which is governed by hereditary rule. Even though ADIA does not represent the personal fortunes of the ruling Al Nahyan family, the explicit and implicit control the family has over ADIA suggests it is not completely inseparable.
2. There are several, mostly commercial, organizations that track the number of SWFs in operation and their assets under management, such as Institutional Investor's Sovereign Wealth Center (www.sovereignwealthcenter.com). Academic organizations tracking SWFs include: ESADEgeo-Center for Global Economy and Geopolitics (www.esadegeo.com/global-economy) and the Sovereign Investment Lab at Bocconi (http://www.baffi.unibocconi.it/wps/wcm/connect/Cdr/Centro_BAFFIen/Home/Sovereign+Investment+Lab/).
3. Pension reserve funds have not been included in this figure given the debate over whether they actually constitute a sovereign wealth fund (see Monk, 2011).

REFERENCES

Auty, R.M. (2001), *Resource Abundance and Economic Development*, Oxford, UK: Oxford University Press.

Auty, R.M. and R.F. Mikesell (1998), *Sustainable Development in Mineral Economies*, Oxford, UK: Clarendon.

Bagnall, S. and E. Truman (2011), 'IFSWF report on compliance with the Santiago Principles: admirable but flawed transparency', *Peterson Institute for International Economics Policy Brief No. 11–14*.

Balding, C. (2012), *Sovereign Wealth Funds: The New Intersection of Money and Politics*, New York: Oxford University Press.

Barbary, V., A. Monk and T. Triki (2011), 'The new investment frontier: SWF investment in Africa', in V. Barbary and B. Bortolotti (eds), *Braving the New World: Sovereign Wealth Fund Investment in the Uncertain Times of 2010*, London: Monitor Group, pp. 54–60.

Bassan, F. (2011), *The Law of Sovereign Wealth Funds*, Cheltenham, UK and Northampton, MA, USA: Edward Elgar Publishing.

Beck, R. and M. Fidora (2008), 'The impact of sovereign wealth funds on global financial markets', *Intereconomics*, **43** (6), 349–58.

BP (2013), *BP Statistical Review of World Energy June 2013*, accessed 23 August 2015 at https://www.bp.com/content/dam/bp/pdf/statistical-review/statistical_review_of_world_energy_2013.pdf.

Bremmer, I. (2010), *The End of the Free Market: Who Wins the War Between States and Corporations?*, New York: Portfolio.

CIC (2012), *China Investment Corporation Annual Report 2012*, accessed 23 August 2015 at http://www.china-inv.cn/wps/wcm/connect/f61dc0a7-31ec-4fb8-8add-1b4e8984f837/CIC_2012_annualreport_en.pdf?MOD=AJPERES&CACHEID=f61dc0a7-31ec-4fb8-8add-1b4e8984f837.

Clark, G.L., A.D. Dixon and A.H.B. Monk (2013), *Sovereign Wealth Funds: Legitimacy, Governance and Global Power*, Princeton, NJ: Princeton University Press.

Collier, P., R. van der Ploeg and M. Spence et al. (2010), 'Managing resource revenues in developing economies', *IMF Staff Papers*, **57** (1), 84–118.

Corden, W. (1984), 'Booming sector and Dutch disease economics: survey and consolidation', *Oxford Economic Papers*, **36** (3), 359–80.

Corden, W. and J. Neary (1982), 'Booming sector and de-industrialisation in a small open economy', *The Economic Journal*, **92** (368), 825–48.
Das, U., Y. Lu and C. Mulder et al. (2009), 'Setting up a sovereign wealth fund: some policy and operational considerations', *IMF Working Paper No. 09/179*.
Davis, J.M., R. Ossowski and A. Fedelino (2003), *Fiscal Policy Formulation and Implementation in Oil-Producing Countries*, Washington, DC: International Monetary Fund.
Dixon, A.D. and A.H.B. Monk (2011), 'What role for sovereign wealth funds in Africa's development?', *Center for Global Development Oil-to-Cash Initiative Background Paper October 2011*.
Dixon, A.D. and A.H.B. Monk (2012), 'Rethinking the sovereign in sovereign wealth funds', *Transactions of the Institute of British Geographers*, **37** (1), 104–17.
Dooley, M., D. Folkerts-Landau and P. Garber (2004), 'The revived Bretton Woods system', *International Journal of Finance and Economics*, **9** (4), 307–13.
Eichengreen, B. (2008), *Globalizing Capital: A History of the International Monetary System*, 2nd edition, Princeton, NJ: Princeton University Press.
Engelen, E., I. Ertürk and J. Froud et al. (2011), *After the Great Complacence: Financial Crisis and the Politics of Reform*, Oxford, UK: Oxford University Press.
Gibson-Graham, J.K. (1996), *The End of Capitalism (As We Knew It): A Feminist Critique of Political Economy*, Cambridge, MA/Oxford, UK: Blackwell.
Gilligan, G. and M. Bowman (2014), 'State capital: global and Australian perspectives', *Seattle University Law Review*, **37** (2), 597–38.
Griffith-Jones, S. and J.A. Ocampo (2011), 'The rationale for sovereign wealth funds: a developing country perspective', in P. Bolton, F. Samama and J. Stiglitz (eds), *Sovereign Wealth Funds and Long-Term Investing*, New York: Columbia University Press.
Gylfason, T. (2011), 'Natural resource endowment: a mixed blessing?', *CESifo Working Paper Series No. 3353*.
Haberly, D. (2011), 'Strategic sovereign wealth fund investment and the new alliance capitalism: a network mapping investigation', *Environment and Planning A*, **43** (8), 1833–52.
Haberly, D. (2014), 'White knights from the Gulf: sovereign wealth fund investment and the evolution of German industrial finance', *Economic Geography*, **90** (3), 293–320.
Hatton, K.J. and K. Pistor (2012), 'Maximizing autonomy in the shadow of great powers: the political economy of sovereign wealth funds', *Columbia Journal of Transnational Law*, **50** (1), 1–81.
Karl, T.L. (1997), *The Paradox of Plenty: Oil Booms and Petro-States*, Berkeley, CA/London: University of California Press.
Khalaf, S.N. (1992), 'Gulf societies and the image of the unlimited good', *Dialectical Anthropology*, **17** (1), 53–84.
Kirshner, J. (2009), 'Sovereign wealth funds and national security: the dog that will refuse to bark', *Geopolitics*, **14** (2), 305–16.
Langland, E. (2009), 'Misplaced fears put to rest: financial crisis reveals the true motives of sovereign wealth funds', *Tulane Journal of International & Comparative Law*, **18** (1), 263.
Markowitz, H. (1952), 'Portfolio selection', *Journal of Finance*, **7** (1), 77–91.
Momani, B. (2008), 'Gulf cooperation council oil exporters and the future of the dollar', *New Political Economy*, **13** (3), 293–314.
Monk, A.H.B. (2011), 'Sovereignty in the era of global capitalism: the rise of sovereign wealth funds and the power of finance', *Environment and Planning A*, **43** (8), 1813–32.
Norton, J.J. (2010a), 'The "Santiago Principles" and the international forum of sovereign wealth funds: evolving components of the new Bretton Woods II post-global financial crisis architecture and another example of ad hoc global administrative networking and related "soft" rulemaking?', *Review of Banking and Finance Law*, **29**, 465–529.
Norton, J.J. (2010b), 'The Santiago Principles for sovereign wealth funds: a case study on international financial standard-setting processes', *Journal of International Economic Law*, **13** (3), 645–62.

Rodrik, D. (2006), 'The social cost of foreign exchange reserves', *International Economic Journal*, **20** (3), 253–66.
Rose, P. (2009), 'Sovereign wealth fund investment in the shadow of regulation and politics', *Georgetown Journal of International Law*, **40** (4), 1–33.
Rose, P. (2011), 'American sovereign wealth', *Ohio State Public Law Working Paper No. 161*, accessed 27 January 2015 at http://dx.doi.org/10.2139/ssrn.1960706.
Ross, M.L. (2012), *The Oil Curse: How Petroleum Wealth Shapes the Development of Nations*, Princeton, NJ: Princeton University Press.
Rozanov, A. (2005), 'Who holds the wealth of nations?', **15** (4), 52–7.
Santiso, J. (2009), 'Sovereign development funds: key actors in the shifting wealth of nations', *Revue d'Economie Financière*, **9** (1), 291–315.
Scammell, W.M. (1980), *The International Economy Since 1945*, London: Macmillan.
Storper, M. and R. Walker (1989), *The Capitalist Imperative: Territory, Technology, and Industrial Growth*, Oxford, UK: Basil Blackwell.
Truman, E.M. (2010), *Sovereign Wealth Funds: Threat or Salvation?*, Washington, DC: Peterson Institute for International Economics.
Wallerstein, I. (1983), *Historical Capitalism*, London: Verso.
Yeung, H. (2011), 'From national development to economic diplomacy? Governing Singapore's sovereign wealth funds', *The Pacific Review*, **24** (5), 625–52.

20. Super-rich capitalism: managing and preserving private wealth management in the offshore world

Jonathan V. Beaverstock and Sarah Hall

INTRODUCTION

In the post-2008 era of managing the political economy of austerity, offshore financial centres (OFCs) have once again come under the spotlight, but this time scrutiny has been from both the academy and national governments, mainly in OECD countries, who are critically examining their 'secrecy' and role as 'tax havens' (see Hampton and Christensen, 2002; Maurer, 2008; Palan et al., 2010; Sikka and Willmott, 2010; Shaxson, 2012). Importantly, national governments are not only putting organizations that use OFCs as 'corporate taxation havens' under the microscope (see *The Guardian*, 2013), but they are also exposing the secrecy, opaqueness and non-compliance in personal taxation that have shrouded the OFCs' nexus of private banking and private wealth management. From the late 2000s, the USA and UK especially, have put the issue of 'super-rich' personal tax minimalism high on the political agenda as their respective revenue agencies, the Internal Revenue Service and Her Majesty's Revenue & Customs, seek to collect personal income tax from their citizens who have investments in private banks or other wealth management institutions. The introduction of the USA's Foreign Account Tax Compliance Act (FATCA) in 2010, which mandated banks to share the personal taxation information on their US citizens, has paved the way for other OECD countries to seek global data sharing on banking secrecy (*The Financial Times*, 2014a). But, unlike the role of OFCs for the corporate world[1] there is a real dearth of conceptual writing and empirical data on the role of OFCs – 'tax havens' – in managing the offshore private wealth of the global super-rich. This is not a surprise given that to the best of our knowledge, no national government, supranational organization (like the International Monetary Fund) or non-governmental organization (e.g., Oxfam) has access to official statistics and data on the stock of private wealth piled up in OFCs.

Thus, the specific aim of this chapter is twofold: first, to explore the role of OFCs specifically in managing private wealth; and second, to reveal, from a variety of non-official data sources, the position of OFCs in managing the flow and stock of private wealth, assets under management (AUM), deposited in these offshore jurisdictions. Following this introduction, the remainder of this chapter is organized into four distinctive parts. First, we explain the role of the specialist banking and advanced business services (ABSs) sector in managing private wealth, drawing on our own previous work (Beaverstock, 2012; Beaverstock et al., 2013a, 2013b), and the specialist writing on private banking (see Maude, 2006; Cassis and Cottrell, 2009) and OFCs (Haberly and Wójcik, 2013; Wójcik, 2013). Second, we draw on a range of private official data sources (like Capgemini Merrill Lynch, the Royal Bank of Canada and the Boston Consulting Group) to expose the flow and stock of private wealth in selected OFCs, noting in particular the rise of the Asia-Pacific region as both a global source of, and destination for, private wealth. Third, we look at the rise of Singapore as a global centre for private wealth management given that it is predicted to race ahead of Switzerland as the major global centre for booking AUM offshore by the end of this decade (see Deloitte, 2013; PwC, 2013). Finally, we close the chapter with several concluding points highlighting the increasingly important role of OFCs for the management of private wealth in the world.

THE OFFSHORE FINANCIAL CENTRE AND THE 'NEXUS' FOR MANAGING PRIVATE WEALTH MANAGEMENT

There has been a rich seam of research on the role of OFCs in the international financial system, classically small island economies or Principalities often with close colonial links with the United Kingdom and other European economies (Roberts, 1994; Cobb, 1998; Hampton and Abbot, 1999; Hampton and Christensen, 1999; Hudson, 2000; Warf, 2002; Palan, 2006; Cameron, 2008). But, in most of these studies, and more recently (e.g., Haberly and Wójcik, 2013; Wójcik, 2013) there has been an absence of analysis on the role of OFCs in the management of *private* wealth. But, we do acknowledge that there has been critically acclaimed research and intervention, from an interdisciplinary perspective, which has made extremely visible the role of OFCs as 'tax havens' and places where corporations especially are exceedingly efficient at minimizing their corporate taxation compliance.[2] We now argue that in order to better understand the operation of the private wealth management industry and its reliance upon offshore jurisdictions it is instructive to integrate understanding of ABSs (including management consultancy, accountancy and legal services) more fully into work on the geographies of finance.

Research on the role of ABSs and their 'spatial complex' in the global financial centre has been at the forefront of the intersections between economic geography and world cities studies (see Cook et al., 2007; Faulconbridge et al., 2007; Beaverstock, 2012; Sassen, 2013). In this respect, we extend work on the ABS offshore nexus (Wójcik, 2013) to include the neglected importance of private wealth. This framework is valuable because it demonstrates the importance of attending to the range of different financial and ABSs that are central to making up the private wealth management industry *offshore* and reveals the ways in which this nexus articulates the operation of the private wealth management industry across the on- and offshore jurisdictions.

In order to develop our argument on the offshore private wealth management nexus we advance Wójcik's (2013) work on what he terms the ABS offshore nexus. Whilst this approach has developed primarily in relation to corporate offshore practices, we extend it in our analysis here to include the case of private wealth management. As part of a wider set of arguments concerning the need to integrate finance with economic geographies of globalization, Wójcik (2013) argues that ABSs (including wholesale financial services but also law, accountancy and management consultancy) are critical to understanding the role of offshore jurisdictions in the global economy because they are central to the establishment of financial and legal instruments (special purpose entities – SPEs) that are aimed at overcoming or escaping governmental regulations on issues such as tax. As such, Wójcik (2013) argues that ABSs are vital in governing the global economy by using such entities to bring together the onshore world, primarily through leading world cities where the dense networks of ABS firms allow them to undertake the formation of such entities, offshore spaces through which capital flows within such SPEs and the rest of the world that serves as clients for these products (Figure 20.1). This approach is instructive for our focus on the private wealth management industry because it draws attention to the range of ABS firms involved in choreographing on- and offshore financial networks (Figure 20.1). It is a particularly useful approach for understanding the transformation of the private wealth management industry because, as the sector has grown out of its private banking history, the range of service providers has increased dramatically, reflecting the growth and changing global distribution of private wealth requiring management both on- and offshore (see Maude, 2006; Cassis and Cottrell, 2009; Beaverstock et al., 2013a, 2013b).

Historically, the landed gentry and wealthy were predominantly served by the private banking system. Jurisdictions like Switzerland, Luxembourg,

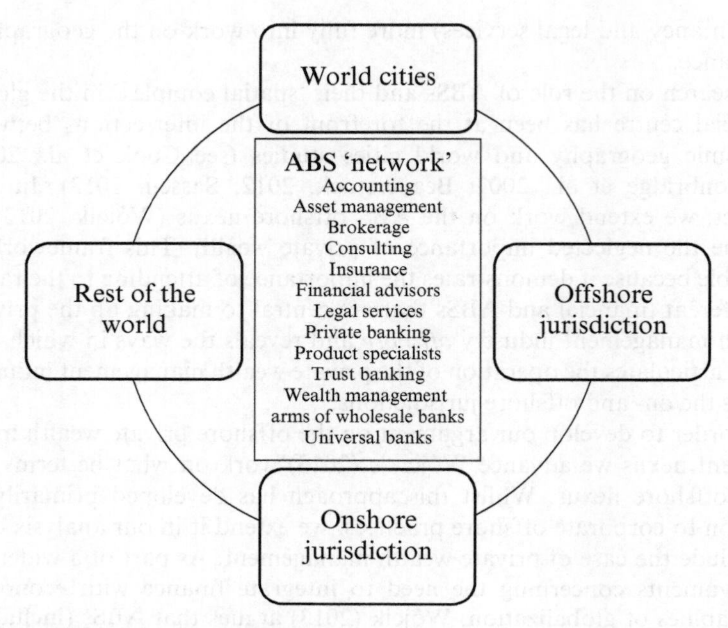

Source: Adapted from Wójcik (2013).

Figure 20.1 The ABS–offshore practice network and nexus for private wealth management

Liechtenstein, Andorra, Monaco, Gibraltar and the Channel Islands in Europe, small tropical islands (like the Cayman Islands, Bermuda, British Virgin Islands) and further east, Singapore and Hong Kong, developed as OFCs as they became the strongholds for private banks to establish offshore office networks (see Bicker, 1996). For much of the twentieth century, the OFC nexus for managing the private wealth of the super-rich was the domain of the private banking milieu, whether that be in the form of taking deposits, offering current accounts, and savings and brokerage, for example (Maude and Molyneux, 1996; Laulajainen, 1998; Cassis and Cottrell, 2009). From the late 1970s onwards, there began the explosion in the growth of the on- and offshore ABS nexus for managing and preserving private wealth (Maude, 2006; Beaverstock et al., 2013a). The advent of wealth produced by 'new money' sources, including, for example, exorbitant executive remuneration, the 'bonus culture' in banking, finance and professional services, selling or the stock market listing of private business, or receiving one-off wealth events (Frank, 2007; Irvin, 2008; Armstrong, 2010), led to a significant growth in the number of wealthy customers who would require retail wealth management products and advisory services (Atkinson and Piketty, 2010; Hay and Muller, 2012; Beaverstock et al., 2013a; Hay, 2013). During this period, successive OECD country governments reduced the tax burden for high earners (e.g., in the UK, the top rate of personal taxation was reduced from 83 per cent in 1980 to 40 per cent in 2004: Irvin, 2008; Sayer, 2012) and from the 2000s, the super-rich effectively became the beneficiaries of 'financialization' as they were able to accumulate wealth from assets and speculation, playing the casino of the international financial system (Folkman et al., 2007; Froud and Williams, 2007; Hall, 2009; Sayer, 2012), becoming 'active investors' in the global economy (Langley, 2006).

As a result of the rapid growth of wealthy individuals in North America (Frank, 2007), Europe (Freeland, 2012) and the Asia-Pacific (Long and Tan, 2010; *The Economist*, 2013), and the accompanying demand for more specialist financial products and advisory services, traditional private banks were truly opened out to competition from the entry of global ABS firms in accounting, legal, insurance, and real estate, and wholesale investment, commercial and retail banks, and specialist brokerages (Maude, 2006; Beaverstock et al, 2013a). Thus was born the global private wealth management industry, a collective of private, wholesale and universal banks, and financial and professional services sectors that quickly established office networks and advisory services throughout the on- and offshore worlds of international finance (as depicted in Figure 20.1). According to Scorpio Partnership, the London-based private wealth management research strategy consultancy, the global top ten 'mega-wealth managers' (Scorpio Partnership, 2013, p. 1) are dominated by universal and wholesale banks' global wealth management and private banking subsidiaries and/or divisions (Table 20.1), all having significant office representation in OFCs across Europe, Caribbean and Panama, Hong Kong and Singapore, and Switzerland.

GLOBAL PRIVATE WEALTH AND THE OFFSHORE JURISDICTION

The market for global private wealth has become a very lucrative segment of total assets under management (AUM) in OFCs. In 2012, the Boston Consulting Group (BCG) (2013) estimated that there was approximately US$8.5 trillion of wealth booked in the offshore domain, deposited almost entirely by the global super-rich, the so-called ultra high net worth individual (UHNWI) and high net worth individual (HNWI) market.[3] As the private wealth management industry is dependent on the demand for its

Table 20.1 Global top ten mega-wealth managers by assets under management, 2012

Institution	AUM (US$ Billion)	Reporting Currency	HQ
UBS	1705.0	CHF	Zurich & Basel
Bank of America[a]	1673.5	USD	Charlotte, NC, USA
Wells Fargo[b]	1400.0	USD	San Francisco, USA
Morgan Stanley	1308.0	USD	New York City, USA
Credit Suisse	854.6	CHF	Zurich
Royal Bank of Canada	628.5	CAD	Toronto
HSBC	398.0	USD	London, Hong Kong
Deutsche Bank	387.3	EUR	Frankfurt
BNP Paribas	346.9	EUR	Paris
Pictet	322.2	CHF	Geneva

Notes:
a. Data for Bank of America include its Global Wealth Management Division, including Merrill Lynch Global Wealth Management, US Trust, Bank of America Private Wealth Management and its Retirement Services business.
b. Wells Fargo data also include Retirement Service.

Source: Adapted from Scorpio Partnership (2013).

products and advisory services from these HNW markets, defined precisely for the wealth management sector (Beaverstock et al., 2013a), the private wealth management industry has become a truly global financial marketplace as the number of millionaires and billionaires, and the value of their global wealth, have increased on a global scale. But, official data on investible private wealth are difficult to acquire from the public domain unlike government statistics on personal income.

The global super-rich have entered the public lexicon through popular 'rich lists' (e.g., *Forbes*' World Billionaires List) and specialist intelligence on the HNWI market published by global private wealth management organizations like Capgemini, Merrill Lynch, the Royal Bank of Canada Wealth Management and the BCG (Beaverstock et al., 2004). For example, Capgemini Royal Bank of Canada Wealth Management's *2013 World Wealth Report* (CRBCWM, 2013) estimates that in 2012 there were 12.0 million HNWIs, with investable assets greater than US$1 million, holding US$42.2 trillion of global wealth. Back in the mid-1990s, when Capgemini collected this data with Merrill Lynch, they estimated the figure in 1996 to be 4.5 million HNWIs with US$16.6 trillion of global wealth. Thus, between 1996 and 2012, the number of HNWIs increased by

Table 20.2 Global population of HNWIs and value of private wealth, 1996–2012

Year	Number (Millions)	Change (%)	Wealth (US$ Trillions)	Change (%)
1996	4.5	–	16.6	–
2000	7.2	+60.0	27.0	+62.7
2005	8.8	+22.2	33.4	+23.7
2006	9.5	+8.0	37.2	+11.4
2007	10.1	+6.3	40.7	+21.9
2008	8.6	−14.9	32.8	−19.4
2009	10.0	+17.1	39.0	+18.9
2010	10.9	+8.3	42.7	+9.7
2011	11.0	+0.8	42.0	−1.7
2012	12.0	+9.2	46.2	+10

Source: Adapted from CMLGWM (2008, 2009, 2010, 2011); CML (2002, 2007); CRBCWM (2012, 2013).

+167 per cent (+7.5 million) and more importantly, their share of global wealth went up by +US$25.6 trillion (+154 per cent) (Table 20.2). It is this 'golden age' of wealth creation from the burgeoning HNWI population that has created the demand conditions, and importantly supply of investible assets, for the private wealth ABS nexus to manage in both on- and offshore jurisdictions.

Moreover, given that the concentration and relative regional growth in the HNWI population, as estimated by CRBCWM (2013), was in North America, Europe and the Asia-Pacific (including China and Japan), those private wealth management organizations in their respective on- and offshore locales – New York, London, Singapore, Hong Kong, Zurich/Geneva, British Channel Islands and Caribbean and Panama – benefited from assets being booked in these locations (BCG, 2013; Deloitte, 2013), as 'local bias' dominated the HNWI 'active investor' (Langley, 2006), particularly those from the Asia-Pacific (PwC, 2013). As for the empirical evidence, in 2012, North America and the Asia-Pacific each had 3.7 million HNWIs, with global wealth valued at US$12 trillion and US$12.7 trillion respectively, but the rise of the Asia-Pacific has been meteoric since 2000, experiencing a +131 per cent increase in the number of HNWIs (from 1.6 to 3.7 million) and a +150 per cent increase in the value of private wealth, from US$4.8 trillion to US$12 trillion (CML, 2002; CRBCWM, 2013). The major trends and regional nuances in these CRBCWM (2013) and CML (2002) data for the HNWI population and their global wealth trends are corroborated in many other independent analyses of millionaires and billionaires (see BCG, 2013; Deloitte, 2013; PwC, 2013; Knight Frank, 2014). But, what is less well known in the public domain is the role of private wealth in OFCs, and it is to this that we turn now.

Between 2006 and 2012, the BCG (2007, 2013) has estimated that the stock of wealth held in OFCs has increased by just over a quarter, from US$6.7 trillion to US$8.5 trillion. The fallout of the global financial crisis in 2008 for the HNWI market was short-lived as the stock of wealth reduced by only −9 per cent (–US$0.6 trillion) between 2007 and 2008, before returning to growth in 2009 to a level of US$7.4 trillion (BCG, 2008, 2009, 2010). The BCG's (2013) analysis of the destination of offshore wealth shows that Switzerland remains top of the pile, attracting US$2.2 trillion of offshore wealth in 2012, followed by the United Kingdom (London), Channel Islands and Dublin (US$2.0 trillion) (Figure 20.2). But what are of significance in these data are the relative positions of Singapore and Hong Kong as destinations for offshore wealth. In combination, between 2009 and 2012, both of these OFCs recorded more than a +50 per cent growth rate (from US$0.7 trillion to US$1.2 trillion) as destinations for offshore wealth, benefiting from the Asia-Pacific region's wider +40 per cent growth rate in the origin of offshore wealth (from US$1.5 trillion to US$2.1 trillion), where the 'local bias' to invest was disproportionately directed at Singapore and Hong Kong. For example, in 2011, the BCG (2012) estimated that 76 per cent of Singapore and Hong Kong's US$1 billion offshore bookings were sourced from the Asia-Pacific region (US$0.76 billion). It is no wonder that Singapore especially is becoming tagged with the strapline, the 'Switzerland of Asia' (Allen, 2006; Pow, 2013; *The Economist*, 2013). We now turn our attention to Singapore's rising status as one of the premier OFCs for the management of private wealth.

THE LION CITY: SINGAPORE'S RISING STATUS AS A 'HAVEN' FOR PRIVATE WEALTH

Singapore is the '[r]ising star of wealth management' (*The Economist*, 2013) not only in the Asia-Pacific, but also on a global stage. PwC (2013, p. 18) predicts that the outlook for Switzerland is challenging, as '[it]. . .is expected. . .to be overtaken by Singapore', primarily as the Asian-Pacific rapidly growing HNWI population look to invest their private wealth locally. This is not a surprising feat over very recent times given that Singapore, and other countries in the wider Asia region like India and China, was relatively unaffected by the fallout of the North American and European subprime financial crisis (Khor and Kee, 2008). The Monetary

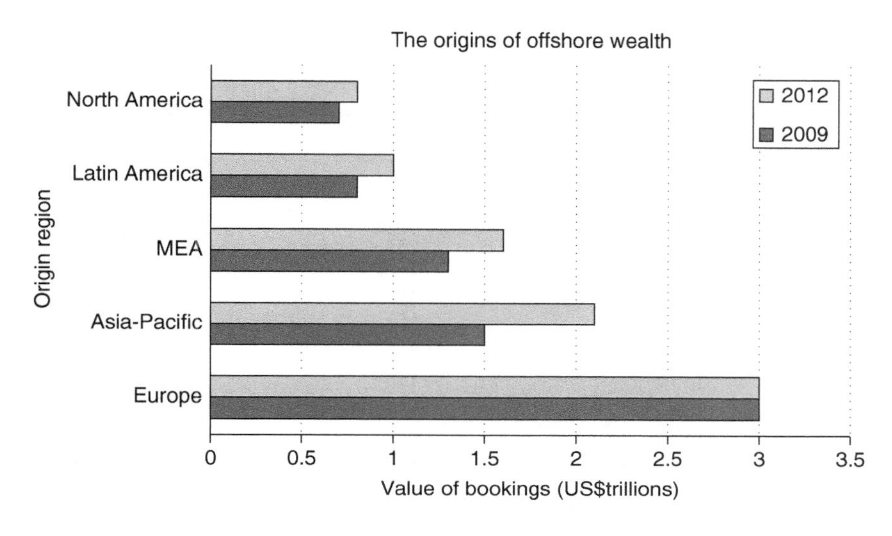

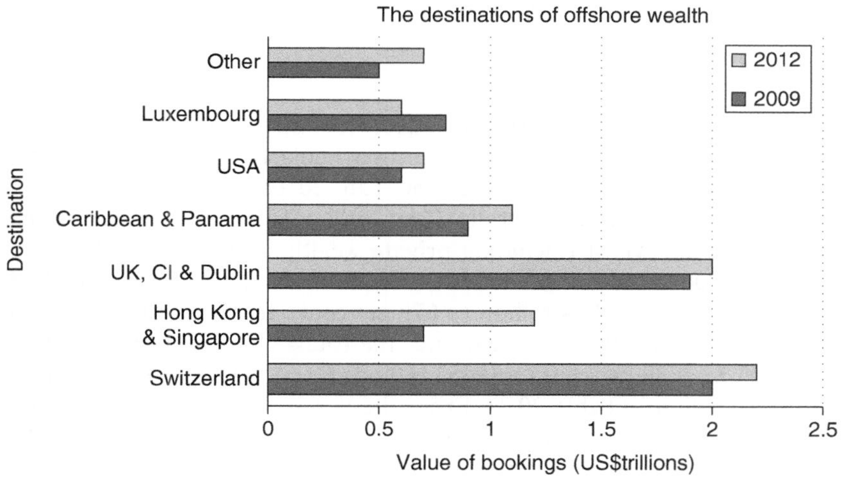

Note: MEA represents the Middle East and Africa; Asia-Pacific includes Japan; CI represents the Channel Islands (Jersey and Guernsey); the USA is mainly Miami and New York; 'Other' includes Dubai and Monaco.

Source: Adapted from BCG data (2010, 2013).

Figure 20.2 The origins and destinations of offshore wealth, 2009 and 2012

Table 20.3 Growth of total assets under management in Singapore, 1998–2012

Year	AUM (U$ Billion)	Growth Rate (S$ Billion)	Percentage Change
1998[a]	150.6	–	–
2000	276.2	+125.6	+46
2005	720.4	+444.2	+161
2006	891	+170.6	+24
2007	1173	+452.6	+63
2008	864	−309	−26
2009	1208	+344	+40
2010	1354	+146	+13
2011	1338	−16	−1
2012	1626	+288	+22

Note: a. Pre-1998, AUM were recorded as Total Discretionary only, which stood at S$18.1 billion in 1990 and S$86.4 billion in 1995 (MAS, 1998). All figures for 1999 onwards included Total Discretionary and Total Advisory (Non-Discretionary) Funds.

Source: MAS (1998, 2005, 2012).

Authority of Singapore (MAS) actively promotes the competiveness of the city-state's financial centre, and particularly the wealth management and insurance sectors (Long and Tan, 2010) alongside other retail financial services more generally (Lai, 2013). Table 20.3 indicates the growth of total AUM (including private wealth) in Singapore. Between 1998 and 2007 (pre-financial crisis), the total AUM in Singapore had increased by +U$1022.4 billion (+679 per cent), from S$150.6 billion to S$1173 billion. At the 2012 year-end, the total AUM in Singapore had increased by +39 per cent (+S$453 billion) from 2007, to S$1626 billion (US$1.33 trillion) (Table 20.3). From the depths of the global financial crisis in 2008 when AUM were S$864 billion, Singapore's total AUM have experienced a +88 per cent growth rate (+S$762 billion) up to 2012 (S$1626 billion). MAS (2013a, p. 1) suggests that in 2012, 'approximately 77% of the total AUM was sourced from outside Singapore . . . demonstrating Singapore's primary role in serving regional and international investors'. In 2012, the distribution of AUM in Singapore by region was: Asia-Pacific (70 per cent); Europe (10 per cent); and the US (9 per cent) (and Other 1 per cent) (MAS, 2013a).

The development of Singapore's private wealth and asset management industry has been at the forefront of contributing to the growth in the volume of total AUM in Singapore. As Long and Tan (2010, p. 107) note, 'this growth has primarily come from the development of the private wealth management sector, with an increasing number of Swiss and international private banks and wealth managers setting up shop in Singapore'. But, as we shall discuss now, the reasons for Singapore's rise and pre-eminence as one of the global premier private wealth management offshore domains is underpinned by significant 'boosterist' state promotion for the sector, and the entirety of Singapore's status and reputation as a truly global financial centre, promoting both political and economic stability, enhanced by numerous interlinking sector-specific factors.

The attractiveness of Singapore as an enclave for private AUM, drawn from the local Asia-Pacific region, Europe and the USA is five-fold. First, and most significantly, Singapore is a global financial centre with a well-established banking (including wholesale, retail, commercial, private), financial and professional services sector (especially in accounting, legal services, insurance). Singapore also has a global reputation for trust, a stable political regime, and post the 'Nick Leeson affair', has a strong and robust regulatory structure. The Z/Yen (2013) Global Financial Centre Index consistently ranks Singapore fourth in its world ranking of global financial centres, often in close competition with its local rival, Hong Kong (Table 20.4). According to MAS (2013b), in 2013, Singapore had approximately 117 foreign banks, 41 merchant banks, 347 holders of capital markets services licences and 162 registered insurers, from about 2700 financial and closely allied institutions. Significantly, Singapore as a global financial centre has a population of the major international branch offices (including Asian-Pacific headquarters) of many leading global banks from North America, Europe, the Middle East and the Asia-Pacific. Moreover, since the mid-1990s, Singapore's financial centre, the 'Golden Shoe' district, has also become a prime location for the leading global players in professional services, particularly in accounting, consulting, legal services, insurance and real estate (Beaverstock, 2002). Z/Yen

Table 20.4 The top five global financial centres in the Global Financial Centre Index

GFCI 1 – 2007	Score	GFCI 8 – 2010	Score	GFCI 13 – 2013	Score
1 London	765	1 London	772	1 London	807
2 New York	760	2 New York	770	2 New York	787
3 Hong Kong	684	3 Hong Kong	760	3 Hong Kong	761
4 Singapore	660	4 Singapore	728	4 Singapore	759
5 Zurich	656	5 Zurich	699	5 Zurich	723

Source: Adapted from the City of London (2007) and Z/Yen (2010, 2013).

(2013, p. 11) deems Singapore to be a 'global leader', in close competition with New York, London and Hong Kong, primarily because like these other centres, it has such 'broad and deep financial services activities and. . .[is]. . .connected with many other financial centres'.

Second, Singapore has a globally recognized specialist banking and financial services cluster focused on private banking, and associated wealth and asset management, which has grown rapidly in size, and in reputational terms, in the Asia-Pacific and North America and Europe since the late 1990s/early 2000s onwards (Long and Tan, 2010). In 2012, Singapore's asset management industry had over 500 registered players, managing a total of S$1363 billion AUM (MAS, 2013a, 2013b). But, it is also the advent of the private banking sector that has given Singapore, like Hong Kong, its competitive advantage in the network of global financial centres. Nearly all of the leading European-based British, Swiss and Dutch, New York-based US and Asia-Pacific-based private banks and private banking or private wealth management subsidiaries of wholesale and commercial banks, are in Singapore (Box 20.1). Since its inception as an entrepôt in the nineteenth century, Singapore has attracted a private banking sector focused on the management of an elite, private clientele. Entry into Singapore has been historic, like, for example, ABN AMRO Private Banking associated with its trading legacy in the nineteenth century (see ABN AMRO, 2014), and current, like Swiss (e.g., Julius Baer Group) and US banks (e.g., Citibank, Goldman Sachs, Morgan Stanley)

BOX 20.1 LEADING PRIVATE BANKS IN SINGAPORE, 2013

ABN AMRO Private Banking
ANZ PTE Bank Singapore
Barclays Wealth
BNP Paribas Wealth Management
China Construction Bank Corporation
Citi Private Bank
Coutts & Co. Ltd
Credit Industriel et Commercial
Credit Suisse
DBS PTE Bank
Deutsche Bank PTE Wealth Man.
Goldman Sachs PTE Wealth
HSBC PTE Bank (Suisse) SA
Industrial & Commercial Bank of China
JP Morgan PTE Clients
Julius Baer & Co. Ltd
Merrill Lynch International Bank
Morgan Stanley
RHB Bank BHD
Rothschild
Schroder & Co Asia Ltd
Société Générale PTE Banking (S)
UBS AG Singapore
Union BanCaire Privée (S) Ltd

Note: PTE = private bank.

Source: *ST Directory*, accessed 24 August 2015 at http://directory.stclassifieds.sg/singapore-directory/Private%20Banks/c/.

expanding or entering during the late 1990s and 2000s (Long and Tan, 2010). For the local Singaporean high net worth population, Long and Tan (2010) have charted the importance of the establishment of private trust companies for planning family finances associated with inheritance and trusts.

Third, Singapore has a profound talent pool of highly skilled bankers, financial and professional services specialists and advisories (accountants, lawyers, insurance experts) drawn from the local Singaporean labour market and from foreign workers, expatriates, whose nationalities are primarily North American (US), European (British, French, Dutch and German), Chinese and Japanese (Beaverstock, 2011; Ye and Kelly, 2011). Between 1991 and 2008, there were +322000 jobs created in financial services, real estate and leasing and professional services (Ministry of Manpower, 2009). In 2010, the Ministry of Manpower reported that there were 110000 expatriates in Singapore across a range of sectors, including banking and financial services (*The Financial Times*, 2010). In June 2013, 21 per cent of all service sector employment (339900 from a total of 1.647 million) was in financial and insurance services (including banking), real estate and professional services (including accounting, legal) (MOM, 2013). The presence of a third of a million workers in these financial and professional occupations ensures that there is a significant supply of knowledge and expertise in the financial centre, which supports the markets for private banking and private wealth and asset management. The private banking and wealth management talent pool has the technical, advisory and relationship-banking experience to manage private wealth, and seek specialist services within Singapore's private wealth community and between the Lion City and other financial centres (Lai, 2012). Moreover, almost all of the different nationality foreign private retail and wholesale banks (with wealth management functions), and professional services firms, would have employed home-country nationals alongside expatriates (Beaverstock, 2011). Since the rapid improvement in economic conditions in Asia-Pacific and Singapore post-2008, there has been a resurgence in employment growth in private banking and wealth management as Singapore's private wealth 'Asian hub' status has been enhanced with the expansion and influx of European banks (e.g., Bank Sarasin, Switzerland) (*The Financial Times*, 2010).

Fourth, and very much on the demand side, Singapore has a high proportion of S$ millionaires, which Jetley (2013) estimates is nearly one in every 30 individuals, and BCG (2013) suggests as being almost one in ten of all households (8.2 per cent). The 'home bias' for investment (PwC, 2013), draws in Singaporean and the local Asia-Pacific HNWI populations' private wealth (Table 20.5) into the city-state for management and

Table 20.5 Millionaire households in proximity to Singapore, 2012

Country	Number of Households (000s)		Proportion of Households (%)	
	Millionaire	UHNW	Millionaire	UHNW
Japan	1460	n.a.	2.2	n.a.
China	1304	851	n.a.	n.a.
Taiwan	312	n.a.	4.0	n.a.
Hong Kong	231	323	9.4	13
Australia	178	231	n.a.	n.a.
India	164	n.a.	n.a.	n.a.
Indonesia	n.a.	221	n.a.	n.a.
Qatar	n.a.	n.a.	14.3	8
Kuwait	n.a.	n.a.	11.5	7
Singapore	n.a.	n.a.	8.2	7
Bahrain	n.a.	n.a.	4.9	n.a.
UAE	n.a.	n.a.	4.0	3
Oman	n.a.	n.a.	3.3	n.a.

Source: Boston Consulting Group (2013).

preservation (Jetley, 2013). To corroborate the significance of AUM in Singapore originating from the local Asia-Pacific region, as already reported, an analysis of MAS's (2013a, 2013b) official data for total AUM in 2012 (totalling S$1626 billion) does indicate that appoximately 70 per cent, S$1138 billion, originated from other Asia-Pacific countries.

Fifth, Singapore is one of the most attractive cities for the super-rich to 'live, work and play' in (Long and Tan, 2010, p. 116). Knight Frank's (2014) *World Wealth Report 2014* ranks Singapore third in the cities that matter to the world's wealthy, after London and New York, and predicts that by 2024 the city will still be ranked in the top five most desirable cities for the wealthy (ranked fourth, after London, New York and Hong Kong). The 'financial elites' resident in Singapore, across its corporate economy, including banking and finance, have a ready-made private wealth management 'complex' on their doorstep, composed of an array of domestic and foreign institutions and they can '"live and bank" quite literally "onshore" in the country' (Pow, 2013, p. 1). Pow (2011, 2013) has noted that the City's ability to attract wealth and the super-rich is very much aligned to Singapore's reputation as a 'liveable city for the super-rich' (Pow, 2013, p. 61). On the one hand, the Singaporean government has pursued a pro-private banking and wealth management policy for the country to become the 'Switzerland of Asia' (Allen, 2006; Pow, 2013), including a relatively low and competitive personal income tax regime (up to a maximum of 20 per cent; Long and Tan, 2010). On the other hand, an array of state elites, planners, policy-makers and developers, have invested in an infrastructure to attract and retain the super-rich from Singapore and the region, most notably China. Singapore's 'super-rich' infrastructure not only includes an array of exclusive gated condominiums, waterfront apartments and serviced transport facilities for private yachts, jets and helicopters in places like Marina Bay and Sentosa, but also access to prize-winning restaurants, six-star hotels, casinos, exclusive luxury goods and international events (Pow, 2013).

Projecting forward, Singapore is expected to retain its position as one of the leading global centres for private wealth management. Moreover, as the market for private wealth grows significantly in the Asia-Pacific region, it will ultimately sustain its place over Switzerland and, perhaps London, to become the pre-eminent offshore private wealth domain. In many ways, China holds the key to Singapore's future success, particularly if it is to become the second wealthiest nation on earth by the end of the decade (BCG, 2013). Returning to Knight Frank's (2014) research, given that nine of the projected top 20 global city UHNWI populations by 2023 are in the Asia-Pacific region (Table 20.6), and following on from PwC (2013) and Deloitte's (2013) estimations, it will be no surprise if Singapore's private wealth sector continues to grow exponentially if a high proportion of these assets are allocated as AUM in its offshore jurisdiction.

CONCLUSIONS

In this chapter, we have addressed two main aims. First, we have brought in to the academic realm a conceptual and empirical understanding of the role of private wealth management in the functioning of the offshore jurisdiction of financial centres, OFCs. Whilst quite rightly the eyes of academia and the public consciousness has been focused on the ways in which multinational corporations use offshore jurisdictions and OFCs in their global business models to minimize their corporate tax (see *The Financial Times*, 2012), we have made sure that the wealth of the private individual has not been left out of this debate. Our argument shows quite clearly that as the number of HNWIs and the value of their global wealth has increased since the early 1990s, so too has the significance of the emergence of the private wealth management sector in both on- and offshore jurisdictions, particularly beyond the historical centres of private banking like Switzerland, former British Empire territories and protectorates, and European Principalities like Luxembourg and Monaco.

Table 20.6 Global cities ranked by projected UHNWI population, 2023

City	Region	UHNWI Population 2013	2023	Ten-year Growth (%)
London	Europe	4226	4940	+17
Singapore	Asia-Pacific	3154	4878	+55
New York	North America	2929	3825	+31
Tokyo	Asia-Pacific	3525	3818	+8
Hong Kong	Asia-Pacific	2560	3502	+37
Frankfurt	Europe	1868	2091	+12
Beijing	Asia-Pacific	1318	1872	+42
Sao Paulo	Latin America	1310	1843	+41
Osaka	Asia-Pacific	1450	1716	+18
Paris	Europe	1500	1656	+10
Munich	Europe	1113	1651	+48
Seoul	Asia-Pacific	1302	1595	+23
Taipei	Asia-Pacific	1255	1576	+26
Shanghai	Asia-Pacific	1028	1542	+50
Istanbul	Europe	1110	1531	+38
Zurich	Europe	1314	1521	+16
Mexico City	Latin America	1088	1431	+32
Toronto	North America	1184	1456	+23
Geneva	Europe	1156	1394	+21
Mumbai	Asia-Pacific	577	1302	+126

Source: Knight Frank (2014).

Second, and perhaps more importantly, we have mined data on private wealth management flows and AUM to present a brief geography of the financial terrain for the private wealth. The advent of 'new money' and the financialization of the rich and wealthy has contributed to the astronomical rise of the HNWI, particularly from Europe, North America and the Asia-Pacific, which in turn has created selected havens like London, New York, Switzerland, the Caribbean and Panama, and Singapore and Hong Kong, for private wealth to be managed, preserved and accumulated through a nexus of private banks, global accounting, legal and insurance companies, stockbrokers and the like. Post-2008, the rise and stock of private wealth, AUM, cannot be underestimated as a major financial market for global financial institutions and ABS in both on- and offshore jurisdictions, and OFCs as the number of HNWIs keep on rising in the world and the value of their global investible wealth reaches record levels up to the end of the decade and beyond.

Looking to the future, with the USA's FATCA 2010 legislation in place and similar schemes under review from the European Union nations in particular to lift the lid on personal tax avoidance across banking and financial institutions mainly doing business in the offshore jurisdiction (*The Financial Times*, 2014b), the role of OFCs' private wealth management nexuses will increasingly come under the microscope from nation-states, supranational organizations and global regulators alike. Unveiling the secrecy and opaqueness of private banking and the offshore jurisdiction can only be the right and just way forward for nation-states to begin to tackle inequality through the fiscal system, and for the industry itself to enhance its moral and social responsibility for advancing fiscal fairness in global society.

NOTES

1. See Tax Justice Network at http://www.taxjustice.net/ and Shaxson (2012).
2. See note 1.
3. The BCG (2013, p. 11) definition of offshore wealth reads, 'assets booked in a country where the investor has no legal residence or tax domicile'.

REFERENCES

ABN AMRO (2014), 'ABN AMRO private banking at a glance', accessed 24 August 2015 at https://www.abnamroprivatebanking.com/en/about-us/about-abn-amro/index.html.

Allen, M. (2006), 'Swiss financial blue print inspires Singapore', *Swissinfo.ch*, 31 October, accessed 30 August 2013 at http://www.swissinfo.ch/eng/archive/Swiss_financial_blueprint_inspires_Singapore.html?cid=682448.

Armstrong, S. (2010), *The Super-Rich Shall Inherit the Earth: The New Global Oligarchs and How They are Taking Over Our World*, London: Constable and Robinson Limited.

Atkinson, A.B. and T. Piketty (eds.) (2010), *Top Incomes: A Global Perspective*, Oxford, UK: Oxford University Press.

Beaverstock, J.V. (2002), 'Transnational elites in global cities: British expatriates in Singapore's financial district', *Geoforum*, **33** (4), 525–38.

Beaverstock, J.V. (2011), 'Serving British expatriate talent in Singapore: exploring ordinary transnationalism and the role of the "expatriate" club', *Journal of Ethnic and Migration Studies*, **37** (5), 709–28.

Beaverstock, J.V. (2012), 'The privileged world city: private banking, wealth management and the bespoke servicing of the global super-rich', in B. Derudder, M. Hoyler and P.J. Taylor et al. (eds), *International Handbook on Globalization and World Cities*, Cheltenham, UK and Northampton, MA, USA: Edward Elgar Publishing, pp. 378–89.

Beaverstock, J.V., P.J. Hubbard and J.R. Short (2004), 'Getting away with it? Exposing the geographies of the super-rich', *Geoforum*, **35** (4), 401–7.

Beaverstock, J.V., S. Hall and T. Wainwright (2013a), 'Servicing the super-rich: new financial elites and the rise of the private wealth management retail ecology', *Regional Studies*, **47** (6), 834–49.

Beaverstock, J.V., S. Hall and T. Wainwright (2013b), 'Private wealth management and the

City of London', in I. Hay (ed.), *Geographies of the Super-Rich*, Cheltenham, UK and Northampton, MA, USA: Edward Elgar Publishing, pp. 43–60.
Bicker, L. (1996), *Private Banking in Europe*, London: Routledge.
Boston Consulting Group (BCG) (2007), *Global Wealth* 2007, accessed 24 August 2015 at https://www.bcg.com/documents/file15105.pdf.
Boston Consulting Group (BCG) (2008), *Global Wealth 2008*, accessed 24 August 2015 at http://www.bcg.com.cn/export/sites/default/en/files/publications/reports_pdf/Global_Wealth_ES_Sept_2008.pdf.
Boston Consulting Group (BCG) (2009), *Global Wealth 2009*, accessed 24 August 2015 at http://www.bcg.com.cn/en/files/publications/reports_pdf/BCG_Global_Wealth_Sep_2009_tcm42-28793x1x.pdf.
Boston Consulting Group (BCG) (2010), *Global Wealth 2010*, accessed 24 August 2015 at http://www.bcg.com.cn/en/files/publications/reports_pdf/BCG_Regaining_Lost_Ground_Global_Wealth_Jun_10_ENG.pdf.
Boston Consulting Group (BCG) (2012), *Global Wealth 2012*, accessed 24 August 2015 athttps://www.bcg.com/documents/file106998.pdf.
Boston Consulting Group (BCG) (2013), *Global Wealth 2013*, accessed 24 August 2015 at http://www.bcg.de/documents/file135355.pdf.
Cameron, A. (2008), 'Crisis? What crisis? Displacing the spatial imaginary of the fiscal state', *Geoforum*, **39** (3), 1145–54.
Capgemini Merrill Lynch (CML) (2002), *World Wealth Report 2001*, accessed 6 January at www.ml.com; see also http://www.valentinecapitalassetmanagement.com/worldwealth.html.
Capgemini Merrill Lynch (CML) (2007), *World Wealth Report 10th Anniversary 1997–2006*, accessed 24 August 2015 at https://www.no.capgemini.com/resource-file-access/resource/pdf/World_Wealth_Report_2006.pdf.
Capgemini Merrill Lynch Global Wealth Management (CMLGWM) (2008), *World Wealth Report 2007*, accessed 24 August 2015 at https://www.capgemini.com/resources/world_wealth_report_2007.
Capgemini Merrill Lynch Global Wealth Management (CMLGWM) (2009), *World Wealth Report 2008*, accessed 24 August 2015 at https://www.capgemini.com/resources/world-wealth-report-2008.
Capgemini Merrill Lynch Global Wealth Management (CMLGWM) (2010), *World Wealth Report 2010*, accessed 6 January 2010 at www.ml.com.
Capgemini Merrill Lynch Global Wealth Management (CMLGWM) (2011), *World Wealth Report 2010*, accessed 24 August 2015 at https://www.capgemini.com/resources/world-wealth-report-2010.
Capgemini and Royal Bank of Canada Wealth Management (CRBCWM) (2012), *World Wealth Report 2012*, accessed 21 June 2012 at http://www.capgemini.com/insights-and-resources/by-publication/world-wealth-report-2012--spotlight/.
Capgemini and Royal Bank of Canada Wealth Management (CRBCWM) (2013), *World Wealth Report 2013*, accessed 24 August 2015 at https://www.capgemini.com/resource-file-access/resource/pdf/wwr_2013_0.pdf.
Cassis, Y. and P. Cottrell (2009), *The World of Private Banking*, Farnham, UK: Ashgate.
City of London (2007), *The Global Financial Centres Index (GFCI) 1*, accessed 24 August 2015 at http://www.zyen.com/PDF/GFCI.pdf.
Cobb, S. (1998), 'Global finance and the growth of offshore financial centres: the Manx experience', *Geoforum*, **29** (1), 7–21.
Cook, G., N. Pandit and J.V. Beaverstock et al. (2007), 'The role of location in knowledge creation and diffusion: evidence of centripetal and centrifugal forces in the City of London financial services agglomeration', *Environment and Planning A*, **39** (6), 1325–45.
Deloitte (2013), *The Deloitte Wealth Management Centre Rankings 2013*, accessed 26 August 2015 at http://www2.deloitte.com/content/dam/Deloitte/ch/Documents/financial-services/ch-en-financial-services-banking-wealth-management-centre-ranking.pdf.
Faulconbridge, J., E. Engelen and M. Hoyler et al. (2007), 'Analysing the changing landscape

of European financial centres: the role of financial products and the case of Amsterdam', *Growth and Change*, **38** (2), 297–303.
Folkman, P., J. Froud and S. Johal et al. (2007), 'Working for themselves? Capital market intermediaries and present day capitalism', *Business History*, **49** (4), 552–72.
Frank, R. (2007), *Richi$tan. A Journey through the 21st Century Wealth Boom and the Lives of the New Rich*, New York: Piatkus.
Freeland, C. (2012), *Plutocrats. The Rise of the New Global Super-Rich*, London: Penguin.
Froud, J. and K. Williams (2007), 'Private equity and the culture of value extraction', *New Political Economy*, **12** (3), 405–20.
Haberly, D. and D. Wójcik (2013), 'Tax havens and the production of offshore FDI: an empirical analysis', *University of Oxford, School of Geography and the Environment Working Paper in Employment, Work and Finance No. 13-02*, accessed 1 October 2013 at http://dx.doi.org/10.2139/ssrn.2252431.
Hall, S. (2009), 'Financialised elites and the changing nature of finance capitalism: investment bankers in London's financial district', *Competition and Change*, **13** (2), 173–89.
Hampton, M.P. and J. Abbott (eds) (1999), *Offshore Financial Centres and Tax Havens: The Rise of Global Capital*, Basingstoke, UK: Palgrave Macmillan.
Hampton, M.P. and J.E. Christensen (1999), 'Treasure Island revisited. Jersey's offshore finance centre crisis: implications for other small island economies', *Environment and Planning A*, **31** (9), 1619–37.
Hampton, M.P. and J.E. Christensen (2002), 'Offshore pariahs? Small island economies, tax havens and the re-configuration of global finance', *World Development*, **30** (9), 1657–73.
Hay, I. (ed.) (2013), *Geographies of the Super-Rich*, Cheltenham, UK and Northampton, MA, USA: Edward Elgar Publishing.
Hay, I. and S. Muller (2012), '"That tiny, stratospheric apex that owns most of the world" – exploring geographies of the super-rich', *Geographical Research*, **50** (1), 75–88.
Hudson, A.C. (2000), 'Offshoreness, globalization and sovereignty: a postmodern geo-political economy?', *Transactions, Institute of British Geographers*, **25** (1), 269–83.
Irvin, G. (2008), *Super Rich. The Rise of Inequality in Britain and the United States*, London: Polity Press.
Jetley, N. (2013), 'Singapore's multimillionaires', *Forbes*, 27 September, accessed 1 October 2013 at http://www.forbes.com/sites/neerjajetley/2013/09/27/anatomy-of-a-singapore-multi-millionaire-a-new-wealth-report-busts-many-myths/.
Khor, H.E. and R.X. Kee (2008), 'Asia: a perspective on the sub-prime crisis', *Finance & Development*, **45** (2), accessed 28 May 2014 at http://www.imf.org/external/pubs/ft/fandd/2008/06/khor.htm.
Knight Frank (2014), *The World Wealth Report 2014*, accessed 28 May 2014 at http://www.thewealthreport.net/#sthash.xp0S8Xhz.dpbs.
Lai, K. (2012), 'Differentiated markets: Shanghai, Beijing and Hong Kong in China's financial centre network', *Urban Studies*, **49** (6), 1275–96.
Lai, K. (2013), 'The Lehman mini-bond crisis and financialisation of investor subjects in Singapore', *Area*, **45** (3), 273–82.
Langley, P. (2006), 'The making of investor subjects in Anglo-American pensions', *Environment & Planning D: Society and Space*, **24** (6), 919–34.
Laulajainen, R. (1998), *Financial Geography. A Banker's View*, Gothenburg: Gothenburg School of Economics and Commercial Law, Gothenburg University.
Long, J.A. and D. Tan (2010), 'The growth of the private wealth management industry in Singapore and Hong Kong', *Capital Markets Law Journal*, **6** (1), 104–26.
Maude, D. (2006), *Global Private Banking and Wealth Management. The New Realities*, Chichester, UK: Wiley.
Maude, D. and P. Molyneux (1996), *Private Banking*, London: Euromoney Books.
Maurer, B. (2008), 'Re-regulating offshore finance?', *Geography Compass*, **2** (1), 155–75.
Ministry of Manpower (MOM) (2009), Data extracted from the Annual Employment Change by Industry statistics, accessed 19 January at www.mom.gov.sg/publish/momportal/en/com

munities/others/mrsd/statistics/Employment.html; for an up to date analysis of employment and labour market trends, see http://stats.mom.gov.sg/Pages/ExploreStatisticsPublications.aspx#PublicationSearch?topic=employment, accessed 26 August 2015.
Ministry of Manpower (MOM) (2013), *Labour Force in Singapore 2013*, accessed 28 May 2014 at http://stats.mom.gov.sg/Pages/Labour-Force-In-Singapore-2013.aspx.
Monetary Authority of Singapore (MAS) (1998), *MAS: Survey of Fund Management Activities in Singapore*, accessed 28 May 2014 at http://www.mas.gov.sg/~/media/resource/eco_research/surveys/1998_Survey_of_F.pdf.
Monetary Authority of Singapore (MAS) (2005), *2005 Singapore Asset Management Industry Survey*, accessed 28 May 2014 at http://www.mas.gov.sg/~/media/resource/eco_research/surveys/2005%20Asset%20Management%20Survey.pdf.
Monetary Authority of Singapore (MAS) (2012), *2012 Singapore Asset Management Industry Survey*, accessed 2 May 2014 at http://www.mas.gov.sg/~/media/MAS/News%20and%20Publications/Surveys/Asset%20Management/2012%20AM%20Survey%20Public%20Report%20FINAL.pdf.
Monetary Authority of Singapore (MAS) (2013a), *2013 Singapore Asset Management Industry Survey*, accessed 24 August 2015 at http://www.mas.gov.sg/~/media/MAS/News%20and%20Publications/Surveys/Asset%20Management/2013%20AM%20Survey%20Public%20Report_25072014_Final%20revised.pdf.
Monetary Authority of Singapore (MAS) (2013b), *Directory of Financial Institutions*, accessed 24 August 2015 at https://masnetsvc.mas.gov.sg/FID.html.
Palan, R. (2006), *The Offshore World. Sovereign Markets, Virtual Places and Nomad Millionaires*, New York: Cornell University Press.
Palan, R., R. Murphy and C. Chavagneux (2010), *Tax Havens: How Globalization Really Works*, Ithaca, NY: Cornell University Press.
Pow, C.-P. (2011), 'Living it up: super-rich enclave and elite urbanism in Singapore', *Geoforum*, **42** (3), 382–93.
Pow, C.-P. (2013), '"The world needs a second Switzerland": onshoring Singapore as the liveable city for the super-rich', in I. Hay (ed.), *Geographies of the Super-Rich*, Cheltenham, UK and Northampton, MA, USA: Edward Elgar Publishing, pp. 61–76.
PricewaterhouseCoopers (PwC) (2013), *Global Private Banking Wealth Management Survey 2013*, accessed 6 September 2013 at http://www.pwc.com/gx/en/banking-capital-markets/private-banking-wealth-management-survey/index.jhtml?WT.ac=vt-wealth.
Roberts, S.M. (1994), 'Fictitious capital, fictitious spaces: the geography of offshore financial flows', in S. Corbridge, R. Martin and N. Thrift (eds), *Money, Power and Space*, Oxford, UK: Blackwell, pp. 91–15.
Sassen, S. (2013), *Cities in a World Economy*, 4th edition, London: Sage.
Sayer, A. (2012), 'Facing the challenge of the return of the rich', in W. Atkinson, S. Roberts and M. Savage (eds), *Class Inequality in Austerity Britain*, Basingstoke, UK: Palgrave Macmillan, pp. 163–79.
Scorpio Partnership (2013), 'Global private banking benchmark 2013', press release, 10 July, accessed 26 August 2015 at http://www.scorpiopartnership.com/press/global-private-banking-benchmark-report-launch/.
Shaxson, N. (2012), *Treasure Islands. Tax Havens and the Men Who Stole the World*, London: Vintage/Random House.
Sikka, P. and H. Willmott (2010), 'The dark side of transfer pricing: its role in tax avoidance and wealthy retentiveness', *Critical Perspectives on Accounting*, **21** (4), 342–56.
The Economist (2013), 'Storm survivors', 14 February, accessed 6 August 2013 at http://www.economist.com/blogs/schumpeter/2013/02/special-report-offshore-finance.
The Financial Times (2010), 'Singapore: an island of opportunity for the qualified outsider', 9 December, accessed 11 April 2014 at http://www.ft.com/cms/s/0/2f93c7fc-0280-11e0-ac33-00144feabdc0.html?siteedition=uk#axzz330HITWdB.
The Financial Times (2012), 'Multinationals face pressure on tax', 3 December.
The Financial Times (2014a), 'Global tax standard attracts 42 countries', 13 February, accessed

20 May 2014 at http://www.ft.com/cms/s/0/f6cd39e6-94b0-11e3-9146-00144feab7de.html?siteedition=uk#axzz330HITWdB.
The Financial Times (2014b), 'The cost of guilt for Credit Suisse', 20 May, accessed 28 May 2014 at http://www.ft.com/cms/s/0/29b05da8-e014-11e3-9534-00144feabdc0.html?siteedition=uk#axzz330HITWdB.
The Guardian (2013), 'G20 summit: states chase tax evaders with plan to swap data globally', 4 September, accessed 1 October 2003 at http://www.theguardian.com/world/2013/sep/04/g20-states-tax-evaders-swap-data.
Warf, B. (2002), 'Tailored for Panama. Offshore banking at the crossroads for the Americas', *Geografiska Annaler B*, **84** (1), 47–61.
Wójcik, D. (2013), 'Where governance fails. Advanced business services and the offshore world', *Progress in Human Geography* [online], accessed 24 August 2015 at http://phg.sagepub.com/content/early/2012/10/09/0309132512460904.full.
Ye, J. and P.F. Kelly (2011), 'Cosmopolitanism at work: labour market exclusion in Singapore's financial sector', *Journal of Ethic and Migration Studies*, **37** (5), 691–708.
Z/Yen (2010), *GFCI 8*, accessed 24 August 2015 at http://www.zyen.com/GFCI/GFCI%208.pdf.
Z/Yen (2013), *GFCI 13*, accessed 24 August 2015 at https://masnetsvc.mas.gov.sg/FID.html.

21. Troubling tax havens: multi-jurisdictional arbitrage and corporate tax footprint reduction

Ronen Palan and Giovanni Mangraviti

INTRODUCTION

It is well known among tax experts that corporate tax planning schemes are typically organized through a multitude of jurisdictions (Western et al., 2011; Palan, 2014a). Companies, large and small, are normally seen as unitary entities. They are referred to by their trademark names – Google, Amazon, IBM, British Telecom, BMW or Toyota, and so on – and are commonly thought of as American, British, German or Japanese, as the case may be. In reality, the vast majority of such companies consist of a multitude of companies, veritable ecologies in some cases, typically numbering in the hundreds or even thousands. Goldman Sachs, a dual entity combining a bank holding company (BHC) with a financial holding company (FHC), consists of 3115 separate legal entities, 1670 of which are registered outside the USA. JPMorgan Chase and Co, another BHC, consists of 3389 separate legal units, 451 of which are registered outside the USA (Avraham et al., 2012). BP, supposedly a British company,[1] consists of 1180 affiliates in 84 countries going 12 tiers deep (that is, 12 tiers of affiliates holding other affiliates and so on) (OpenOil, 2014). They are registered in various jurisdictions, among which notable tax havens, such as the Cayman Islands, British Virgin Islands, Jersey, Ireland, the Netherlands or Luxembourg, feature heavily.

Within these ecologies, each corporate structure provides its own contribution, as it were, to the tax planning scheme of the parent company, insofar as it brings with it a unique bundle of corporate tax laws and regulations that are particular to the jurisdiction of its location. Considered in isolation, entities of this sort may appear innocuous. It is only when intertwined with other entities of different jurisdictions that such schemes reveal their true purpose – that is, the exploitation of jurisdictional arbitrage for the purposes of minimizing individual and corporate tax footprints. The exploitation of jurisdictional arbitrage under such schemes may be blatant, but is hardly conspicuous. The industrial complex built around the business of selling tax minimization strategies is contingent as well as dependent upon a thick welter of intertwined jurisdictions where arbitrage schemes can grow and proliferate out of sight.

To investigate the nature and mechanisms of tax arbitrage, this chapter explores the linkages between wealth and multi-jurisdictionality: how and why the corporate world operates as a set of global business ecologies on top of an undergrowth of jurisdictional networks. We start with a brief discussion of the concept of wealth. Our analysis will proceed from the premise that the lion's share of global wealth is in fact 'intangible'. Distinguishing between tangible and intangible wealth is crucial to operating another important analytical distinction at the level of tax reduction and wealth: that between tax minimization schemes aimed at *accumulated* wealth, or 'wealth-already-accumulated', and those aimed at *accumulating* wealth, or 'wealth-to-be-accumulated'. The opaque relationship between tax minimization schemes and wealth accumulation is at the heart of what makes tax havens just so troublesome.

TROUBLESOME WEALTH AND TAX HAVENS

Troublesome wealth is, perhaps ironically, 'protected' wealth. As the principal strategy of wealth protection, the otherwise mundane activity of individual and corporate tax planning has become the kingpin of the most extraordinarily opaque system of wealth transfer in the world. The bulk of this system of wealth protection is aimed at escaping the clutches of public fiscal authorities. In a world with no common tax rules, the business of protecting, or otherwise concealing wealth plays out mostly via clever strategies of capital transfers: complex international transactions can elude tax supervision through a complex maze of legal trap doors and escape hatches, to the point where the difference between avoidance and evasion – guile and crime – becomes all but immaterial.

In pursuance of either elusive or avoidant tax planning, the bulk of this enormous traffic of wealth takes place in and out of special tax zones – that is, tax havens (Palan et al., 2010; Henry, 2012; Stewart, 2012; The Permanent Subcommittee on Investigations, 2013). Also referred to as 'offshore financial centres' by their professionals, or 'secrecy jurisdictions' by their detractors, tax havens are sovereign states or suzerain jurisdictions that offer international capital a system of eased tax regimes in direct competition with the regulatory provisions of other states. They can be defined broadly as 'any jurisdiction that satisfies two criteria. First, it has tax laws that are attractive to global investors and entrepreneurs. Second, it protects its fiscal sovereignty by choosing, in at least some cases, not to enforce the bad tax laws of other nations' (Mitchell, 2009). With the progressive expansion of state expenditure as well as state coffers across most advanced industrialized countries over the past century, the minimization of both individual and corporate tax footprint has become an expanding and lucrative business. Albeit an important strategy of financial growth and global competition, this industrial complex is less subservient to the economic interests of tax havens themselves than to those of the larger community of resident and non-resident accountants, lawyers, bankers, financiers and businesses who purchase the protection of offshore jurisdictions – essentially legal sovereignty – for the purposes of tax arbitrage.

There are many ways the super-rich can avail themselves of tax havens.[2] Perhaps the easiest one, for both individuals and businesses, is to establish an offshore trust or foundation. Exempt from registration in most jurisdictions, trusts separate the beneficiary of an asset from a trustee, often an offshore legal owner, who is assigned fiduciary duty to manage the asset for the benefit of the former. Alternatively, businesses, and increasingly also international financial institutions, can also establish a subsidiary, affiliate or independent company directly in a tax haven. Often set up as limited liability companies, or international business corporations (IBCs), these subsidiaries are employed to store profitable capital, trade in financial markets, hold property rights and manage investment funds in complete anonymity with respect to the parent company, whose transactions with subsidiaries are concealed under the company's 'consolidated' accounts. Aside from transferring capital to one of the above-mentioned offshore instruments, individuals have also the option of relocating their domicile or non-domiciled residency to a tax haven, or more drastically, renouncing both domicile and residency to live as PTs, or 'permanent tourists' exempt from tax and other legal obligations (Maurer, 1998). The basic principle of both individual and business tax planning is to spread one's assets across a multitude of jurisdictions so that most profit and capital transfers disappear from the tax radar.

The beneficiaries of these offshore tax-planning schemes are the global elite. The super-rich, in particular, hold the rough equivalent of the US annual GDP parked in offshore jurisdictions in direct competition with public fiscal authorities (Henry, 2012). The withdrawal of the product and the process of wealth accumulation from both fiscal and legal supervision spells trouble all round. Not only do tax minimization schemes clear the way for money laundering and other illegal activities, but they also skew the global distribution of income to the detriment of those individuals and businesses who lack the financial wherewithal, or moral proclivity, to engage in offshore tax minimization schemes. As will be explained below, what makes such schemes particularly troublesome for the distribution of global wealth is that the intangible nature of most assets makes multi-jurisdictional tax arbitrage an instrument of wealth creation as much as of wealth concealment (Beaverstock et al., 2013).

WEALTH PROTECTION AND INTANGIBLE WEALTH

The concept of aggregate global household wealth, despite a long intellectual heritage, has generated surprisingly little interest among economists.[3] The focus of most writing in political economy falls on GDP trends, often treated as growth projections, as well as debt, trade data or even income distribution. Wealth accumulation, one of the primary objectives of most civilizations, seems to play a curiously small role in mainstream understandings of political economy.

This attitude is changing. Since 2009, the Swiss bank Credit Suisse began issuing trend reports on the direction of global aggregate wealth. Its latest report, *Global Wealth 2013*, suggests that aggregated household wealth in 2013 stood at US$242 trillion, approximately three times the world's annual GDP (Keating et al., 2013). Credit Suisse additionally goes on to estimate a 40 per cent rise in global wealth over the course of the next five years. That figure is extraordinarily out of kilter with the current estimates of growth in the world economy, which most economists predict will rise very modestly over the same period.[4] What accounts for such a widening gap between global wealth and global growth?

Behind this apparent contradiction is a fundamental accounting magic. Economists tend to identify household wealth with 'tangibles', such as stock holding and real estate, in contrast to what they consider 'intangibles', such as human capital. Data on household wealth are typically referred to as 'net household wealth' and, accordingly, wealthy individuals are known as 'high net worth individuals' (HNWIs). The prefix 'net', here, is conventionally understood to refer to assets minus liabilities. But there is another element that distinguishes estimates of individual net worth from those of corporate 'net worth'. Whereas corporate accounts include the category of 'goodwill', individual ones do not. The assumption is that individuals 'own' two types of properties: 'tangible' properties (e.g., shares, real estate, yachts, art), and 'intangible' properties, also known as 'goodwill'. Wealth data generally only account for net assets – that is, accumulated wealth, as opposed to intangible assets.

The notion of goodwill and intangible assets is best understood in connection with the concept of 'futurity' (Palan, 2012, 2014b). The term, in John R. Commons's original reformulation of Böhm-Bawerk and MacLeod's definitions of futurity value, refers to a nineteenth-century legal innovation that marked a watershed in the development of capitalist institutions in the West (Commons [1924] 1959; 1961). The idea of futurity, with its particular understanding of valuation, is the keystone concept that holds together the notions of wealth-already-accumulated and wealth-to-be-accumulated.

Historically, businesses were valued on the basis of two modes of valuations. One was based on replacement value, equal to the current market prices of a firm's assets, and the other was based on liquidation value, which amounts to the value of a firm's assets upon cessation of business. In both cases, valuation is backward looking, in the sense that it reflects past business records and temporarily assumes the present company to have ceased operations. These valuations represent the power of creditors, who are seeking solid securities for their investment. Yet, businesses are not set up to fail; they are set up to capture future income streams. A different mode of valuation was therefore developed in the late nineteenth century, primarily in the USA, to better capture this reality. This mode was modelled after the principle of goodwill. In this view, firms are considered organic and future-oriented entities that only stand for what businesses are ultimately after: claims on future income streams. Accordingly, firms are treated as 'going concerns' (i.e., institutional complexes embedded in time and invested in the future). The pervasiveness of increasingly highly valued goodwill, notably in the case of big corporations such as the Coca-Cola Company, is testament to the fact that futurity, as a forward-looking mode of valuation, is truer to the everyday logic of business enterprises and particularly suited to explain the relationship between intangible wealth-already-accumulated (e.g., goodwill) and tangible wealth-to-be-accumulated (e.g., future income).

This distinction is of particular importance when investigating the dynamics between wealth accumulation and tax planning. In the business of wealth protection, intangible assets often call for special treatment. While some wealth schemes are primarily intended to protect wealth already accumulated, generally referred to as 'net household wealth', others are intended to protect future or potential income-generating titles. The academic literature on the subject seems long on the former and short on the latter. Protection schemes for future or potential income are usually only mentioned in the context of what are described in the trade as 'sophisticated' schemes, where the objectives of both types of wealth protection are mixed together.

The intangible dimension of wealth looms even larger. That is because conventional classifications of tangible and intangible assets are conceptually misleading. A number of tangible assets commonly conceived as part of wealth-already-accumulated, such as stock, shares and even bonds, contain in fact an inherent degree of futurity value. As noted in a different context (Palan, 2014b), even the seemingly tangible assets of real estate are regularly valued as claims on future income streams. In other words, the current value of those assets factors in future profits. As their value is based largely on projections of future profits, those seemingly tangible assets would deserve inclusion in corporate accounts as part of their goodwill.

Contrary to common impression, therefore, it is reasonable to conclude that most 'wealth' is made up of non-tangible assets that represent claims on future income streams. However, by their intangible nature, claims on future income streams are nomadic, and their place of registration – and taxation – is easy to manipulate.

Goodwill and Tax Havens

Let us briefly consider goodwill management with reference to individual wealth. There is little doubt that a renowned footballer of the likes of David Beckham, for instance, will have accumulated a great stock of wealth in his lifetime on the back of his many professional achievements.[5] What is often neglected, however, is that his profile and visibility will have also earned him a commensurate amount of goodwill among the public. Though intangible, goodwill is a considerable addition to the footballer's personal wealth because, as promise of future profit, it constitutes an asset for sale. Commercial establishments will seek to obtain this asset – his 'endorsement' in this context – in order to associate his goodwill to their products, thereby attracting potential customers and increasing sales. Companies are therefore prepared to purchase Beckham's goodwill at a price. They purchase the right to use his image, or words attributed to him (both under strict conditions), to endorse their products. David Beckham, to pursue the example further, can therefore be said to be in the business of selling his goodwill, which he can also do to more than one company at a time. There are a few restrictions: for example, he may only be allowed to support, say, one brand of fizzy drink at any one time. That is because a goodwill persona is for all intents and purposes a business proposition and, as such, goes by a different set of rules to the person itself. In our example, 'goodwill Beckham' is not allowed to indulge in both Coca-Cola and Pepsi at the same time, whereas 'real Beckham' may well enjoy both (or neither).

This intangible side to personal wealth lays itself wide open to tax minimization schemes. Spotting an opportunity in the management of celebrity goodwill, for instance, the island of Guernsey has recently introduced a new law, the Image Rights (Bailiwick of Guernsey) Ordinance

(States of Guernsey, 2012), which enables the registration of personalities and images associated to them. The advantage of a Guernsey registration is twofold. First, as a UK jurisdiction, Guernsey can project its laws overseas and hence ensure (to an as yet untested degree) the maintenance of such rights throughout the world. Second, Guernsey corporate taxation is 20 per cent as opposed to the 50 per cent that top earners, such as David Beckham, would pay in the UK. Hence, by the simple provision of registering image rights in Guernsey, a well-known celebrity can already reduce tax liabilities to 20 per cent. As we will see later, that 20 per cent can be chipped further away.

Celebrities like David Beckham (and Sean 'Diddy' Combs as discussed by Watson in Chapter 9 in this volume) are not alone is selling their goodwill; in principle, we all do. When academic lecturers, such as the authors of this chapter, sign a contract with a university, the university is purchasing something about the future accomplishment and services of said academics, typically based on projections of past performances. In that sense, not only do we all 'own' something that may generate future income, but the potential for future income may also be valuable in itself. Insofar as both sides to the contract make promises to the future, these contracts are essentially futures.

What mainstream accounts of 'net household wealth' or HNWI leave out of their calculations, however, are two variables: (1) estimates of aggregate potential earnings accruing from individual goodwill, and (2) potential for future gains that may or may not materialize. These items are usually missing from data logs on HNWI because they are scarcely employed for the purposes of traditional accounting. In practice, however, these intangible variables make a considerable difference in future-sensitive contexts. For example, they are playing an increasing role in divorce proceedings, especially of celebrities and the super-rich, who may be ordered to award former spouses a 'percentage ownership interest in the future income stream [generated from] professional goodwill' in the interest of equitable distribution.[6] This analysis therefore alerts us to the existence of three types of personal wealth:

- wealth that has already been accumulated;
- income streams generated against personal goodwill; and
- future income streams that may be generated against personal goodwill.

Real Estate and Tax Havens

That most wealth is largely intangible, and hence nomadic, may appear counterintuitive. Let us take the example of what has a good claim to be the least mobile of assets, real estate – also known, appropriately enough, as 'immovable property'. The vast majority of household 'net' wealth is residential and commercial real estate, which is subject to various layers of taxation during the ownership cycle as well as at point of sale. There is a business case, therefore, to find ways of grabbing income from real estate without claiming direct ownership. Different tax planning strategies can be used to achieve this, such as through financial derivatives (which cannot be explored at sufficient length here) or through schemes that prevent the registration of the sales of real estate in a taxing country. In all these cases, multi-jurisdictional arrangements are necessary, preferably through tax havens.

Let us take the example of a London property. In the UK, realized appreciation in the value of properties is subject to capital gains tax, currently standing at 18 per cent. In addition, the UK imposes a 7 per cent stamp duty on purchase of assets worth more than £2 million (risen to 12 per cent in late 2014), the value of a small flat in central London. The figure for properties owned by companies goes up to 15 per cent. Duties on the sales of such properties can therefore reach up to 33 per cent of the value of the sale. How can taxes be avoided on such sales? These are, after all, immobile assets.

Let us consider the following scenario:

1. London Property A is owned by a Spanish Company B. Company B has only one asset, the London property.
2. Company B is owned by Company C, registered in the Cayman Islands.
3. Company C is owned by Company D, registered in Bermuda.
4. A legal person wishes to purchase London property A. To do so, they set up a company in Luxembourg (Company E).

The scheme can work in a number of ways. For instance, Company C sells Company B to Company E registered in Luxembourg. Alternatively, Company D sells Company C to Company E. In both of these cases, as far as the UK is concerned, no sale took place: Property A is still owned by company B, and hence there are no tax issues arising from this sale. Taxation on sales of companies in the Cayman Islands or Bermuda is minimal. In fact, the sale may have no tax implications at all (known in the trade as 'tax neutral' transaction).

At face value, each company in the above scenario appears innocuous; these are entirely legitimate Spanish, Cayman, Bermudan and Luxembourgian companies. Furthermore, in this scenario, each of the jurisdictions may even pride itself – as they increasingly do – in being highly regulated domestically. But what is the scope of isolated regulation in the midst of such intangible transactions? It is only in combination that the true purpose of the scheme comes to light. In fact, the sole objective of this multi-jurisdictional arrangement is to remove the location of the sale of a property from the location of the property itself, so as to avoid the sales tax in the original location of the property. It is difficult to see any other purpose to these arrangements.

The net results of these multi-jurisdictional schemes is that the brunt of property tax and stamp duties is borne by the worse-off and the middle classes, whereas the better-off and large corporations have the de facto privilege of avoiding them, if they so wish. These are, in other words, wealth concentration devices. Tax havens stand to benefit the most from this, in that they are in a position to design and create the legal loopholes for such schemes to their own advantage.

To take another example, in 2010 Luxembourg, under pressure from the EU, withdrew its 1929 Holding Company legislation. These types of holding companies were used pervasively for tax avoidance purposes because they were exempt from corporate income tax and could withhold tax on dividends and certain other Luxembourgian tax. As the backbone of many corporate tax minimization techniques, they had come in for widespread condemnation. On the face of it, Luxembourg appeared to have bowed to criticisms by terminating the 1929 Holding Company legislation. In reality, in anticipation of the withdrawal, Luxembourg had already established a number of new structures to replace it. Among them are the Private Asset Management Companies (SPFs) introduced in 2007. The Luxembourg Consulting Group helpfully lists the tax liabilities of these types of companies as follows:

- a one-off registration tax of €75 that is payable at the formation of an SPF and when the articles of association are amended;
- subscription tax of 0.25 per cent annually on the deposited capital (+ issuing bonuses);
- no DBA authorization; (database authorization, i.e., data provided by the company is taken at face value);
- no VAT registration;
- complete exemption from corporate income tax, excise tax, and assets tax;
- no withholding tax on interest payments (restrictions apply to individuals);
- no withholding tax on dividend payments (non-residents);
- no taxation of capital profit arising from the sale of SPF shares (non-residents);

- no taxation of liquidation revenues from the SPF (non-residents) (Luxembourg Consulting Group, 2014).

One can see how SPFs can support what the same document, perhaps unadvisedly, refers to as 'opportunistic real estate investment'. The specific recommendation of the Luxembourg Consulting Group is for individuals and companies to set up SPFs 'just in case'. SPFs are considered legal and above board. Indeed, Luxembourg is considered a highly regulated financial environment. But it is difficult to deny that SPFs, and a few other provisions like it, are intended to perpetrate tax avoidance schemes routed through Luxembourg.

The scenario presented above also tells us that the data on assets registered offshore are highly dubious for two related reasons. First, there is no inherent reason why any of the holding companies, of whatever type or denomination, should revise asset values as they rise and fall. In the case of the Luxembourg SPFs, which is rather typical, there is an implicit promise on the part of the Luxembourg authorities not to relinquish the data provided by the SPFs to the national authorities. Most likely, declared asset values in such entities count as business propositions, and may be subject to considerations of taxation, leverage, collaterals and the like. Second, it is unclear whether there is any occurrence of double, triple or quadruple accounting, as some assets are linked together in chains of ownership spanning different jurisdictions. Each of these jurisdictions is keen to boast the aggregate value of all the registered assets. Indeed, the practice is to create at least one or two additional shell companies in different locations, 'just in case' a future sale scenario would benefit from a slightly different organization of the chain.

The Firm and Tax Havens

The intangible dimension of wealth is expressed to the fullest in the context of business companies. A surprising amount of business assets fall under the rubric of goodwill, and companies take full advantage of this flexibility for tax minimization purposes. Let us start with some clarification about the nature of the firm and corporations. Dominant theories of the firm, as summed up by Jean-Philippe Robé, 'are built around the notions of agency, property rights and contracts. . . Firms are assumed to be operating within perfect legal and political environments [where] all externalities within the firms' production prices [and] all interests affected by the firm's activities' are internalized and protected' (2011, p. 2). This view, however, tends to conflate the ownership structures of assets (the capital) with institutions (such as multinational corporations), and legal institutions (corporations) with economic units (firms). The corporation is a legal entity that is licensed by a sovereign entity and can only operate within the bounds of one national space at a time.[7] The firm, in contrast, is an economic entity. Firms often control strings of legal entities or corporations – at times numbering in the thousands – and in that sense can operate in many jurisdictions. The firm, on the other hand, lacks legal existence. In other words, legally speaking, there is no such thing as a 'multinational corporation'.

In this context, tax minimization techniques exploit the differences between the economic control of firms and the legal foundations of territorially bounded companies. All legal companies are bound by national rules and regulations that, if duly observed, are sufficient to raise companies above legal suspicions, as in the case of Luxembourg's SPFs described above. Firms, on the other hand, are organizations put together by accountants, such as the Big Four accounting firms (Deloitte, PwC, Ernst and Young, KPMG). They typically span a number of territories and are used to link companies from different jurisdictions in such a way as to minimize their overall tax footprint.

A STORY OF RUSSIAN DOLLS

A telling example of firms' tax minimization techniques was revealed in 2012 by Emily Yiolitis, partner at Harneys, Cyprus. In her piece, Yiolitis describes a deal organized by her firm for a proposed US$500 million restructuring of a Russian individual shareholding with the view 'of maximising the tax efficiency of the corporate structure and with a view to a prospective sale of part of the operations' (Yiolitis, 2012). Harneys proposed the following steps:

- Step one: The Russian company (RI) sets up a Cypriot holding company (CY). The company can only access the double tax treaty network with Cyprus as a local tax resident, so they shift management and control to the island by setting up an office in Limassol, where they move three Russian representatives and hire two administrative staff.
- Step two: The Russian shareholder contributes 100 per cent of the shares of their Russian corporation (RusCo) to the CY in return for further shares in CY. The exchange of shares is carried out with no tax implication as such qualifying reorganizations are exempt from stamp duty (otherwise applicable at 0.2 per cent per cent of the value of the transaction). As a result, RI becomes the shareholder of an

increased number of shares in CY, and CY becomes the shareholder of RusCo.
- Step Three: Profits derived by the manufacturing operations of RusCo are sent as dividends to the sole shareholder, CY.

The double tax treaty between Cyprus and Russia stipulates that the dividends payable from Russia to Cyprus are subject to 5 per cent withholding tax in Russia provided the investment exceeds €100 000. As RusCo was engaged in active manufacturing operations, CY, according to Cyprus law, was not subject to any tax in Cyprus. Hence, Cyprus did not levy any withholding tax on CY. Cyprus income tax law provides for a tax exemption from profits realized by Cyprus companies upon the sale of securities. Therefore the sale by CY of 30 per cent of its shares in RusCo to a purchaser situated in the British Virgin Islands (BVI) did not attract any income tax in Cyprus as a trading gain. As Cyprus does not tax capital gains, there was no incidence of capital gains tax in the sale of 30 per cent of RusCo by CY.

Here we have a case of a deal that operates through three different jurisdictions, taking advantage of tax loopholes in each to ensure that only the minimal amount of tax is paid. Cyprus is therefore used by the Russian individual to minimize his tax footprint in Russia.

WHEN APPLES FALL FAR FROM THE TREE

Another common way a firm's intangible assets can be exploited for tax minimization purposes is by recourse to the futurity value of those intangibles formally recognized in English Common Law as 'goodwill'. These include all the usual business assets that are valued on the basis of the firm's future earning capacity, such as trademarks and brand names as well as the organizational and managerial business structure.

One way firms can utilize these assets in multi-jurisdictional tax minimization schemes is by separating the tangible sources of income, such as the sale of hardware, from the intangible sources of income embedded therein, such as the sale of patent rights or intellectual property rights associated with the hardware. The different sources of income can be subsequently apportioned to different companies. A large company of the size of BMW, for example, could hypothetically set up a string of offshore entities, each with a claim to different portions of a car: one owning the rights of income from the use of the label, another owning the rights of income from the sale of physical assets, and yet another owning the rights of income from the patents embedded in the cars. In this model, BMW car sales would have to pay royalties to each of those separate companies. Since they are legally separated, they are each treated as separate entities for tax purposes. Needless to say, those companies will be registered in different jurisdictions, chosen largely for tax purposes.

Let us take the example of Apple Inc. The following is taken from a detailed study of the Levin Congressional Committee (Apple, 2013; The Permanent Subcommittee on Investigations, 2013). It appears that Apple Inc. has created three offshore corporations that receive tens of billions of dollars in income, but which have no tax residence – neither in Ireland, where they are incorporated, nor in the USA, where the Apple executives who run them are located. 'Apple has arranged matters so that it can claim that these ghost companies, for tax purposes, exist nowhere. One has paid no corporate income tax to any nation for the last 5 years; another pays tax to Ireland equivalent to a tiny fraction of 1 per cent of its total income' (The Permanent Subcommittee on Investigations, 2013, p. 3).

One of Apple's shell companies is Apple Operations International (AOI). AOI directly or indirectly owns most of Apple's other offshore entities. Under Irish law, only companies that are managed and controlled in Ireland are considered residents for tax purposes. Since AOI is only incorporated, but not managed or controlled, in Ireland it does not count as an Irish tax resident. Under US law, on the other hand, a company is generally taxed on the basis of where it is incorporated, not where it is managed and controlled. Since AOI is not incorporated in the USA, it is not tax resident in the USA either. AOI, therefore, is tax resident nowhere. In fact, AOI has as many as zero employees.

The second corporate shell set up by Apple in Ireland is Apple Sales International (ASI). ASI holds the economic rights to Apple intellectual property rights outside of the USA. From 2009 to 2012, its sales income amounted to US$74 billion. Similarly to AOI, the company is incorporated in Ireland but operated from the USA. ASI only paid a minimal amount of tax to Ireland. For example, in 2011 it paid US$10 million against US$22 billion in revenue. Apple's third subsidiary, Apple Operations Europe (AOE), sits between ASI and AOI. It, too, has no tax home.

Not unlike many firms of its kind, Apple is taking advantage, in other words, of discrepancies in incorporation and tax residency rules between different countries. Compared with more sophisticated techniques used by other household name firms, this is one of the simplest schemes of tax minimization, but one that makes full use of multi-jurisdictional tax arbitrage. Needless to say, it is all legal.

TROUBLESOME TAX HAVENS: AGGLOMERATIONS AND NICHE-SEEKING STRATEGIES

The cases mentioned above offer a flavour of how multi-jurisdictionality works in practice. It is a world that relies heavily on financial services, the Big Four accounting firms, as well as a myriad of smaller law, banking and financial firms. A UK parliamentary committee estimates that this industry generates about US$25 billion of annual income (House of Commons, 2013).

One interesting question that arises from this discussion is whether those countries colloquially known as tax havens are actively encouraging the development of tax avoidance strategies or whether they have developed their own taxation rules independently and are just being exploited by unscrupulous accounting firms. There are two possible answers to this question. One is based on the theory of the captured state or captured elites. This theory suggests that lacking the necessary manpower and tertiary education facilities, many small island jurisdictions and their governments are not capable of developing successful offshore financial centres (OFCs) on their own. They are effectively 'captured' by powerful foreign finance and legal firms who write the laws of these countries that they then exploit. There is good evidence to this effect (Sagar et al., 2012). The other theory suggests that tax havens evolved in jurisdictions that were traditionally outward looking and dominated by commercial and trading interests. Many tax havens, such as Switzerland or Singapore, were originally known as entrepôt centres for regional trading activities. Their local elites were therefore strongly predisposed to develop OFCs.[8] The two theories are not incompatible. Indeed, the second may add nuance to the first.

The latter theory points to the historical evolution of two broad types of tax haven agglomerations. This argument is founded on an analysis of the Bank for International Settlements (BIS) locational statistics on international lending and borrowing data (BIS, 2014). The data reveal that one agglomeration of tax havens has a distinct British imperial flavour. It consists, first and foremost, of the City of London, and includes, in addition, the British Crown dependencies of Jersey, Guernsey and the Isle of Man; a few British Overseas Territories, including the Cayman Islands, Bermuda, British Virgin Islands, Turks and Caicos, and Gibraltar; and recently independent British colonies such as Hong Kong, Singapore, the Bahamas, Cyprus, Bahrain and Dubai.[9] The British imperial pole accounted for a combined average of 38.3 per cent of all outstanding international loans and deposits by March 2010.

The other, far looser agglomeration consists of a string of mid-size Western European states known for an odd coupling of welfare and tax haven provisions. They include the Benelux countries, Belgium, Netherlands and Luxembourg, as well as Ireland and Switzerland. The European agglomeration accounted for a combined 14.9 per cent of all outstanding international loans and deposits by March 2010, exactly the same as the entire USA (BIS, 2014).[10]

What explains the emergence of these two agglomerations of international financial centres? Why do so many of the world's leading international financial centres have a British imperial link? The root cause behind this differentiation is found in the distinction, operated since the legal establishment of futurity valuation in the late nineteenth century, between two types of incorporeal properties, financial instruments and goodwill instruments (or intangibles).

London's rise took place in the midst of the City's attempt to survive the period of imperial decline. The tax haven agglomeration linked to it evolved as a group of centres designed to trade in incorporeal assets and therefore geared to operate as one gigantic offshore financial centre with the City of London at its core. The subsequent re-emergence of the City of London as the world's premier financial centre was in no small part due to the emergence of the Euromarket in London in 1957 (Burn, 2006). According to this theory, the Bank of England came to an informal agreement with London merchant banks that it would treat certain types of financial transactions, those between non-resident parties and those denominated in foreign currencies, as if they did not take place in London. In doing so, the bank effectively created a new regulatory space outside of its own jurisdiction – as well as a new concept, that of offshore finance. As the transactions taking place in London were deemed by the Bank of England to be taking place elsewhere, they ended up under no regulation at all, and therefore 'offshore'. This new, unregulated locus of transaction came to be known as the Euromarket, or offshore financial market (Burn, 2006).

The Euromarket remained small and practically unknown for three or four years until American banks discovered it in the early 1960s. They quickly developed branch networks so that they could avoid restrictive domestic regulations through their London subsidiaries. Once the facilities of the Euromarket were discovered, corporate clients also began to bypass the banks and to tap directly into the offshore financial market to earn higher rates of interest while their clients, too, learnt to tap in the Euromarket to fund their operations (Sylla, 2002; Burn, 2006).

London emerged as an offshore financial market as a result of what could be seen as an administrative accident. All other areas under the jurisdiction of the UK at the time, including Hong Kong, the Channel Islands, the Cayman Islands and other British Caribbean Islands, happened to enjoy the same legal provisions, and spontaneously developed as offshore centres as a result. It did not take long, of course, for banks and other financial institutions to appreciate the useful synergies between tax havens and OFCs, particularly if located in the same place. In dual-status tax havens/OFCs, banks and other financial institutions could not only circumvent stringent financial regulations, but also find 'tax neutral' ways of conducting their business. This is in fact what drove some tax havens to develop as OFCs.

Some smaller US and Canadian banks, faced with the high infrastructural costs of a London base, 'realized that the Caribbean OFCs offered a cheaper and equally attractive regulatory environment – free of exchange controls, reserve requirements and interest rate ceilings, and in the same time zone as New York' (Hudson, 1998, p. 541). According to various reports (Sylla, 2002), the early spillover of OFCs activities into the Bahamas and Cayman was motivated, like the London Euromarket, not by tax advantages, but by the cheaper transaction costs of setting up branches there.

The London Euromarket, while effectively unregulated, was still heavily taxed. However, as tax was levied at point of maturation, it became common practice to register syndicated loans and, later on, many other financial activities in commensurate offshore centres in British dependencies (although for reasons that go beyond the scope of this chapter, until 1974, Euromarket operations could develop only in British overseas territories and not in the three Channel Islands). Due to its historical roots, the British imperial pole of tax havens came to specialize in financial affairs, such as syndicated loans, derivatives, forex, insurance hedge funds and 'off the shelf' companies.

The European agglomeration, on the other hand, preceded the British one. A recent study by Christophe Farquet (2013) demonstrates that concerns over Swiss and Belgian support of tax avoidance and evasion strategies were already expressed in the 1920s, and then extended to Luxembourg in the 1930s. European centres, primarily the Benelux countries and Ireland, and to a lesser extent Switzerland, emerged as tax havens for international capital harvested from intangible assets. They developed rules and regulations aimed at attracting holding companies that serve as repositories of international incomes from logos, goodwill, trademarks and brand names. Analysis of German inbound and outbound foreign direct investment (FDI), for instance, shows that the Netherlands and Switzerland serve as the two leading conduit jurisdictions for most German businesses over the past 20 years (Weichenrieder and Mintz, 2007). The study also found that, typically for cases of this type, British-linked tax havens, such as Barbados, Bermuda and the Caymans, played no significant role in hosting German conduit entities. Whatever their different historical origins, the Imperial and European agglomerations combined accounted, by March 2010, for approximately 53.3 per cent of all international banking assets and liabilities.

CONCLUSION

Tax footprint reduction makes business sense. Businesses, as economists are fond of repeating, are by their nature profit oriented. What plain economic accounts tend to overlook, however, is the extent to which profit orientation varies along two conceptual and logistical boundaries: profit before and after taxes, and profit within and without national borders. Cross-border differences in tax structures incentivize the proliferation of tax arbitration schemes that push well beyond the legal and economic boundaries of profit orientation and wealth accumulation as traditionally understood.

Corporate tax footprint reduction makes wealth accumulation troublesome on two fronts. On the one hand, tax minimization schemes hinge on the legal dismemberment of corporate entities into ecologies of ancillary businesses operating in multiple jurisdictions. This scale of multi-jurisdictionality, insofar as it clouds both actors and processes, is apt to provide cover for transactions of all but indiscernible legality. On the other hand, multi-jurisdictional arbitrage takes full advantage of tax reduction schemes directed not just at traditional notions of wealth, but also at intangible aspects of wealth, such as goodwill, that are treated as claims on future income streams. When multi-jurisdictional arbitrage operates on assets with such futurity value, the practice of tax minimization no longer involves just the protection of *accumulated* wealth, or wealth-already-accumulated, but also the process of *accumulating* wealth, or wealth-to-be-accumulated. Corporate restructuring practices aimed at the multi-jurisdictional arbitrage of capital gains tax, stamp duties or income tax all play on this troublingly elusive principle.

At the service of the huge cross-border traffic of capital resulting from this process of wealth accumulation is a global industry whose explicit purpose and product is the reduction of individual and corporate tax footprint. Emerged through an organic process of agglomerations and competitive niche-seeking strategies, tax havens have an estimated turnover of US$50 billion as well as around 10–12 per cent of global aggregate wealth parked within their shores. It is certainly not wealth per se, but the unaccountable generation and accumulation of wealth across these tax jurisdictions that spells most trouble.

NOTES

1. 'BP', the official name of the company since 2001, was short for 'British Petroleum' (*The Economist*, 2010; BP, 2014).
2. See Beaverstock and Hall, Chapter 20 in this volume for further discussion.
3. Notable exceptions include the likes of Tony Atkinson, Joseph Stiglitz, Emmanuel Saez and Thomas Piketty, who have led the vanguard of economic analyses of wealth accumulation, particularly in connection with inequality (Atkinson, 2000; Piketty and Saez, 2003; Atkinson and Piketty, 2007; Stiglitz et al., 2010).
4. Most forecasts hover between a 2 per cent and a 3 per cent rise (United Nations, 2013; World Bank, 2014).
5. For a discussion of celebrity goodwill, see Walzer and Gabrielson (1986).
6. This is driving lively discussions in specialist law journals on the subject (see Walzer and Gabrielson, 1986; Kelly, 1999; Bartow, 2001).
7. 'Corporations are apart among the legal instruments used to legally structure firms. The reason for this is that they are treated by the legal systems as if they were "real" persons (with some adaptations), i.e. they can participate in the legal systems through the phenomenon of "juridical personality". They can own property, have debts, contract, sue and be sued in courts, get bankrupt, etc. – i.e. they can "function" in the economy like human beings because they are treated by the legal system as if they were "persons"' (Robé, 2011, p. 9).
8. See Beaverstock and Hall, Chapter 20 in this volume.
9. Bermuda, the largest captive insurance centre in the world in spite of its relatively small banking centre, can be included too, and so can Cyprus and the numerous but less significant former British colonies in the Pacific. For a discussion of Bermuda's financial centre, see Crombie (2008). For a discussion of the Pacific offshore centres and their relationship to the UK, see Sharman and Mistry (2008).
10. The USA, in contrast, accounted for 12.4 per cent and 12.9 per cent of all outstanding international loans and deposits, while Japan accounted for 4.5 per cent and 3.8 per cent respectively in March 2009. The European havens were about 2 per cent higher only a year before. The USA appears to be the only large net gainer during the crisis of 2007 up to this day.

REFERENCES

Apple (2013), 'Testimony of Apple Inc. before the Permanent Subcommitee on Investigations (US Senate)', 21 May, accessed 25 August 2015 at https://www.apple.com/pr/pdf/Apple_Testimony_to_PSI.pdf.

Atkinson, A.B. (2000), *Handbook of Income Distribution*, Amsterdam/New York: Elsevier.

Atkinson, A.B. and T. Piketty (2007), *Top Incomes over the 20th Century: A Contrast between Continental European and English-Speaking Countries*, Oxford, UK/New York: Oxford University Press.

Avraham, D., P. Selvaggi and J. Vickery (2012), 'A structural view of U.S. bank holding companies', *FRBNY Economic Policy Review*, July, 65–76.

Bank for International Settlements (BIS) (2014), 'Locational banking statistics', accessed 5 December 2014 at http://www.bis.org/statistics/bankstats.htm.

Bartow, A. (2001), 'Intellectual property and domestic relations: issues to consider when there is an artist, author, inventor, or celebrity in the family', *Family Law Quarterly*, **35** (3), 383–424.

Beaverstock, J., S. Hall and T. Wainright (2013), 'Servicing the super-rich: new financial elites and the rise of the private wealth management retail ecology', *Regional Studies*, **47** (6), 834–49.

BP (2014), 'Our brands', accessed 25 August 2015 at http://www.bp.com/en/global/corporate/about-bp/our-brands.html.
Burn, G. (2006), *The Re-Emergence of Global Finance*, Basingstoke, UK: Palgrave Macmillan.
Commons, J.R. ([1924] 1959), *The Legal Foundations of Capitalism*, Madison, WI: University of Wisconsin Press.
Commons, J. (1961), *Institutional Economics*, Madison, WI: University of Wisconsin Press.
Crombie, R. (2008), 'Bermuda in-depth series. Part I: lighting and fire', *Risk and Insurance*, 1 January.
Farquet, C. (2013), 'Tax avoidance, collective resistance and international negotiations: foreign tax refusal by Swiss banks and industries between the two World Wars', *Journal of Policy History*, **25** (3), 334–53.
Henry, J. (2012), 'The price of offshore revisited: a review of methods and estimates for "missing" global private wealth, income, inequality, and lost taxes', *Tax Justice Network*, July, accessed 25 August 2015 at http://www.taxjustice.net/cms/upload/pdf/Price_of_Offshore_Revisited_120722.pdf.
House of Commons (2013), *Tax Avoidance: The Role of Large Accountancy Firms*, London: The Stationery Office.
Hudson, A.C. (1998), 'Reshaping the regulatory landscape: border skirmishes around the Bahamas and Cayman offshore financial centers', *Review of International Political Economy*, **5** (3), 534–64.
Keating, G., M. O'Sullivan and A. Shorrocks et al. (2013), *Global Wealth Report 2013*, Zurich: Credit Suisse AG.
Kelly, A.B. (1999), 'Sharing a piece of the future post-divorce: toward a more equitable distribution of professional goodwill', *Rutgers Law Review*, **51** (3), 569–635.
Luxembourg Consulting Group (2014), 'Private Asset Management Company (SPF) in Luxembourg', accessed 16 May 2014 at http://www.lcg-luxembourg.com/Private-Asset-Management-Compa.472+M52087573ab0.0.html.
Maurer, B. (1998), 'Cyberspatial sovereignties: offshore finance, digital cash, and the limits of liberalism', *Indiana Journal of Global Legal Studies*, **5** (2), 493–519.
Mitchell, D.J. (2009), 'In praise of tax havens', *The Freeman*, **59** (6), 23–7.
OpenOil (2014), *Mapping BP – Using Open Data to Track Big Oil*, accessed 5 December 2014 at www.openoil.net: http://openoil.net/mapping-bp-using-open-data-to-track-big-oil/.
Palan, R. (2012), 'The financial crisis and intangible value', *Capital and Class*, **37** (1), 65–77.
Palan, R. (2014a), 'Where the struggle against international tax avoidance is heading: multi-jurisdictionality', *The FSC Report 2014*, pp. 11–14.
Palan, R. (2014b), 'Futurity, pro-cyclicality and financial crises', *New Political Economy*, **20** (3), 367–85.
Palan, R., R. Murphy and C. Chavagneux (2010), *Tax Havens: How Globalization Really Works*, Ithaca, NY: Cornell University Press.
Piketty, T. and E. Saez (2003), 'Income inequality in the United States, 1913–1998', *Quarterly Journal of Economics*, **118** (1), 1–39.
Robé, J.-P. (2011), 'The legal structure of the firm', *Accounting, Economics, and Law*, **1** (1), Art. 5.
Sagar, P., J. Christensen and N. Saxhson (2012), 'British government attitudes to British tax havens: an examination of Whitehall responses to the growth of tax havens in British dependent territories from 1967–75', in A.L. Waris and J. Leaman, *Why Tax Justice Matters in Global Economic Development*, New York: Bergam.
Sharman, J. and P.S. Mistry (2008), *Considering the Consequences: The Development Implications of Initiatives on Taxation, Anti-Money Laundering and Combating the Financing of Terrorism*, London: Commonwealth Secretariat.
States of Guernsey (2012), *The Image Rights (Bailiwick of Guernsey) Ordinance, 2012. Ordinance of the States XLVII*, Guernsey: Royal Court House.
Stewart, J. (2012), 'Low tax financial centres and the financial crisis: the case of the Irish Financial Services Centre', *IIIS Discussion Paper No. 420*, Dublin: Trinity College.

Stiglitz, J., A. Sen and J.-P. Fitoussi (2010), *Mismeasuring Our Lives: Why GDP Doesn't Add Up*, New York: New Press.

Sylla, R. (2002), 'United States banks and Europe: strategy and attitudes', in S. Battilossi and Y. Cassis (eds), *European Banks and the American Challenge: Competition and Cooperation in International Banking under Bretton Woods*, Oxford, UK: Oxford University Press.

The Economist (2010), 'BP and British Petroleum: what's in a name?' *The Economist*, 16 June, accessed 5 December 2014 at http://www.economist.com/node/16373115.

The Permanent Subcommittee on Investigations (2013), 'Offshore profit shifting and the U.S. tax code – Part 2 (Apple Inc.)', *Permanent Subcommittee on Investigations*, Washington, DC: US Congress.

United Nations (2013), *World Economic Situation and Prospects 2013*, New York: United Nations.

Walzer, S.B. and J.C. Gabrielson (1986), 'Celebrity goodwill', *Journal of the Academy of Matrimonial Lawyers*, **2**, 35–44.

Weichenrieder, A.J. and J. Mintz (2007), 'What determines the use of holding companies and ownership chains?' *Oxford University Centre for Business Taxation, Working Paper No. WP08 (03)*.

Western, M., D. Maughan and A. Quinn (2011), 'Best of both worlds – Cayman Incorporated, Irish tax resident', *AirFinance Journal*, September, accessed 25 August 2015 at http://www.maplesandcalder.com/fileadmin/uploads/maples/Documents/PDFs/Airfinance%20Journal_The%20Best%20of%20Both%20Worlds.pdf.

World Bank (2014), *Global Economic Prospects*, Washington, DC: World Bank.

Yiolitis, E. (2012), 'Russia–Cyprus: case study on Project M', *International Financial Centre Review*, 1 November, accessed 25 August 2015 at http://www.ifcreview.com/restricted.aspx?articleId=5698&areaId=20.

22. No change there! Wealth and oil

*Isaac 'Asume' Osuoka and Anna Zalik**

INTRODUCTION

It was the business of oil that created the world's first US dollar billionaire and changed the fortunes of many more. Since John D. Rockefeller's Standard Oil became one of the world's first mega corporations in the late nineteenth century, controlling the bulk of the market share of consumer petroleum fuels in the USA, the oil industry has demonstrated a predilection to amass. To sustain multibillion-dollar profits, industry–state alliances compete to control the global resource base, and its power over peoples. Pursuit of oil reserves has been a significant factor shaping global geopolitical dramas over the twentieth century. But in these dynamics, and the millionaires created as a consequence, millions of people have been placed on the margins of power, as the mega transnational corporations, national oil companies and smaller (but not insignificant) players seem to share the spoils.

The extractive sector broadly and the oil and gas industry make up a huge section of the global economy. Business sources observe that the world's top 20 extractive industries recorded profits of US$211 trillion[1] in 2005, whereas US GDP in the same year totalled US$11 trillion (Cortese et al., 2009, p. 27). According to the US Congressional Research Service, in 2011 the revenues of the top five oil majors alone totalled 10 per cent of US GDP, US$1.8 trillion in revenues for the companies and US$15 trillion GDP (Pirog, 2012, p. 1). In 2013, Chatham House, the Royal Institute of International Affairs, cited a total value of mining and fossil fuels as surpassing US$10 trillion (Stevens et al., 2013).

Beyond the wealth created for a few, we are all affected by petro-capital; not only through climate change and the more localized pollution associated with the production and transportation of oil and gas, but also as regular consumers of the varied products derived from petroleum. In purchasing food and fuel we shape and are shaped by a broader capitalist economy and global oil complex, which reproduces its own logic. It is a multifaceted order in which present and future profits from the sector encompass not only oil and gas – as fuel – but also a broad range of petroleum subproducts and infrastructures. These include the pipeline infrastructure for distribution, whose construction is subcontracted to a vast number of firms, as well as petrochemicals and plastics that are central facets of daily life.

An important global oil elite competes and cooperates to shape relations in the global oil market. As petroleum corporations navigate investment terrains that transcend the financial markets into new frontiers and compradorial strongholds, old stakeholders seek to consolidate and expand, even as new players continue to emerge. These old and new elites, some more localized than others, all have a material interest in maintaining the flow of crude oil and natural gas in ways that reproduce the global capitalist system. From the push toward producing US shale oil for export that benefits the new oil barons of the Bakken (a prolific source of oil in North America), to the oil industry indigenization policies in the West African context, new patterns are emerging that consolidate oil industry rents in elite hands.

Through a multiscalar examination of accumulation in the global oil and gas industry, this chapter examines the increasing concentration of wealth that accrues through a combination of executive compensation, inherited (familial) wealth, and revenues from the sector in North America and West Africa. The varied ways whereby compensation (to key corporate executives and, in some cases, the national elite) and avenues for global accumulation are effected constitute overall industry strategies to access and/or restrict markets. Despite global initiatives promoting transparency in the extractives sectors, opaque arrangements, including in a parallel form of high-level cronyism, create avenues for enrichment of members of the political class in some petroleum exporting countries such as those in the Gulf of Guinea region of Africa. In North America, where calls for transparency and accountability in executive pay increased after 2008, by 2012 reported compensation to executives of oil majors exceeded pre-crisis figures (Herbst, 2008).

Based on a broadly conceived rentierism associated with access to oil and gas royalties, and the not insignificant number of elites that have benefited from the rising inequality associated with this phenomenon, we use examples from the West African and North American regions to assert our central argument: the present conjuncture witnesses not only growing inequality through rentierist relations associated with extractive oil and gas capital, but the contours of this accumulation benefit a sizeable global elite while disenfranchising a much larger population through volatile fuel and food prices and accompanying socioecological contradictions.[2]

CONCEPTUALIZING OIL AND ACCUMULATION

As we prepare this chapter for publication, oil and gas prices are falling, in part the result of the investment in new oil and gas frontiers prompted by dynamics of the past two decades, as well as the turn to alternative energy sources in the wake of rising concern over climate change. Prior to this recent downturn, and since the turn of the millennium, the oil and gas sector saw windfall profits. Business sources attributed this to the uncertainty prompted by the events of 9/11 – particularly the concerns regarding the insecurity of Persian Gulf and North African energy sources they entailed; rising attention to political change and instability in OPEC exporters like Nigeria and Venezuela; and increased demand for energy from large industrializing countries, notably China and India. Despite the fall in prices following the 2008 financial crash, the strength of the sector was buttressed as global markets saw a flood of investment into commodities, notably minerals, oil and gas, as a relatively safe haven from the financial storm (Yergin, 2008a, 2008b). Oil prices recovered within a year. Although the share of investment in commodities may have shifted over the subsequent six years, the relative importance of hydrocarbons in global capital is underscored by the listing of companies like ExxonMobil, Royal Dutch Shell and Petro China along with big banks and investment houses among the top ten of the *Forbes* Global 2000 (*Forbes*, 2014b). Indeed, all but one of the top ten of the world's biggest corporations is either a financial services or petroleum corporation. General Electric, the exception, actually has an 'extremely profitable' financial arm (Van Arnum and Naples, 2013, p. 1160).

In citing such data, we must point out that we take seriously economist Thomas Piketty's methodological caution on the use of data sourced from the business press (Piketty, 2014b; Piketty and Zucman, 2014). As he notes in a 2014 interview:

> It amounts to a real abdication of responsibility when researchers and public institutions fail to describe existing inequalities in accurate terms. It leaves the field open to wealth rankings by magazines like *Forbes*, or the Global Wealth Reports put out by big banks, who take on the role of 'knowledge producers'. But the methodological basis for their data remains unclear; the results are largely ideological, a hymn to entrepreneurship and well-deserved fortunes. Moreover, the simple fact of focusing on the 'richest five hundred' is a way of depoliticizing the issue of inequality. The number is so small that it becomes meaningless. It appears to show extreme inequalities, but in reality it gives a mollifying picture. Inequalities have to be grasped in a more extensive fashion. For example, if one takes fortunes of over €10 million, rather than over €1 billion, they amount to a very significant proportion of total wealth. We need the right tools to represent inequality. The American movement of the

> 99 per cent was one way of doing this. Focusing on the richest 1 per cent makes it possible to compare different societies that would otherwise seem incommensurate. Talking about 'top executives' or 'rentiers' may seem more accurate, but these terms are historically specific. (Piketty, 2014b)

While gazing at the oil billionaires, we take note of a much larger number of beneficiaries of petroleum rents, through royalties to landowners, and the varied licit and illicit transactions that both sustain intergenerational wealth, and create the nouveau riche. These groups, in different ways, have interests that are intricately linked to the need to maintain global crude flows, and the order of accumulation that so often tends to exacerbate inequalities. In this case, Piketty's point regarding the salience of *inherited wealth and accumulated capital* to wealth stratification is highly pertinent. Analyses such as Piketty's build on a longer set of scholarly studies of elite power that also shed light on the present conjuncture's extreme inequalities (Kees van der Pijl, 1998, 2004; Sklair, 2001) while also offering a lens through which to explore the history and contemporary operations of the oil and gas industry.

An analysis of how accumulation in the twenty-first-century petroleum sector reproduces intergenerational wealth and power necessarily considers how old and new elites are endowed with particular access to capital windfalls. Both intergenerational wealth transfers and contemporary windfall profits/executive compensation should be seen as attributes of a broad theory of 'rentierism'. A nuanced theory of rentierism understands it as not only concerning how individual agents access state coffers, as has been the case in conventional literatures critiquing 'developing state' corruption and the resource curse. Rather, rentierism as we understand it here (also) centres on how intergenerational wealth, amassed through the exploitation of living labour (and nature), is reproducible and indeed augmentable over time through political dynamics that privilege those holding assets of fixed capital. These holdings include access to land and resources, mineral rights to highly valuable territories, as well as controlling public assets in trust – with the latter pertinent to state officials in some of the countries that we examine below.

OIL WEALTH UNDER ANGLO-AMERICAN IMPERIALISM

The United States remains exceptional internationally because private landowners hold subsoil rights in many jurisdictions. The promise that came with producing oil on a commercial scale resulted in a mad rush, as under the so-called 'rule of capture', neighbouring landowners competed to extract resources more quickly than their neighbours. Between the mid-nineteenth century to the early twentieth century, such 'extractive anarchy' is said to have acted as an obstacle to efficient development of oil and gas resources (Libecap and Smith, 2002). The USA addressed this to some extent by mandated unitization agreements (Daintith, 2010) and protections on its production but nevertheless created a significant tranche of landowners and firms profiting from the sector (Bina, 2006). Outside the USA, the country's oil industry came to dominate global production during this period as its oil corporations redefined the capitalist enterprise via international cartels controlling the oil trade. US participation in these cartels helped produce some of the wealthiest and most powerful individuals in history. John D. Rockefeller's Standard Oil stood out among the oil and gas oligarchies. Founded in 1870, Standard Oil controlled over 80 per cent of the market share of refined petroleum products in the USA by 1907. While in 1911 a US antitrust law resulted in the breaking up of Standard Oil and John D. Rockefeller's control in the market domestically, he went on to become the world's first US dollar billionaire with significant stakes in the successor companies. Moreover, imperial relations internationally allowed for the continuous dominance of US and other Western capitals in the global market.

In popular conceptions of the oil industry, the fact of Western cartels' domination of the global market has been masked by a focus on the oil producers of the Middle East and the creation of the Organization of the Petroleum Exporting Countries (OPEC). The attention on this region offered an external outlet for US populist demands for cheap fuel (Huber, 2013), and was served by racist Islamophobia directed at the OPEC states. Thus substantive control of oil and gas markets by Western private firms was veiled by attention to OPEC states as a so-called cartel, whose interests in the past four decades in fact mirror those of these firms very closely (Bichler and Nitzan, 2004). Indeed, contrary to the dominant misperception of the inverse, the OPEC period from the 1970s onward has been labelled as one of *decartelization* (Mitchell, 2002; Bina, 2006).

This notion of decartelization requires an examination of the earlier period, 1870–1910, which marked the first US period of 'early cartelization' wherein oil trusts dominated the market. This period was followed by what has been referred to as the 'era of colonial concessions in the Middle East', from roughly 1910 to 1950 (Bina, 2013). In this period, the 1928 'As-is' Agreement, signed at Achnacarry Castle in Scotland, consolidated global cartelization in the oil industry. The agreement aimed to address the global price wars that threatened the private profit rate in the sector during the 'Roaring Twenties'. At Achnacarry, Sir Henri Deterding of

Royal Dutch Shell invited then president of Standard Oil of New Jersey/Exxon, Walter C. Teagle, and Sir John Cadman of Anglo-Persian Oil (BP) to hunt grouse. There, along with representatives of Standard Oil New Jersey in Germany, and Standard Oil Indiana, they signed an agreement 'assigning a status quo in market share to members and restricting production so as to prevent competition from affecting profitability'. Achnacarry's terms complemented the control over Iraqi oil by the oil majors under the 'Red Line Agreement' of the same year. Covering coal and chemical industries and strengthening patents on synthetic fuels (Mitchell, 2011, p. 97), the Red Line Agreement similarly instituted favourable terms for private industry.

The post-war period and decolonization, which allowed for the rise of the Global South under the fractured hegemony of the Cold War (Girvan, 1975), was transformative for the global oil industry. As described above, in contrast to popular understanding, the case has been made that cartelized practice dominated by the private, global oil industry maintained hegemony over much of the global market until 1972. As such, the 1950–72 era should be seen as a *transitional decartelization* period marked by 'the spread of competition against the prearranged production, captive oil concessions, "gentlemen's agreements", and arbitrary accounting of oil royalties (and rents) according to fictitious "posted pricing"', as market forces became more preponderant (Bina, 2013, p. 103).

This cartelization and decartelization process emerged from the fact, emphasized by critical scholars of hydrocarbon capitalism, that a key dilemma for the oil and gas sector through the twentieth century has been that of ensuring *scarcity* of global oil on the market in the midst of an otherwise plentiful resource, so as to sustain profit rates in an industry with high fixed capital costs (Bichler and Nitzan, 2004; Labban, 2008; Mitchell, 2011; Bina, 2013). Again, this runs counter to popular views on oil supply, which assume limited resources (Goodstein, 2005), including some variants of peak oil theory (Bridge and Wood, 2010). Rather, although intercapitalist competition among transnational and national oil and gas corporations for reserves and markets is intense, the sector as a whole has protected its profits through cartel-like behaviour (Bina, 2013).

We see then that OPEC arose as a regional response to the cartelized actions of the oil majors in the Persian Gulf. Such actions consisted of the cuts to the posted prices of oil implemented by the 'International Petroleum Cartel' in the late 1950s.[3] This was accompanied by the decision of the Global South to exert its collective role, setting the stage for what has been described by Bina (2013) as the 'post-cartelization' period, since 1974. OPEC's challenge to the majors, which some understand as their ability to exert 'absolute rent', as a *sovereigntist territorialized* exertion of power by Southern states (Coronil, 1997; Mommer, 2002), subsequently prompted the rise of market relations (including in financialized form), in a way that disabled the sovereigntist elements of their action. In doing so, market relations also displaced the explicit cartel behaviour of the oil elites in the previous period. Yet, although with new attributes, the influence of oil industry elites persists under these market relations, they just take a neoliberal form, marking a shift in the operations of the transnational elite classes from the earlier part of the twentieth century.

To reiterate, then, while the assertion of OPEC power is understood conventionally as expressing the interests of a 'cartel', critical views present the emergence of the organization as marking decartelization, as Achnacarry was undone, and opening up the global oil industry for a more robust interplay of global market forces (Bina, 2013). These forces did not spell the end of the power of the oil industry elite, but rather the 'market' and elite interests were realigned in ways that would become increasingly entrenched with the rise of global futures markets in oil and their coincidence with the dismantling of state regulations, so-called 'neoliberalism'. Under these conditions, but with contours that differed from the earlier period, contemporary oil and gas markets remain dominated by corporate elites that fund and use significant lobby groups like the American Petroleum Institute and the Canadian Association of Petroleum Producers to shape public and regulatory policy (Cortese et al., 2010; Cayley-Daoust and Girard, 2012; Stewart, 2013; Rowell, 2014).

THE OIL INDUSTRY AND INDIVIDUAL WEALTH UNDER NEOLIBERALISM AND POST-NEOLIBERALISM

One way of understanding the dominance of corporate elites under the period of 'decartelization' and the broader market dynamics surrounding the oil industry, is to consider the point made by economist Tae Hee Jo (2013) and others (De Kuijper, 2009): competition and cooperation among business elites are two sides of the same coin. Thus, even as conglomerates of the 'Seven Sisters'[4] oil companies have restructured, divided and merged once again, the influence of these elites has in no way been diminished. Indeed, despite the rise of new players on the global field, we can trace the line of dominance of the oil majors back to the early oil cartels. This is underlined by the role of familial oil and gas oligarchies in significantly influencing late nineteenth- and twentieth-century hydrocarbon capitalism. As shown earlier, these oligarchies shared a significant objective: that of protecting profits by restricting oil and gas availability while ensuring scarcity on the global market. As one observer remarked in a 2010 letter to *Fortune* magazine:

> While reading the Fortune 500 issue (May 3), I could not help but notice the tremendous impact that John D. Rockefeller still has in the 21st century. Counting just three of the surviving entities (Exxon, Mobil, and Chevron) of the original 34 companies that came out of the 1911 breakup of Standard Oil, we would have a company that would be the indisputable No. 1 on the Fortune 500. I would love to see Rockefeller's place in business history restored to where it belongs. (Hernandez, 2010)

The enduring impact of Rockefeller sheds light on the power of inheritance. About 200 of Rockefeller's descendants are collectively worth about US$10 billion, according to *Forbes* magazine. Other super-wealthy families are associated in some form with the oil and gas industry. The 'Top Ten' wealthiest individuals in the US oil and gas industry compiled by Wealth X, a research firm on the ultra-wealthy (which provides services to that group), ranks the controversial Koch brothers first and second at US$41.5 billion each.[5] While they clearly belong to the club of oil industry magnates, the source of their wealth is identified by Wealth-X as 'inheritance/self-made'. Their inherited company, Koch Industries, is described by Standard and Poors as among the largest privately held participants in the refining and marketing sector of the oil industry and they pursue exploration globally, including in the Canadian Tar Sands. The Charles G. Koch Charitable Foundation has reportedly retained formal capacity to shape hiring priorities and decisions in higher education institutions such as Florida State University, through large financial endowments (Hundley, 2011). Supporters of the US Tea Party, the Koch brothers have also financed various conservative think-tanks deriding climate science and their policy reach is wide, including within the right-wing Cato Institute. However, analysts of climate denial point out that too great a focus on the Koch brothers diverts attention from a larger group of ultra-conservatives who fund climate-denial think-tanks (Goldenberg, 2013), a point that squares with Piketty's critique of the tendency of the business press and larger public to fetishize individuals among the super-rich (for a discussion, see Koh, Wissink and Forrest, Chapter 2 this volume).

On the Wealth-X ranking the Koch Brothers prove to be somewhat wealthier than many of their business counterparts. In next place, 'self-made' Harold Hamm of Continental Resources trails each of them by more than US$25 billion, listed with wealth of US$14.1 billion. Yet, the enormity of these figures for a few individuals belies the fact that perhaps 8 million landowners in the USA receive oil and gas royalties (Huber, 2013). Thus, a fuller analysis of the wealth generated under neoliberalism requires attending not simply to billionaires but also the varied ways of accumulating 'ultra-wealth'. Conservative news sources are now quick to point out that North Dakota, the heart of Bakken shale oil and gas development, moved up 15 spots in the individual wealth ranking of US states with the largest number of billionaires per capita in 2013. While the amounts garnered will clearly be much lower for smaller landowners, these nevertheless directly link the interests of a not insignificant sector of the US population to oil wealth and windfalls.

An exaggerated focus on the ultra-rich stakeholders also distracts attention from salaried executives of the financialized firms who receive significant compensation and fat bonuses. To buttress this point, Wealth X's 'Top Ten' list does not include ExxonMobil's Rex Tillerson nor indeed any of the top ten highest-paid oil industry executives in the USA. A more complete picture of big earners from the oil and gas sector would include people like Tillerson who, according to *Forbes*, received over US$28 million as compensation in 2013. While a reduction from over US$40 million in compensation in 2012 (Morningstar, 2014) the figure nevertheless denotes ongoing significant holdings among the global rich. In Europe, Shell's CEO Peter Voser's total compensation in 2013 was €3.6 million (or US$4.9 million) (according to *Forbes*), down from US$6.5 million in 2012 and US$7.2 million in 2011. The implications of these salaries for inequality are highlighted by Labban who indicates that corporate executives announcing large numbers of layoffs have seen higher absolute and median pay than other executives (Labban, 2014, p. 9). In a recent report concerning executive compensation in Canada, an institute at a Montreal University's business school reviewed the massive divide between CEO compensation and average salary in private firms. Under financialization, skyrocketing executive compensation, through stock options to CEOs and salary increases, has intensified the divide between upper management and employees. While initially US executive compensation far exceeded that of Canadian firms, by the end of 2010 Canadian executive compensation roughly equalled that of the USA. Most significantly, the 'relationship of Canadian CEOs' compensation to the average salary of Canadian private-sector employees jumped from 60 times in 1998 to some 150 times in 2010' (Allaire, 2012, p. 11). In the oil industry today, Canadian chief executive compensation is certainly competitive with US rates, and higher than European figures. Talisman Energy's John Manzioni received the highest 2012 compensation of Canadian oil and gas companies by far, at over CAN$18 million, followed by Suncor's Steve Williams at CAN$11.8 million, Cenovus's Brian Ferguson at CAN$10.8 million and TransCanada's Russell Girling at CAN$8.6 million.

The immense executive returns in the oil and gas sector are an expression of the broader trends toward CEO compensation that have accompanied financialization in the past three decades, including the vicious debt cycle it has entailed among the economically subjuguated. In this regard, dynamics in the oil and gas sector reflect those seen across capitalist firms globally.

THE CONTOURS OF OIL WEALTH IN WEST AFRICA

Unlike the USA, in most countries, subsoil resources are owned by the state. And in some of these countries, the location of individuals within interstices of state power reveals how oil wealth is accumulated domestically. The emergence of OPEC tended to obfuscate the hegemony of US and Western European corporations, but the oil majors continued to be influential through varied partnership with OPEC members. Those North–South liaisons, being a constant feature of the global oil business, are not without tensions, as transnational capital and indigenous power each seek advantages. In the Gulf of Guinea basin of West and Central Africa, transnational corporations have largely maintained a missionary position over the countries of the region, while the national political elite tends to play the role of holding down the local populations. However, the narrative of the 'rape of Africa', which captures the story of how over a century of imperial relations with the continent has shaped its current predicament, must now be nuanced given the increasing agency of indigenous capitalism, and indigenous accumulation in joint ventures with global powers. Nigeria, Angola and Equatorial Guinea and other major oil exporting countries in the African Gulf of Guinea showcase how revenues from the commodity have driven colossal accumulation among the power elite amidst mass impoverishment.

In January 2013, Isabel Dos Santos became the first African woman to be acknowledged as a US dollar billionaire by *Forbes* magazine, two years after the same magazine rated her fortune as 'at least $50 million' (*Forbes*, 2011). The phenomenal rise in the estimation of Dos Santos's wealth over such a short period throws into doubt the legitimacy of her investments, primarily in Portugal and Angola. What is clear is that she is the eldest daughter of the man who has ruled Angola since 1978. In this country much of the oil revenues from the 1980s, when the country was embroiled in civil war, ended up in private bank accounts of regime functionaries. As conditions normalized in the country, state functionaries were involved in buying public assets in privatization programmes (Hammond, 2011) and using stolen oil wealth to finance equity stakes in a wide range of businesses. Sonangol, the national oil company, continues to make huge cash transfers to members of the political elite (CNN, 2012). Dos Santos is seen widely as a mere front for the president as accumulators seek outlets to normalize loot as legitimate capital. Even as she was being advertised as a billionaire, 70 per cent of the population of Angola remained in abject poverty.

Isabel Dos Santos is in the league of the 'royal' families of oil-producing countries, which includes the Biyas in Cameroon, the Bongos in Gabon, the Sassou-Nguessos in Republic of Congo, and Obiangs in Equatorial Guinea. In all of these countries, the oil-fuelled opulence of the politically connected stands in stark contrast to the impoverished conditions under which the majority live.[6] Emblematic of the contrast is Equatorial Guinea, a country dubbed the 'Kuwait of Africa' because of its small size and large oil and gas reserves. New investments to exploit the country's natural resources have generated one of the most positive economic growth records in the past decade. But the entrenched corruption in the oil industry has ensured that Western corporations, rather than Equatorial Guinea's fewer than a million people, benefit from the country's riches. In the 1990s, during the initial years of oil production, Equatorial Guinea retained only 12 per cent of the value of oil exports as revenue due to contracts skewed in favour of foreign companies. Concurrently, US companies were paying hundreds of millions of dollars into the private US bank accounts of Equatorial Guinea's dictator and family members (Leung, 2003), continuing a trend of illicit payments by northern corporations to officials of other jurisdictions such as Nigeria (BBC, 2009; Baltimore, 2012). In one case, Kellogg Brown & Root (KBR), a subsidiary of Halliburton, appears to have paid over US$180 million in bribes to Nigerian officials between 1994 and 2004 to secure contracts related to natural gas projects worth US$6 billion. Following investigations and prosecutions in the USA, KBR officials who participated in the scheme were ordered to serve time in prison and pay fines. Former Vice President of the USA, Dick Cheney, who was the head of Halliburton during periods when the crimes were committed, was not indicted (Rudolf, 2012). Other Western company bribery of Nigerian officials has involved Shell (Rubenfeld and Palazzolo, 2010) and Noble (Viswanatha, 2012).

The liaisons between Western corporations and Africa's power elite are uncovered in occasional 'scandals'. But such scandals have not succeeded in changing the way business is done. French investigations into 'ill-gotten gains' by families of mainly francophone African dictators followed revelations of hundreds of millions of dollars in properties and 'bling' obtained through the loot of oil rents by state executives (Global Witness, 2009). At one point, French authorities auctioned over US$3 million worth of sports cars confiscated from 'Teodorin' Obiang, Vice

President of Equatorial Guinea, the 'playboy' son of President Teodoro Obiang Nguema Mbasogo (Freymeyer, 2014). In the United States, the Justice Department ordered Teodorin to forfeit US$30 million worth of illegally obtained assets in the country as part of a settlement in a civil forfeiture case involving over US$80 million in alleged loot (Lazupa, 2014). It has been suggested that from 2004 to 2011, Teodorin may have spent over US$300 million on the most expensive cars in the world, exotic property and pop culture memorabilia in the United States and Europe (Freymeyer, 2014). While the Teodorin case may suggest that 'France has gotten tough on corruption' (Chrisafis, 2012), the reality is that 'corruption' has been the conscious mechanism through which the African oil complex is being concretized via an emergent class of indigenous super-rich oil barons. The French government worked in tandem with its transnational, Elf Aquitaine, to support the promotion of accumulation by the local power elite of Gabon and the Republic of Congo through illicit financial payments and laundering of millions of dollars from the 1960s to 1990s (Henley, 2003; Heilbrunn, 2005; Aljazeera, 2014), offering one example of what appears to be more than just an arrangement of convenience.

Yet condemnations of the lifestyles of the dictators from Gabon and the Republic of Congo mask how French officials colluded to launder millions of dollars in North–South private sector joint ventures. Through these arrangements indigenous oil companies, owned by ex-dictators, politicians, retired civil servants and their family members and acolytes, obtain oil blocks, which are subsequently operated by Western companies. Nigeria may have the majority of such arrangements in the subregion. Supported by the government's 'local content' policies, Nigerian indigenous oil companies have been expanding as active players in upstream and downstream operations. Nearly all the indigenous oil companies in Nigeria that have obtained oil blocks have ex-government officials or their family members as shareholders. Among them are at least three former federal ministers of petroleum, and a former defence minister, General Theophilus Danjuma, one of the richest men in Nigeria, who confessed to earning US$1 billion from selling an oil block awarded to him by a military regime (Nairaland.com, 2010). That is why a look at the list of the nouveau riche in Africa shows that 'oil remains the surest route to an African billion' (Popham, 2013). Among the billionaires is Folorunsho Alakija who is worth an estimated US$7 billion. The one-time clothes maker sewed gorgeous blouses and headgear for the wife of Ibrahim Babangida, Nigeria's longest serving military dictator of the 1980s and 1990s. Alakija, personal friend to the first family, 'is believed to have ridden on the crest of this relationship to acquire an oil block in 1993 at a relatively inexpensive price' (Ventures Africa, 2013). To exploit oil, her new company Famfa entered into a joint venture arrangement in which it ceded a 40 per cent stake to subsidiaries of Chevron and Petrobras.

Participation of Nigerians in the oil business is linked to localized neoliberal reforms in some countries of Sub-Saharan Africa, including via the privatizing of public assets. Such privatization, promoted vigorously by Western institutions, has seen the increasing transfer of equity of energy assets to indigenous businesses, which, as shown above, are often owned by the politically connected – some of whom are local beneficiaries of loot. Such individuals and the commercial entities linked to them are also positioned to benefit from local content legislation in countries like Nigeria and Angola that has been enacted to promote the domiciling of production processes and business opportunities for locals. In Nigeria, local content legislation has been designed to promote the national economic interest in hydrocarbon extraction by stipulating minimum numbers of local businesses in different processes in the value chain of oil and gas production. Activities such as welding, engineering design, fabrications, supply of specialized tools and equipment and other oil field, administrative and communication services are supposed to be regulated by local content laws (Ovadia, 2013). However, beyond the stated objective of mitigating the enclave nature of the oil industry and adding value to the local economy generally, comprador capitalism is consolidated. This mirrors aspects of Black Economic Empowerment in Southern Africa. Introduced by the post-apartheid government as a way of granting economic opportunities to previously disadvantaged Africans, it has contributed fundamentally to promoting the 'maintenance of the status quo: [as] new economic groups are co-opted by old groups. This is essentially the generic liberal preference, where "trickle down" development is held out as a longer-term possibility, thus keeping a lid on any outbursts of popular demands for improvements in living conditions' (Andreasson, 2007, p. 17).

In a similar vein, Nigerian local content policies, which facilitate elite accumulation, along with other privatization policies, may enable the institutionalization and legalization of the loot of oil rents. Here mainstream capital – indeed global capital – employs joint venture arrangements between indigenous owners of new oil blocks and transnational corporations to seek more favourable tax and royalty schemes. The new local content scheme is similar to the 1970s' indigenization programme where local business operated as mere fronts for foreign capital (Ovadia, 2012). The overall result is that we have an emergent class of local stakeholders whose material interests are tied to maintaining the flow of oil for the benefit of a global capitalist system. While they are clearly a minority of the population, they are hardly a tiny population.

Indeed, in Nigeria, 'trickle-down' economics is now official. Despite realities of mass impoverishment, President Goodluck Jonathan is adamant in his claim that there is no poverty in the country. Disputing classification by the World Bank, which places Nigeria as a poor country, the President pointed out that:

> [i]f you talk about ownership of private jets, Nigeria will be among the first 10 countries, yet they are saying that Nigeria is among the five poorest countries. Some of you will experience that there is an amount of money you will give to a Nigerian who needs help and [the person] will not even regard it and thank you but if you travel to other countries and give such an amount, the person will celebrate. But the World Bank statistics shows that Nigeria is among the five poorest countries. Our problem is not poverty, our problem is redistribution of wealth. (*The Nation*, 2014)

CONCLUSION: OIL AND WEALTH, HEGEMONY AND RESISTANCE

As reported by a major Nigerian newspaper, the wealthy in Nigeria, oil magnates among them, spent US$6.5 billion on private jets between 2007 and 2012 (*Punch*, 2012). The use of private and 'executive' jets by the rich, including government officials, increased in subsequent years as Nigeria competed with China for the position of fastest-growing market for the high-flying lifestyle. While the conscious 'empowerment' of local entrepreneurs may have contributed to the boom in private jets, as even soon-to-be-former President Goodluck Jonathan proudly acknowledges, the redistribution of wealth is hampered by a lack of political commitment. A plan by a federal government agency to introduce a luxury tax on private jets was immediately rebuffed by the Senate in 2013 (BBC, 2014).

Today, the nouveau riche, whether their wealth is a product of oil via indigenization policies in the West African context or the push toward producing US shale oil from the Bakken, share the material interests of the ultra-wealthy billionaires. There are many more of the former group than those who rank among the top ten, or top 100 lists of the ultra-wealthy. Thus, the number of those with a direct interest in maintaining neoliberal regulation is much greater than those who attract the attention of *Forbes* magazine. While their existence is purported to spur the trickle-down growth that has long been the mantra of economic liberals, the increasing number of permanently unemployed and those, especially oil industry workers, who work under precarious conditions belies this trickle-down promise. In the case of Nigeria, local content policies, while indicating a commitment to drastically increasing the number of Nigerian employees as against expatriates, have done little or nothing to address the casualization of employment, which has become the norm in both foreign and locally owned companies (Solidarity Center, 2010; Eroke, 2014). There is a feeling among oil workers in Nigeria that, even by poor oil industry labour standards, remuneration and other benefits are worse in many of the indigenous companies than in the transnational firms. As Isaac Aberare, General Secretary, Nigerian Union of Petroleum and Natural Gas Workers, complained:

> You can't be using people to do your job for years without regularising their employment. The provision in our law that says after three months, regularise employment, employers are not following it. They employ people and within three months they sack the person and still reemploy the person again and again. (Cited in Vanguard, 2013)

The Nigerian setting manifests a similar context of labour precarity to that seen in the US domestic context and internationally. Here the oil sector mirrors trends across the world's most profitable industries, with accumulation of wealth associated with increasing social stratification and inequality, including hyper-extractive labour relations (Harvey, 2011; Labban, 2014; Piketty, 2014a). Globally, rising CEO compensation and the pressure toward paying dividends to shareholders has been accompanied by serial layoffs, labour repression and reduced trade unionism in the oil sector. Even for full-time workers, income inequality increased 26 per cent (based on the Gini index) in the USA between 1980 and 2007, as a result of financialization (Lin and Tomaskovic-Devey, 2013). Such rising inequality has been shown to emanate from the diminishing power of workers relative to rentiers (Van Arnum and Naples, 2013). To compound the problem, oil workers globally are controlled through the spatial organization of their labour and the provision of contracts in forms that prevent unionization (Pérez, 2009; Appel, 2012). Today, with the turn to relatively labour-intensive unconventional production on offshore platforms, fracking work camps and tar sands mining, pressure on workers is evident at the site of production in the form of repressive subcontracting relations and shiftwork that militates against collective organization.

In this context, and despite calls for transparency and accountability in executive pay after 2008, by 2012 reported executive compensation among the oil majors exceeded pre-crisis figures (Smith and Kuntz, 2013; Sweet, 2014). In 2013, Bloomberg reported that the ratio of CEO to worker pay had in fact tripled since 2009 (Smith and Kuntz, 2013).

Critiques of these developments from a Marxist position centre on accumulation resulting from the theft of common property through its privatization and enclosure (Harvey, 2003). More liberal analyses, as per Wilkinson and Pickett (2010), point out that greater social inequality leads to a diminished quality of life for all people, including those in the wealthiest stratum. And so, countermovements of the past few years such as the 'Occupy Nigeria' movement have demanded greater access to wealth and redistribution, or those such as the 'Idle No More' movement in Canada have claimed historic compensation for land and resources stolen and enclosed through colonialism and imperialism for extractive purposes. In the meantime, an attribute of one of the contradictions of capital, its tendency to destroy the natural and human means via which it generates profit, looms large.

NOTES

* The authors thank Halah Akash and Stephen Agboaye for their research assistance.

1. The variation between this figure and that offered by Chatham House, the Royal Institute of International Affairs, is notable. The variation depicts the considerable debate over accounting standards in this sector and others – with particular attention post the 2008 financial crash. For a review of these processes prior to 2008, see Cortese et al. (2009).
2. While this chapter will not treat this attribute at length, clearly the socioecological deterioration associated with the so-called 'second contradiction of capitalism' – where the drive to accumulate propels capital to degrade the conditions necessary for its own reproduction – is also important to rising inequality. Ecologically this is today associated with climate change, but various human aspects of overextraction of surplus – including threats to human labour – are also attributes of this contradiction.
3. Bina (2006) attributes this to various factors including the recession of the late 1950s, the growth of Russian oil production, and the US import quota imposed on its domestic oil market in 1959; the latter hurt both US domestic consumers and the Persian Gulf states.
4. Anglo-Persian Oil Company (now BP), Gulf Oil, Standard Oil of California (SoCal), Texaco (now Chevron), Royal Dutch Shell, Standard Oil of New Jersey (Esso/Exxon) and Standard Oil Company of New York (Socony).
5. On the *Forbes* list of the ten wealthiest Americans, the Koch brothers tie one another in fourth place, after Bill Gates, Larry Ellison and Warren Buffett.
6. Nigeria's petroleum exports were valued at US$61.8 billion in 2010 (OPEC, 2010/2011). In the same year, the poverty rate in the country, relative to the living standard of the majority of the population, was 69 per cent. Those living on less than US$1 per day constituted 61.2 per cent of the entire population, according to the government (National Bureau of Statistics, 2012).

REFERENCES

Aljazeera.com (2014), 'The French African connection', *Aljazeera.com*, 7 April, accessed 5 May 2014 at http://www.aljazeera.com/programmes/specialseries/2013/08/201387113131914906.html.

Allaire, Y. (2012), 'Pay for value: cutting the Gordian Knot of executive compensation', *IGOPP Innovative Policy Papers*, Montreal: Institute for Governance of Public and Private Corporations, accessed 1 September 2015 at https://igopp.org/en/pay-for-value/.

Andreasson, S. (2007), 'The resilience of comprador capitalism: "new" economic groups in Southern Africa' in A.E. Fernández Jilberto and B. Hogenboom (eds), *Developing Countries and Transition Economies Under Globalization*, Abingdon, UK/New York: Routledge, pp. 274–96.

Appel, H. (2012), 'Offshore work: oil, modularity, and the how of capitalism in Equatorial Guinea', *American Ethnologist*, **39** (4), 692–709.

Baltimore, C. (2012), 'Ex-KBR CEO gets 30 months for Nigeria scheme', *Reuters*, 23 February, accessed 4 December 2014 at http://www.reuters.com/article/2012/02/23/us-kbr-bribery-idUSTRE81M1NX20120223.

BBC (2009), 'Swiss hold "$150m Nigeria bribes"', *BBC News*, 9 April, accessed 4 December 2014 at http://news.bbc.co.uk/2/hi/africa/7991447.stm.

BBC (2014), 'Nigeria's jet set: how the super-rich travel', *BBC News*, 5 March, accessed 6 April 2014 at http://www.bbc.com/news/world-africa-26439657.

Bichler, S. and J. Nitzan (2004), 'Differential accumulation and Middle East wars – beyond neo-liberalism', in K. van der Pijl, L. Assassi and D. Wigan (eds), *Global Regulation: Managing Crises after the Imperial Turn*, London: Palgrave Macmillan.

Bina, C. (2006), 'The globalization of oil: a prelude to a critical political economy', *International Journal of Political Economy*, **35** (2), 4–34.

Bina, C. (2013), *A Prelude to the Foundation of Political Economy: Oil, War and Global Polity*, New York: Palgrave Macmillan.

Borras, S. and J.C. Franco. (2011), 'Global land grabbing and trajectories of agrarian change: a preliminary analysis', *Journal of Agrarian Change*, **12** (1), 34–59.

Bridge, G. and A. Wood (eds) (2010), *Geographies of Peak Oil: Geoforum Special Issue*, July, **41** (4), 519–666.

Cayley-Daoust, D. and R. Girard (2012), *Big Oil's Oily Grasp: The Making of Canada as a Petro-State and How Oil Money is Corrupting Canadian Politics*, Polaris Institute, accessed 1 September 2015 at https://d3n8a8pro7vhmx.cloudfront.net/polarisinstitute/pages/31/attachments/original/1411065312/BigOil'sOilyGrasp.pdf?1411065312.

Chrisafis, A. (2012), 'France has finally got tough on corruption by seizing a dictator's Paris mansion', *The Guardian*, 6 August, accessed 5 May 2014 at http://www.theguardian.com/commentisfree/2012/aug/06/france-tough-corruption-dictators-mansion.

CNN (2012), 'The billion-dollar question: where is Angola's oil money?', *CNN*, 29 November, accessed 6 April 2015 at http://edition.cnn.com/2012/11/28/business/angola-oil-revenues/.

Coronil, F. (1997), *The Magical State: Nature, Money and Modernity in Venezuela*, Chicago, IL: University of Chicago Press.

Cortese, C.L., H.J. Irvine and M.A. Kaidonis (2009), 'Extractive industries accounting and economic consequences: past, present and future', *Accounting Forum*, **33** (1), 27–37.

Cortese, C.L., H.J. Irvine and M.A. Kaidonis (2010), 'Powerful players: how constituents captured the setting of IFRS 6, an accounting standard for the extractive industries', *Accounting* Forum, **34** (2), 76–88.

Daintith, T. (2010), *Finders Keepers?: How the Law of Capture Shaped the World Oil Industry*, Washington, DC: REF Press.

De Kuijper, M. (2009), *Profit Power Economics: A New Competitive Strategy for Creating Sustainable Wealth*, New York: Oxford University Press.

Eroke, L. (2014), 'Casualisation and the spectre of underemployment', *Thisdaylive.com*, 28 January, accessed 20 November 2014 at http://www.thisdaylive.com/articles/casualisation-and-the-spectre-of-underemployment/169888/.

Forbes (2011), 'Africa's richest women', *Forbes.com*, 7 May, accessed 6 April 2015 at http://www.forbes.com/sites/mfonobongnsehe/2011/05/02/africas-richest-women/.

Forbes (2014a), 'America's richest families', *Forbes.com*, accessed 6 April 2015 at http://www.forbes.com/families/list/2/.

Forbes (2014b), 'Global 2000 leading companies', *Forbes.com*, accessed 20 November 2014 at http://www.forbes.com/global2000/.

Freymeyer, C. (2014), 'Illicit financial flows: the elephant in the room at the EU-Africa summit', *Afronline*, 3 April, accessed 25 August 2015 at http://www.afronline.org/?p=33583.

Girvan, N. (1975), 'Economic nationalism', *Daedalus*, **104** (4), 145–58.
Girvan, N. (1978), *Corporate Imperialism: Conflict and Expropriation: Transnational Corporations and Economic Nationalism in the Third World*, New York: Monthly Review Press.
Global Witness (2009), 'The secret life of a shopaholic: how an African dictator's playboy son went on a multi-million dollar shopping spree in the U.S.', *Global Witness.com*, 17 November, accessed 5 May 2014 at https://www.globalwitness.org/campaigns/corruption-and-money-laundering/banks/secret-life-shopaholic/.
Goldenberg, S. (2013), 'Conservative groups spend up to $1bn a year to fight action on climate change', *The Guardian*, 20 December, accessed 1 September 2015 at http://www.theguardian.com/environment/2013/dec/20/conservative-groups-1bn-against-climate-change.
Goodstein, D. (2005), *Out of Gas: The End of the Age of Oil*, New York: W.W. Norton.
Hammond, J.L. (2011), 'The resource curse and oil revenues in Angola and Venezuela', *Science and Society*, **75** (3), 348–78.
Harvey, D. (2003), *The New Imperialism*, New York: Oxford University Press.
Harvey, D. (2011), *The Enigma of Capital and the Crises of Capitalism*, London: Profile Books.
Heilbrunn, J.R. (2005), 'Oil and water? Elite politicians and corruption in France', *Comparative Politics*, **37** (3), 277–96.
Henley, J. (2003), 'Gigantic sleaze scandal winds up as former Elf oil chiefs are jailed', *The Guardian*, 12 November, accessed 4 December 2014 athttp://www.theguardian.com/business/2003/nov/13/france.oilandpetrol.
Herbst, M. (2008), 'Oil CEOs: high prices, fat paychecks', *Bloomberg Business News*, 17 June.
Hernandez, D. (2010), 'Oil companies still top of the 500', *Fortune*, **161** (8), 15.
Huber, M.T. (2013), *Lifeblood: Oil, Freedom, and the Forces of Capital*, Minneapolis, MN: University of Minnesota Press.
Hundley, K. (2011), 'Billionaire's role in hiring decisions at Florida State University raises questions', *Tampa Bay Times*, 9 May, accessed 1 May 2015 at http://www.tampabay.com/news/business/billionaires-role-in-hiring-decisions-at-florida-state-university-raises/1168680.
Jo, T.-H. (2013), 'Uncertainty, instability, and the control of markets', *MPRA*, accessed 1 September 2015 at https://mpra.ub.uni-muenchen.de/47936/1/MPRA_paper_47936.pdf.
Labban, M. (2008), *Space, Oil and Capital*, New York: Routledge.
Labban, M. (2010), 'Oil in parallax: scarcity, markets, and the financialization of accumulation', *Geoforum*, **41** (4), 541–52.
Labban, M. (2014), 'Against value: accumulation in the oil industry and the biopolitics of labour under finance', *Antipode*, **46** (2), 477–96.
Lazupa, J. (2014), 'US seizes more than $30m from Guinean official', *Voice of America*, 12 October, accessed 4 December 2014 at http://www.voanews.com/content/us-seizes-property-from-guinea-vice-president-in-corruption-case/2481179.html.
Leung, R. (2003), 'Kuwait of Africa?: Equatorial Guinea has vast oil reserves, but poverty still prevalent', *60 Minutes*, 14 November, accessed 5 May 2014 at http://www.cbsnews.com/news/kuwait-of-africa/.
Libecap, G.D. and J.L. Smith (2002), 'The economic evolution of petroleum property rights in the United States', *The Journal of Legal Studies*, **31**(S2), S589–S608.
Lin, K.H. and D. Tomaskovic-Devey (2013), 'Financialization and US income inequality, 1970–2008', *American Journal of Sociology*, **118** (5), 1284–329.
Mitchell, T. (2002), 'McJihad: Islam in the US global order', *Social Text*, **20** (4), 1–18.
Mitchell, T. (2011), *Carbon Democracy: Political Power in the Age of Oil*, New York: Verso.
Mommer, B. (2002), *Global Oil and the Nation State*, Oxford, UK: Oxford University Press.
Morningstar (2014), 'Exxon Mobil: key executive compensation', accessed 2 September 2015 at http://quote.morningstar.ca/Quicktakes/Insiders/ExecutiveCompensation.aspx?t=XOM.
Nairaland.com (2010), 'I made $500 million from oil block, but I didn't know how to spend it – T.Y. Danjuma', *Nairaland Forum*, 18 February, accessed 25 August 2015 at http://www.nairaland.com/400110/made-500million-oil-block-didn't.

National Bureau of Statistics (2012), *Annual Abstract of Statistics*, Government of Nigeria, accessed 3 September 2015 at www.nigerianstat.gov.ng/pages/download/253.
Obi, C.I. and S.A. Rustad (2011), *Oil and Insurgency in the Niger Delta: Managing the Complex Politics of Petroviolence*, London: Zed Books.
OPEC (2010/2011), *OPEC Annual Statistical Bulletin*, accessed 3 September 2015 at http://www.opec.org/opec_web/static_files_project/media/downloads/publications/ASB2010_2011.pdf.
Osuoka, I.A. (2013), 'Privatizing dissent: community, civil society, and contested hegemonies in southern Nigeria', PhD dissertation, Faculty of Environmental Studies, York University, Canada.
Ovadia, J.S. (2012), 'The dual nature of local content in Angola's oil and gas industry: development vs. elite accumulation', *Journal of Contemporary African Studies*, **30** (3), 395–417.
Ovadia, J.S. (2013), 'The making of oil-backed indigenous capitalism in Nigeria', *New Political Economy*, **18** (2), 258–83.
Pérez, A.L. (2009), 'Campeche Basin: paradigm of labour exploitation', *Coordination marée noire*, accessed 25 January 2015 at http://coordination-maree-noire.eu/spip.php?article13651.
Piketty, T. (2014a), *Capital in the Twenty-First Century* [French edition published 2013 as *Le capital au XXI siècle*, Editions du Seuil], Cambridge, MA: Belknap Press of Harvard University Press.
Piketty, T. (2014b), 'Interview: dynamics of inequality', *New Left Review*, **85**, 103–16.
Piketty, T. and G. Zucman (2014), 'Capital is back: wealth–income ratios in rich countries, 1700–2010', *The Quarterly Journal of Economics* [online], accessed 25 August 2015 at http://qje.oxfordjournals.org/content/early/2014/05/21/qje.qju018.
Pirog, R. (2012), *Financial Performance of the Major Oil Companies 2007–2011*, Washington, DC: Congressional Research Service, accessed 2 September 2015 at https://www.fas.org/sgp/crs/misc/R42364.pdf.
Popham, P. (2013), 'Africa rises: continent triples its number of billionaires', *The Independent*, 8 October, accessed 5 May 2014 at http://www.independent.co.uk/news/world/africa/africa-rises-continent-triples-its-number-of-billionaires-8866758.html.
Punch (2012), 'Wealthy Nigerians spend $6.5bn on 130 private jets', *Punch*, 17 September, accessed 6 April 2015 at http://www.punchng.com/news/wealthy-nigerians-spend-6-5bn-on-130-private-jets/.
Rowell, A. (2014), 'How companies anonymously influence climate policy', *Oil Change International*, 16 January, accessed 1 September 2015 at at http://priceofoil.org/2014/01/16/companies-anonymously-influence-climate-policy/.
Rubenfeld, S. and J. Palazzolo (2010), 'Panalpina settlements announced, with $236.5 million in penalties', *The Wall Street Journal*, 4 November, accessed 4 December 2014 at http://royaldutchshellplc.com/2010/11/06/panalpina-settlements-announced-with-236-5-million-in-penalties/.
Rudolf, J. (2012), 'Albert Stanley, former Halliburton exec, sentenced in bribery scheme', *Huffington Post*, 24 February, accessed 4 December 2014 at http://www.huffingtonpost.com/2012/02/24/albert-stanley-halliburton-kbr-bribery-sentence_n_1299760.html.
Sklair, L. (2001), *The Transnational Capitalist Class*, Oxford, UK: Blackwell.
Smith, E.B. and P. Kuntz (2013), 'CEO pay 1,795-to-1 multiple of wages skirts U.S. law', *Bloomberg Business News*, 30 April.
Solidarity Center (2010), *The Deregulation of Work: Oil and Casualization of Labor in the Niger Delta*, Washington, DC: Solidarity Center.
Stevens, P., J. Kooroshy and G. Lahn et al. (2013), *Conflict and Coexistence in the Extractive Industries*, London: Chatham House, accessed 3 September 2015 at http://www.chathamhouse.org/publications/papers/view/195670.
Stewart, K. (2013), 'Confidential documents detail oil industry lobbying to weaken greenhouse gas rules', *Greenpeace* Canada, 8 November accessed 1 September 2015 at http://www.greenpeace.org/canada/en/blog/Blogentry/confidential-documents-detail-oil-industry-lo/blog/47293/.

Sweet, K. (2014), 'Fat cats getting fatter as stocks power CEO pay to record high', *Associated Press*, 28 May.
The Nation (2014), 'Jonathan to World Bank: Nigeria not poor', *The Nation*, 2 May, accessed 5 May 2014 at http://thenationonlineng.net/new/jonathan-world-bank-nigeria-poor-2/.
Van Arnum, B.M. and M.I. Naples (2013), 'Financialization and income inequality in the United States, 1967–2010', *American Journal of Economics and Sociology*, **72** (5), 1158–82.
Van der Pijl, K. (1998), *Transnational Classes and International Relations, Vol. 1*, London: Routledge.
Van der Pijl, K. (2004), 'Two faces of the transnational cadre under neo-liberalism', *Journal of International Relations and Development*, **7** (2), 177–207.
Vanguard (2013), 'Government is not doing anything to curb casualisation in Nigeria – Aberare', *Vanguard*, 19 October, accessed 5 May 2014 at http://www.vanguardngr.com/2013/10/government-anything-curb-casualisation-nigeria-aberare/#sthash.I0a10Cbj.dpuf.
Ventures Africa (2013), 'The richest people in Africa, 2013', *Ventures Africa.com*, 9 October, accessed 5 May 2014 at http://www.ventures- africa.com/2013/10/richest-people-africa-2013/.
Viswanatha, A. (2012), 'U.S. charges former, current Noble execs over bribes', *Reuters*, 24 February, accessed 4 December 2014 at http://fr.reuters.com/article/idUKL2E8DOBPF20120224?pageNumber=2andvirtualBrandChannel=0.
Wilkinson, R. and K. Pickett (2010), *The Spirit Level: Why More Equal Societies Almost Always Do Better*, London: Allen Lane.
Yergin, D. (2008a), 'Oil has reached a turning point', *Financial Times*, 27 May.
Yergin, D. (2008b), 'What lower oil prices mean for the world', *Financial Times*, 10 November.

Index